Advances in
PARASITOLOGY

VOLUME 32

Editorial Board

Advances in
PARASITOLOGY

Edited by

J. R. BAKER

Cambridge, England

and

R. MULLER

International Institute of Parasitology
St Albans, England

VOLUME 32

ACADEMIC PRESS

Harcourt Brace & Company, Publishers
London San Diego New York Boston
Sydney Tokyo Toronto

ACADEMIC PRESS LIMITED
24/28 Oval Road
LONDON NW1 7DX

United States Edition published by
ACADEMIC PRESS INC.
San Diego CA 92101

A CIP record for this book is available from the British Library

ISBN 0-12-031732-X

This book is printed on acid-free paper

Filmset by Bath Typesetting Ltd., London Road, Bath
Printed in Great Britain by T. J. Press (Padstow) Ltd, Padstow, Cornwall.

CONTRIBUTORS TO VOLUME 32

S. J. BALL, *Department of Life Sciences, University of East London, Romford Road, London, E15 4LZ, UK*

P. F. L. BOREHAM, *Queensland Institute of Medical Research, The Bancroft Centre, 300 Herston Road, Herston, Brisbane, Queensland 4029, Australia*

A. J. DAVIES, *School of Life Sciences, Kingston University, Penrhyn Road, Kingston upon Thames, Surrey, KT1 2EE, UK*

F. Y. LIEW, *Department of Immunology, University of Glasgow, Western Infirmary, Glasgow, G11 6NT, UK*

A. H. W. MENDIS, *School of Biomedical Sciences, Curtin University of Technology, Bentley, Western Australia, 6001, Australia*

C. A. O'DONNELL, *Department of Immunology, University of Glasgow, Western Infirmary, Glasgow, G11 6NT, UK*

A. RAIBAUT, *Laboratoire de Parasitologie Comparée, URA CNRS 698, Université Montpellier II, Sciences et Techniques du Languedoc, Place Eugène Bataillon, 34095 Montpellier Cédex 5, France*

J. A. REYNOLDSON, *Institute for Molecular Genetics and Animal Disease and School of Veterinary Studies, Murdoch University, Murdoch, Western Australia, 6150, Australia*

D. J. STENZEL, *Analytical Electron Microscopy Facility, Queensland University of Technology, George Street, Brisbane, Queensland 4001, Australia*

R. C. A. THOMPSON, *Institute for Molecular Genetics and Animal Disease and School of Veterinary Studies, Murdoch University, Murdoch, Western Australia, 6150, Australia*

J. P. TRILLES, *Laboratoire d'Ecophysiologie des Invertébrés, Université Montpellier II, Sciences et Techniques du Languedoc, Place Eugène Bataillon, 34095 Montpellier Cédex 5, France*

D. ZILBERSTEIN, *Department of Biology, Technion–Israel Institute of Technology, Haifa 32000, Israel*

PREFACE

Unusually for the series as a whole, this volume of *Advances in Parasitology* has a predominantly protozoological flavour, which we hope may go some way to redressing the overall helminthological bias.

The first contribution deals with what the authors, Drs Boreham and Stenzel, refer to as an enigma: the well known but little understood *Blastocystis*. Both this organism's taxonomic position and its possibly pathogenic role remain in doubt, but Boreham and Stenzel conclude that *Blastocystis* is indeed a protist, but is probably *sui generis*—although perhaps most closely related to the amoeboflagellates. Since its first recognition by Alexieff in 1911, this organism has been variously regarded as a fungus, yeast, the cyst of a flagellate, or a degenerate vegetable cell. Opinion about its pathogenicity has been, and still is, similarly divided; the authors conclude that it is at present premature to attempt to reach a conclusion.

Drs Thompson, Reynoldson and Mendis next present a masterly review of the knowledge concerning *Giardia* which has accumulated since this parasite was last reviewed in this series in 1979 (Vol. 17). The pathogenicity of *Giardia* is not in doubt, it now being recognized as "one of the ten major parasites of humans", and nor is its taxonomic position among the flagellates. However, the increasing and intriguing (and sometimes conflicting) evidence that *Giardia* represents a very early branching of the eukaryotic stem is discussed, and the authors conclude that "some caution may be prudent" in reaching a conclusion about this. The review also deals with the relatively newly discovered intranuclear RNA virus, GLV or Giardiavirus, about which much remains to be discovered, and other aspects of the parasite's morphology, the vexed question of speciation, the life cycle and transmission (concluding that definitive proof of zoonotic transmission in nature has yet to be obtained), biochemistry (which is discussed in considerable detail), and control.

Following the review of the interaction between *Leishmania* and its macrophage host cells in the previous volume, this volume contains a full overview of current knowledge of the immunology of leishmaniasis by Professor Liew and Dr O'Donnell. Considerable advances in this subject have been made over the last decade or so, including knowledge of the genetic regulation of the response to infection in human and murine hosts, the cellular response (especially the roles of various T cell populations), cytokines and the effector mechanisms by means of which the parasites may be killed as a result of the host's response to infection. This leads to a discussion of vaccination (which has, of course, been practised for centuries in the form of "leishmanization" for oriental sore). The authors conclude that leishmaniasis is now perhaps one of the infectious diseases best understood from an immunological viewpoint.

Dr Zilberstein reviews, in a compact and concentrated form, current knowledge of the means by which trypanosomatids transport nutrients and ions across their membranes, including proton transport and the proton motive force, and transport of glucose, amino acids and calcium. These mechanisms, which have until relatively recently been a neglected topic, are clearly of great importance in maintaining the parasites' intracellular homeostasis and thus ensuring their survival within their hosts. Equally clearly, it is possible that a fuller understanding of the mechanisms of transport could lead to the development of directed chemotherapeutic agents, aimed at blocking or disrupting these essential processes.

Dr Davies and Professor Ball next review the biology of the coccidian parasites of fish. This very comprehensive, fully and beautifully illustrated chapter complements the review of eimeriid coccidia by Professor Ball and others in Volume 28. This interesting group of organisms, which contains, in the authors' words, a "bewildering array" of parasites, has until recently been much less studied than the equivalent parasites of mammals and birds. However, the currently growing interest in fish farming and increasing awareness of the potential of these coccidia to cause disease, especially under the intensive conditions inseparable from farming, has led to a tendency to redress this imbalance—a process which will be much helped by the present review. All aspects of the parasites' biology are covered: life cycles, transmission, structure and host–parasite interactions. The authors conclude by summarizing fields in which our knowledge of these parasites has increased considerably and, perhaps more importantly, those in which it has not much increased. This latter category includes the possibility of autoinfection, immunity and pathogenicity, and taxonomy. There is still confusion and argument over the number of valid genera; alongside the well established genera *Eimeria* and *Goussia*, should the newer genera *Epieimeria*, *Epigoussia* and *Nucleogoussia* be accepted? The mechanism of oocyst wall formation also remains incompletely understood. The authors conclude optimistically that the significant recent advances in knowledge should provide the basis for controlled experiments to provide answers to these questions.

Finally Drs Raibaut and Trilles review the sexuality of parasitic crustaceans. This diverse and sometimes bizarre group of parasitic organisms is not well known to the majority of parasitologists but can supply fascinating insights into the evolutionary aspects of comparative parasitology. Parasitism has led to a multiplicity of sexual modes with separate sexes and various degrees of hermaphroditism often reflecting the motile or sessile habits of their free-living ancestors. The problems of mate encounter which occur when one of the partners is fixed, and particularly when both are, have been solved in various ways, and these are discussed and amply illustrated.

J. R. Baker
R. Muller

CONTENTS

Transport of Nutrients and Ions across Membranes of Trypanosomatid Parasites

D. ZILBERSTEIN

The Biology of Fish Coccidia

A. J. DAVIES AND S. J. BALL

The Sexuality of Parasitic Crustaceans

A. RAIBAUT AND J. P. TRILLES

Blastocystis in Humans and Animals: Morphology, Biology, and Epizootiology

PETER F. L. BOREHAM

Queensland Institute of Medical Research, The Bancroft Centre, 300 Herston Road, Herston, Brisbane, Queensland 4029, Australia

AND

DEBORAH J. STENZEL

Analytical Electron Microscopy Facility, Queensland University of Technology, George Street, Brisbane, Queensland 4001, Australia

ADVANCES IN PARASITOLOGY VOL. 32
ISBN 0-12-031732-X

I. Introduction

Blastocystis has been described as an enigma among the protists, with a number of unique features (Zierdt, 1991). This contention has certainly been true since the first accurate descriptions of this parasite (Alexeieff, 1911; Brumpt, 1912), and remains so today, largely because of the lack of critical research. *Blastocystis* has a controversial history, and progress in our knowledge has been hindered by sweeping generalizations, made from too little scientific evidence, which have been repeated without being seriously questioned. Thus, many of the dogmas currently held are based on minimal factual data, often collected more than 20 years ago, and must now be challenged. In this review an attempt will be made to differentiate between fact and speculation, with a view to putting the biology of this organism on a scientific footing and indicating the major deficiencies in our knowledge. The literature of *Blastocystis* is dominated by the work of Charles Zierdt, who must be credited with bringing this organism to the notice of medical scientists. However, his work has not been subjected to rigorous scrutiny by other scientists and does contain many inconsistencies.

The history of *Blastocystis* has recently been reviewed in detail (Zierdt, 1991), and the possible contributions of Brittan (1849), Swayne (1849), Lösch (1875) and Perroncito (1899) to the discovery of the organism appraised. However, none of the descriptions given in these early reports is entirely satisfactory and, like a number of later publications, does not exclude the possibility that the authors were looking at artefacts, such as degenerate vegetable, tissue or yeast cells. The history of the parasite will not be considered further in this review except where it impinges on current interpretation of data, and the reader is referred to Zierdt's excellent historical account for further details (Zierdt, 1991).

II. Taxonomy

Our current knowledge of the taxonomy of *Blastocystis* is grossly deficient. The history of this organism illustrates the confusion that has always existed concerning its taxonomic position. Early workers were unable to classify *B. hominis*, and variously described it as the cyst of a flagellate, vegetable material, yeast and fungus (see Zierdt, 1978, 1991). It was not until 1967 that evidence was provided to assign *B. hominis* to the subkingdom Protozoa (see Zierdt *et al.*, 1967), based on morphological and physiological criteria. Ultrastructurally it resembles the protists as it lacks a cell wall, but contains nuclei, smooth and rough endoplasmic reticulum, Golgi complex and mitochondria. Physiologically it is anaerobic, sensitive to oxygen, fails to grow on fungal media, grows optimally at 37°C and neutral pH, and is not killed by antifungal agents such as amphotericin (Zierdt *et al.*, 1967; Tan and Zierdt, 1973; Zierdt, 1973; Tan *et al.*, 1974; Zierdt and Williams, 1974).

B. hominis was subsequently classified in the subphylum Sporozoa, in a separate suborder Blastocystina (Zierdt, 1978), and more recently in the subphylum Sarcodina (Zierdt, 1988). Molecular sequencing studies on a single human isolate (Netsky), utilizing small subunit ribosomal ribonucleic acid sequencing techniques, have shown that *B. hominis* is not monophyletic with *Saccharomyces* nor with any of the sarcodines or sporozoans, suggesting that *B. hominis* is not closely related to any of these groups (Johnson *et al.*, 1989) (Fig. 1). When the Apicomplexa were examined, only *Sarcocystis* and *Toxoplasma* were found to be monophyletic, and are separated from *Plasmodium* and *Blastocystis* by the ciliates and the dinoflagellate *Prorocentrum*. Thus, the exact taxonomic position of *B. hominis* remains undetermined, although it seems likely that eventually it will be shown to form a new group, possibly closely related and analogous to the amoebo-flagellates (see Section IV.H).

Two non-human species have recently been described: *B. galli* from the caecum of chickens in the Commonwealth of Independent States (Belova and Kostenko, 1990), and *B. lapemi* from the sea snake, *Lapemis hardwickii*, collected in Singapore (Teow *et al.*, 1991). The former was identified on morphological criteria but, due to the great variation seen between individual organisms, care should be exercised in interpreting this result until further confirmatory evidence is obtained. *B. lapemi* was differentiated from *B. hominis* on its different optimal culture requirements (*B. lapemi* growing best at 24°C rather than 37°C) and different electrophoretic karyotype.

In this review we will refer to the parasite isolated from humans as *B. hominis*, and use *Blastocystis* sp. for organisms isolated from other hosts, since there are currently no data on their interrelationship.

Intraspecific variation has been examined in *B. hominis* (Kukoschke and Müller, 1991; Boreham *et al.*, 1992). Analysis of 10 stocks of *B. hominis* isolated from human stools revealed two discrete groups of organisms. Proteins of the two groups were immunologically distinct (Figs 2, 3), and hybridization with random probes generated from the deoxyribonucleic acid

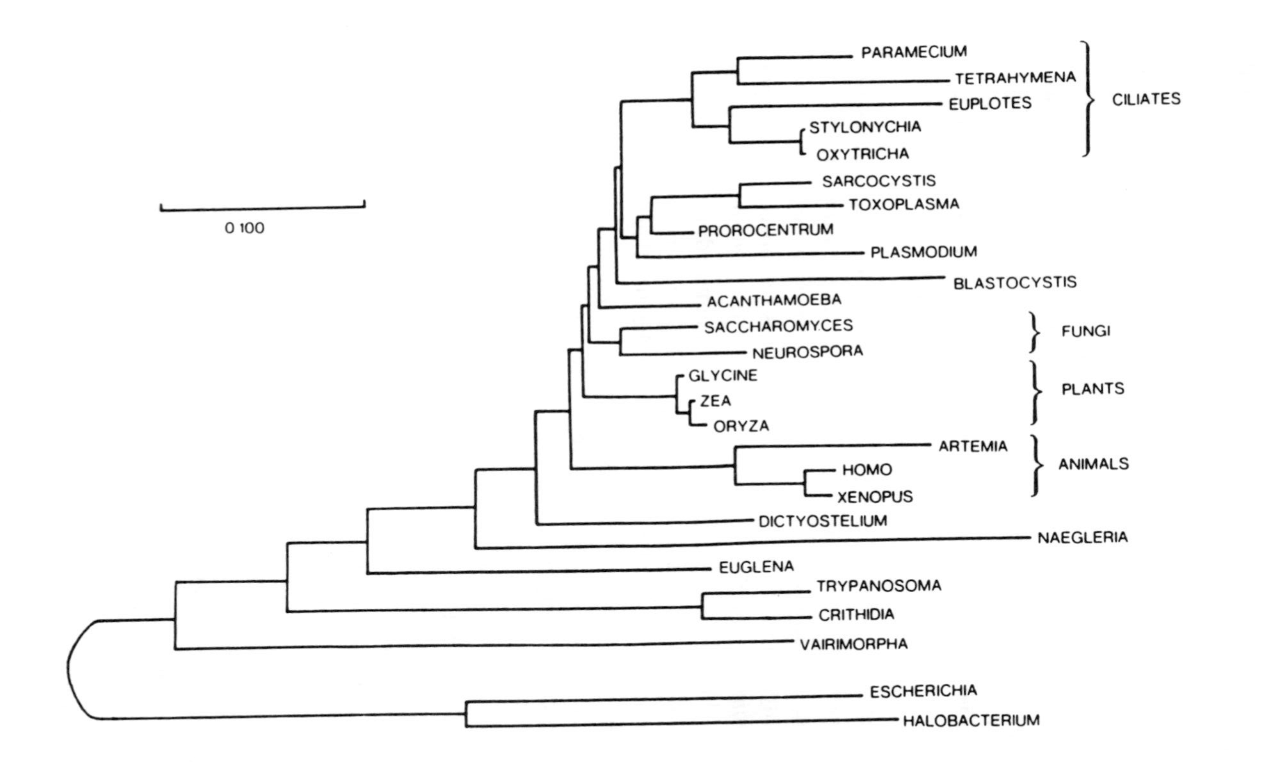

FIG. 1. Phylogenetic tree for 25 genera of eukaryotes to show the relationship of *Blastocystis* to other species. The tree is based on the degree of nucleotide divergence for 215 semi-conserved sites of the srRNA gene. Branch lengths shown are proportional to the amount of evolutionary change along each branch. (Reproduced by permission from Johnson and Baverstock, 1989.)

(DNA) of one stock showed that the DNA content of the two groups was also different (Fig. 4) (Boreham *et al.*, 1992). This raises the possibility that there are more than one species of *Blastocystis* in humans. However, without further epidemiological data it is not appropriate to designate a new species. Rather, it is germane to regard these two groups as demes in accordance with the nomenclature developed for trypanosomes (World Health Organization, 1978). Demes are defined as 'Populations that differ from others of the same species or subspecies in a specified property or set of properties'. In this case, protein and DNA criteria are used to identify demes.

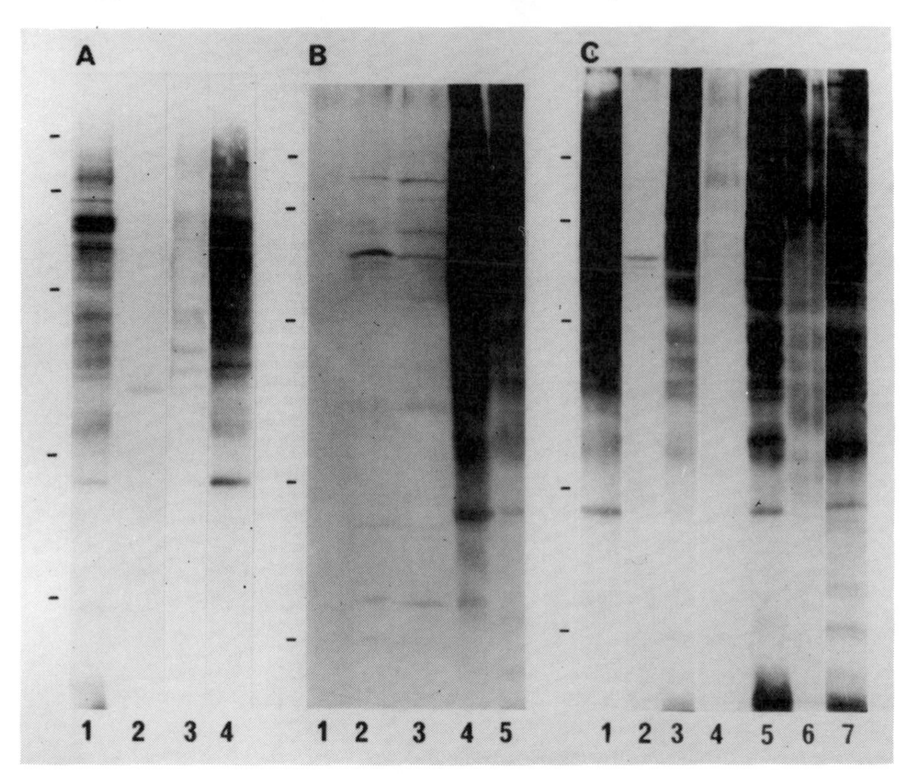

FIG. 2. Immunoblot of *B. hominis* stocks reacted with antisera raised against Netsky stock. Panel A. Lane 1, BRIS/88/HEPU/34; Lane 2, BRIS/88/HEPU/28; Lane 3, BRIS/87/HEPU/11; Lane 4, Netsky. Panel B. Lane 1, BRIS/87/HEPU/2; Lane 2, BRIS/87/HEPU/11; Lane 3, BRIS/87/HEPU/12; Lane 4, Netsky; Lane 5, BRIS/88/HEPU/23. Panel C. Lane 1, Netsky; Lane 2, BRIS/87/HEPU/11; Lane 3, BRIS/88/HEPU/23; Lane 4, BRIS/88/HEPU/28; Lane 5, BRIS/88/HEPU/32; Lane 6, BRIS/88/HEPU/33; Lane 7, BRIS/88/HEPU/34. Marker proteins are indicated in each panel and represent proteins of molecular mass 21.5, 30, 46, 69 and 92.5 kDa in ascending order. (Reproduced with permission of the Australian Society for Parasitology from Boreham *et al.*, 1992.)

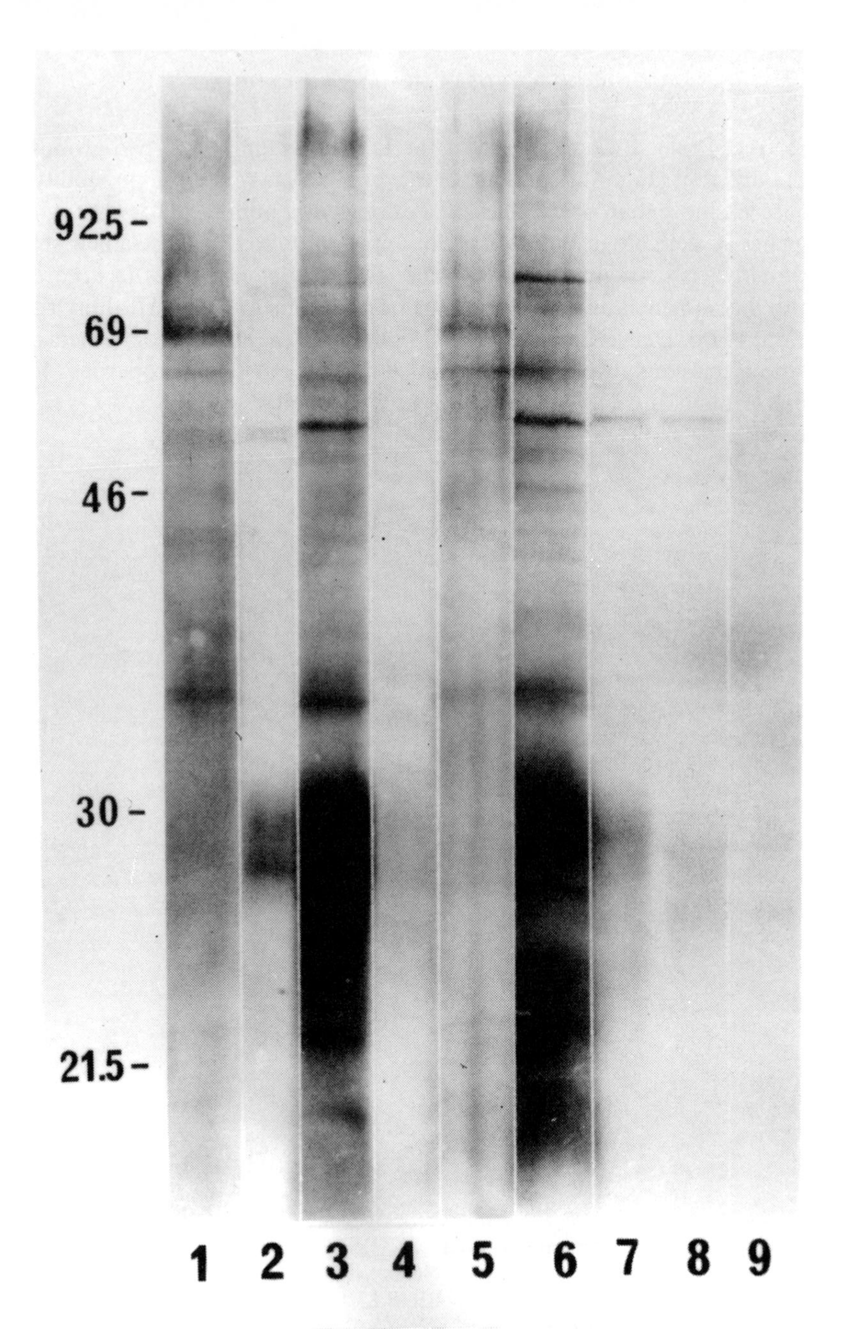

FIG. 3. Immunoblot of *B. hominis* stocks reacted with antisera raised against stock BRIS/87/HEPU/12. Lane 1, BRIS/87/HEPU/12; Lane 2, BRIS/88/HEPU/34; Lane 3, BRIS/88/HEPU/28; Lane 4, BRIS/88/HEPU/23; Lane 5, BRIS/87/HEPU/12; Lane 6, BRIS/87/HEPU/11; Lane 7, BRIS/87/HEPU/6; Lane 8, BRIS/87/HEPU/2; Lane 9, Netsky. Protein molecular masses are given in kDa. (Reproduced with permission of the Australian Society for Parasitology from Boreham *et al.*, 1992.)

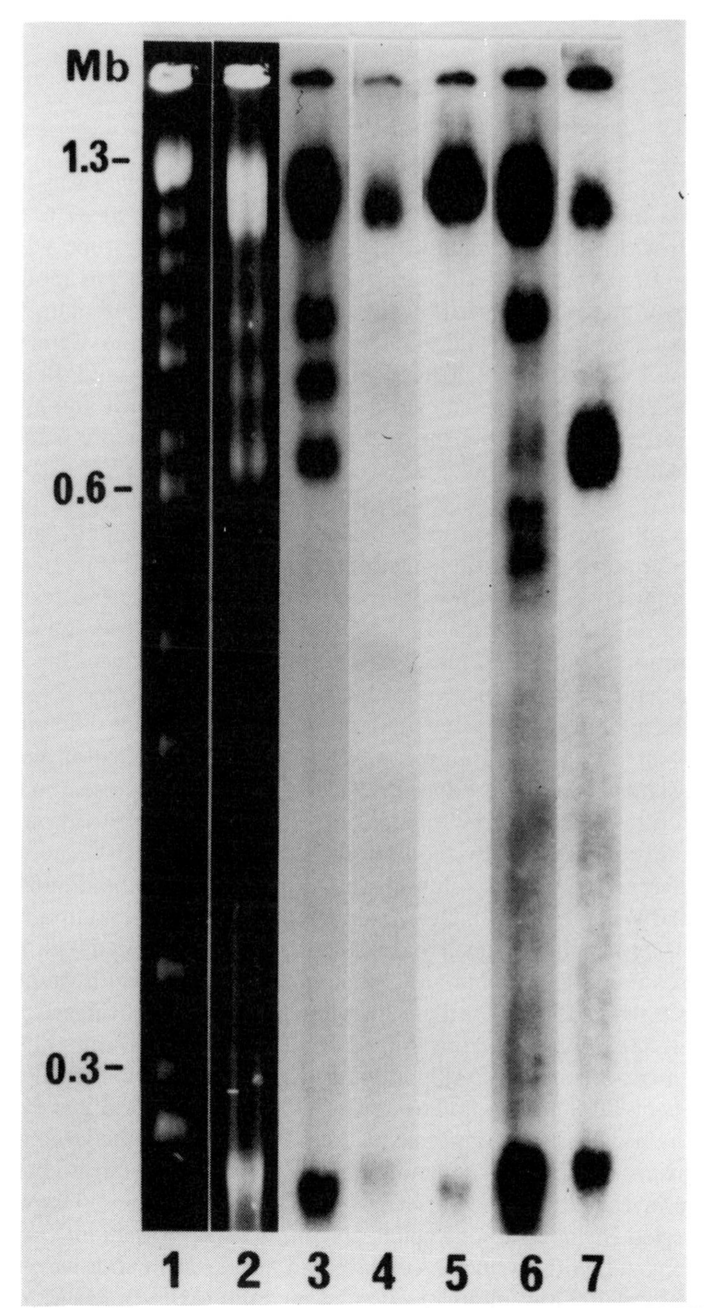

FIG. 4. Hybridization of chromosomes of *B. hominis* stock BRIS/88/HEPU/23 with DNA probes generated from the Netsky stock. Lane 1, yeast chromosome markers; Lane 2, chromosomes of stock BRIS/88/HEPU/23 separated by field inversion gel electrophoresis and stained with ethidium bromide. Lanes 3–7, chromosomes of stock BRIS/88/HEPU/23 Southern-transferred and hybridized with Netsky DNA probes. Lane 3, probe B3; Lane 4, probe B4; Lane 5, probe B37; Lane 6, probe B38; Lane 7, probe B7. Chromosome sizes are included in megabases (Mb). (Reproduced with permission of the Australian Society for Parasitology from Boreham *et al.*, 1992.)

III. Culture of *Blastocystis hominis*

The first report of the successful culture *in vitro* of *B. hominis* was in 1921, using 10% human serum in 0.5% saline with incubation at 37°C (Barret, 1921). Growth occurred mainly in the lower part of the tube where the oxygen content was reduced. Subsequently, the technique was modified to include ovarian cyst and peritoneal exudate fluids rich in albumin (Lynch, 1922). Attempts at growth on solid and semi-solid media were initially unsuccessful (Lynch, 1922; Ciferri and Redaelli, 1938), until Boeck and Drbohlav's (1925) inspissated egg medium was used (Zierdt and Williams, 1974). Growth was optimal on medium which had been pre-reduced for 48 h, at neutral or slightly alkaline pH, with incubation at 37°C. Growth did not occur at 30°C or room temperature. Axenization of stocks has been achieved with antibiotics over a period of at least 1 month, using, in particular, ampicillin, colistin and streptomycin (Zierdt and Williams, 1974). Ceftizoxime and vancomycin have been used to inhibit bacteria resistant to the former antibiotics (Zierdt, 1991). Other workers have been unsuccessful in axenization (Dunn, 1992), suggesting that in some instances the bacteria may be essential for the survival of the organism.

There have been several attempts experimentally to induce different forms of *B. hominis* in culture. Initially it was found that axenization using egg medium slants generally produced cells larger than those present in conventional cultures (Zierdt and Williams, 1974). Short-term cultivation of axenized strains on TPN broth and brain-heart infusion with 10% horse blood and an overlay of Hank's solution was found to favour the granular form (Zierdt and Williams, 1974). Other media reported to be useful for growth of *Blastocystis* sp. include Dobell and Laidlaw's medium covered with Ringer's solution containing 20% human serum and supplemented with streptomycin sulphate (Silard, 1979), Loeffler's medium covered with Ringer's solution containing 20% human serum (Silard *et al.*, 1983), and Diamond's trypticase-panmede-serum (TP-S-1) monophasic medium (Molet *et al.*, 1981). The usefulness of TP-S-1 medium has not been confirmed (Dunn and Boreham, 1991). The need for a monophasic medium for culture of the large numbers of organisms required for chemotherapeutic and molecular studies led to extensive investigation of a variety of media and conditions (Dunn, 1992). Media tested included thioglycollate broth, TYI-S-33 medium (Diamond, 1987) alone and with pyruvate, catalase and horse serum additives, medium 1640, medium 199 and minimal essential medium (MEM). MEM, containing 10% horse serum (MEMS), which had been pre-reduced for 48 h was found to be the best monophasic medium to maintain growth of *B. hominis*, although some growth occurred in medium 199.

Standardization of culture conditions enabled the measurement of doubling times of *B. hominis* in culture. Zierdt and Swan (1981) found that the doubling times of eight axenic strain of *B. hominis* in biphasic egg medium ranged from 6 to 16 h, and the time required for an individual *B. hominis* cell to divide was 30–60 min. The doubling time for the Netsky strain of *B. hominis* was 19.7 ± 3.3 h in MEMS, compared to 23.4 ± 5.4 h on egg slants (Dunn and Boreham, 1991). These doubling times were considerably longer than the 8.5 h previously reported for this stock grown on an egg slant medium (Zierdt and Swan, 1981). This discrepancy may reflect selection of the parasite through continuous growth over several years, or possibly the different time period over which measurements were made.

Although *B. hominis*, unlike bacteria, cannot be stored by freeze drying (Zierdt, 1991), it may readily be cryopreserved using 7.5% dimethylsulphoxide as a cryoprotectant. Best results are obtained by controlled slow cooling and subsequent maintenance in liquid nitrogen. On retrieval from liquid nitrogen, the cells must be rapidly thawed at 37°C, washed twice in Locke's solution and inoculated into a fresh culture tube of pre-reduced medium. This allows stocks to be stored for extended periods and to be transported to other laboratories. In addition, reference material can be maintained for antigenic analysis and to ensure that continuous culture does not lead to genetic selection.

IV. Morphology

Reports of the morphology of *B. hominis* have consistently noted several major forms (vacuolar, granular and amoeboid), and a number of less common forms, of the organism (see Zierdt, 1991). It is currently unknown whether these differences in morphology reflect differences in the biochemistry and general cell biology of the organism. The true life cycle has not been conclusively elucidated, and the differentiation pathways are uncertain. The stage of the parasite responsible for transmission is also undefined. This review on the current status of the morphology of *Blastocystis* is written with these deficiencies in mind, in order to identify those areas where research is most needed and to separate actuality from conjecture.

A. Light Microscopy

Many reports have described in detail the morphology of *B. hominis* as determined by light microscopy. Many of the early reports were meticulous in detail, and in general closely correlate with recent electron microscopic descriptions. Most reports described a spherical cell, with a large central body (also called a central vacuole, internal body or reserve body), and a thin peripheral rim of cytoplasm (Wenyon, 1910; Chatton and Lalung Bonnaire, 1912; Macfie, 1915; Kofoid *et al.*, 1919; Wenyon, 1926; Ciferri and Redaelli, 1938). Many also reported the presence of a thick, mucilaginous coat surrounding the organism (Kofoid *et al.*, 1919; Beaurepaire Aragão, 1922a,b; Knowles and Das Gupta, 1924; Ciferri and Redaelli, 1938; Lavier, 1952) and the presence of multiple nuclei (Chatton and Lalung Bonnaire, 1912; Macfie, 1915; Matthews, 1918; Beaurepaire Aragão, 1922a,b; Wenyon, 1926). The original description by Alexeieff (1911) even included what is now considered to be the characteristic nucleus of *Blastocystis*, with its cap of condensed chromatin. This nuclear structure has been noted in almost all reports of *Blastocystis*. The cell diameter was generally reported to be in the range of 5–20 µm (see Matthews, 1918; Wenyon, 1926). Several reports also noted a separate smaller cell or "spore" (Alexeieff, 1911; Lynch, 1917; Kofoid *et al.*, 1919; Lavier, 1952). This form was surrounded by a thick wall, and thought to be a resistant stage. No vacuole was present in the smaller form, but there were reports of glycogen and lipid inclusions (see Lavier, 1952).

Size variations of *Blastocystis* in faecal samples, including a smaller form, were noted but not detailed (Lynch, 1917; Wenyon and O'Connor, 1917; Matthews, 1918; Beaurepaire Aragão, 1922a,b). Lynch (1917) noted a small form in formed stools and older specimens, compared with larger forms found in the stools of patients with diarrhoea.

Light microscopy offers a rapid method of detection of *Blastocystis* in faecal samples, but the limited resolution restricts its use as a research tool.

B. SCANNING ELECTRON MICROSCOPY

Very few studies have used scanning electron microscopy (SEM) to investigate the structure of *B. hominis* cells. A severe limitation to the data obtained with conventional SEM is the limited resolution attainable at higher magnifications. With increasing interest and developments in high resolution SEM new data should become available in the future.

At present, SEM allows observation of the three-dimensional structure of *B. hominis*, at a level far beyond the capabilities of light microscopy. The study by van Saanen-Ciurea and El Achachi (1983) included a number of scanning electron micrographs of *B. hominis* from short-term culture. The cells appeared spherical or elongated with a rough surface, and large pores and indentations were present. No difference was noted between several samples. Bacteria were often seen adherent to the surface of *B. hominis* and this has been confirmed by transmission electron microscopy (TEM), by

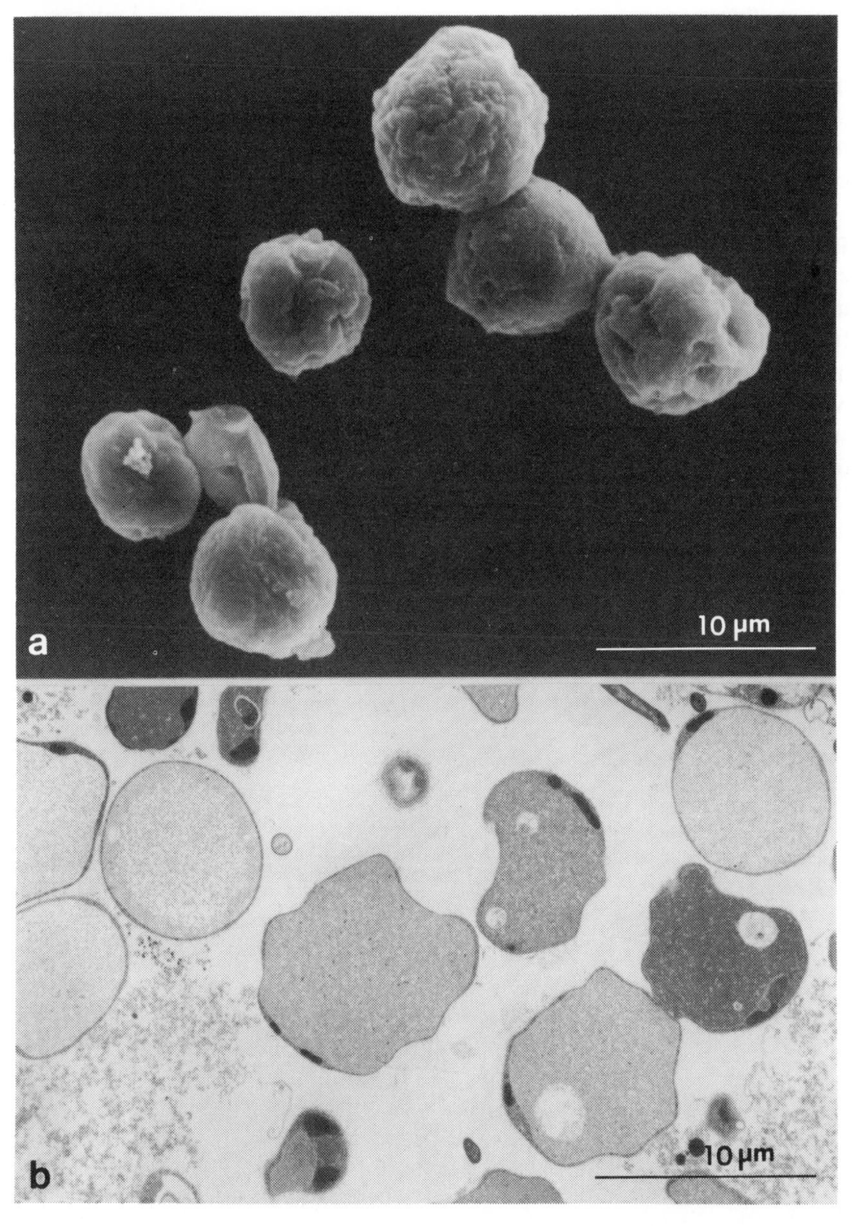

FIG. 5. Micrographs of *B. hominis* from laboratory culture of Netsky stock. (a) Scanning electron micrograph showing shape and surface morphology. Although generally rounded in shape, many indentations are seen on the cell surface. (b) Transmission electron micrograph of cells from the same sample. Variations in profile are seen, and considerable variation in appearance of internal structures is visible. (Reprinted with permission from *Comparative Biochemistry & Biology*, Pergamon Press, Oxford.)

means of which bacteria have been seen deeply embedded in the surface coat (Silard and Burghelea, 1985; Dunn *et al.*, 1989). The only other published report included a single micrograph of cultured *Blastocystis* sp., and concluded that the organism was spherical in shape (Matsumoto *et al.*, 1987).

Work in our laboratory (D. G. Skelly, M. F. Cassidy, D. J. Stenzel and P. F. L. Boreham, unpublished observations) confirmed the generally spherical or elongated shape of *B. hominis* from culture (Figs 5, 6), and showed indentations and deep pores extending into the organism (Fig. 5a). The function of these pores has not been elucidated, although they may correspond to cytoplasmic openings seen by TEM. Two types of outer surface were discernible in organisms from culture: (i) a very rough and ruffled surface, which probably corresponds to the surface coat seen by TEM, or its remnants, and (ii) a smooth surface, broken only by deep grooves and pores (Fig. 6). This second type of surface may correspond to organisms without a distinct surface coat. Bacteria were seen adhering to the surface of both rough- and smooth-surfaced organisms. No evidence of phagocytosis of bacteria has yet been noted by SEM.

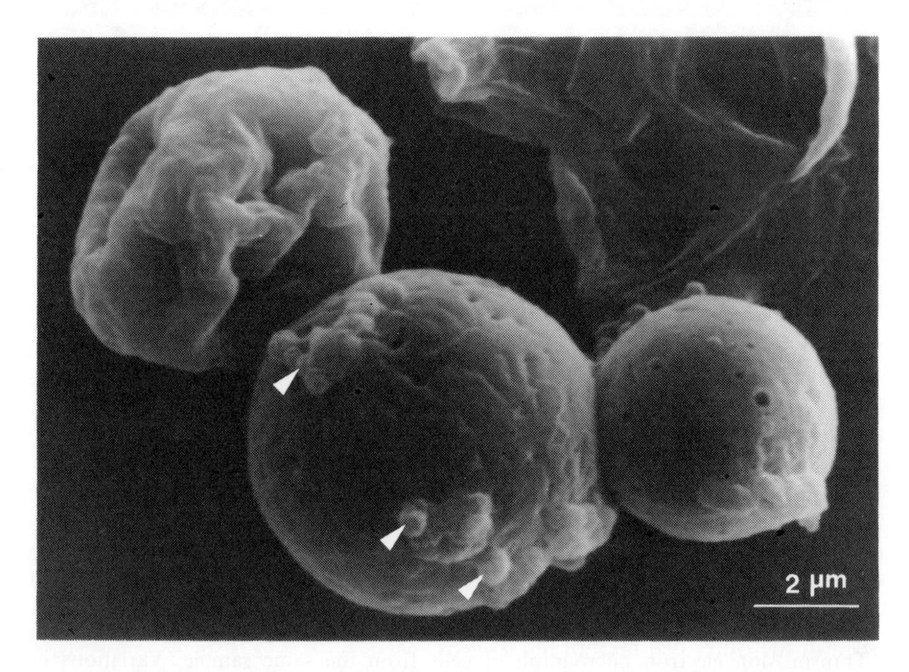

FIG. 6. Scanning electron micrograph of *B. hominis* from laboratory culture of Netsky stock. Openings and infoldings of the cell surface are seen. Bacteria (indicated by arrows) are associated with the surface coat of the cells.

C. TRANSMISSION ELECTRON MICROSCOPY

TEM has provided much of the evidence of the protist nature of *B. hominis*. Almost all the new information on the life cycle and cell biology of *Blastocystis* has resulted from the use of TEM (see Sections IV.F and G). Major variations in ultrastructure have been reported, but it is not known if these relate to morphological differences in various stages of the life cycle, or whether morphologically distinct species of *Blastocystis* exist. It is unlikely that the morphological differences noted in recent reports were due to isolation or preparative techniques, although this is probably the case with some earlier reports, in which many discrepancies can be attributed to artefacts resulting from the techniques used (see Magaudda-Borzì and Pennisi, 1961). Should *Blastocystis* eventually prove to consist of several different species, with distinct host ranges, this would account for some of the observed differences.

D. FREEZE ETCH AND FREEZE FRACTURE ELECTRON MICROSCOPY

Two studies of *B. hominis* used freeze fracture techniques (Tan *et al.*, 1974; Yoshikawa *et al.*, 1988). Freeze fracture techniques have extended the results of conventional TEM studies in regard to cell shape, distribution of granules in the central vacuole and the presence of membrane-bound organelles. The study by Tan *et al.* (1974) confirmed the presence of pores (approximately 50 nm in diameter) in the outer membrane and demonstrated that they were not present in the membrane surrounding the central vacuole. However, a later report, using conventional TEM, stated that pores were present on the outer membrane and the central vacuole membrane (Zierdt and Swan, 1981). The central vacuole membrane and cell membrane occasionally appeared to be continuous. Perhaps the only novel information obtained from these studies was that the outer cell membrane and that surrounding the central vacuole appeared to be of different composition as assessed by the distribution of intermembranous particles. However, intramembranous particle numbers and spatial distribution should be interpreted with considerable caution, as rearrangements are now known to occur when chemical fixatives and cryoprotectants are used in sample preparation.

The report by Yoshikawa *et al.* (1988) presented information on the appearance of membranous formations and components. Again, pores were seen in the outer cell membrane of *B. hominis*, although their size (approximately 75–80 nm diameter) was larger than that reported by Tan *et al.* (1974). Pores were not seen in the central vacuole membrane. Additional irregularities in the membrane, extending outwards from the cell, were also noted. Protrusions of membrane into the central vacuole occurred, and vacuoles with similar internal composition to the central vacuole were present in the cytoplasm. Although some vacuoles appeared to be connected to the central vacuole, others appeared to be completely enclosed by their own membrane. This work has now been confirmed by TEM studies (Dunn *et al.*, 1989).

E. LIFE CYCLE

A life cycle for *Blastocystis* was first proposed by Alexeieff in 1911 (Fig. 7). He described a complex cycle, involving binary fission of a binucleate stage (plasmotomic division), and autogamy, a sexual phenomenon, to form primary cysts. These cysts produced spores by multiple budding. The spores, or secondary cysts, were thought to be resistant forms, and were smaller than the primary cysts, uninucleate and surrounded by a thick membrane. Merogony ("schizogony"), an asexual mode of reproduction, and sporogony, a sexual phenomenon, were also described. The validity of this life cycle was questioned by other authors (see Lynch, 1917; Ciferri and Redaelli, 1938; Lavier, 1952). Binary fission of *Blastocystis*, also known as plasmotomic division (Beaurepaire Aragão, 1922 a,b; Knowles and Das Gupta, 1924), seems to have been readily accepted, and described in a number of studies (Chatton and Lalung Bonnaire, 1912; Lynch, 1917; Matthews, 1918; Knowles and Das Gupta, 1924; Wenyon, 1926; Tan and Zierdt, 1973; Zierdt and Swan, 1981; Yamada *et al.*, 1987a; Zierdt, 1988; Dunn *et al.*, 1989; Zierdt, 1991). Confusion exists, since a more recent report indicated that binary fission and plasmotomy were two distinct modes of reproduction (Zierdt, 1988). Detailed descriptions of the division modes and life cycle of *B. hominis* were also provided by Beaurepaire Aragão (1922a,b), Knowles and Das Gupta (1924), Ciferri and Redaelli (1938) and Lavier (1952). These reports debated the process of sporulation, with both endosporulation and exosporulation proposed and refuted. Budding was also questioned and merogony ("schizogony") suggested (Ciferri and Redaelli, 1938).

The life cycles and modes of reproduction described in these reports were extremely varied, and often based on very little evidence. It is always possible that some of these reports reflected division of organisms other than *Blastocystis*. Conclusive information on the life cycle of *B. hominis* cannot be gained from the study of these early reports, although they do provide good descriptions of the general morphology of the organism. Commonly, two distinct cell types were reported: a large cell with a large central vacuole, and a smaller cell type, usually described without a vacuole, and thought to be a resistant form. It is interesting to note that, until the very recent publications by Mehlhorn (1988a) and Stenzel and Boreham (1991), a smaller form of *B. hominis* had not been described in the modern literature.

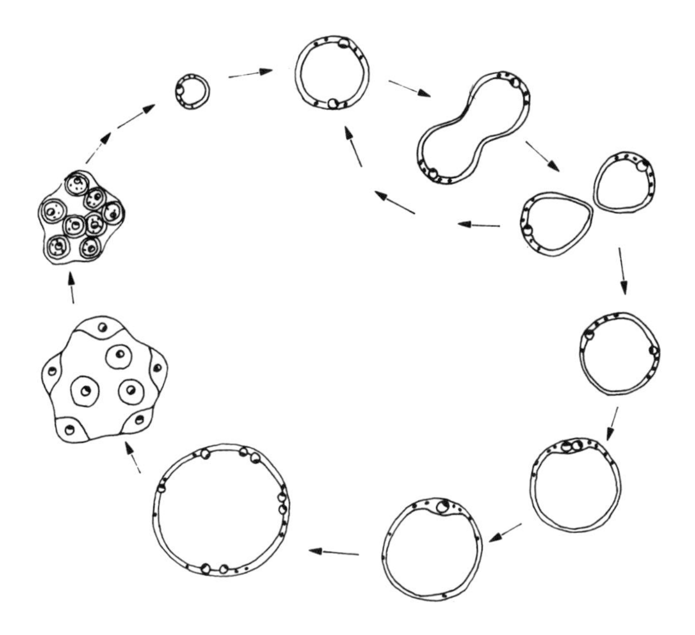

FIG. 7. Life cycle of *Blastocystis* proposed by Alexeieff (1911). Redrawn from original article; for description see text.

A new life cycle for *B. hominis* was proposed by Zierdt (1973), based on his light microscopical observations, and this has been used, unmodified and unquestioned, in the literature since that date (Fig. 8). This life cycle assumed the vacuolar form differentiated into either the granular form, which subsequently released vacuolar daughter cells from the central body, or to the amoeboid form which produced prospective vacuolar daughter cells by budding. Division by at least four modes, all asexual, was suggested (Zierdt, 1991). These were described as binary fission, plasmotomy, endodyogeny and "schizogony" (merogony), but were not illustrated in the proposed life cycle and little evidence was presented for their existence. "Schizogony" was reported, particularly in axenic cultures (see Zierdt and Tan, 1976b). The central vacuole membrane in the "schizont" (meront) was said to be lost, and the contents seen to merge with the cytoplasm. The viable progeny released by this process were thought to be resistant to air exposure, drying and suboptimal temperatures (Zierdt, 1991). Additionally, sporulation was said to occur (Zierdt and Williams, 1974), but no detail was provided. Zierdt and Williams (1974) observed the apparent union of two cells, and raised the possibility of sexual conjugation in *B. hominis*. In contrast, Zierdt (1991) indicated that sexual reproduction did not occur. A reproductive form not visually distinguishable from bacteria by light microscopy has been suggested to exist (Zierdt *et al.*, 1967), but has not been described further.

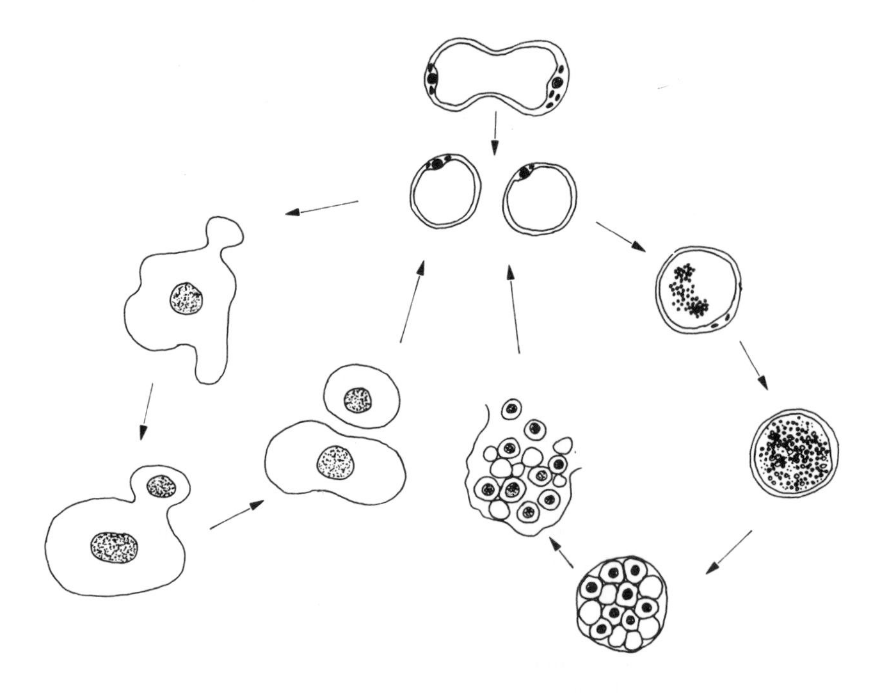

FIG. 8. Life cycle of *B. hominis* proposed by Zierdt (1973). Redrawn from original article; for description see text.

Extensive electron microscopical studies have failed to reveal sporogony, endodyogeny or merogony. Membrane-bound projections and cytoplasmic bridges which appear to divide the central vacuole (Dunn, 1992) may have given the appearance of these supposed means of division when examined by light microscopy. Thus, binary fission is the only proven method of reproduction for *B. hominis* at the present time. *B. hominis* cells dividing by binary fission are seen commonly by light microscopy, and less frequently by electron microscopy (McClure *et al.*, 1980; Matsumoto *et al.*, 1987; Yamada *et al.*, 1987a; Dunn *et al.*, 1989). The cells divide into two approximately equal portions, with the distribution of organelles into both halves. Division of organelles has not been observed, but multiple nuclei, mitochondria and other organelles are present in *B. hominis* cells.

F. DESCRIPTIONS OF FORMS

1. *Form seen* in vivo

Only two reports have given descriptions of the form of *B. hominis* thought to be present in the intestines of the host. Zierdt and Tan (1976a) described a patient with severe enteric disease, discharging copious volumes of diarrhoeal fluid. These authors believed that the unusual cells found in the diarrhoeal fluid were the trophozoites of *B. hominis*. A more recent report by Stenzel *et al.* (1991) described *B. hominis* cells obtained from a patient by colonoscopy. The morphology reported in both studies was similar. The cells were rounded to ovoid, approximately 5 μm in diameter, with a plasma membrane, but no surface coat (Fig. 9). Small vacuoles and vesicles were noted within the cytoplasm, but no larger vacuoles. The nucleus showed a crescentic band of condensed chromatin, as found in other forms of *B. hominis*, but was often larger than the nuclei previously described. An additional "spot" of condensed chromatin was often seen (Fig. 9), and Zierdt and Tan (1976a) indicated that this was the nucleolus (but see Section IV. G.3). When two nuclei were noted within these cells they were sometimes enclosed within the same perinuclear envelope (Stenzel *et al.*, 1991).

The mitochondria of this form of *B. hominis* differed morphologically from those of other forms. The mitochondrial matrix was relatively electron-lucent, with numerous cristae. The cristae were described as sacculate (Zierdt and Tan, 1976a) or lamellar (Stenzel *et al.*, 1991). In comparison, all other forms of *B. hominis* have been reported to have a simple mitochondrial structure, with few cristae, and usually an electron-opaque matrix (see Section IV.G.2). The significance of this difference in structure remains unknown, although it may indicate that different biochemical pathways operate in the different forms. Zierdt and Tan (1976a) successfully cultured organisms from the diarrhoeal samples of their patient and only vacuolar forms were present in culture.

2. *Fresh faecal form*

Light microscopy of faecal samples suggested that the form seen before inoculation into culture was generally smaller (4–15 μm) than that recorded for cultured *B. hominis* (15–25 μm) (Zierdt, 1991). Few reports, however, have detailed the ultrastructure of *B. hominis* in fresh faecal

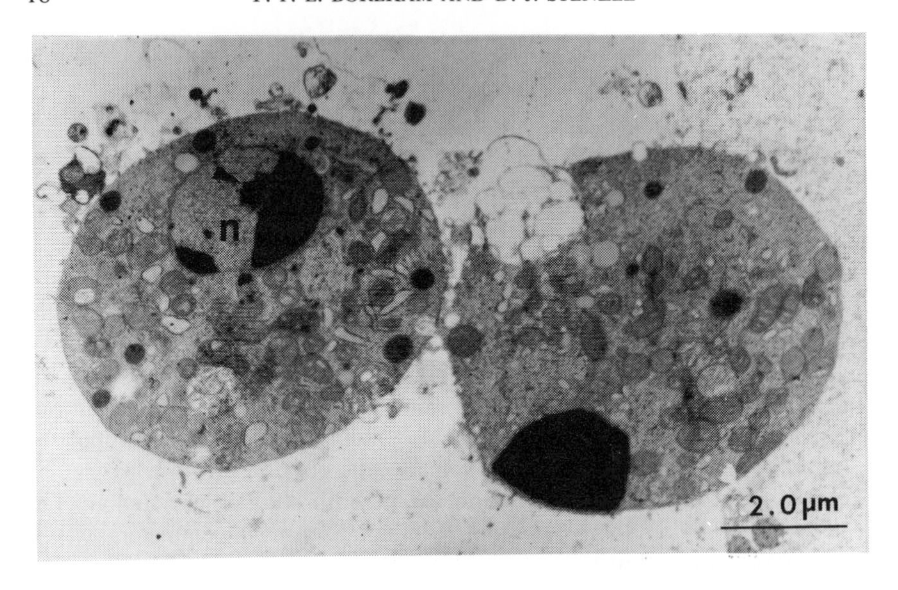

FIG. 9. Transmission electron micrograph of *B. hominis* cells obtained at colonoscopy. These cells are without a central vacuole. In addition to crescentic areas, a spot of condensed chromatin (indicated by arrow) is often seen within the nucleus (n).

samples, choosing instead to examine the forms found in culture. It is now known that cultivation *in vitro* modifies the morphology of *B. hominis* (see Stenzel *et al.*, 1991), resulting in descriptions not representative of the form found in the faeces or the host. Some earlier work also suggested that culture conditions could be used to select particular forms of *B. hominis* (see Zierdt, 1973), but these observations have been largely ignored.

The organisms present in human faecal material (Fig. 10) have been shown by TEM to be predominantly multivacuolar rather than containing the single large vacuole seen in cultured cells (Stenzel *et al.*, 1991). A very thick surface coat, often up to 0.5 µm, surrounded the cells, and bacteria were frequently seen associated with, or embedded in, this fibrillar matrix (Fig. 10). The vacuolar form, with the large central vacuole, was rarely present in faecal samples and, when found, was generally smaller than similar cells in culture.

Changes in morphology resulted when organisms from faeces were placed into culture (Stenzel *et al.*, 1991). Within 3 days of culture on egg slants, typical vacuolar cells were present. Granular forms were initially seen when faecal forms were placed into MEMS cultures, but after about 12 days vacuolar forms predominated.

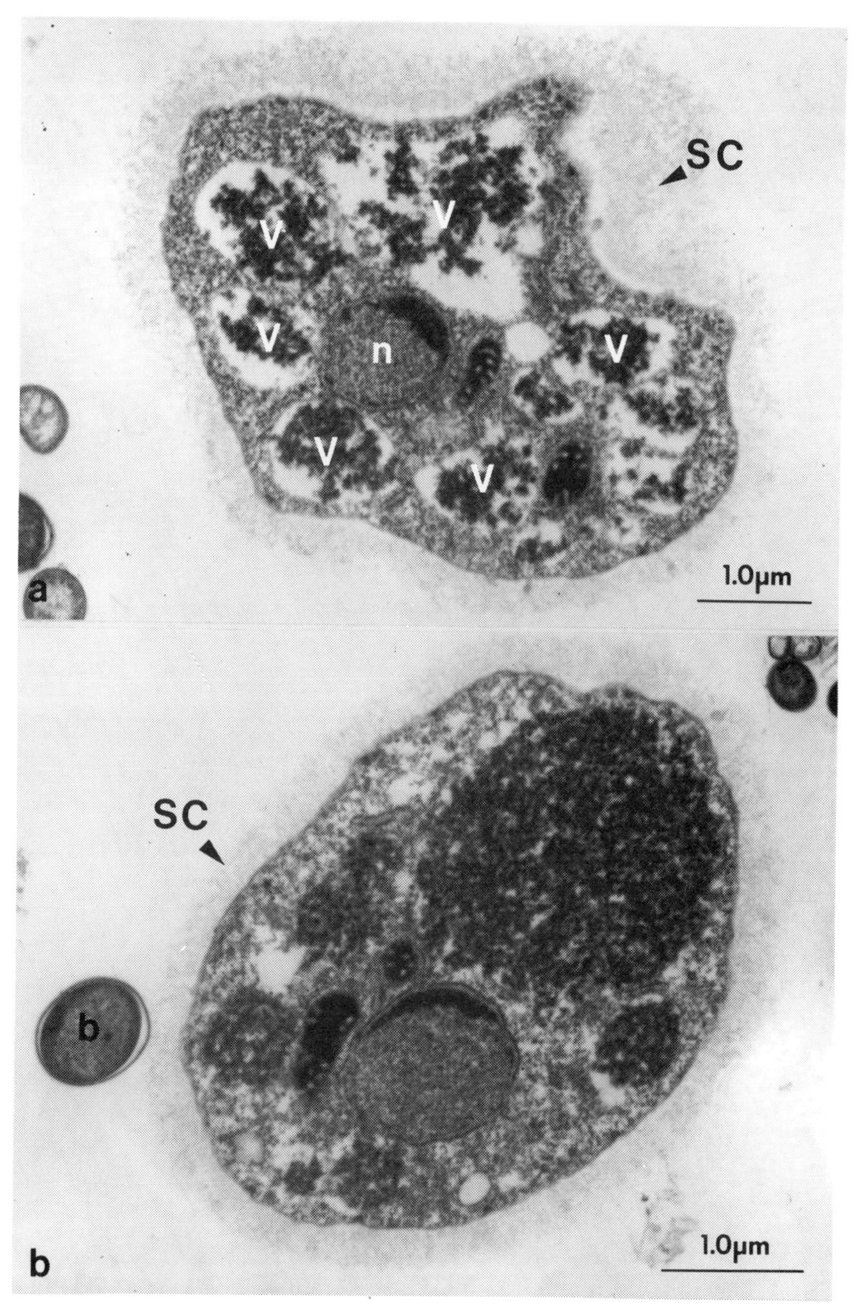

FIG. 10. Transmission electron micrographs of *B. hominis* from fresh human faecal sample. Both cells are from the same sample, but the morphology and vacuolar contents differ. (a) Multiple vacuoles (V), rather than a large, single central vacuole, are seen. The nucleus (n) shows a crescentic band of condensed chromatin. A very thick surface coat (SC) surrounds the cell. (b) A bacterium (b) is seen in close association with the surface coat (SC).

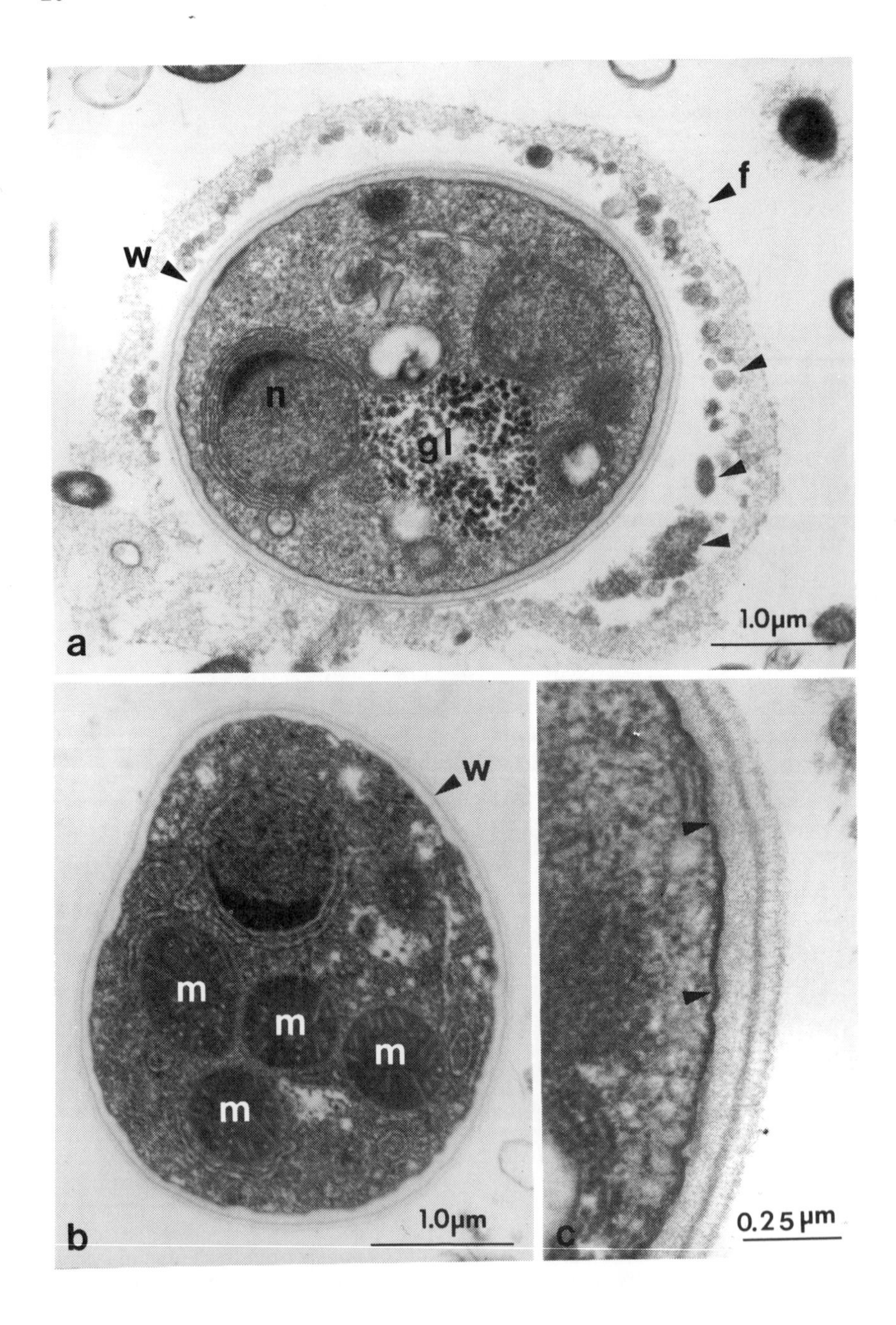
f
w
n
gl
1.0μm
a
w
m
m
m
m
1.0μm
b
c
0.25 μm

3. *Cyst form*

Although *Blastocystis* was originally considered to be the cyst of other protists, notably *Trichomonas* (see Bohne and Prowazek, 1908; Bensen, 1909; Prowazek, 1911; James, 1914), the presence of a cyst as part of the life cycle of *Blastocystis* has been extensively debated. The granular form was suggested to be the cyst (Zierdt *et al.*, 1967), but later reports describing the life cycle refuted this suggestion, claiming that a cyst does not exist (Zierdt, 1988, 1991). Only recently has conclusive evidence been presented that *Blastocystis* does have a cyst form (Fig. 11).

Mehlhorn (1988a) described the ultrastructure of cyst-like *B. hominis* cells obtainedfrom the fresh faeces of a patient with the acquired immune deficiency syndrome (AIDS) and, more recently, Stenzel and Boreham (1991) reported the presence of cyst-like forms of *B. hominis* in fresh human faeces and in long-term cultures. These cyst-like forms were more common in stored faecal specimens, suggesting that the form may be a mechanism of survival in the external environment.

The cyst-like organisms were smaller than the typical cultured forms, with a thick wall and condensed cytoplasm, as is typical for other protozoan cysts (Mehlhorn, 1988b). Many small vacuoles were present in the cytoplasm, rather than the large single vacuole typical of the cultured organism, and the typical nuclear structure of *B. hominis* was seen (Mehlhorn, 1988a; Stenzel and Boreham, 1991). Cysts have been reported to be 5–10 μm (Mehlhorn, 1988a) and 3.7–5 μm in diameter (Stenzel and Boreham, 1991). There are other differences between the two reports which need to be resolved. Whereas Mehlhorn (1988a) was unable to distinguish mitochondria, Stenzel and Boreham (1991) reported mitochondria with an electron-opaque matrix and few cristae (see Fig. 11b). Lipid inclusions were frequently present (Stenzel and Boreham, 1991), and glycogen deposits were also present. In organisms isolated from monkey faeces, these glycogen deposits may be

---

FIG. 11. Transmission electron micrographs of *B. hominis* cysts from fresh human faecal samples. (a) Cyst with a thick wall (w), showing an outer fibrillar layer (f) with similar morphology to the surface coat. This layer may surround the entire cyst and enclose dense material (indicated by arrows). The layer appears to dissociate later from the cyst wall to result in a bare cyst as shown in (b). Long strands of rough endoplasmic reticulum partially enclose the nucleus (n) in this micrograph, and glycogen (gl) is present in the cytoplasm. (b) Micrograph of a cyst, surrounded by a wall (w), but without a fibrillar layer. Mitochondria (m) are present, often with rough endoplasmic reticulum closely associated. (c) High magnification of cyst wall, showing multilayered composition. A membrane (indicated by arrows) is seen surrounding the cytoplasm, but beneath the cyst wall.

extensive (D. J. Stenzel, unpublished observations); possibly they function as metabolic reserve material, as has been described for other protozoan cysts (Mehlhorn, 1988b).

In the cyst-like forms, up to four nuclei were described by Mehlhorn (1988a), whereas only one nucleus was seen by Stenzel and Boreham (1991). However, cysts from monkey faeces were multinucleate and larger than those reported from humans (D. J. Stenzel, P. F. L. Boreham and K. Lai Peng Foon, unpublished observations). In *Macaca* monkeys cysts were very common, and in one animal accounted for about 90% of the *Blastocystis* organisms seen. It remains to be determined if these different cyst morphologies represent different species of *Blastocystis*.

The cyst wall is a thick, multilayered fibrillar structure (Fig. 11c). An additional outer layer, morphologically similar to the surface coat described on vacuolar and granular forms of *B. hominis*, was present on some cells (Stenzel and Boreham, 1991, 1993) (Fig. 11a). This suggests that the cyst wall may be formed beneath the surface coat and, after maturation of the wall, the surface coat is shed from the organism to reveal a typical cyst. The composition of the wall is unknown, and its ability to protect the organism from environmental factors has not been investigated. Silard and Burghelea (1985) presented electron micrographs of morphologically altered cells in a drug-resistant strain of *B. hominis* which, although not identified as such, appeared closely to resemble the cyst-like forms of *B. hominis* described above. Perhaps this explains why this isolate was able to survive in the presence of antiprotozoal drugs.

It now seems apparent that a cyst form of *B. hominis* does exist and provides a possible mechanism of survival outside the host, and the means of transmission between hosts. The vacuolar and granular forms are both known to be sensitive to environmental conditions (Zierdt, 1991), and appear unable to survive temperature changes, hypertonic or hypotonic environments, or exposure to air (Matsumoto *et al.*, 1987), and hence seem unlikely to provide a mode of transmission. Experimental proof of transmission via the cyst form must await development of a suitable animal model.

4. *Vacuolar form*

The vacuolar (or vacuolated) form (Fig. 12), also referred to as the "central body" form (Zierdt, 1991), was first defined by light microscopy (Zierdt *et al.*, 1967). Until recently, it has been considered to be the "typical" *Blastocystis* cell form.

Vacuolar cells are usually spherical, and thus show rounded profiles in transmission electron micrographs, although some irregularly shaped cells

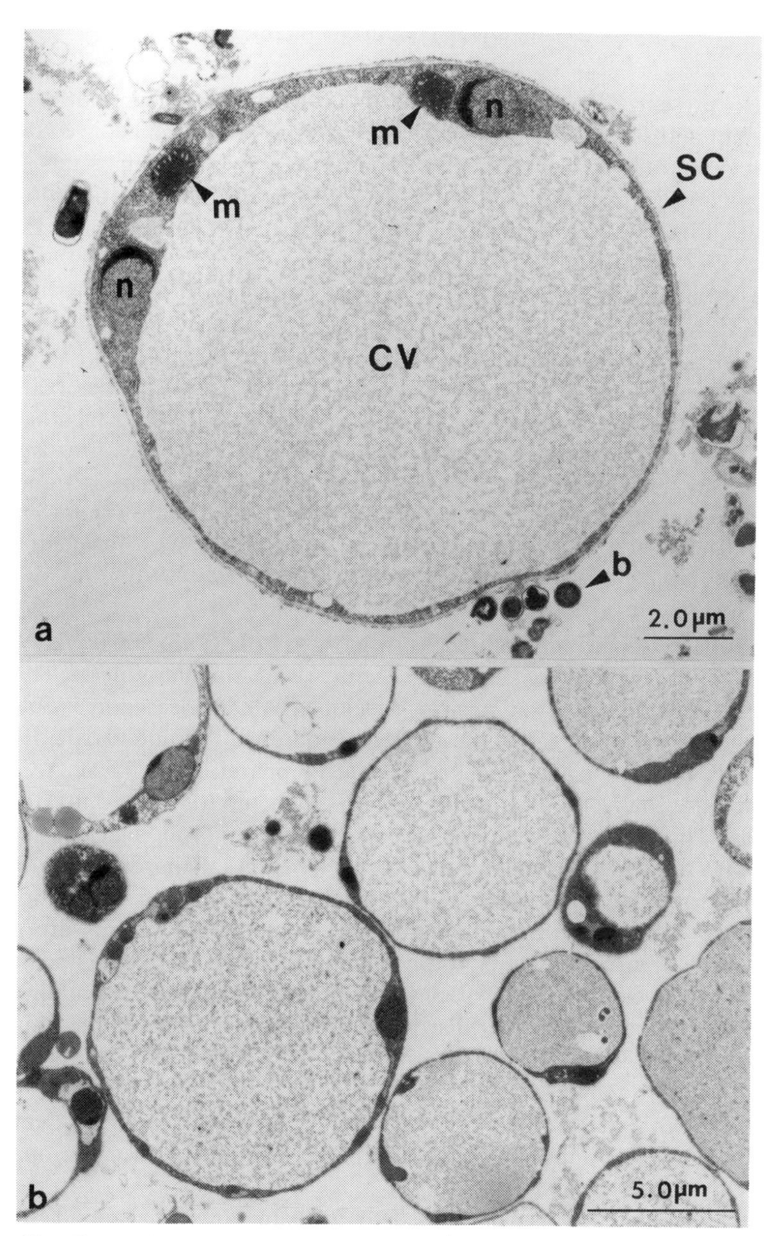

FIG. 12. Transmission electron micrographs of the vacuolar form of *B. hominis* from laboratory culture. (a) High magnification micrograph of stock BRIS/87/HEPU/8 showing cellular morphology. A thin band of cytoplasm, containing organelles, surrounds a large central vacuole (CV). Two nuclei (n) are present in this section of the cell, both showing the crescentic band of chromatin typical of *B. hominis*. Bacteria (b) are closely associated with the surface coat. m, Mitochondrion. (b) Low magnification micrograph showing variability of the cell morphology of Netsky stock vacuolar forms in culture. Central vacuole contents vary in appearance.

are also present (Fig. 12). Variations in ultrastructure occur, but appear consistent and stable for each isolate, even during prolonged culture *in vitro* (Dunn *et al.*, 1989). Cells vary greatly in size, ranging from 2 μm (van Saanen-Ciurea and El Achachi, 1985) to 200 μm in diameter (Zierdt and Tan, 1976b). The average diameter of cells is usually between 4 and 15 μm (Zierdt, 1991), but it does vary between isolates.

The vacuolar cell appears as a thin peripheral band of cytoplasm surrounding a large central vacuole (Fig. 12). A surface coat of varying thickness, sometimes called a slime layer or capsule (Zierdt *et al.*, 1967), is present on some cells. The limiting cell membrane surrounding the organism has been shown to contain coated pits (Tan and Zierdt, 1973) which have a function in endocytosis (Stenzel *et al.*, 1989). The organelles are usually gathered in thickened areas of cytoplasm, which often appear at opposite poles of the cell. These may protrude into the central body (Dunn *et al.*, 1989; Zierdt, 1991), or extend outwards (Dunn *et al.*, 1989) to give the cell an irregular outline. Most of the organelles seen in *Blastocystis* are simply representative of their type.

Multiple nuclei are present in cells from most isolates (Dunn *et al.*, 1989), with four being commonly reported (Zierdt, 1973; Matsumoto *et al.*, 1987). They are surrounded by a nuclear envelope, with nuclear pores. Several reports have noted the lack of a nuclear membrane (Tan and Zierdt, 1973), but this is likely to be an artefact of specimen preparation for TEM. A Golgi complex is usually present (Dunn *et al.*, 1989), in close proximity to the nucleus. This structure generally appears to be a morphologically simple type, with few cisternal stacks and associated vesicles. Ribosomes are found in the cytoplasm, and appear to be of eukaryotic size. Short strands of endoplasmic reticulum are often seen, although longer strands are rare. There have been reports of ribosomes lining the central vacuole and small vesicles in the cytoplasm, together with rough endoplasmic reticulum surrounding the mitochondria (Silard and Burghelea, 1985). Lipid droplets have been seen in the cytoplasm and vacuole of some cells (Dunn *et al.*, 1989).

Mitochondria are present, varying in morphology and number (Dunn *et al.*, 1989). Cristae appear as simple structures within an electron opaque matrix, typical of protists.

The central vacuole is surrounded by a bilaminar membrane (Tan and Zierdt, 1973; Matsumoto *et al.*, 1987; Dunn *et al.*, 1989) and shows considerable variation in the morphology of its contents, ranging from completely electron-lucent to electron-opaque, fully distended vacuoles. The contents are usually finely granular or flocculent and unevenly distributed within the vacuole (Dunn *et al.*, 1989). Vacuoles may show development of numerous granules of many morphological types (Tan and Zierdt, 1973).

Hence, Zierdt and Williams (1974) stated that the structure was not a true vacuole, but was the site of development of granules during the transition to the granular form. The vacuole was also reported to function in endogeny and merogony ("schizogony") (Zierdt, 1988, 1991), but this has not been supported by further studies.

5. *Granular form*

The granular form (Figs 13, 14) has a similar ultrastructure to the vacuolar form, apart from the nature of its central vacuolar contents (Tan and Zierdt, 1973; Dunn *et al.*, 1989) and often being slightly larger. Diameters of 10–60 µm (Zierdt *et al.*, 1967), 15–25 µm (Zierdt, 1973), 3–80 µm (Zierdt and Williams, 1974), and 6.5–19.5 µm (Dunn *et al.*, 1989) have been reported.

The granules in the central vacuole were first described in detail by Zierdt *et al.* (1967), although descriptions by Lavier (1952) included a cell type in which the vacuole was entirely filled with refringent spheres. Zierdt *et al.* (1967) proposed two granule types based on light microscopy studies. One type was thought to develop into *B. hominis* cells, and another, located at the cell periphery, was suggested to have a role in metabolism. This was later extended by TEM studies (Tan and Zierdt, 1973), as a result of which three types of granules were referred to as metabolic, lipid and reproductive granules respectively. Metabolic granules were cytoplasmic, lipid granules found in the central vacuole and cytoplasm, and the reproductive granules were present in the central vacuole (Zierdt, 1973). These observations were not confirmed by Dunn *et al.* (1989), who noted myelin-like inclusions, small vesicles, and crystalline granules and lipid droplets in the central vacuole of the granular form. Lipid droplets were also present in the cytoplasm. Small vacuoles and vesicles in the cytoplasm of granular cells also contained granules of similar appearance to those in the central vacuole (Yoshikawa *et al.*, 1988; Dunn *et al.*, 1989).

Subsequent development of "reproductive granules" within the central vacuole must be considered as unproven, and the descriptions of the granules present are unlike those expected for viable entities. It is much more likely that they are the result of metabolism or storage within the cell (Dunn *et al.*, 1989). A suggestion that the granules are mitochondria (Zierdt, 1991) must similarly be rejected.

Cytochemical studies have indicated that many of the granules are composed of lipids, in particular phospholipids and fatty acids (Zierdt, 1973; Dunn *et al.*, 1989), and probably represent a form of energy storage. Small crystalline granules present in some cells were suggested to be protein (Dunn, 1992). The differing morphology of the granules may reflect their differing composition.

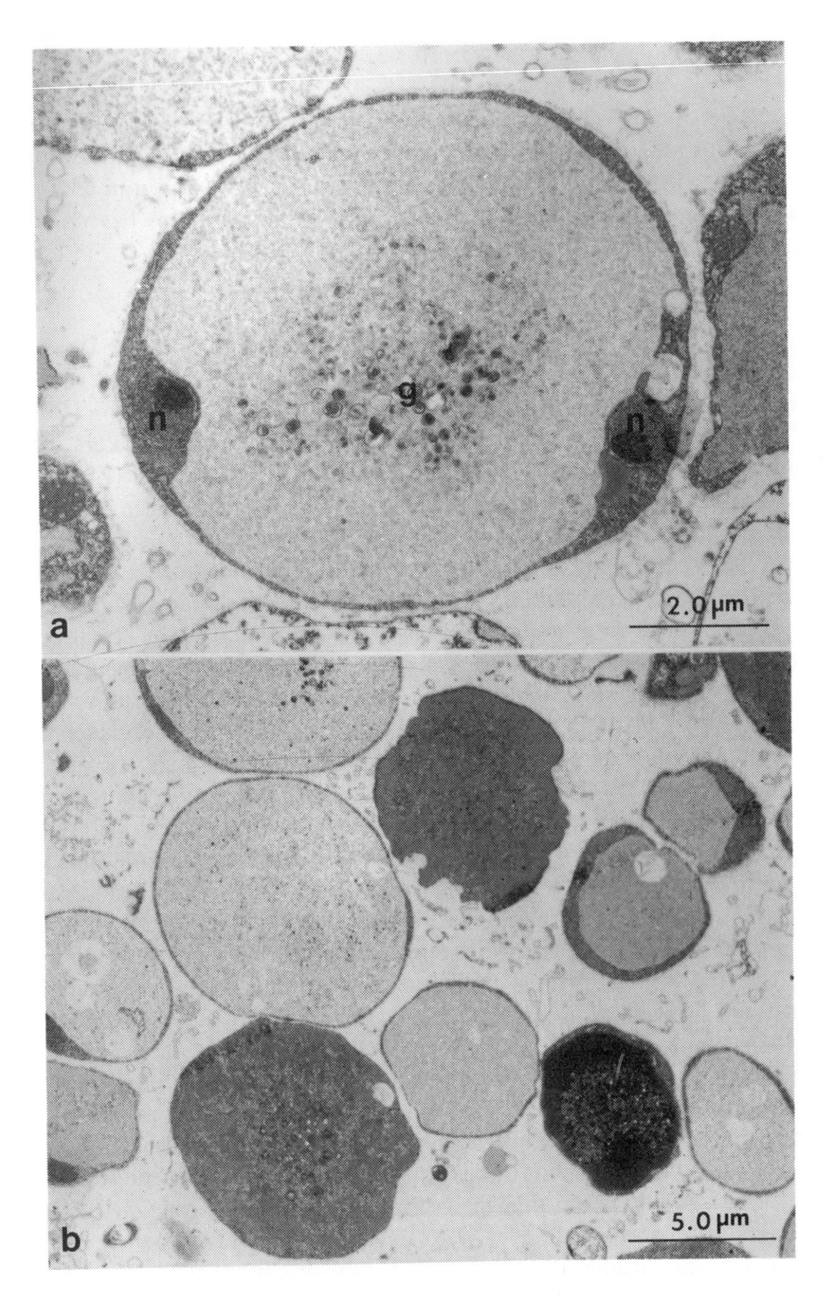

FIG. 13. Transmission electron micrographs of the granular form of *B. hominis* Netsky stock induced in laboratory culture. (a) Granular cell, induced by culture in MEMS, with similar morphology to the vacuolar forms, apart from the granules (g) present in the central vacuole. n, Nucleus. (b) Granular cells (induced by increasing the concentration of horse serum in the culture medium) show differing morphology; the appearance of the vacuolar contents varies greatly. Granules may be concentrated in the central portion or be dispersed throughout the central vacuole.

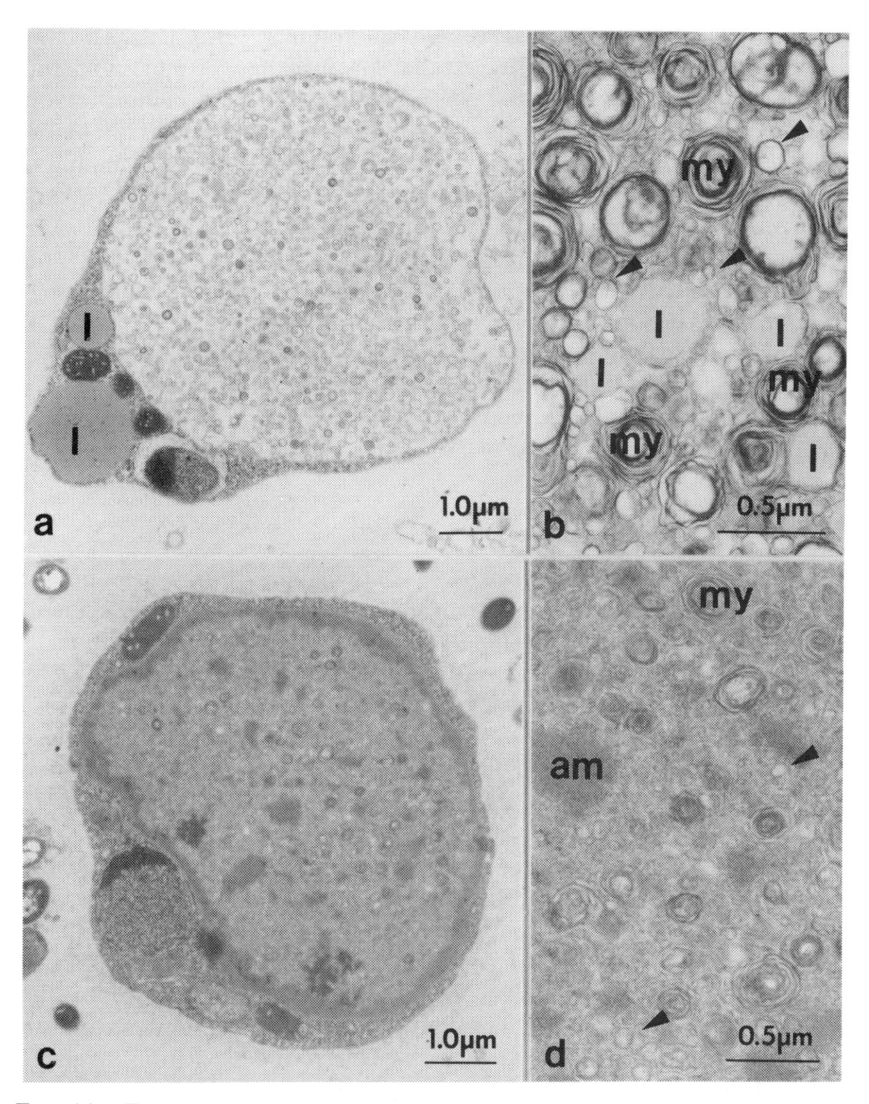

FIG. 14. Transmission electron micrographs of granular forms of *B. hominis* induced by altered culture conditions. (a) Granular form of Netsky stock showing a large accumulation of lipid (l) in the cytoplasm, and myelin-like granules in the central vacuole. This form was induced by the addition of high concentrations of horse serum to MEMS culture. (b) High magnification of granules from granular form. Small vesicular structures (indicated by arrows), lipid droplets (l) and myelin-like granules (my) are present. (c) Granular form of stock BRIS/87/HEPU/11 induced by the addition of norfloxacin to the culture medium. (d) High magnification of the central vacuole seen in (c). In addition to vesicular structures (arrows) and myelin-like accumulations (my), electron-opaque amorphous material (am) is present.

The frequency and occurrence of the granular form has also been disputed. Conflicting reports suggest that this form may be more numerous in older cultures (Zierdt *et al.*, 1967), or that short-term cultures favour its development (Zierdt and Williams, 1974). Various conditions are known to induce granular form formation from the vacuolar form, including (i) increased serum concentrations in the culture medium (Lavier, 1952, Zierdt, 1973; Silard, 1979; Dunn *et al.*, 1989) (Figs 13b, 14a,b), (ii) transfer from egg slant medium into MEMS media (Stenzel *et al.*, 1991) (Fig. 13a), (iii) axenization (Zierdt and Williams, 1974), and (iv) addition of some antibiotics, in particular norfloxacin and amphotericin B (Dunn, 1992) (Fig. 14c,d). It is apparent that the vacuolar form transforms into the granular form under a variety of adverse conditions. Thus there seems little evidence that the granular form should be considered a distinct form, separate from the vacuolar form, in the life cycle of *B. hominis*.

6. *Amoeboid form*

The amoeboid form (Fig. 15) of *B. hominis* requires further definition, as there are conflicting reports on its morphology (see Dunn *et al.*, 1989; Zierdt, 1991). Clarification of the terminology is also required, as the terms amoeba-like form (Zierdt *et al.*, 1967), amoeba form (Tan and Zierdt, 1973, Zierdt, 1991), amoebiform (Zierdt, 1988) and amoeboid form (Dunn *et al.*, 1989) have all been used. We propose to use the term amoeboid in this review. Distinct morphological differences within the amoeboid cell need to be determined, and variation in cell shape alone should not be used as the sole criterion for distinguishing a separate form.

Zierdt (1973) described an amoeboid form which appeared in small numbers in older cultures and those treated with antibiotics, and occasionally in faecal samples. The form was described as lacking a central vacuole, and was irregular and lobed in outline. Pseudopods were present, but the report did not indicate if they were involved in motility of the organism. The cells were reported to feed on bacteria. One or two nuclei were seen, at the centre of the cell. Stalked appendages were seen to detach from the parent cell, and this was presumed to be a form of reproduction.

Tan and Zierdt (1973) described a quite different morphology. Again, these cells were present in low numbers only, in stools and in culture. A central body occupying about one-third of the cell was reported, but it did not appear to be surrounded by a limiting membrane. It was also noted that, in organisms from cultures with a higher serum content, this central body, like the granular form, contained electron-lucent granules. The cells were oval in outline, with one or two large pseudopods. A distinct cell wall or membrane was not seen, but dense pockets, which probably corresponded to

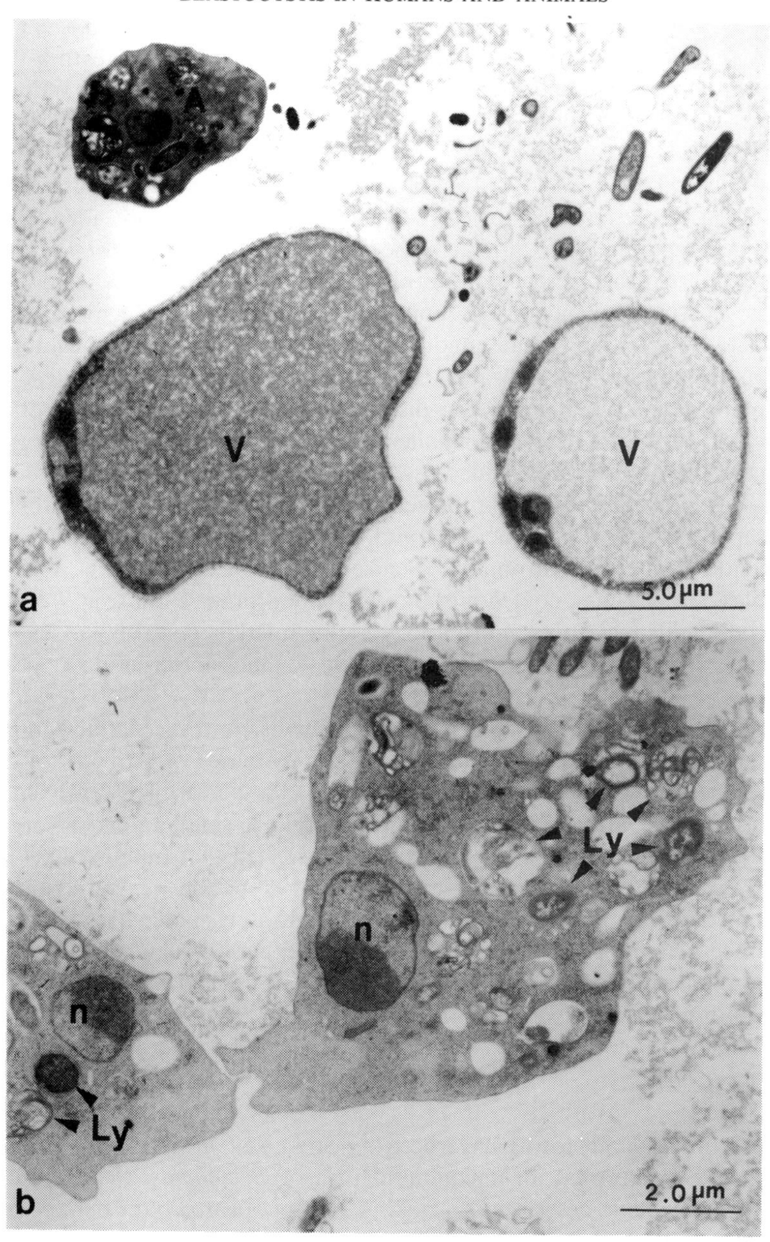

FIG. 15. Transmission electron micrographs of the amoeboid form of *B. hominis* stock BRIS/87/HEPU/1 from laboratory culture. (a) Vacuolar (V) and amoeboid (A) forms, demonstrating the size differences between the two forms. (b) High magnification of amoeboid forms, showing nuclear structure. Bacteria are present in lysosome-like inclusions (Ly) in the cell cytoplasm. n, Nucleus.

the coated pits described by Dunn *et al.* (1989), were seen beneath a thick filamentous surface layer. The nucleus was not surrounded by a double nuclear membrane, but was limited by the cytoplasm. The only criterion for defining these organisms as the amoeboid form appears to be the lack of membranes. This was most probably an artefact of sample preparation or sectioning of the material for electron microscopy, and is not adequate to describe a separate form. Similarly, the description by McClure *et al.* (1980), which reported the presence of a central body surrounded by a membrane, but no cell-limiting membrane, appears to be the result of a preparation artefact. It is unlikely that an organism would remain viable without a surrounding membrane or wall.

Dunn *et al.* (1989) noted the discrepancies in the descriptions of the amoeboid form and described smaller cells (2.6–7.8 μm diameter), irregular in shape and often with extended pseudopodia. Engulfed bacteria were seen in lysosome-like bodies within the cell (Fig. 15b). The nuclear structure was the same as that of the vacuolar and granular forms, with the band of crescentic condensed chromatin (Stenzel and Boreham, 1993). A Golgi complex, surface coat, coated pits and mitochondria were not seen (Fig. 15). The possibility that this morphology represented a distinct species was considered, but in view of the characteristic nuclear structure it seems unlikely.

The mode of division among amoeboid cells is controversial and unproven. It has been suggested that reproduction is by budding, with one to three smaller cells developing around the parent cell's periphery (Zierdt *et al.*, 1967), which separate after several hours to release small amoeboid organisms. However, Zierdt and Williams (1974) asserted that budding does not occur. Sporulation was reported by Sun (1988), and reproduction by plasmotomy, forming progeny without a central vacuole, has also been suggested (Zierdt, 1991). There appears to be no good evidence to support any of these reproductive modes. No conclusive information is available on the position of the amoeboid form in the life cycle of *Blastocystis*.

7. *Other forms*

Several unusual cell forms have been reported, but little detail was given. Zierdt (1991) reported an uncommon form of *B. hominis*, the "schizont" form. An adequate description is lacking, but it appeared to be derived from vacuolar cells. The cell was reported to fill with progeny by an asexual reproductive process which caused it to burst and release its progeny. Another cell type was reported to have no cytoplasm, but this was unlikely to have been a living cell, but rather the residual central vacuole often found with other cell debris, resulting from degenerating cells in culture. This debris often appears to contain minute *B. hominis* cells when examined by light microscopy. Electron microscopical examination reveals the true nature of the debris, which consists of membranous material, degenerating and disrupted organelles, and flocculent clumps of cytoplasmic and vacuolar material (D. J. Stenzel, unpublished observations). Little is known of the effects of drugs on the morphology of *Blastocystis*, and this may have accounted for some of the anomalies reported in the literature.

G. STRUCTURES OF UNCERTAIN FUNCTION

1. *Central vacuole*

The central vacuole of *Blastocystis* was first described by Alexeieff (1911), and even this report suggested an alternative name of "internal body" for the structure. Since that time, the vacuole of *B. hominis* has also been known as the reserve body and *Innenkörper* (Lavier, 1952), and the central body (Zierdt, 1991). It is more important to maintain a consistent terminology for this structure than to debate the most appropriate term on the limited evidence of function currently available. We have chosen to describe the structure as the central vacuole, defined as a membrane-bound structure with chemically and functionally unidentified contents.

Although Zierdt *et al.* (1967) indicated that the central vacuole had a role in metabolism and reproduction, evidence for this was not given. Cytochemical stains have been used to examine the central vacuole, in an attempt to determine its function, but the results have often been contradictory.

Although often considered to play a role in reproduction, the central vacuole has not been shown to contain nucleic acids. A study by Tan and Zierdt (1973), using methyl green–pyronin staining, indicated that RNA was not present in the central vacuole, although staining was seen in the peripheral cytoplasm. Similarly, no fluorescence was seen in the vacuole when DAPI (4,6-diamidino-2-phenylindole) staining for DNA was performed, although the nucleus was strongly fluorescent (Matsumoto *et al.*, 1987).

Zierdt (1973) asserted that the central vacuole had no role in storage in *B. hominis*, although data from other studies suggested that, at least in some circumstances, material may accumulate or be stored in the vacuole (Yamada *et al.*, 1987a; Stenzel *et al.*, 1989). Iodine staining for starch has usually given negative results (Zierdt, 1973; McClure *et al.*, 1980), although occasionally stained cells have been seen (Yamada *et al.*, 1987a). This may reflect accumulation of starch under adverse culture conditions (Kofoid *et al.*, 1919; Silard 1979). Glycogen and lipids were not found in the central vacuole (Zierdt, 1973; McClure *et al.*, 1980), although Tan and Zierdt (1973) were able to detect lipids in the central vacuole of the vacuolar form. Lipids are present in the granular form (Lavier, 1952; Tan and Zierdt, 1973; Dunn *et al.*, 1989) and in older cells (Zierdt *et al.*, 1967). These results suggest that the central vacuole may have a function in the storage of some metabolic products, but further studies are required.

Cytochemical analysis has given negative results for acid phosphatase, a marker for lysosomes (Stenzel *et al.*, 1989), and alkaline phosphatase (Zierdt *et al.*, 1988). Further analysis of enzyme activity within the central vacuole may provide evidence of its function.

Endocytosis is known to occur, at least in the vacuolar and granular forms (Stenzel *et al.*, 1989). Cationized ferritin, an electron-opaque marker for endocytosis, was used to follow uptake from the external environment. Tracer ferritin was noted in small vesicles and vacuoles within the cytoplasm and, ultimately, deposited in the central vacuole (Fig. 16). It was assumed that these cytoplasmic structures fuse with the central vacuole to release their contents, and a similar mechanism has been suggested for the deposition of granules into the central vacuole (Yoshikawa *et al.*, 1988). Conclusive evidence for this fusion has yet to be obtained, and due to the dynamic and rapid nature of membrane fusion it is unlikely that electron microscopy alone will provide this evidence. Fluorescent dyes to trace endocytotic events using confocal microscopy offer the best prospects for following dynamic events in living cells.

The passage of material from the exterior through coated pits during endocytosis appears similar to that described for other eukaryotic cells, including a number of protists. Antibodies to clathrin bind to the coated pits of *B. hominis*, suggesting that their composition is similar or identical to that of mammalian cells (Stenzel *et al.*, 1989).

Endosymbionts have been reported in the central vacuole of *B. hominis* (see Zierdt and Tan, 1976b). Bacillary and ovoid forms, enclosed by a membrane, were reported in close proximity to the central vacuole membrane. Other workers have not confirmed this observation, and these structures could be explained by the projection of membrane-bound cytoplasm into the central vacuole, as described by Dunn *et al.* (1989). Sectioning cells containing these projections for TEM often results in circular or elongated profiles, apparently separate from the cytoplasm. Other profiles reveal a continuation with the peripheral cytoplasm. The report by Zierdt and Tan (1976b) indicated that only a few cells in culture displayed endosymbionts, and that axenization with antibiotics increased their frequency. This correlates with the data of Dunn (1992), which indicated that the presence of cytoplasmic bridges and projections into the central vacuole increased during unfavourable culture conditions, and particularly in cul-

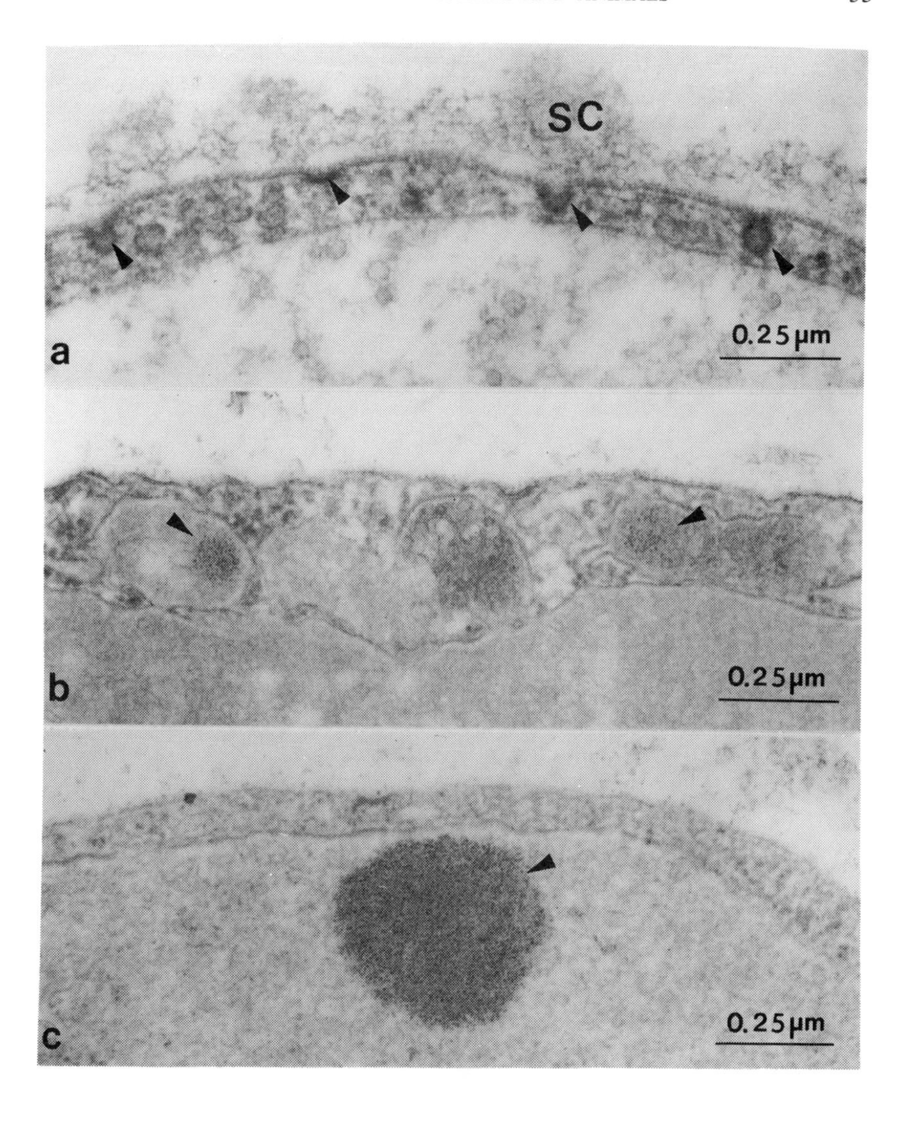

FIG. 16. Transmission electron micrographs of endocytotic structures in the vacuolar forms of *B. hominis*. (a) Coated pits (indicated by arrows) are seen on the outer membrane of stock BRIS/87/HEPU/5, beneath the surface coat (SC). (b) The electron-opaque tracer for endocytosis, cationized ferritin, can be used to follow endocytotic events. Cells (Netsky stock) sampled after 5 min incubation with cationized ferritin showed the tracer (indicated by arrow) present in vesicles within the cell cytoplasm. (c) After 30 min incubation, the cationized ferritin is found as large aggregations (shown by arrow) in the central vacuole, indicating a role for this structure in uptake of external substances.

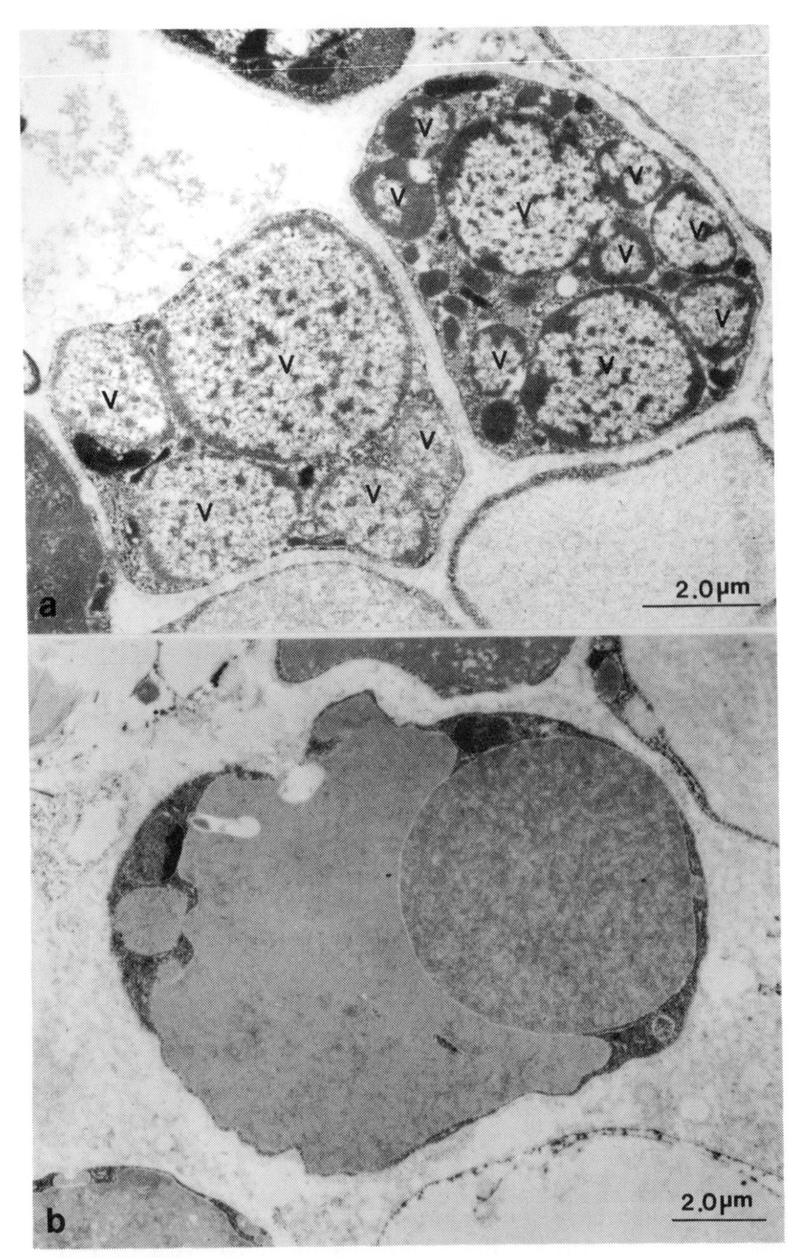

FIG. 17. Transmission electron micrographs of atypical *B. hominis* forms from Netsky stock, induced by altered culture conditions. By light microscopy, these forms may be easily misconstrued as undergoing sporulation or endodyogeny. The altered appearance is due to the segregation of the central vacuole into smaller, multiple components. (a) Apparent multiple vacuoles (v), as shown in these cells, are common in cultures with increased concentrations of horse serum. (b) Cell from culture in medium 199 with division of the central vacuole, often by cytoplasmic inclusions containing organelles.

tures containing high concentrations of antibiotics. The presence of cytoplasmic strands in the central vacuole, often appearing to divide the vacuole into compartments (Fig. 17), has probably also been mistaken for division and reproduction of *B. hominis*.

The true nature of the central vacuole and its contents, and their function in the biology of the organism, remain unresolved. The evidence does suggest that the central vacuole has a storage function and, given the observed diversity of ultrastructural appearances, it is not inconceivable that storage products of many different compositions exist. Whether the central vacuole has further digestive functions or is merely a repository for waste or storage materials is unknown.

2. *Mitochondria*

The mitochondria of *B. hominis* show considerable morphological variation (Fig. 18), both within isolates and within individual cells. Although generally appearing rounded or slightly elongated in profile, the mitochondria may be more elongated or sinuous in some cells and occasionally irregular profiles may be seen. Mitochondria are generally 0.5–1 μm in diameter, and although they may be found around the entire rim of peripheral cytoplasm, they tend to be concentrated near the nucleus.

The small size of the mitochondria in *B. hominis* makes light microscopy observations of limited value. TEM studies show that the mitochondria (Fig. 18d) of *B. hominis* are surrounded by a bilaminar outer membrane, and have an inner bilaminar membrane from which the cristae arise (Zierdt *et al.*, 1988). This is the classical structure of mitochondria (see Ghadially, 1988). The cristae appear as short tubules or bulbs (Zierdt *et al.*, 1988), typical of protists but unlike those of metazoa and yeasts (see Ghadially, 1988). Circular profiles of cristae have also been reported (Dunn *et al.*, 1989). Ribosomes and rough endoplasmic reticulum are often seen in close contact with the mitochondria (Silard *et al.*, 1983). Freeze fracture studies by Yoshikawa *et al.* (1988) confirmed that the mitochondria are usually rounded, and of a size similar to, or slightly smaller than, the nucleus.

Variations in the apparent electron-opacity of the mitochondria occur (Silard *et al.*, 1983) (Fig. 18a,c), often within the same cell (Dunn *et al.*, 1989). This may relate to the energy state of the organelle. However, these differences do not correlate with the ultrastructural changes related to differing energy states described by Hackenbrock (1968). Crystalline inclusions (Dunn, 1992) (Fig. 18d) and osmophilic inclusion bodies (Matsumoto *et al.*, 1987) are present in some mitochondria, and may represent protein storage, crystallized enzymes, or degenerative stages. Although the numbers of mitochondria per cell have not been accurately determined, considerable variation can occur, ranging from two to four in an individual cell to hundreds in giant and old cells (Zierdt *et al.*, 1988).

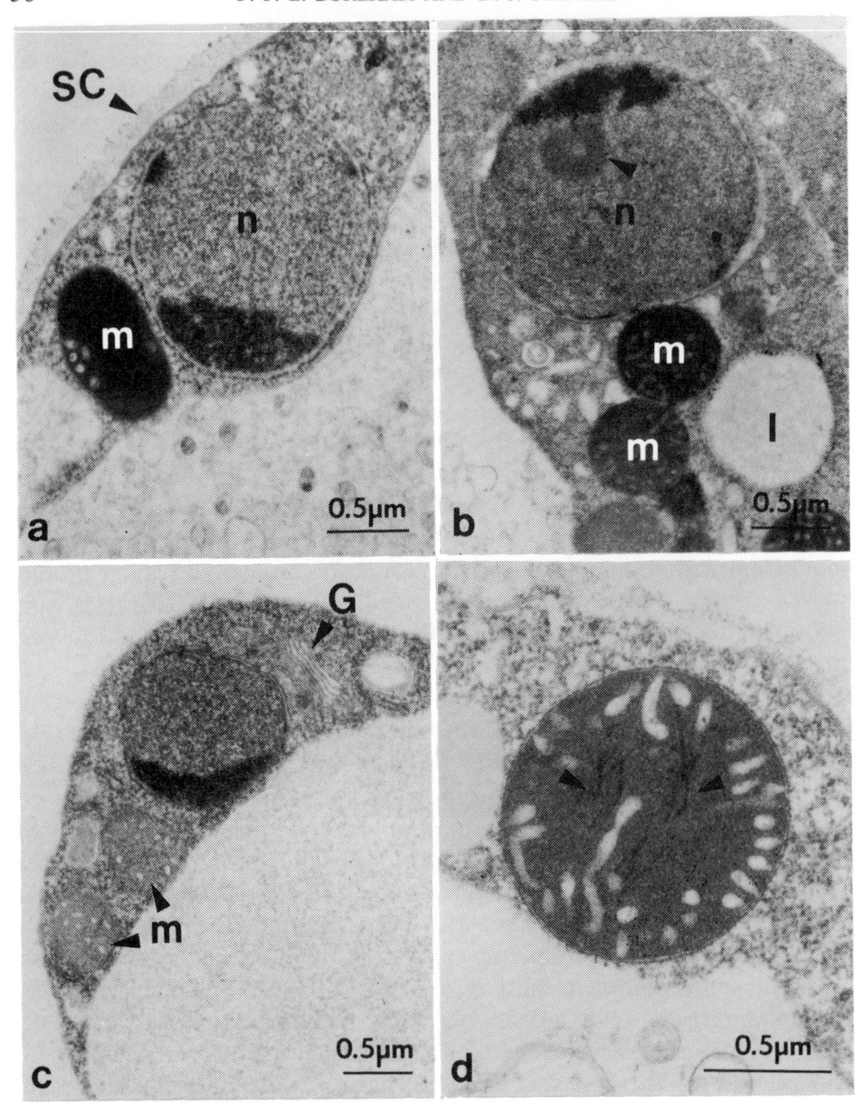

FIG. 18. High magnification transmission electron micrograph showing morphology of organelles in *B. hominis* cells. (a) Nucleus (n) showing typical crescentic band of condensed chromatin. Mitochondria (m) are electron-opaque, and generally have few cristae. SC, surface coat. (b) Nucleus (n) with an additional "spot" (indicated by arrow) of condensed chromatin. This may represent the nucleolus. m, Mitochondria; I, lipid inclusion. (c) Mitochondria (m) with electron-lucent matrices. Golgi complex (G) is present, but is not seen in all sections of *B. hominis* cells. (d) Mitochondrion showing crystalline inclusions (indicated by arrows). Cristae appear bulbous or as short tubules which rarely extend deeply into the mitochondrial matrix.

The mitochondria of *B. hominis* stain selectively with Janus green (Lavier, 1952; Zierdt and Tan, 1976a; Zierdt, 1986; Zierdt *et al.*, 1988; Dunn, 1992), and rhodamine 123 is taken up exclusively by the mitochondria (Zierdt, 1986; Zierdt *et al.*, 1988). The accumulation of this cationic dye is dependent upon the presence of a mitochondrial transmembrane potential (see Divo *et al.*, 1985). Thus, the mitochondria of *B. hominis* are able to generate energy and maintain this potential, indicating that they are functional, and not vestigial remnants of non-functional organelles. This is supported by the fact that high concentrations of rhodamine 123 (100 $\mu g\,ml^{-1}$) inhibit or block division of *B. hominis* (see Zierdt *et al.*, 1988). Metabolic inhibitors and uncouplers have not yet been used to determine functional pathways within the mitochondria. DAPI, a fluorescent dye which complexes specifically with DNA, shows some staining of the mitochondria, suggesting that mitochondrial DNA is present (Matsumoto *et al.*, 1987). This has not been confirmed.

Biochemical analysis indicated that a number of typical mitochondrial enzymes associated with energy metabolism were absent from *B. hominis*. Cytochrome oxidase, catalase and peroxidase were not present (Zierdt, 1986), and the pyruvate dehydrogenase complex, ketoglutarate dehydrogenase complex, isocitrate dehydrogenase, glutamate dehydrogenase, and cytochrome *c* oxidase have also been shown to be absent (Zierdt *et al.*, 1988). These enzymes are considered to be marker enzymes for mitochondria in many other organisms (see Ghadially, 1988). Conflicting data on the existence of the enzymes diaphorase, lactate dehydrogenase and aldolase are given in the literature (Zierdt, 1988; Zierdt *et al.*, 1988), and these results must be regarded circumspectly until substantiated by further analysis.

The presence of mitochondria in this anaerobic organism is an enigma, and further research should be devoted to ascertaining their true function and metabolic capabilities. It has been suggested that the mitochondria may have a role in lipid synthesis (Zierdt *et al.*, 1988), but this remains to be tested experimentally. Recently, a number of anaerobic ciliate protists have been found to possess hydrogenosomes which morphologically resemble mitochondria (Finlay and Fenchel, 1989; Fenchel and Finlay, 1991). Since it has been reported that hydrogenosomes lack catalase and cytochromes (Benchimol and de Souza, 1983; Müller, 1991), it would be of interest to determine whether the organelles thought to be mitochondria in *B. hominis* were, in reality, hydrogenosomes. Given the strict requirements for an anaerobic environment for the survival and growth of *B. hominis*, the presence of hydrogenosomes would be more easily explained.

3. *Nucleus*

The structure of the nucleus of *B. hominis* is less controversial than that of other organelles. Both light and electron microscopy have revealed a crescentic band of condensed chromatin (Figs 18a–c), usually at one pole of the nucleus. The composition and function of this structure is unknown. It has been described as the nucleolus (Zierdt *et al.*, 1967; Zierdt, 1988), but other reports have refuted this designation (Tan and Zierdt, 1973; Zierdt and Tan, 1976a). This crescent has been found in all forms of the organism, and appears to have no relationship to cellular division. Rarely, an additional "spot" of condensed material (Fig. 18b) has been seen within the nucleus (Zierdt and Tan, 1976a; Stenzel *et al.*, 1991). This spot appeared to be more common in the *in vivo* form of *B. hominis* and was considered by Zierdt and Tan (1976a) to be the nucleolus. The nucleus appears spherical to oval in shape, and approximately 1 μm diameter (Fig. 18a–c). It is surrounded by a membrane with nuclear pores, and an outer nuclear envelope. The perinuclear space may be dilated, and this is more frequent in organisms grown under stress conditions, such as unfavourable temperatures or unsuitable media (Dunn, 1992). Several nuclei are often found within the same perinuclear space, indicating that nuclear division may precede the enclosure of nuclei by individual perinuclear envelopes.

The method of nuclear division in *B. hominis* is unknown, although nuclear fission (Lavier, 1952), fusion by autogamy (Alexeieff, 1911) and mitosis (Knowles and Das Gupta, 1924) have been proposed. The freeze fracture study by Yoshikawa *et al.* (1988) indicated that the nucleus may divide by fission and, while TEM studies support this, conclusive evidence is lacking. Microtubules have not been observed in or adjacent to the nucleus.

The occurrence of DNA within the nucleus has been confirmed by the use of specific stains, including Hoechst 33258, acridine orange, methylene green and brilliant cresyl blue (Dunn, 1992), and DAPI (Matsumoto *et al.*, 1987). RNA appears to be concentrated around the nucleus (Dunn, 1992).

4. *The surface coat*

A fibrillar surface coat has been reported to surround most *B. hominis* cells (Figs 10, 12). This structure was first noted in early reports (Alexeieff, 1911; Lynch, 1917; Kofoid *et al.*, 1919), and has been described as a mucilaginous envelope, layer or coat, and as a gelatinous or slime capsule. Only one report specifically noted the absence of such a layer surrounding *B. hominis* (Macfie, 1915), although some recent reports have noted its absence from one or more forms of *B. hominis* (see Dunn *et al.*, 1989). The surface coat may be up to half the thickness of the cell diameter (Zierdt *et al.*, 1967).

A dense, fibrillar surface coat, approximately 0.25–0.5 μm in width, was found on all *B. hominis* examined from fresh human faeces (Stenzel *et al.*, 1991) (Fig. 10), whereas the surface coat in cultured cells was much thinner.

Unpublished results (D. J. Stenzel, P. F. L. Boreham, K. Lai Peng Foon and M. F. Cassidy) indicated that a very thick fibrillar surface coat is also present on all *Blastocystis* sp. examined from the faeces of monkeys and birds. The surface coat varies in thickness and density between isolates and may slough off some cells in culture, so that portions of the cell surface appear bare. A minority of cells in culture lacks a surface coat. The amoeboid forms described by Dunn *et al.* (1989), cells obtained at colonoscopy (Stenzel *et al.*, 1989), and cells from a patient with severe enteric disease (Zierdt and Tan, 1976a), all lacked a surface coat. The role of the surface coat is not known in the *in vivo* environment, but the loss of surface coat *in vitro* indicates that this role is not a contributing factor to the survival of the organism in laboratory culture.

Bacteria, often causing an indentation, have been seen in close association with the surface coat (Silard and Burghelea, 1985; Dunn *et al.*, 1989) (Figs 10b, 12a). The adherence of bacteria to the surface of *B. hominis* has been confirmed by SEM studies (van Saanen-Ciurea and El Achachi, 1983; D. J. Stenzel and M. F. Cassidy, unpublished observations) (Fig. 6), but its significance is undetermined. The surface coat has been suggested to have a function in the non-specific recognition involved in the attachment of bacteria before phagocytosis (Dunn *et al.*, 1989), or to have a possible toxic effect on bacteria (Silard and Burghelea, 1985). The surface coat probably acts as a mechanical and chemical protective barrier against host responses, in a similar manner to the surface coats of other protozoa (Mehlhorn, 1988b). Further studies are required to determine the chemical composition of the surface coat, and to ascertain its role in the survival and pathogenicity of *B. hominis*.

H. PROPOSED LIFE CYCLE

From the data currently available we propose the life cycle shown in Fig. 19. The predominant form present in the colon of humans appears to be a small non-vacuolated cell without a surface coat. As the non-vacuolated form passes through the colon, the small vesicles probably coalesce to form the multivacuolar stage seen in faeces. This form has a thick surface coat, which is sloughed off under as yet undefined conditions, once a cyst wall has formed underneath. We hypothesize that the resultant cyst is resistant to external environmental conditions and is the dominant infective stage of the life cycle. This thesis requires experimental verification. Excystation possibly occurs as a result of exposure to gastric acid and intestinal enzymes, as has been described for *Giardia* (Schaefer, 1990). Differentiation to the amoeboid form is the least understood part of the cycle. However, since there are some morphological similarities, it is possible that the amoeboid form arises *in vivo* from the non-vacuolated form. A precedent for this is the amoeba–flagellate transformation described for some soil amoebae (Schuster, 1979). The mechanism of this transformation in the amoeboflagellates appears to be dependent upon host physiological factors which are poorly understood. As with *Naegleria*, the transformation of the non-vacuolated form to the amoeboid form is probably reversible.

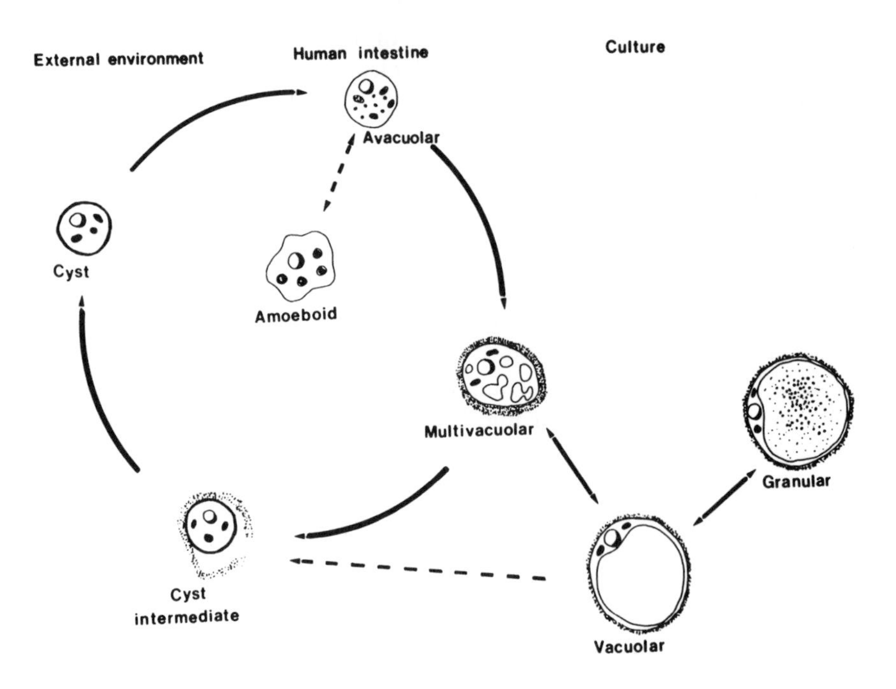

FIG. 19. Proposed life cycle of *B. hominis*. The broken lines indicate hypothetical pathways.

The vacuolar form has been shown to arise in culture from the multivacuolar form, apparently by coalescence of the smaller vacuoles. Based on current evidence, the vacuolar form does not appear to play a major role in the life cycle of *B. hominis*. It remains to be determined whether the vacuolar form is infective to humans. Since cysts are occasionally seen in culture, there is a possibility that the vacuolar form may also develop into a cyst. Similarly, culture conditions may occasionally cause the vacuolar form to revert to the multivacuolar form. Adverse conditions have been demonstrated to initiate the transformation of the vacuolar form to the granular cell. We believe it is not necessary to invoke the concept of a distinct form for the granular cell, but rather to consider it as a vacuolar form in the central vacuole of which granules arise. Information on the organism isolated from animals is so sparse that at this time it is not possible to speculate whether identical life cycles occur in all hosts.

V. *Blastocystis* spp. in Animals other than Humans

Many studies have reported the existence of *Blastocystis* sp. in invertebrates (Alexeieff, 1911; Lavier, 1937a,b, 1952; Ciferri and Redaelli, 1938), amphibia (Alexeieff, 1911; Beaurepaire Aragão 1922a,b; Lavier, 1937a,b, 1952), reptiles (Wenyon, 1920; Lavier 1937a,b, 1952), birds (Alexeieff, 1911; Ciferri and Redaelli, 1938; Lee, 1970; Yamada *et al.*, 1987a; Belova and Kostenko, 1990), and a variety of mammals including rats (Alexeieff, 1911; Lynch, 1917; Knowles and Das Gupta, 1924), guinea-pigs (Alexeieff, 1911; Knowles and Das Gupta, 1924; Molet *et al.*, 1981), cats (Knowles and Das Gupta, 1924), monkeys (Macfie, 1915; Knowles and Das Gupta, 1924; McClure *et al.*, 1980; Yamada *et al.*, 1987a), and apes (Alexeieff, 1911; Zierdt, 1988).

Many of these reports cannot be verified because of confusion over the identification and classification of *Blastocystis* sp. during the early part of this century. Thus, only more recent reports will be discussed here. *Blastocystis* sp. seems capable of infecting at least some of the common domestic animals, although comprehensive survey data are lacking.

A. Pigs

Large numbers of *Blastocystis* sp. were found in 36 (60%) of 60 pigs sampled from four farms in England (Burden, 1976; Burden *et al.*, 1978/1979). Greater numbers of *Blastocystis* sp. were present in pigs suffering from diarrhoea or dysentery, but no evidence was found to suggest that *Blastocystis* sp. had a pathogenic role. Burden (1976) questioned whether *Blastocystis* sp. caused the disease, or if the diarrhoea produced conditions which were suitable for its growth. The organisms were small (about 5 μm diameter), with a central vacuole surrounded by peripheral cytoplasm containing nuclei, and prominent mitochondria with bulbous cristae. The morphology was found to be similar in faecal and cultured samples.

Blastocystis sp. isolated from pigs were easily grown in horse serum diluted with saline, in which a variety of forms was seen, and in egg slant medium, where an amoeba-like stage was seen. These forms were not described further, and so cannot be compared with descriptions from human hosts (see Section IV.F). Binary fission was noted, and possibly endosporulation, in which daughter cells were thought to grow within the central vacuole.

As the supposed endosporulation was seen only in the horse serum cultures, it seems more likely that it was the formation of granules in the granular form, and not true endosporulation.

These results were recently confirmed (Pakandl, 1991) when pigs more than 2 days old from five farms in southern Bohemia were found to be infected with *Blastocystis* sp. Twenty-two of 30 faecal samples from wild pigs (*Sus scrofa*) were also found to harbour *Blastocystis* sp. While vacuolar and granular forms were present in culture, amoeboid forms were not seen. No difference was noted from previous ultrastructural studies of *B. hominis*.

B. MONKEYS

Blastocystis sp. infections in monkeys have been reported in recent literature on several occasions (McClure *et al.*, 1980; Yamada *et al.*, 1987a). The report by McClure *et al.* (1980) described a case of diarrhoea in a pig-tailed macaque (*Macaca nemestrina*), with large numbers of *Blastocystis* sp. in the faeces. Only the vacuolar form was seen by light microscopy and bacteria were occasionally reported in the central vacuole. TEM revealed predominantly vacuolar forms, and rarely amoeboid forms. The amoeboid form was characterized by a large central vacuole surrounded by a distinct membrane. An outer, limiting, cell membrane was not visible, although a filamentous layer surrounded the cell. Pseudopodial projections were present. The vacuolar form displayed a limiting membrane. Rarely, organisms contained lipid or glycogen inclusions in the cytoplasm.

Yamada *et al.* (1987a) found 15 of 26 monkeys in Japan to be infected with *Blastocystis* sp. These included *Macaca fuscata fuscata*, *M. mulatta*, *M. radiata* and *Saimiri sciurea*. The authors compared the organisms from monkeys with those from humans and fowls, and *Blastocystis* sp. from faeces, lumenal contents and culture were reported to be morphologically similar for all three species, except for variations in size and the contents of the central vacuoles. Vacuolar cells were the most common form detected in fresh faeces, and had diameters of 8–18 μm in monkeys, compared to 9–32 μm for those from fowls and cultures. Up to four nuclei were seen, either located at opposite poles of the cell, or adjacent to each other. Occasionally, a large single granule was present within the central vacuole. Our own studies on *Blastocystis* sp. from *Macaca* and *Presbytis* monkeys from Malaysia also indicated basic morphological similarities with human isolates (D. J. Stenzel, P. F. L. Boreham and K. Lai Peng Foon, unpublished observations). However, some significant differences were detected. Vacuolar (Figs 20, 21a), rather than multivacuolar, cells predominated in the faeces of monkeys. These frequently contained electron-opaque inclusions in the central vacuole. Cysts, in particular multinucleated cysts (Fig. 21b), were frequently found in the samples from *Macaca* but have not been identified from *Presbytis* faecal samples.

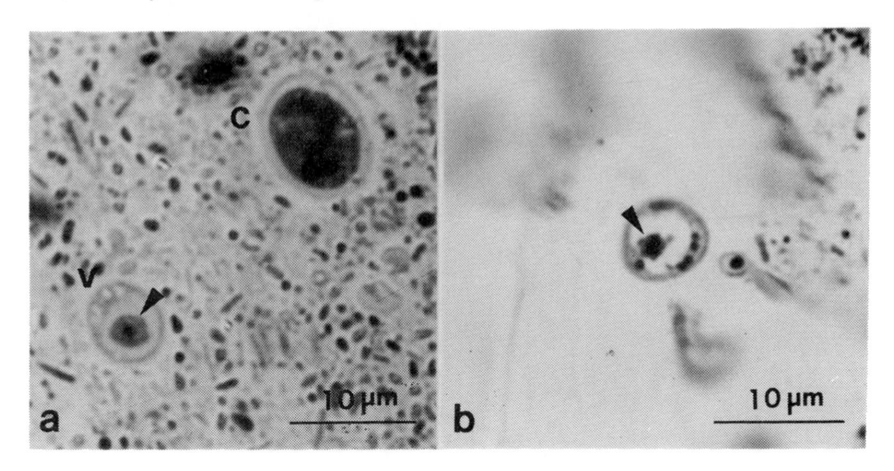

FIG. 20. Light micrograph of *Blastocystis* sp. from monkey faecal samples. Sections (1 µm) stained with toluidine blue. (a) Cyst (C) and vacuolar (V) forms from *Macaca* monkey. A dense inclusion (indicated by arrow) is seen in the central vacuole of the vacuolar form. (b) Vacuolar form from *Presbytis* monkey. An inclusion (indicated by arrow) is seen within the central vacuole.

C. GUINEA-PIGS

Apart from a short report noting the isolation and culturing of *Blastocystis* sp. from one animal (Molet *et al.*, 1981), no recent report of naturally acquired infections of guinea-pigs has been published. However, *B. hominis* infections have been produced experimentally in germ-free guinea-pigs (Phillips and Zierdt, 1976). *B. hominis* from cultured human isolates was inoculated orally or intracaecally to produce infections in approximately 30% of animals. Guinea-pigs infected with large numbers of *B. hominis* were symptomatic, with watery diarrhoea. *B. hominis* was most numerous in the caecum, with smaller numbers occasionally found in the ileum and colon, adjacent to the caecum. No ulceration or other lesion was found in the intestinal tract of infected animals, but amoeba-like forms were reported in the epithelial cells. There was no inflammation or other symptom associated with this invasion, if in fact it did occur. The low rates of infection of the guinea-pigs suggests that unidentified factors may be involved in the transmission and growth *in vivo* of *B. hominis*. Possibly the stages of *B. hominis* usually seen in culture, and hence those used to inoculate animals in this study, are not the infective forms of the organism.

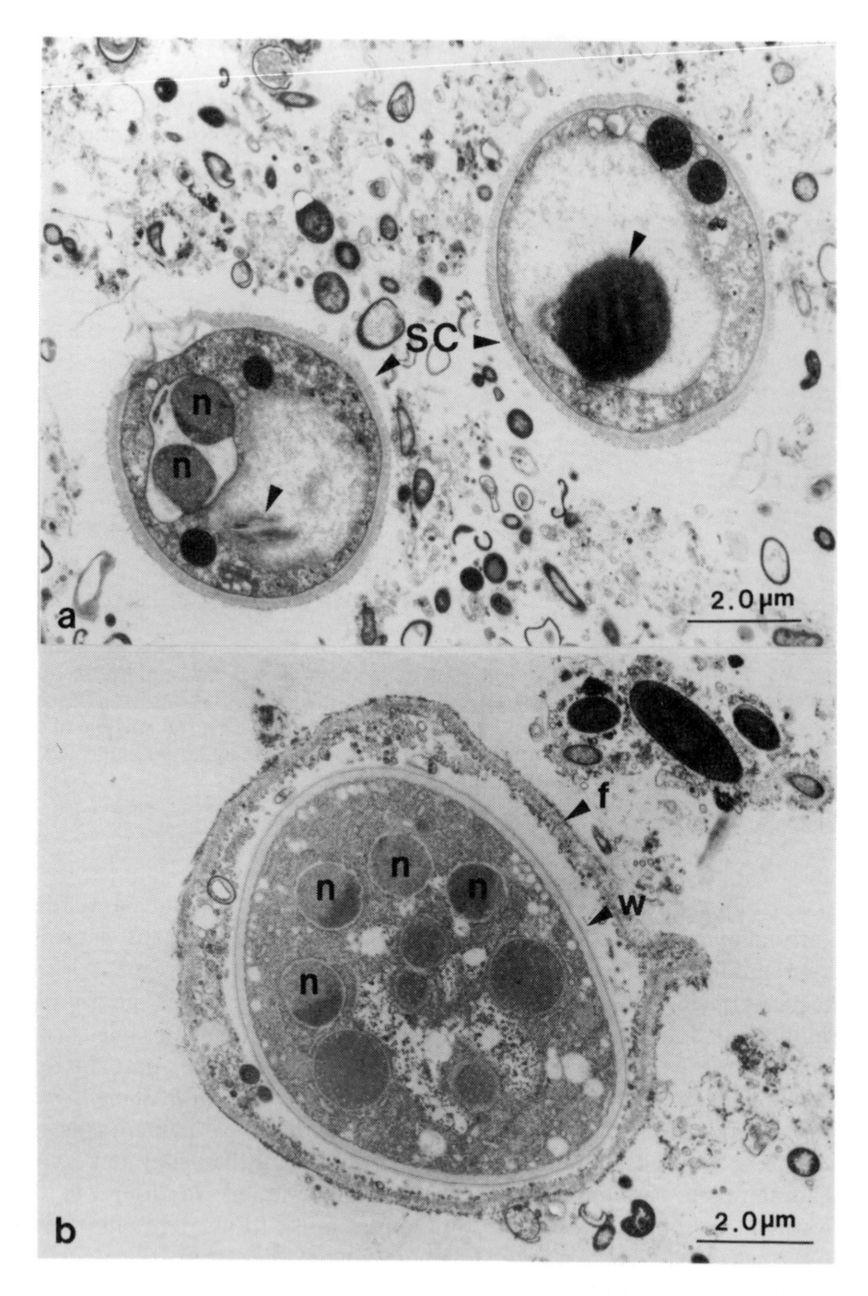

FIG. 21. Transmission electron micrographs of *Blastocystis* sp. from *Macaca* monkey faecal sample. (a) Vacuolar forms show inclusions (indicated by arrows) within the central vacuole. Nuclear morphology is similar to that of *B. hominis*. A thick surface coat (SC) surrounds the cells. n, Nucleus. (b) Multinucleated cyst, showing outer fibrillar layer (f) separating from cyst wall (w). n, Nucleus.

D. BIRDS

Blastocystis sp. infections appear to be common in birds (Lee, 1970; Yamada *et al.*, 1987a; Belova and Kostenko, 1990; Pakandl and Pecka, 1992; D. J. Stenzel, M. F. Cassidy and P. F. L. Boreham, unpublished observations). Lee (1970) described *Blastocystis* sp. from the caecum of turkeys, believing the organism to be a fungus. It was described as having a thick coat of "mucilage" rather than a typical fungal cell wall. The organism was found in tissue sections, in cultures and in smears. The organism from the caecal contents appeared as a round or oval cell, varying in size from 5 to 10 µm in diameter, with a mucilaginous coat, a central vacuole, several mitochondria, and one or more nuclei. The central vacuole contained finely granular material and occasionally a crystalline inclusion. The cell membrane was covered with a coat of fine filamentous material, probably mucoprotein or mucopolysaccharide, about 300 nm thick. Similarities were noted with the description by Lavier (1952), who thought *Blastocystis* to be the vegetative stage of a fungus belonging to, or closely related to, the phycomycetes. Lee (1970) often found *Blastocystis* sp. in superficial lesions in the caecal wall which had been disrupted by *Histomonas* sp., but concluded that it was almost certainly a secondary invader of the lesions. However, since the organism was able to survive and multiply in these areas, it may have aggravated existing lesions.

A light microscopical study of *Blastocystis* sp. in Japan found 100% infection in the limited number of domestic fowls and ostriches examined (Yamada *et al.*, 1987a). At autopsy, *Blastocystis* sp. organisms were found in the caecum, either distributed freely in the lumen or attached to the epithelium, but not within epithelial cells. A large variation in size was noted (9–32 µm diameter).

The incidence of *Blastocystis* sp. infection in domestic hens (*Gallus gallus domesticus*) from the Commonwealth of Independent States was high, ranging from 80 to 100% (Belova and Kostenko, 1990). Only very young birds were not infected. In samples obtained from the large bowel the organisms were rounded to ellipsoid, approximately 18 µm in diameter, with one to four nuclei. A well-defined glycocalyx was present external to the plasma membrane. The large electron-transparent vacuole was described as a reproductive organelle, and was divided by cytoplasmic membranes. These cytoplasmic protrusions and divisions of the vacuole sometimes contained mitochondria and ribosomes. The nucleus contained a nucleolus, and blocks of chromatin concentrated at one of its poles, but a crescentic band of condensed chromatin was not observed within the nucleus. Belova and Kostenko (1990) have proposed a new species, *B. galli*, based on the morphological observations in this study. Preliminary studies of *Blastocystis*

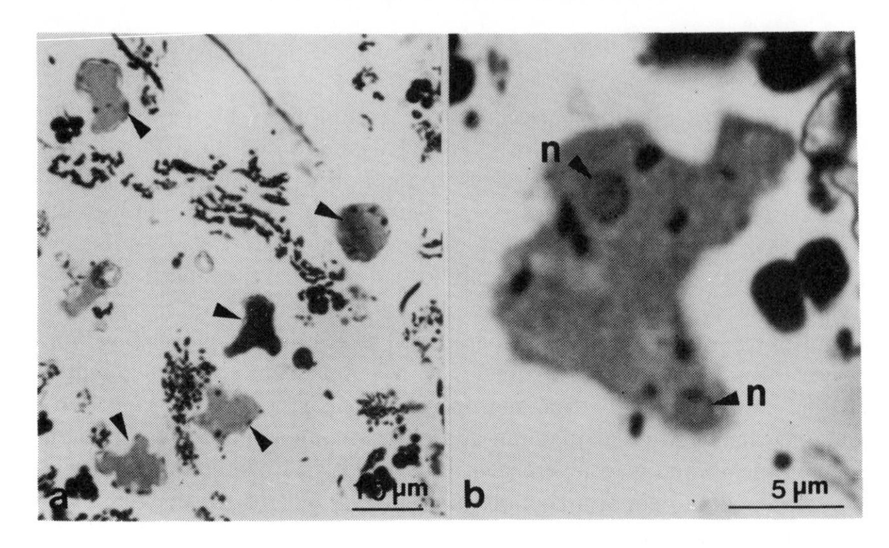

FIG. 22. Light micrographs of *Blastocystis* sp. from chicken faeces. Sections (1 µm) stained with toluidine blue. (a) Low magnification showing irregularly shaped cells (indicated by arrows). (b) High magnification of cell with two nuclei (n).

sp. isolated from domestic chickens in our laboratory (D. J. Stenzel, M. F. Cassidy and P. F. L. Boreham, unpublished observations) (Figs 22, 23) confirm the differing nuclear morphology (Fig. 23b) of such isolates.

Recently *Blastocystis* sp. has been isolated from the domestic duck (*Anas platyrhynchos* f. *domestica*) with an infection rate of 80% in adults and 25% in ducks 1 month old (Pakandl and Pecka, 1992). Pakandl and Pecka were also able to infect domestic ducks with the caecal contents of mallards, and found *Blastocystis* sp. in the caecum and cloaca after 1 year. The ultrastructure of *Blastocystis* from ducks was similar to that of the human parasite.

E. REPTILES

A new species, *B. lapemi*, isolated from the sea snake (*Lapemis hardwickii*), was recently designated. Although morphologically similar to human isolates, it grew at 26°C only, instead of 37°C (Teow *et al.*, 1991). The electrophoretic karyotype also differed from that of human isolates, but in view of the considerable differences detected among human isolates (Upcroft *et al.*, 1989; Boreham *et al.*, 1992) caution is required in using electrophoretic karyotypes alone to distinguish species.

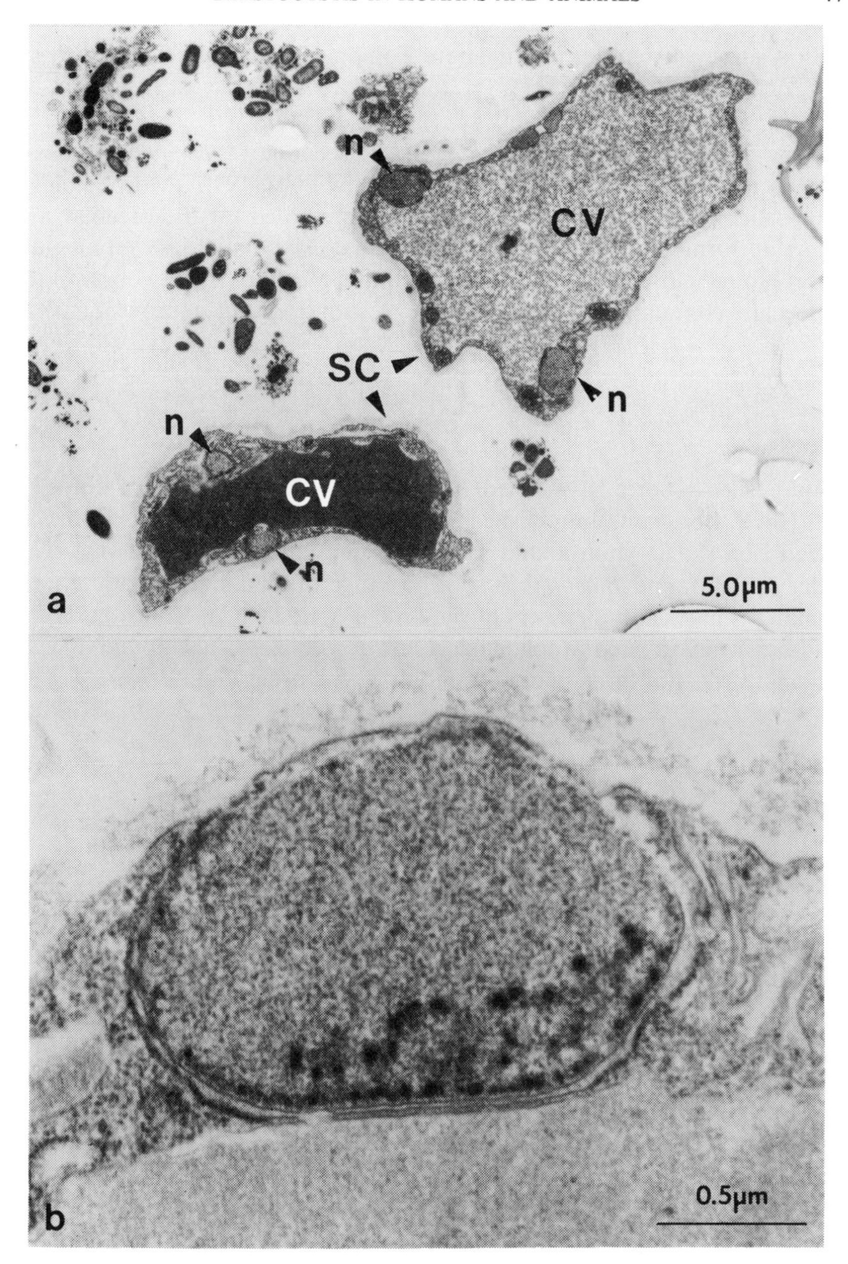

FIG. 23. Transmission electron micrographs of *Blastocystis* sp. from chicken faeces. (a) Two binucleated, irregularly shaped cells showing variation in central vacuolar contents. n, Nucleus; SC, surface coat; CV, central vacuole. (b) High magnification showing detail of nucleus.

In a survey of reptiles in Singapore Zoological Gardens, *Blastocystis* sp. was detected in three species of tortoises (*Geochelone elephantopus*, *G. elegans* and *G. carbonaria*), three species of snake (*Boiga dendrophilla*, *Python reticulatus* and *Elaphe radiata*), a crocodile (*Crocodylus porosus*), and an iguana lizard (*Cyclura cornuta*) (Teow *et al.* 1992). Morphological similarities of these isolates to *B. hominis* were noted for the vacuolar and granular forms. The workers suggested that, since all these reptiles are always found close to water, they may acquire infections by the consumption of faecally contaminated water. This suggestion is as yet unproven.

F. OTHER HOSTS

Unpublished reports have noted *Blastocystis* sp. in the faeces of domestic cats (R. E. Boreham, unpublished observations) and in mammals from San Diego Zoo (L. S. Munroe and M. Lewis, unpublished observations). This latter survey found *Blastocystis* sp. in a range of mammals including non-human primates and members of the orders Artiodactyla, Rodentia, Hyracoidea, Carnivora, and Marsupialia. *Blastocystis* sp. was also found in eight different orders of birds, most commonly in the order Galliformes.

G. SUMMARY

It is evident that *Blastocystis* sp. is widespread throughout the animal kingdom, and it appears to be particularly common in monkeys, pigs and chickens. However, since the data are so sparse, this conclusion may merely reflect the species chosen for study. The major question outstanding is whether *Blastocystis* from these hosts represent a single species or multiple species. Since it appears relatively easy to establish *Blastocystis* in culture, although it is considerably more difficult to produce axeni cultures, the application of molecular techniques such as immunoblotting (Upcroft *et al.*, 1989; Kukoschke and Müller, 1991; Boreham *et al.*, 1992), electrophotetic karyotyping (Upcroft *et al.*, 1989; Teow *et al*, 1991; Boreham *et al.*, 1992) and DNA fingerprinting (Upcroft *et al.*, 1990; Archibald *et al.*, 1991) should assist in resolving this question. Almost nothing is known about the pathogenicity of *Blastocystis* sp. from animals and future research should perhaps concentrate initially on animal hosts which have the potential to transmit infection to humans, and on those which are of commercial importance.

VI. CLINICAL ASPECTS OF *B. HOMINIS* INFECTIONS

A. EPIDEMIOLOGY

Nowhere is our knowledge of *B. hominis* more lacking than in the study of its epidemiology. There are very few publications where the topic is even mentioned and we can only speculate about such important practical issues as the mode of transmission, importance of animal reservoirs and prevalence in symptomatic and asymptomatic individuals.

Until the recent report of a cyst form (Stenzel and Boreham, 1991), it was hard to envisage the mode of transmission, although some protists, for example *Pentatrichomonas hominis* and *Dientamoeba fragilis* (see Ockert, 1990), have developed successful strategies for transmission without a cyst. It is generally assumed that *B. hominis* is transmitted by the faecal–oral route in a similar manner to *Giardia* and *Entamoeba*, but there is no corroborating evidence. Reasons for this include the lack of an appropriate animal model, and misinformation and confusion over the status of *B. hominis*. Waterborne and foodborne transmission are other possible routes, but these have not been seriously examined. There is a single report of a statistical association between *B. hominis* infection and the consumption of raw water (Kain *et al.*, 1987). The literature contains the following snippets of information which may be relevant to the epidemiology of this infection, but at this stage cannot be considered as conclusive.

(i) While there may be a slight tendency for an increased female to male sex ratio (Doyle *et al.*, 1990), most authors report no difference between the sexes (Guirges and Al-Waili, 1987; Qadri *et al.*, 1989).

(ii) Most authors agree that adults are more likely to be infected than children. The median age of persons infected with *B. hominis* was reported as 31 years (Kain *et al.*, 1987) and 37 years (Doyle *et al.*, 1990). In another survey of 515 persons, 72% of infected people were aged 13–50 years, while 19% were over 50 and only 9% were under 13 years of age (Qadri *et al.*, 1989). In a study of parasites present in the residents of the Asaro Valley in Papua New Guinea, the age prevalence curves rose to a peak at about age 10 years, then fell slowly, rising again in late life. This pattern was similar to that observed for *Entamoeba coli* and *Endolimax nana* (Ashford and Atkinson, 1992), possibly suggesting a similar mode of transmission.

(iii) Familial transmission (Guglielmetti *et al.*, 1989; Bratt and Tikasingh, 1990) and transmission within institutes for the mentally retarded (Yamada *et al.*, 1987b; Libanore *et al.*, 1991) have been reported. This would be consistent with faecal-oral transmission.

(iv) Travel history has been implicated as an important risk factor in three studies, in which 52%, 57.5% and 74% of infected people had a history of recent travel either overseas or to wilderness areas (Sheehan *et al.*, 1986; Kain *et al.*, 1987; Doyle *et al.*, 1990). However, other reports do not substantiate this conclusion (Garcia *et al.*, 1984; Sun *et al.*, 1989).

(v) It has been suggested that the incidence of *B. hominis* infections may be related to the weather conditions, being higher in hot weather (Knowles and Das Gupta, 1924) and during the pre-monsoon months (Babcock *et al.*, 1985). Data from Brisbane, Australia, indicate a greater prevalence in the hot, wet summer months than the cold, dry winter months (R. McDougall, unpublished observations).

(vi) Forty-four per cent of 121 patients infected with *B. hominis* had a history of exposure to pets or farm animals (Doyle *et al.*, 1990). Garavelli and Scaglione (1989) suggested that *B. hominis* is a zoonotic infection carried via food and water. Insufficient evidence exists to support these assertions.

(vii) The age prevalence curve for *B. hominis* in Papua New Guinea, obtained by Ashford and Atkinson (1992), suggested that there was no evidence of the acquisition of protective immunity or pathogenicity at the community level.

B. DIAGNOSIS

B. hominis can be identified by light microscopy of wet mounts of either fresh stools (Fig. 24a) or concentrates. Staining with iodine or permanent trichrome mounts may aid in diagnosis. Care must be taken not to confuse *Blastocystis* with leucocytes. Many diagnostic laboratories have taken the vacuolar form of 5–8 μm as the standard morphology, but in the light of recent observations of the form *in vivo* (Stenzel *et al.*, 1991) it is essential to look for smaller and multivacuolar forms in fresh stools (see Fig. 24) or else infected patients may be missed.

Kukoschke *et al.* (1990) compared the effectiveness of microscopy and culture in the identification of infected stools, and found that culture had no benefit over microscopy. Invasive techniques have occasionally been used in diagnosis, but have not been thoroughly evaluated and generally are not required. Matsumoto *et al.* (1987) found that endoscopic examinations revealed numerous *B. hominis* organisms in the lumen of the ileum and caecum. Narkewicz *et al.* (1989) reported a 16-year-old male haemophiliac with acquired immune deficiency syndrone (AIDS)-related complex, in whom *B. hominis* was found in duodenal secretions taken by the Enterotest® (string test), as well as in stools. These reports indicate that *B. hominis* may not always be restricted to the large intestine, as is generally accepted dogma.

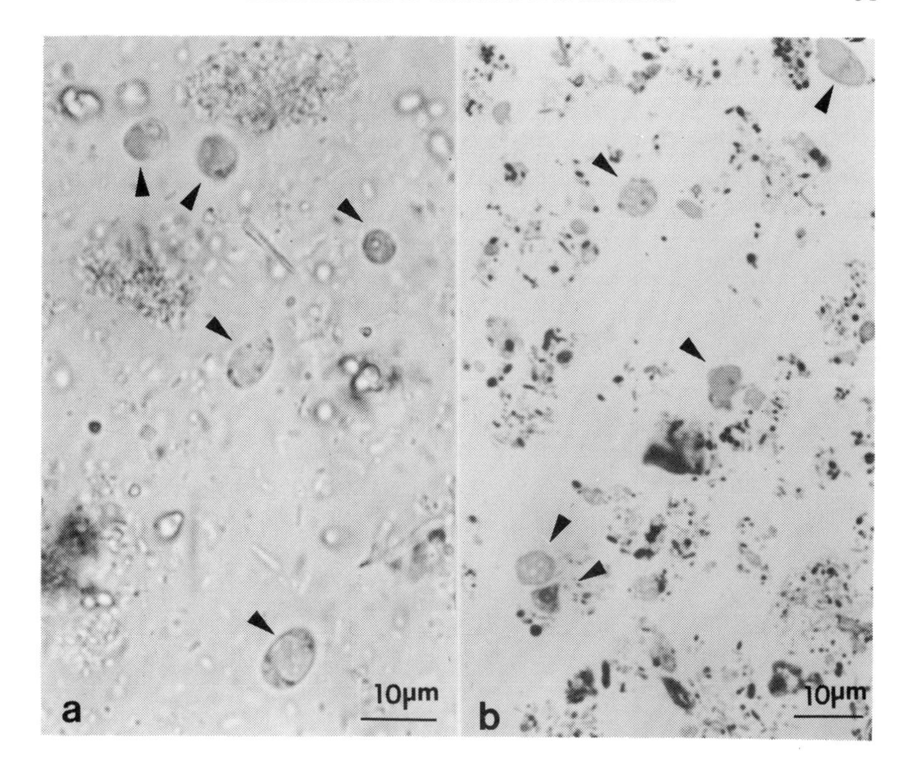

FIG. 24. Light micrographs of *B. hominis* (indicated by arrows) from human faecal samples. (a) Unstained wet mount. (b) Section (1 µm) stained with toluidine blue.

C. PREVALENCE

While there are many surveys published in the literature which report prevalence data for *B. hominis*, it is difficult to interpret and compare surveys. Lack of detail is common in published accounts and, while some studies report on the number of infected stools, others detail the number of patients. Most surveys are biased in that they report data from parasitology laboratories, rather than random surveys, and thus largely reflect symptomatic patients. Table 1 reports the results from some major studies undertaken. We have arbitrarily chosen only those studies which included more than 100 stool samples or patients. These prevalence data give little indication of risk factors involved in acquiring *B. hominis* infection, and all that can be concluded is that *B. hominis* is a ubiquitous parasite reported from both tropical and temperate regions.

TABLE 1 *Prevalence of* Blastocystis hominis *infections*

Country	*Group*[a]	*Number*[b]	*Sample*[c]	*Percentage infected*	*Reference*	*Comment*
Germany	?	832	s	14.2	Hahn and Fleischer, 1985	People returning from tropics
	?	1934	s	3.6	Kukoschke *et al.*, 1990	Patients with diarrhoea
Switzerland	?	1460	p	4.7	Nguyen and Krech, 1989	Patients with gastroenteritis
Italy	C	938	p	1.9	Guglielmetti *et al.*, 1991	Symptomatic and asymptomatic patients
Yugoslavia	?	276	p	14.1	Pikula, 1987	Symptomatic patients
USA	C,A	2360	p	12.2	Garcia *et al.*, 1984	Pathology laboratory records
	C,A	389	p	11	Sheehan *et al.*, 1986	Pathology laboratory records
	?	2700	p	3	Babb and Wagener, 1989	Pathology laboratory records
	?	5505	s	0.3	Diaczok and Rival, 1987	Routine hospital stools
	C,A	1442	s	4.4	Narkewicz *et al.*, 1989	Symptomatic and asymptomatic patients
	C,A	6262	s	1.6	Sun *et al.*, 1989	Pathology laboratory records
	C,A	2744	s	9.5	Zuckerman *et al.*, 1990	Pathology laboratory records
	A	3001	?	34.1	Kofoid and Swezy, 1921	Returned servicemen
	?	?	?	11.6	Markell and Udkow, 1986	Hospital patients
Canada	?	1496	p	12.7	Kain *et al.*, 1987	Pathology laboratory records
	C,A	16545	s	3.2	Doyle *et al.*, 1990	Pathology laboratory records
	C,A	8399	s	8	Senay and MacPherson, 1990	Pathology laboratory records
Chile	C	100	p	45	Figueroa *et al.*, 1990	Children with diarrhoea
	C	3061	p	31.7	Puga *et al.*, 1991	Daycare centre children
	C,A	6162	p	30.4	Mercado and Arias, 1991	Ambulatory Santiago persons

Venezuela	C,A	2009	s	10.3	Castrillo de Tirado *et al.*, 1990	Pathology laboratory records
Saudi Arabia	C,A	12136	s	17.5	Qadri *et al.*, 1989	Pathology laboratory records
Kuwait	C	1960	p	2	Zaki *et al.*, 1991	Hospital patients
India	?	1849	s	29	Knowles and Das Gupta, 1924	Pathology laboratory records
Pakistan	?	800	s	12	Zafar, 1988	Pathology laboratory records
Nepal	C,A	1813	s	17.4	Babcock *et al.*, 1985	Mainly foreign residents and tourists
Vietnam	?	100	p	5	Picher and Aspöck, 1980	Refugees
Japan	?	1251	p	3.8	Yamada *et al.*, 1987b	Symptomatic and asymptomatic patients
Egypt	C,A	1435	p	31	El Masry *et al.*, 1990	Pathology laboratory records
Kenya	C,A	1129	p	17.5	Chunge *et al.*, 1991	Rural community
Nigeria	C,A	479	p	2.5	Reinthaler *et al.*, 1988	Mainly patients with diarrhoea
Australia	C,A	125	p	22.4	Walker *et al.*, 1985	Asymptomatic persons
	C,A	9848	p	5.4	Sullivan, Nicolaides and Partners, unpublished	Pathology laboratory records
Papua New Guinea	C,A	1004	p	54	Ashford and Atkinson, 1992	Village survey

[a]C, Children; A, adults; ? information not supplied.
[b]Some surveys refer to number of patients and others to number of stools.
[c]s, Stools examined; p, patients examined.

D. PATHOGENESIS

The most critical question concerning *Blastocystis* today is whether it is a pathogen or a commensal. Recently, the purported disease in humans caused by *B. hominis* has been called Zierdt–Garavelli disease (Garavelli, 1992; Garavelli *et al.*, 1992). Such a designation seems premature until it is proven that this organism actually causes disease, and the term blastocystosis seems much more appropriate.

Attempts to assign pathogenicity to *B. hominis* by epidemiological studies have been criticized because of the impossibility of eliminating all other causes of symptoms, either infectious or non-infectious, especially when it is considered that approximately 25% of diarrhoeas have no known aetiology (Edmeades *et al.*, 1978). The major problem in deciding whether *B. hominis* is a pathogen has been the inability to fulfil Koch's postulates because of the lack of experimental models, as well as the difficulty of excluding all other potential causes of the symptoms. Several studies, for example, have not considered the possibility of virus infections or non-infectious causes (see Miller and Minshew, 1988). Elimination of the parasite by drugs and subsequent subsidence of symptoms is often taken to indicate that *B. hominis* is pathogenic. This is an unacceptable argument since the drugs used, chiefly nitroimidazoles, are not specific and affect many organisms, including anaerobic bacteria. It is also worth noting that many causes of infectious diarrhoea are self-imiting, and this also may be true of *B. hominis* infections (Sun *et al.*, 1989; Doyle *et al.*, 1990).

Although there have been many suggestions that *B. hominis* causes disease (Garcia *et al.*, 1984; Ricci *et al.*, 1984; Babcock *et al.*, 1985; Le Bar *et al.*, 1985; Vannatta *et al.*, 1985; Sheehan *et al.*, 1986; Guirges and Al-Waili, 1987; Telalbasic *et al.*, 1987, 1991; El Masry *et al.*, 1988; Zierdt, 1988; Qadri *et al.*, 1989; Bories *et al.*, 1990; Dawes *et al.*, 1990; Zaki *et al.*, 1991), there is a similar number of contrary reports (Taylor *et al.*, 1985; Markell and Udkow, 1986; Kain *et al.*, 1987; Miller and Minshew, 1988; Sun *et al.*, 1989; Doyle *et al.*, 1990; Senay and MacPherson, 1990). Symptoms usually attributed to human infection with *B. hominis* include diarrhoea, abdominal discomfort, anorexia, flatulence, and other non-specific gastrointestinal effects.

The most balanced account of pathogenicity was given in the *Lancet* (Editorial, 1991) which concluded:

> Uncontrolled surveys and therapeutic trials, and anecdotal case reports, may encourage uncritical acceptance of pathogenicity without adequate proof. Notwithstanding the limitations of even the extended Koch's postulates, studies so far are generally unconvincing in meeting

such criteria for establishing a primary pathogenic role for the parasite. However, while it may be premature to accept a primary role, it is similarly premature to reject it.

It must also be remembered that *B. hominis*, even if it proves to be primarily a commensal, could be pathogenic under specific conditions such as immunosuppression, poor nutrition or concurrent infections. One way of investigating the pathogenicity of *B. hominis* would be to select patients who are infected with the organism as the sole enteropathogen, and who have been shown to have no other underlying gastrointestinal condition, for treatment with specific therapy and subsequent monitoring of their symptoms. Currently this is not practicable since no effective, specific therapy is available and the selection of patients for such a trial would require very extensive work to exclude all possible causes of diarrhoea. Another approach would be to develop an experimental model for pathogenicity studies. The situation is further complicated by the recent observation, by means of chromosome and protein studies (Boreham *et al.*, 1992), that significant intraspecific variation occurs among *B. hominis* stocks. The implications of this observation need to be determined in respect of sensitivity to drugs.

Several case reports have suggested that *B. hominis* was associated with a variety of diseases, including tropical pulmonary eosinophilia (Enzenauer *et al.*, 1990), reactive arthritis (Lakhanpal *et al.*, 1991), infective arthritis (Lee *et al.*, 1990), irritable bowel syndrome (Drossman, 1979), colitis (Russo *et al.*, 1988; Shikiya *et al.*, 1989), ulcerative colitis (Jeddy and Farrington, 1991), terminal ileitis (Tsang *et al.*, 1989), enteritis (Gallagher and Venglarcik, 1985; García Pascual *et al.*, 1988), leukaemia (Garavelli *et al.*, 1991b), and diabetes (Scaglione *et al.*, 1990; Sheehan and Ulchaker, 1990). The possibility that *B. hominis* causes toxic–allergic reactions leading to non-specific inflammation of the colonic mucosa has been raised (Garavelli *et al.*, 1991a), and oedema of the colonic mucosa with variable presence of lymphocytes and plasmocytes has been noted following sigmoidoscopy and biopsy (Garavelli *et al.*, 1992). The only immunological study so far undertaken did not detect any serum antibody response to *B. hominis* (see Chen *et al.*, 1987), but no attempt has been made to detect a secretory immunoglobulin response.

Several reports have suggested that *B. hominis* may be an opportunistic infection in immunosuppressed patients with AIDS (Henry *et al.*, 1986; Garavelli *et al.*, 1988, 1990; Llibre *et al.*, 1989; Narkewicz *et al.*, 1989; Garavelli and Libanore, 1990; Garavelli and Scaglione, 1990; Libanore *et al.*, 1990; Salavert *et al.*, 1990; Steinmann *et al.*, 1990; Vallano *et al.*, 1991), but whether as a commensal or a pathogen remains to be determined.

E. CHEMOTHERAPY

There is no consensus among physicians whether it is appropriate to treat patients infected with *B. hominis* and, if so, what is the most appropriate drug. Possible reasons for this are its perceived unimportance by many physicians, the fact that the infection may be self-limiting (Sun *et al.*, 1989; Doyle *et al.*, 1990), and the reluctance to use potentially toxic drugs when *B. hominis* is not a proven pathogen. Many physicians believe that when there are symptoms, and no other cause is obvious, treatment should be instigated. The drug of choice is a matter of conjecture and treatment is still largely empirical. Diet was an early recommended form of treatment and, for example, Swellengrebel (1917) treated a patient with the laxative cascara and purgation with magnesium sulphate followed by a diet of eggs and milk. Most authors now believe that either metronidazole (Markell and Udkow, 1986; Guirges and Al-Waili, 1987; Kain *et al.*, 1987; Zafar, 1988; Babb and Wagener, 1989; Garavelli *et al.*, 1989; Guglielmetti *et al.*, 1989; Qadri *et al.*, 1989; Rolston *et al.*, 1989; Zaki *et al.*, 1991) or iodoquinol (diiodohydroxyquinoline) (Babcock *et al.*, 1985; Jarecki-Black *et al.*, 1986; Markell and Udkow, 1986; Zafar, 1988; Rolston *et al.*, 1989) is the most effective treatment, but there are few experimental studies to verify this.

Research on chemotherapy has been hampered by the lack of a monophasic growth medium for studies *in vitro* and of appropriate animal models for work *in vivo*. Only two experimental studies on the effect of drugs on *B. hominis in vitro* have been undertaken. The first study (Zierdt *et al.*, 1983), which was not quantitative, investigated the effect of 10 antiprotozoal drugs on the growth of four strains of *B. hominis*. Drugs were classified into inhibitory compounds, which reduced cell counts at 72 h by 50 to 100-fold (emetine dihydrochloride, metronidazole, furazolidone, trimethoprim-sulphamethoxazole, Enterovioform® (clioquinol), and pentamidine), moderately inhibitory drugs which reduced cell counts by 5 to 50-fold (iodoquinol and chloroquine), and non-inhibitory drugs, reducing cell counts by less than five-fold (diloxanide furoate and paromomycin sulphate).

An assay based on the incorporation of [^{3}H]hypoxanthine allowed quantitative comparison of the efficacy of various drugs (Dunn and Boreham, 1991). The activities of the drugs reported to be useful in the treatment of blastocystosis are shown in Fig. 25. It must be remembered that these were only preliminary studies, based on a single strain of *B. hominis*, and cannot necessarily be extrapolated to their use in humans. Many other factors, particularly pharmacokinetic properties of the drugs, must be taken into account. Most of the drugs which have been used therapeutically do show activity in this test, although iodoquinol, commonly recommended for treatment, was 25 times less active than metronidazole in this assay *in vitro*.

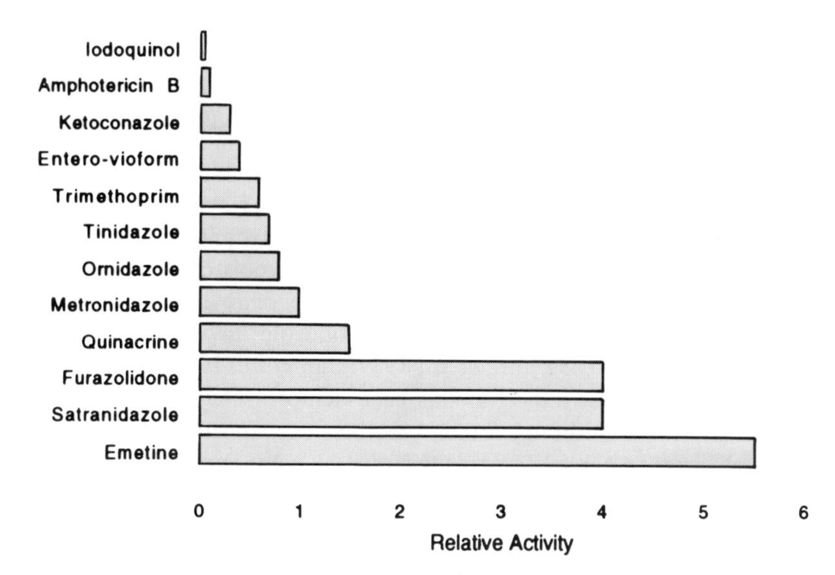

FIG. 25. The activity *in vitro* of drugs against *B. hominis*, relative to that of metronidazole (= 1), determined by means of the [^{3}H]hypoxanthine incorporation assay. (Data from Dunn and Boreham, 1991.)

There is an urgent need to develop an animal model to investigate pathogenesis and to test the effectiveness of potential therapeutic compounds *in vivo*. The only experimental animal work to date is with the gnotobiotic guinea-pig (Phillips and Zierdt, 1976) but, because of the problems and costs of maintaining such animals, this is not an appropriate model. Intrahepatic inoculation of *B. hominis* together with *Entamoeba* into the golden hamster resulted in 12 days survival of *B. hominis* in the liver (Silard *et al.*, 1977). This observation could perhaps be developed into an appropriate experimental model. Other possible models which should be explored are the neonatal mouse, used successfully for *Giardia* chemotherapeutic studies (Boreham *et al.*, 1986), and the chick, which is a natural host.

Double-blind clinical trials have not been conducted, and measurement of the effect of drugs has not been standardized, with reports being based on parasitological cure, reduction in parasite numbers, or alleviation of symptoms. To date only a few studies of series of patients have been reported, none of which has been properly standardized. In one series, 18 Canadian patients infected with *B. hominis* were treated with metronidazole (250–750 mg three times a day for 5–10 days), with 14 becoming asymptomatic or improving clinically (Kain *et al.*, 1987). Similar results were obtained in Kuwait, with all of the 18 patients in the study showing improvement with metronidazole, although additional treatment (ketoconazole) was required to eliminate all the parasites in one patient (Zaki *et al.*, 1991). Two other reports of substantial numbers of patients suggested the effectiveness of metronidazole, but both failed to give details. All of 74 patients receiving metronidazole, 200 mg four times a day for a maximum of 7 days, were cured parasitologically by stool examination on follow-up 1–2 months later (Guirges and Al-Waili, 1987); and treatment of 43 symptomatic patients with metronidazole (500 mg–1 g/day) for 7–10 days resulted in the resolution of all symptoms and absence of parasites from stools on follow-up over a period of 3–6 months (Qadri *et al.*, 1989). To counteract this, there are several reports in the literature of metronidazole having been ineffective (Zierdt and Tan, 1976a; Llibre *et al.*, 1989; Sun *et al.*, 1989). Perhaps this is suggestive of drug resistance, as has been demonstrated for *Giardia* (Boreham *et al.*, 1991), although other possible mechanisms for the apparent failure of drugs to eliminate parasites must be considered, including patient non-compliance, differing pharmacokinetic properties of the drugs in different patients, inaccessibility of the organism to the drug, and the inactivation of the drug by concomitant bacteria or other agents.

Iodoquinol (300–650 mg three times a day for up to 20 days) has also been used by several physicians to treat *B. hominis*, with mixed results (Babcock *et al.*, 1985; Jarecki-Black *et al.*, 1986; Markell and Udkow, 1986; Zafar, 1988; Rolston *et al.*, 1989). Treatment of 14 patients with 300 mg of iodoquinol three times a day for 10 days resulted in the removal of parasites from the stools and cessation of diarrhoea (Zafar, 1988). In contrast, five patients were treated with iodoquinol 650 mg three times a day for 10–20 days and all remained positive for *B. hominis*, with no reduction of symptoms (Markell and Udkow, 1986).

Other drugs that have been reported to be successful in a very limited number of patients include emetine (Low, 1916), furazolidone (Narkewicz *et al.*, 1989; El Masry *et al.*, 1991), ketoconazole (Cohen, 1985; Zaki *et al.*, 1991), ornidazole (Hahn and Fleischer, 1985), quinacrine (Babcock *et al.*, 1985), tinidazole (Babcock *et al.*, 1985; El Masry *et al.*, 1988), trimethoprim-sulphamethoxazole (Miller and Minshew, 1988; Rolston *et al.*, 1989), Entero-vioform® (Zierdt, 1991) and co-trimoxazole (Schwartz and Houston, 1992). Drugs which have been reported to be ineffective include quinacrine (Markell and Udkow, 1986; Sun *et al.*, 1989) and paromomycin (Markell and Udkow, 1986). Recent use of salapyrine appeared to be associated with a reduced carriage of *B. hominis* (see Kain *et al.*, 1987), while previous use of tetracycline decreased the recovery rate of parasites in stools from 25 to 14% in a group of exposed US Peace Corp Volunteers (Schwartz and Houston, 1992).

Again it needs to be emphasized that comprehensive testing of these drugs against *B. hominis* has not been undertaken and their use must still be considered as empirical and unproven until detailed clinical trials are conducted and further experimental studies are undertaken. If drug therapy to eliminate *B. hominis* from stools is thought to be indicated, then metronidazole or another 5-nitroimidazole are reasonable choices in the first instance. It is important to monitor the patient both clinically and parasitologically. If failures occur, further treatment is likely to be empirical, but co-trimoxazole, furazolidone and quinacrine (in the countries where they are available) should be considered. It is, however, important for physicians to realize that at least some *B. hominis* infections are self-limiting and thus no form of therapy may be indicated. This is illustrated by the results of Kain *et al.* (1987), who found that 14 of 18 patients treated with metronidazole became asymptomatic or improved clinically, whereas all of six patients treated with dietary management (details not given), and 13 of 16 receiving no treatment, showed similar improvement.

VII. Conclusions

While there has been some increase in our knowledge of *Blastocystis* in the last few years, there have been many uncritical anecdotal reports which added little to our knowledge. Perhaps the most significant advances have been in our understanding of the life cycle, based on morphological studies, so that now we are able to propose a scheme which fits the available data. However, we remain in a position only to postulate where the amoeboid form fits into this cycle. This review has challenged many of the accepted dogmas and highlighted the major deficiencies in our knowledge of this parasite. A suggestion has been made regarding the age-old problem of the taxonomic position of *Blastocystis*, which can now be tested experimentally. The existence of distinct demes of *B. hominis* has been demonstrated, and may have important epidemiological consequences. Major areas worthy of immediate study include the function of the central vacuole and the mitochondria, the possible role of the cyst form in transmission, the epidemiology and pathogenicity of *B. hominis*, and experimental chemotherapy. *Blastocystis* sp. and blastocystosis will prove to be fruitful areas of research for many years to come, research which is sure to provide many new and exciting insights into the biology of this enigmatic organism.

Acknowledgements

The original research reported in this review has been supported by grants from the Australian National Health and Medical Research Council, the

Australian Research Council, and the Queensland University of Technology. We are especially grateful to Dr Pietro Garavelli, Dr Lee Monroe, Dr Michal Pakandl, Dr Mulkit Singh and Dr Charles Zierdt for providing us with preprints and unpublished data for inclusion in this review, Professor Frank Cox for helpful discussions, and Dr Robyn Boreham for commenting on sections of the review. We also thank Dr Maria O'Sachy for translating original Russian texts.

REFERENCES*

Alexeieff, A. (1911). Sur la nature des formations dites "kystes *de Trichomonas intestinalis". Comptes Rendus des Séances de la Société de Biologie* **71**, 296–298.

Archibald, S.C., Mitchell, R.W., Upcroft, J.A., Boreham, P.F.L. and Upcroft, P. (1991). Variation between human and animal isolates of *Giardia* as demonstrated by DNA fingerprinting. *International Journal for Parasitology* **21**, 123–124.

Ashford, R.W. and Atkinson, E.A. (1992). Epidemiology of *Blastocystis hominis* infection in Papua New Guinea: age prevalence and associations with other parasites. *Annals of Tropical Medicine and Parasitology* **86**, 129–136.

Babb, R.R. and Wagener, S. (1989). *Blastocystis hominis*—a potential intestinal pathogen. *Western Journal of Medicine* **151**, 518–519.

Babcock, D., Houston, R., Kumaki, D. and Shlim, D. (1985). *Blastocystis hominis* in Kathmandu, Nepal. *New England Journal of Medicine* **313**, 1419.

Barret, H.P. (1921). A method for the cultivation of *Blastocystis. Annals of Tropical Medicine and Parasitology* **15**, 113–116.

Beaurepaire Aragão, H. de (1922a). Etudes sur les *Blastocystis. Memorias do Instituto Oswaldo Cruz* **15**, 143–151.

Beaurepaire Aragão, H. de (1922b). Estudos sobre os *Blastocystis. Memorias do Instituto Oswaldo Cruz* **15**, 240–250.

Belova, L.M. and Kostenko, L.A. (1990). [*Blastocystis galli* sp. n. (Protista: Rhizopoda) from the intestine of domestic hens.] *Parazitologiia* **24**, 164–168.

Benchimol, M. and de Souza, W. (1983). Fine structure and cytochemistry of the hydrogenosome of *Tritrichomonas foetus. Journal of Protozoology* **30**, 422–425.

Bensen, W. (1909). Untersuchungen über *Trichomonas intestinalis* und *vaginalis* des Menschen. *Archiv für Protistenkunde* **18**, 115–127.

Boeck, W.C. and Drhohlav, J. (1925). The cultivation of *Entamoeba histolytica. American Journal of Hygiene* **5**, 371–407.

Bohne, A. and Prowazek, S. (1908). Zur frage der Flagellatendysenterie. *Archiv für Protistenkunde* **12**, 1–8.

Boreham, P.F.L., Phillips, R.E. and Shepherd, R.W. (1986). The activity of drugs against *Giardia intestinalis* in neonatal mice. *Journal of Antimicrobial Chemotherapy* **18**, 393–398.

Boreham, P.F.L., Upcroft, J.A. and Upcroft, P. (1991). Biochemical and molecular mechanisms of resistance to nitroheterocyclic drugs in *Giardia intestinalis. In* "Biochemical Protozoology" (G.H. Coombs and M.J. North, eds), pp. 594–604. Taylor and Francis, London.

*Titles in brackets indicate papers in languages other than English which have not been seen in the original version.

Boreham, P.F.L., Upcroft, J.A. and Dunn, L.A. (1992). Protein and DNA evidence for two demes of *Blastocystis hominis* from humans. *International Journal for Parasitology* **22**, 49–53.

Bories, C., Riahl, A. and Leparco, J.C. (1990). Une étiologie rare pour un parasite fréquent: une diarrhée apyrétique à *Blastocystis hominis*. *Bulletin de la Société Française de Parasitologie* **8**, 253–255.

Bratt, D.E. and Tikasingh, E.S. (1990). *Blastocystis hominis* in two children of one family. *West Indian Medical Journal* **39**, 57–58.

Brittan, F. (1849). Report of a series of microscopical investigations on the pathology of cholera. *London Medical Gazette* **9**, 530–542.

Brumpt, E. (1912). Côlite à *Tetramitus mesnili* (Wenyon 1910) et côlite à *Trichomonas intestinalis* Leuchart 1879—*Blastocystis hominis* n. sp. et formes voisines. *Bulletin de la Société de Pathologie Exotique* **5**, 725–730.

Burden, D.J. (1976). *Blastocystis* sp.—a parasite of pigs. *Parasitology* **73**, iv–v.

Burden, D.J., Anger, H.S. and Hammet, N.C. (1978/1979). *Blastocystis* sp. infections in pigs. *Veterinary Microbiology* **3**, 227–234.

Castrillo de Tirado, A., Gonzales Mata, A.J. and Tirado Espinoza, E. (1990). *Blastocystis hominis* infection, frecuencia de infeccion por *Blastocystis hominis*: un ano de estudio. *G E N (Caracas)* **44**, 217–220.

Chatton, E. and Lalung Bonnaire (1912). Amibe limax (*Vahlkampfia* n. gen.) dans l'intestin humain. Son importance pour l'interprétation des amibes de culture. *Bulletin de la Société de Pathologie Exotique* **5**, 135–143.

Chen, J., Vaudry, W.L., Kowalewska, K. and Wenman, W. (1987). Lack of serum immune response to *Blastocystis hominis*. *Lancet* **i**, 1021.

Chunge, R.N., Karumba, P.N., Nagelkerke, N., Kaleli, N., Wamwea, M., Mutiso, N., Andala, E.O. and Kinoti, S.N. (1991). Intestinal parasites in a rural community in Kenya: cross-sectional surveys with emphasis on prevalence, incidence, duration of infection, and polyparasitism. *East African Medical Journal* **68,** 112–123.

Ciferri, R. and Redaelli, P. (1938). A new hypothesis on the nature of *Blastocytis*. *Mycopathologia* **1,** 3–6.

Cohen, A N. (1985). Ketoconazole and resistant *Blastocytis hominis* infection. *Annals of Internal Medicine* **103,** 480–481.

Dawes, R.F.H., Scott, F.D. and Tuck, A.C. (1990). *Blastocystis hominis*: an unusual cause of diarrhoea. *British Journal of Clinical Practice* **44,** 714–716.

Diaczok, B.J. and Rival, J. (1987). Diarrhea due to *Blastocystis hominis*: an old organism revisited. *Southern Medical Journal* **80,** 931–932.

Diamond, L.S. (1987). Lumen dwelling protozoa: *Entamoeba*, Trichomonads, and *Giardia*. In "*In vitro* Cultivation of Protozoan Parasites" (J.B. Jensen, ed.), pp. 65–109. CRC Press, Boca Raton, Florida.

Divo, A.A., Geary, T.G., Jensen, J.B. and Ginsburg, H. (1985). The mitochondrion of *Plasmodium falciparum* visualized by rhodamine 123 fluorescence. *Journal of Protozoology* **32,** 442–446.

Doyle, P.W., Helgason, M.M., Mathias, R.G. and Proctor, E.M. (1990). Epidemiology and pathogenicity of *Blastocystis hominis*. *Journal of Clinical Microbiology* **28,** 116–121.

Drossman, D.A. (1979). Diagnosis of the irritable bowel syndrome. *Annals of Internal Medecine* **90,** 431–432.

Dunn, L.A. (1992). "Variation Among Cultured Stocks of *Blastocytis hominis* (Brumpt 1912)". PhD Thesis, University of Queensland, Australia.

Dunn, L.A. and Boreham, P.F.L. (1991). The *in-vitro* activity of drugs against *Blastocystis hominis*. *Journal of Antimicrobial Chemotherapy* **27,** 507–516.
Dunn, L.A., Boreham. P.F.L. and Stenzel, D.J. (1989). Ultrastructural variation of *Blastocystis hominis* stocks in culture. *International Journal for Parasitology* **19,** 43–56.
Editorial (1991). *Blastocystis hominis*: commensal or pathogen. *Lancet* **337,** 521–522.
Edmeades, R., Halliday, K. and Shepherd, R.W. (1978). Infantile gastroenteritis: relationship between cause, clinical course and outcome. *Medical Journal of Australia* **ii,** 29–32.
El Masry, N.A., Bassily, S. and Farid, Z. (1988). *Blastocystis hominis*: clinical and therapeutic aspects. *Transactions of the Royal Society of Tropical Medicine and Hygiene* **82,** 173.
El Masry, N.A., Bassily, S., Farid, Z. and Aziz, A.G. (1990). Potential clinical significance of *Blastocystis hominis* in Egypt. *Transactions of the Royal Society of Tropical Medicine and Hygiene* **84,** 695.
El Masry, N.A., Bassily, S.B., Farid, Z., Mansour, N.S., Sabry, A.G. and Kilpatrick, M.E. (1991). *Blastocystis hominis*: eradicative therapy for a probable pathogen. *American Journal of Tropical Medicine and Hygiene* **45**, supplement, 95–96.
Enzenauer, R.J., Underwood, G.H., Jr and Ribbing, J. (1990). Tropical pulmonary eosinophilia. *Southern Medical Journal* **83**, 69–72.
Fenchel, T. and Finlay, B.J. (1991). The biology of free-living anaerobic ciliates. *European Journal of Protistology* **26**, 201–215.
Figueroa, L., Moraleda, L. and Garcia N. (1990). Enteroparasitosis en ninos con sindrome diarreico agudo de la ciudad de Valdivia, X region, Chile con especial referencia a *Cryptosporidium* sp. *Parasitología al Dia* **14**, 78–82.
Finlay, B.J. and Fenchel, T. (1989). Hydrogenosomes in some anaerobic protozoa resemble mitochondria. *FEMS Microbiology Letters* **65**, 311–314.
Gallagher, P.G. and Venglarcik, J.S. (1985). *Blastocystis hominis* enteritis. *Pediatric Infectious Disease* **4**, 556–557.
Garavelli, P.L. (1992). Acquisitions récentes sur *Blastocystis hominis* et la blastocystose (maladie de Zierdt et Garavelli). *Bulletin de la Société Française de Parasitologie*, **10**, 21–26.
Garavelli, P.L. and Libanore, M. (1990). *Blastocystis* in immunodeficiency diseases. *Reviews of Infectious Diseases* **12**, 158.
Garavelli, P.L. and Scaglione, L. (1989). Blastocystosis. An epidemiological study. *Microbiologica* **12**, 349–350.
Garavelli, P.L. and Scaglione, L. (1990). [*Blastocystis hominis* infection in AIDS and correlated pathologies.] *Minerva Medica* **81**, supplement, 91–92.
Garavelli, P.L., Orsi, P. and Scaglione, L. (1988). *Blastocystis hominis* infection during AIDS. *Lancet* **ii**, 1364.
Garavelli, P.L., Scaglione, L., Rossi, M.R., Bicocchi, R. and Libanore, M. (1989). Blastocystosis in Italy. *Annales de Parasitologie Humaine et Comparée* **64**, 391–395.
Garavelli, P.L., Scaglione, L., Bicocchi, R and Libanore, M. (1990). Blastocystosis: a new disease in the acquired immunodeficiency syndrome? *International Journal of Sexually Transmitted Diseases and AIDS* **1**, 134–135.
Garavelli, P.L., Scaglione, L., Biococchi, R. and Libanore, M. (1991a). Pathogenicity of *Blastocystis hominis*. *Infection* **19**, 185.
Garavelli, P.L., Scaglione, L., Libanore, M. and Rolston, K. (1991b). Blastocystosis: a new disease in patients with leukaemia. *Haematologica* **79**, 80.

Garavelli, P.L., Scaglione, L., Merighi, A. and Libanore, M. (1992). Endoscopy of blastocystosis (Zierdt–Garavelli Disease). *Italian Journal of Gastroenterology* **24**, 206.

García, L.S., Bruckner, D.A. and Clancy, M.N. (1984). Clinical relevance of *Blastocystis hominis*. *Lancet* **i**, 1233–1234.

García Pascual, L., Bartolomsé Comas, R., Cuenca Luque, R. and San José Laporte, A. (1988). [Enteritis caused by *Blastocystis hominis*] (letter). *Medicina Clinica (Barcelona)* **91**, 797.

Ghadially, F.N. (1988). "Ultrastructural Pathology of the Cell and Matrix. A Text and Atlas of Physiological and Pathological Alterations in the Fine Structure of Cellular and Extracellular Components", 3rd edn. Butterworths, London.

Guglielmetti, P., Cellesi, C., Figura, N. and Rossolini, A. (1989). Family outbreak of *Blastocystis hominis* associated gastroenteritis. *Lancet* **ii**, 1394.

Guglielmetti, P., Sansoni, A., Fantoni, A. and Rossolini, A. (1991). Pathogenesis of blastocystosis. *Lancet* **338**, 57.

Guirges, S.Y. and Al-Waili, N.S. (1987). *Blastocystis hominis*: evidence for human pathogenicity and effectiveness of metronidazole therapy. *Clinical and Experimental Pharmacology and Physiology* **14**, 333–335.

Hackenbrock, C.R. (1968). Ultrastructural bases for metabolically linked mechanical activity in mitochondria II. Electron transport-linked ultrastructural transformations in mitochondria. *Journal of Cell Biology* **37**, 345–369.

Hahn, P. and Fleischer, N.K.F. (1985). *Blastocystis hominis*—is it of clinical importance? *Tropical Medicine and Parasitology* **36**, supplement 2, 7–8.

Henry, M.C., De Clercq, D., Lokombe, B., Kayembe, K., Kapita, B., Mamba, K., Mbendi, N. and Mazebo, P. (1986). Parasitological observations of chronic diarrhoea in suspected AIDS adult patients in Kinshasa (Zaire). *Transactions of the Royal Society of Tropical Medicine and Hygiene* **80**, 309–310.

James, W.M. (1914). A study of the *Entamoebae* of man in the Panama Canal zone. *Annals of Tropical Medicine and Parasitology* **8**, 133–320.

Jarecki-Black, J.C., Bannister, E.R. and Glassmann, A.B. (1986). *Blastocystis hominis*: infection in a 2-yr old. *Clinical Microbiology Newsletter* **8**, 98.

Jeddy, T.A. and Farrington, G.H. (1991). *Blastocystis hominis* complicating ulcerative colitis. *Journal of the Royal Society of Medicine* **84**, 623.

Johnson, A.M. and Baverstock, P.R. (1989). Rapid ribosomal RNA sequencing and the phylogenetic affinities of protists. *Parasitology Today* **5**, 102–105.

Johnson, A.M., Thanou, A., Boreham, P.F.L. and Baverstock, P.R. (1989). *Blastocystis hominis*: phylogenetic affinities determined by rRNA sequence comparison. *Experimental Parasitology* **68**, 283–288.

Kain, K.C., Noble, M.A., Freeman, H.J. and Barteluk, R.L. (1987). Epidemiology and clinical features associated with *Blastocystis hominis* infection. *Diagnostic Microbiology and Infectious Disease* **8**, 235–244.

Knowles, R. and Das Gupta, B.M. (1924). On the nature of *Blastocystis hominis*. *Indian Journal of Medical Research* **12**, 31–38.

Kofoid, C.A. and Swezy, O. (1921). On the prevalence of carriers of *Endamoeba dysenteriae* among soldiers returned from overseas service. *American Journal of Tropical Medicine* **1**, 41–48.

Kofoid, C.A., Kornnhauser, S.I. and Swezy, O. (1919). Criterions for distinguishing the *Endamoeba* of amebiasis from other organisms. *Archives of Internal Medicine* **24**, 35–50.

Kukoschke, K.-G. and Müller, H.E. (1991). SDS-PAGE and immunological analysis of different axenic *Blastocystis hominis* strains. *Journal of Medical Microbiology* **35**, 35–39.

Kukoschke, K.-G., Necker, A. and Müller, H.E. (1990). Detection of *Blastocystis hominis* by direct microscopy and culture. *European Journal of Clinical Microbiology and Infectious Diseases* **9**, 305–307.

Lakhanpal, S., Cohen, S.B. and Fleischmann, R.M. (1991). Reactive arthritis from *Blastocystis hominis*. *Arthritis and Rheumatism* **34**, 251–253.

Lavier, G. (1937a). Sur certaines formes que présentent en culture les *Blastocystis*. *Comptes Rendus des Séances de la Société de Biologie* **125**, 593–595.

Lavier, G. (1937b). Sur la cytologie des protistes du genre *Blastocystis*. *Comptes Rendus de l'Académie des Sciences, Paris* **105**, 340–342.

Lavier, G. (1952). Observations sur les *Blastocystis*. *Annales de Parasitologie Humaine et Comparée* **27**, 339–356.

Le Bar, W.D., Larsen, E.C. and Patel, K. (1985). Afebrile diarrhea and *Blastocystis hominis*. *Annals of Internal Medicine* **103**, 306.

Lee, D.L. (1970). The fine structure of *Blastocystis* from the caecum of turkey. *Transactions of the British Mycological Society* **54**, 313–317.

Lee, M.G., Rawlins, S.C., Didier, M. and De Ceulaer, K. (1990). Infective arthritis due to *Blastocystis hominis*. *Annals of the Rheumatic Diseases* **49**, 192–193.

Libanore, M., Bicocchi, R., Sighinolfi, L. and Ghinelli, F. (1990). Blastocystosis in drug-addicts with HIV-1 infection. *European Journal of Epidemiology* **6**, 108–109.

Libanore, M., Rossi, M.R., Scaglione, L. and Garavelli, P.L. (1991). Outbreak of blastocystosis in institution for the mentally retarded. *Lancet* **337**, 609–610.

Llibre, J.M., Tor, J., Manterola, J.M., Carbonell, C. and Foz, M. (1989). *Blastocystis hominis* chronic diarrhoea in AIDS patients. *Lancet* **i**, 221.

Lösch, F.A. (1875). Massenhafte Entwickelung von Amoeben im Dickdarn. *Archiv für Pathologische Anatomie und Physiologie und für Klinische Medizin* **65**, 196–211.

Low, G. (1916). Two chronic amoebic dysentery carriers treated by emetine, with some remarks on the treatment of *Lamblia*, *Blastocystis* and *E. coli* infections. *Journal of Tropical Medicine and Hygiene* **19**, 29–34.

Lynch, K.M. (1917). *Blastocystis hominis*; its characteristics and its prevalence in intestinal content and feces in South Carolina. *Journal of Bacteriology* **2**, 369–377.

Lynch, K.M. (1922). *Blastocystis* species in culture: a preliminary communication. *American Journal of Tropical Medicine* **2**, 215–222.

Macfie, J.W.S. (1915). A case of dysentery in a monkey in which amoebae and spirochaetes were found. *Annals of Tropical Medicine and Parasitology* **9**, 507–512.

Magaudda-Borzì, L. and Pennisi, L. (1961). A morfologia di *Blastocystis hominis* al microscopio elettronico. *Atti della Società Peloritana di Scienze, Fisiche Matematiche e Naturali, Società Peloritana Messina* **7**, supplement, 575–582.

Markell, E.K. and Udkow, M.P. (1986). *Blastocystis hominis*: pathogen or fellow traveler? *American Journal of Tropical Medicine and Hygiene* **35**, 1023–1026.

Matsumoto, Y., Yamada, M. and Yoshida, Y. (1987). Light microscopical appearance and ultrastructure of *Blastocystis hominis*, an intestinal parasite of man. *Zentralblatt für Bakteriologie, Microbiologie und Hygiene* **A264**, 379–385.

Matthews, J.R. (1918). Observations on the cysts of the common intestinal protozoa of man. *Annals of Tropical Medicine and Parasitology* **12**, 17–26.

McClure, H.M., Strobert, E.A. and Healy, G.R. (1980). *Blastocystis hominis* in a pig-tailed macaque: a potential enteric pathogen for nonhuman primates. *Laboratory*

Animal Science **30**, 890–894.
Mehlhorn, H. (1988a). *Blastocystis hominis*, Brumpt 1912: are there different stages or species? *Parasitology Research* **74**, 393–395.
Mehlhorn, H. (1988b). "Parasitology in Focus. Facts and Trends". Springer, Berlin.
Mercado, R. and Arias, B. (1991). *Blastocystis hominis*: frecuencia de infectión en pacientes ambulatorios del sector norte de Santiago, Chile. *Boletín Chileno de Parasitología* **46**, 30–32.
Miller, R.A. and Minshew, B.H. (1988). *Blastocystis hominis*: an organism in search of a disease. *Reviews of Infectious Diseases* **10**, 930–938.
Molet, B., Werler, C. and Kremer, M. (1981). *Blastocystis hominis*: improved axenic cultivation. *Transactions of the Royal Society of Tropical Medicine and Hygiene* **75**, 752–753.
Müller, M. (1991). Energy metabolism of anaerobic parasitic protists. *In* "Biochemical Protozoology" (G.H. Coombs and M.J. North, eds), pp. 80–91. Taylor and Francis, London.
Narkewicz, M.R., Janoff, E.N., Sokol, R.J. and Levin, M.J. (1989). *Blastocystis hominis* gastroenteritis in a hemophiliac with acquired immune deficiency syndrome. *Journal of Pediatric Gastroenterology and Nutrition* **8**, 125–128.
Nguyen, X.M. and Krech, T. (1989). *Blastocystis hominis*, ein parasitärer Durchfallerreger. *Schweizerische Medizinische Wochenschrift* **119**, 457–460.
Ockert, G. 1990. Symptomatology, pathology, epidemiology and diagnosis of *Dientamoeba fragilis*. *In* "Trichomonads Parasitic in Humans" (B.M. Honigberg, ed.), pp. 384–410. Springer, New York.
Pakandl, M. (1993). Occurrence of *Blastocystis* sp. in pigs. *Folia Parasitologica*, **38**, 297–301.
Pakandl, M. and Pecka, Z. (1992). A domestic duck as a new host of *Blastocystis* sp. *Folia Parasitologica*, **39**, 59–60.
Perroncito, E. (1899). Di un nuovo protozoa dell' uomo e di talune specie animali. *Giornale della Reale Academia di Medicina di Torino* **5**, 36–38.
Phillips, B.P. and Zierdt, C.H. (1976). *Blastocystis hominis*: pathogenic potential in human patients and in gnotobiotes. *Experimental Parasitology* **39**, 358–364.
Picher, O. and Aspöck, H. (1980). Häuffigkeit und Bedeutung parasitärer Infektionen bei vietnamesischen Flüchtlinger. *Wiener Medizinische Wochenschrift* **130**, 190–193.
Pikula, Z.P. (1987). *Blastocystis hominis* and human disease. *Journal of Clinical Microbiology* **25**, 1581
Prowazek, S. (1911). Zur Kenntnis der Flagellaten des Darmtraktus. *Archiv für Protistenkunde* **23**, 96–100.
Puga, S., Figueroa, L. and Navarrette, N. (1991). Protozoos y helmintos intestinales en la población preescolar y escolar de la ciudad de Valdivia, Chile. *Parasitología al Dia* **15**, 57–58.
Qadri, S.M.H., Al-Okaili, G.A. and Al-Dayel, F. (1989). Clinical significance of *Blastocystis hominis*. *Journal of Clinical Microbiology* **27**, 2407–2409.
Reinthaler, F.F., Mascher, F., Klem, G. and Sixl, W. (1988). A survey of gastrointestinal parasites in Ogun State, south-west Nigeria. *Annals of Tropical Medicine and Parasitology* **82**, 181–184.
Ricci, N., Toma, P., Furlani, M., Caselli, M. and Gullini, S. (1984). *Blastocystis hominis*: a neglected cause of diarrhoea? *Lancet* **i**, 966.
Rolston, K.V.I., Winans, R. and Rodriguez, S. (1989). *Blastocystis hominis*: pathogen or not? *Reviews of Infectious Diseases* **11**, 661–662.

Russo, A.R., Stone, S.L., Taplin, M.E., Snapper, H.J. and Doern, G.V. (1988). Presumptive evidence for *Blastocystis hominis* as a cause of colitis. *Archives of Internal Medicine* **148**, 1064.

Salavert, M., Roig, P., Nieto, A., Navarro, V. and Borrás, R. (1990). [Enterocolitis caused by *Blastocystis hominis* and HIV infection.] *Enfermedades Infecciosas y Microbiologia Clinica* **8**, 63–64.

Scaglione, L., Ansladi, E., Troielli, F. and Garavelli, P.L. (1990). *Blastocystis hominis* infection in patients with diabetes mellitus. Clinical case history. *Recenti Progressi in Medicina* **81**, 482–485.

Schaefer, F.W., III. (1990). Methods for excystation of *Giardia*. *In* "Giardiasis" (E.A. Meyer, ed.), pp. 111–136. Elsevier, Amsterdam.

Schuster, F.L. (1979). Small amebas and amoeboflagellates. *In* "Biochemistry and Physiology of Protozoa" (M. Levandowsky and S.H. Hutner, eds), Vol. 1, pp. 215–285. Academic Press, New York.

Schwartz, E. and Houston, R. (1992). Effect of co-trimoxazole on stool recovery of *Blastocystis hominis*. *Lancet* **339**, 428–429.

Senay, H. and MacPherson, D. (1990). *Blastocystis hominis*: epidemiology and natural history. *Journal of Infectious Diseases* **162**, 987–990.

Sheehan, D.J., Raucher, B.G. and McKitrick, J.C. (1986). Association of *Blastocystis hominis* with signs and symptoms of human disease. *Journal of Clinical Microbiology* **24**, 548–550.

Sheehan, J.P. and Ulchaker, M.M. (1990). *Blastocystis hominis*: treatable cause of diabetic diarrhea. *Diabetes Care* **13**, 906–907.

Shikiya, K., Terukina, S., Higashionna, A., Arakaski, T., Kadena, K., Shigeno, Y., Kinjo, F. and Saito, A. (1989). A case report of colitis due to *Blastocystis hominis*. *Gastrointestinal Endoscopy* **31**, 1851–1854.

Silard, R. (1979). Contribution to *Blastocystis hominis* studies. Aspects of degenerescence. *Archives Roumaines de Pathologie Expérimentale et de Microbiologie* **38**, 105–114.

Silard, R. and Burghelea, B. (1985). Ultrastructural aspects of *Blastocystis hominis* strain resistant to antiprotozoal drugs. *Archives Roumaines de Pathologie Expérimentale et de Microbiologie* **44**, 73–85.

Silard, R., Petrovici, M., Panaitescu, D. and Stoicescu, V. (1977). *Blastocystis hominis* in the liver of *Cricetus auratus*. *Archives Roumaines de Pathologie Expérimentale et de Microbiologie* **36**, 55–60.

Silard, R., Panaitescu, D. and Burghelea, B. (1983). Ultrastructural aspects of *Blastocystis hominis*. *Archives Roumaines de Pathologie Expérimentale et de Microbiologie* **42**, 233–242.

Steinmann, E., di Gallo, A., Rüttimann, S., Loosli, J. and Dubach, U.C. (1990). [Aetiology of diarrheal diseases in immunocompetent and HIV-positive patients.] *Schweizerische Medizinische Wochenschrift* **120**, 1253–1256.

Stenzel, D.J. and Boreham, P.F.L. (1991). A cyst-like stage of *Blastocystis hominis*. *International Journal for Parasitology* **21**, 613–615.

Stenzel, D.J. and Boreham, P.F.L. (1993). Ultrastructure of *Blastocystis hominis*. *Journal of Computer Assisted Microscopy*, **5**, 13–16.

Stenzel, D.J., Dunn, L.A. and Boreham, P.F.L. (1989). Endocytosis in cultures of *Blastocystis hominis*. *International Journal for Parasitology* **19**, 787–791.

Stenzel, D.J., Boreham, P.F.L. and McDougall, R. (1991). Ultrastructure of *Blastocystis hominis* in human stool samples. *International Journal for Parasitology* **21**, 807–812.

Sun, T. (1988). "Color Atlas and Textbook of Diagnostic Parasitology". Igaku-shoin, New York.

Sun, T., Katz S., Tanenbaum, B. and Schenone, C. (1989). Questionable clinical significance of *Blastocystis hominis* infection. *American Journal of Gastroenterology* **84**, 1543–1547.

Swayne, J.G. (1849). An account of certain organic cells peculiar to the evacuations of cholera. *Lancet* **ii**, 368–371.

Swellengrebel, N.H. (1917). Observations on *Blastocystis hominis. Parasitology* **9**, 451–459.

Tan, H.K. and Zierdt, C.H. (1973). Ultrastructure of *Blastocystis hominis. Zeitschrift für Parasitenkunde* **42**, 315–324.

Tan, H.K., Harrison, M. and Zierdt, C.H. (1974). Freeze-etch studies of the granular and vacuolated forms of *Blastocystis hominis. Zeitschrift für Parasitenkunde* **44**, 267–278.

Taylor, D.N., Echeverria, P., Blaser, M.J., Pitarangsi, C., Blacklow, N., Cross, J. and Weniger, B.O. (1985). Polymicrobial aetiology of travellers' diarrhoea. *Lancet* **i**, 381–383.

Telabasic, S., Pikula, Z.P. and Drsda, M.A. (1987). Diarrhoea caused by *Blastocystis hominis. Giornale di Malattie Infective e Parassitarie* **39**, 614–615.

Telalbasic, S., Pikula, Z.P. and Kapidzic, M. (1991). *Blastocystis hominis* may be a potential cause of intestinal disease. *Scandinavian Journal of Infectious Diseases* **23**, 389–390.

Teow, W.L., Zaman, V., Ng, G.C., Chan, Y.C., Yap, E.H., Howe, J., Gopalakrishnakone, P. and Singh, M. (1991). A *Blastocystis* species from the sea-snake *Lapemis hardwickii* (Serpentes: Hydrophiidae). *International Journal for Parasitology* **21**, 723–726.

Teow, W.L., Ng, G.C., Chan, P.P., Chan, Y.C., Yap, E.H., Zaman, V. and Singh, M. (1992). A survey of *Blastocystis* in reptiles. *Parasitology Research*, **78**, 453–455.

Tsang, T.K., Levin, B.S. and Morse, S.R. (1989). Terminal ileitis associated with *Blastocystis hominis* infection. *American Journal of Gastroenterology* **84**, 798–799.

Upcroft, J.A., Dunn, L.A., Dommett, L.S., Healey, A. Upcroft, P. and Boreham, P.F.L. (1989). Chromosomes of *Blastocystis hominis. International Journal for Parasitology* **19**, 879–883.

Upcroft, P., Mitchell, R. and Boreham, P.F.L. (1990). DNA fingerprinting of the human intestinal parasite *Giardia intestinalis* with the M13 phage genome. *International Journal for Parasitology* **20**, 319–323.

Vallano, A., Pigrau, C., Hermández, A. and Gavaldá, J. (1991). [*Blastocystis hominis* in an HIV-positive homosexual patient.] *Revista Clínica Española* **188**, 110–111.

Vannatta, J.B., Adamson, D. and Mullican, K. (1985). *Blastocystis hominis* infection presenting as recurrent diarrhea. *Annals of Internal Medicine* **102**, 495–496.

Van Saanen-Ciurea, M. and El Achachi, H. (1983). *Blastocystis hominis*: étude morphologique par microscopie optique et électronique. *Bulletin de la Société de Pathologie Exotique* **76**, 766–776.

Van Saanen-Ciurea, M. and El Achachi, H. (1985). *Blastocystis hominis*: culture and morphological study. *Experientia* **41**, 546.

Walker, J.C., Bahr, G. and Ehl, A.S. (1985). Gastrointestinal parasites in Sydney. *Medical Journal of Australia* **143**, 480.

Wenyon, C.M. (1910). A new flagellate *Macrostoma mesnili* n.sp. from the human intestine, with some remarks on the supposed cysts of *Trichomonas. Parasitology* **3**, 210–216.

Wenyon, C.M. (1920). Observations on the intestinal protozoa of three Egyptian lizards, with a note on a cell-invading fungus. *Parasitology* **12**, 350–365.
Wenyon, C.M. (1926). "Protozoology—A Manual for Medical Men, Veterinarians and Zoologists", Vol. 1. Baillière, Tindall and Cox, London.
Wenyon, C.M. and O'Connor, F.W. (1917). An inquiry into some problems affecting the spread and incidence of intestinal protozoal infections of British troops and natives in Egypt, with special reference to the carrier question, diagnosis and treatment of amoebic dysentery, and an account of three new human intestinal protozoa. *Journal of the Royal Army Medical Corps* **28**, 346–367.
World Health Organization (1978). Proposals for the nomenclature of salivarian trypanosomes and for the maintenance of reference collections. *Bulletin of the World Health Organization* **56**, 467–480.
Yamada, M., Yoshikawa, H., Tegoshi, T., Matsumoto, Y., Yoshikawa, T., Shiota, T. and Yoshida, Y. (1987a). Light microscopical study of *Blastocystis* spp. in monkeys and fowls. *Parasitology Research* **73**, 527–531.
Yamada, M., Matsumoto, Y., Tegoshi, T. and Yoshida, Y. (1987b). The prevalence of *Blastocystis hominis* infection in humans in Kyoto City. *Japanese Journal of Tropical Medicine and Hygiene* **15**, 158–159.
Yoshikawa, H., Yamada, M. and Yoshida, Y. (1988). Freeze-fracture study of *Blastocystis hominis. Journal of Protozoology* **35**, 522–528.
Zafar, M.N. (1988). Morphology and frequency distribution of protozoan *Blastocystis hominis. Journal of the Pakistan Medical Association* **38**, 322–324.
Zaki, M., Daoud, A.S., Pugh, R.N.H., Al-Ali, F., Al-Mutairi, G. and Al-Saleh, Q. (1991). Clinical report of *Blastocystis hominis* infection in children. *Journal of Tropical Medicine and Hygiene* **94**, 118–122.
Zierdt, C.H. (1973). Studies of *Blastocystis hominis. Journal of Protozoology* **20**, 114–121.
Zierdt, C.H. (1978). *Blastocystis hominis*, an intestinal protozoan parasite of man. *Public Health Laboratory* **36**, 147–161.
Zierdt, C.H. (1986). Cytochrome-free mitochondria of an anaerobic protozoan—*Blastocystis hominis Journal of Protozoology* **33**, 67–99.
Zierdt, C.H. (1988). *Blastocystis hominis*, a long misunderstood intestinal parasite. *Parasitology Today* **4**, 15–17.
Zierdt, C.H. (1991). *Blastocystis hominis*—past and future. *Clinical Microbiology Reviews* **4**, 61–79.
Zierdt, C.H. and Swan, J.C. (1981). Generation time and growth rate of the human intestinal parasite *Blastocystis hominis. Journal of Protozoology* **28**, 483–485.
Zierdt, C.H. and Tan, H. (1976a). Ultrastructure and light microscope appearance of *Blastocystis hominis* in a patient with enteric disease. *Zeitschrift für Parasitenkunde* **50**, 277–283.
Zierdt, C.H. and Tan, H. (1976b). Endosymbiosis in *Blastocystis hominis. Experimental Parasitology* **39**, 422–430.
Zierdt, C.H. and Williams, R.L. (1974). *Blastocystis hominis*: axenic cultivation. *Experimental Parasitology* **36**, 233–243.
Zierdt, C.H., Rude, W.S. and Bull, B.S. (1967). Protozoan characteristics of *Blastocystis hominis. American Journal of Clinical Pathology* **48**, 495–501.
Zierdt, C.H., Swan, J.C. and Hosseini, J. (1983). *In vitro* response of *Blastocystis hominis* to antiprotozoal drugs. *Journal of Protozoology* **30**, 332–334.
Zierdt, C.H., Donnolley, C.T., Muller, J. and Constantopoulos, G. (1988). Biochemical and ultrastructural study of *Blastocystis hominis. Journal of Clinical Microbiology* **26**, 965–970.

Zuckerman, M.J., Ho, H., Hooper, L., Anderson, B. and Polly, S.M. (1990). Frequency of recovery of *Blastocystis hominis* in clinical practice. *Journal of Clinical Gastroenterology* **12**, 525–532.

NOTE ADDED IN PROOF

Since submission of this review there have been a number of developments. Jiang and He (1993) have discussed the taxonomic status of *Blastocystis hominis* and concluded that it should not be assigned to the subphylum Sarcodina nor placed in the Apicomplexa; they proposed a new subphylum, Blastocysta, to include class Blastocystea, order Blastocystida, family Blastocystidae and genus *Blastocystis*, of which *B. hominis* is the type species. Two new species have been described, based on light microscopical observations; *B. anatis* (Belova, 1991) from the domestic duck (*Anas platyrhynchos*) and *B. anseri* (Belova, 1992a) from the domestic goose (*Anser domesticus*). *Blastocystis* isolated from turkeys (*Meleagris gallopavo*) in Tadjikistan and Uzbekistan have tentatively been assigned to *B. galli* (see Belova, 1992b). Experimental infection of chickens with cultured *B. galli* did not result in clinical symptoms or pathology of the viscera, and it was concluded that *B. galli* was a commensal (Belova, 1992c).

Blastocystis spp. have been described from single specimens of camel, llama, highland bull and lion from a small travelling circus (Stenzel *et al.*, in press), from a koala, cattle, alpacas and ostriches in Australia (D. J. Stenzel, M. F. Cassidy and P. F. L. Boreham, unpublished observations), and from cockroaches collected in sewers in Singapore (Zaman *et al.*, 1993). Cyst-like stages were found in the cockroach and it was suggested that this may be the mode of transmission to humans.

Burghelea and Radulescu (1991) have described *B. hominis* isolated from human stools, giving an alternative view of differentiation of the cells. They propose that autophagocytic processes within the endoplasmic reticulum result in the formation of the vacuolar form from the amoeboid form. However, their electron micrographs show cells of cyst morphology with obvious cyst walls.

An analysis of the lipids present in *B. hominis* showed that 9 neutral and 11 polar lipids could be resolved (Keenan *et al.*, 1992). Sterol esters, principally esters of cholesterol, were the major neutral lipid constituents while phosphatidylcholine was the major polar lipid present. Of particular interest was the tentative identification of cardiolipin, a characteristic lipid of mitochondria which has a role in cytochrome oxidase function.

Icosahedral double-stranded RNA virus-like particles 30 nm in diameter have been described from *B. lapemi* by Teow *et al.* (1992).

Further data have been published relating to the debate on pathogenicity of *Blastocystis*. A study of a cohort of 49 homosexual men, confirmed to be infected with human immunodeficiency virus (HIV), compared clinical status and enteric parasite load with gastrointestinal structure, function and symptomatology. Although *B. hominis* was the commonest infection (44%) in this group, no correlation was found between the occurrence of any enteric parasitic infection and gastrointestinal symptoms (Church *et al.*, 1992). However, in a prospective cross-sectional study of children in Tanzania, *B. hominis* was detected only in HIV-infected individuals and not in children with chronic diarrhoea but not infected with HIV (Cegielski *et al.*, 1993). Thus, the role of *Blastocystis* as a pathogen in AIDS patients has not been resolved, a view shared by Ayadi *et al.* (1992) in a brief review of the current status of *Blastocystis*. Kukoschke and Müller (1992) reported an increased detection rate of *B. hominis* in faecal samples collected from healthy individuals compared with those from patients suffering from diarrhoea.

The prevalence of *B. hominis* infections in riverside communities of the Valdivia river basin, Chile was 61.8% (Torres *et al.*, 1992). Infection was highest in persons living in houses without sanitary facilities. No sex difference in prevalence was detected.

REFERENCES

Ayadi, A., Dutoit, E. and Camus, D. (1992). *Blastocystis hominis*: à la recherche d'une maladie, un organisme incompris. *La Presse Médicale* **21**, 1677–1679.

Belova, L.M. (1991). *Blastocystis anatis* sp. nov. (Rhizopoda, Lobosea) from *Anas platyrhynchos*. *Zoologicheskii Zhurnal* **70**, 5–10.

Belova, L.M. (1992a). [*Blastocystis anseri* (Protista: Rhizopoda) from domestic goose.] *Parazitologiia* **26**, 80–82.

Belova, L.M. (1992b). [On the occurrence of *Blastocystis galli* (Rhizopoda, Lobosea) in turkey.] *Parazitologiia* **26**, 166–168.

Belova, L.M. (1992c). [*Blastocystis galli* (Rhizopoda, Lobosea)—a parasite or commensal.] *Archiv für Protistenkunde* **144**, 215–218.

Burghelea, B. and Radulescu, S. (1991). Ultrastructural evidence for a possible differentiation way in the life-cycle of *Blastocystis hominis*. *Roumainian Archives of Microbiology and Immunology* **50**, 231–244.

Cegielski, J.P., Msengi, A.E., Dukes, C.S., Mbise, R., Reddinglallinger, R., Minjas, J.N., Wilson, M.L., Shao, J. and Durack, D.T. (1993). Intestinal parasites and HIV infection in Tanzanian children with chronic diarrhea. *AIDS* **7**, 213–221.

Church, D.L., Sutherland, L.R., Gill, M.J., Visser, N.D., Kelly, J.K., Bryant, H.E., Hwang, W.-S. and Sharkey, K.A. (1992). Absence of an association between enteric parasites in the manifestations and pathogenesis of HIV enteropathy in gay men. *Scandinavian Journal of Infectious Diseases* **24**, 567–575.

Jiang, J.-H. and He, J.-G. (1993). Taxonomic status of *Blastocystis hominis*. *Parasitology Today* **9**, 2–3.

Keenan, T.W., Huang, C.M. and Zierdt, C.H. (1992). Comparative analysis of lipid composition in axenic strains of *Blastocystis hominis*. *Comparative Biochemistry and Physiology* **102B**, 611–615.

Kukoschke, K.-G. and Müller, H.E. (1992). Varying incidence of *Blastocystis hominis* in cultures from faeces of patients with diarrhoea and from healthy persons. *Zentralblatt für Bakteriologie/International Journal of Medical Microbiology, Virology, Parasitology and Infectious Diseases* **277**, 112–118.

Stenzel, D.J., Cassidy, M.F. and Boreham, P.F.L. (in press). Morphology of *Blastocystis* sp. isolated from circus animals. *International Journal for Parasitology*.

Teow, W.L., Ho, L.C., Ng, G.C., Chan, Y.C., Yap, E.H., Chan, P.P., Howe, J., Zaman, V. and Singh, M. (1992). Virus-like particles in a *Blastocystis* species from the sea-snake, *Lapemis hardwickii*. *International Journal for Parasitology* **22**, 1029–1032.

Torres, P., Miranda, J.C., Flores, L., Riquelme, J., Franjola, R., Perez, J., Auad, S., Hermosilla, C. and Riquelme, S. (1992). Blastocystosis and other intestinal protozoan infections in human riverside communities from Valdivia River Basin, Chile. *Revista do Instituto de Medicina Tropical de São Paulo* **34**, 557–564.

Zaman, V., Ng, G.C., Suresh, K., Yap, E.H. and Singh, M. (1993). Isolation of *Blastocystis* from the cockroach (Dictyoptera, Blattidae). *Parasitology Research* **79**, 73–74.

Giardia and Giardiasis

R. C. A. THOMPSON AND J. A. REYNOLDSON

Institute for Molecular Genetics and Animal Disease and School of Veterinary Studies, Murdoch University, Murdoch, Western Australia, 6150, Australia

AND

A. H. W. MENDIS

School of Biomedical Sciences, Curtin University of Technology, Bentley, Western Australia, 6001, Australia

ADVANCES IN PARASITOLOGY VOL. 32
ISBN 0-12-031732-X

I. Introduction

It is over 10 years since Meyer and Radulescu (1979) reviewed *Giardia* and giardiasis in *Advances in Parasitology*. In their introduction, they emphasized that "despite their ubiquity and antiquity, the *Giardia* have, until recently, been little studied". In the intervening years, *Giardia* has been extensively studied. The number of papers published has increased enormously, two books on the parasite have been produced (Erlandsen and Meyer, 1984; Meyer, 1990a), and an international conference on *Giardia* has been organized (Wallis and Hammond, 1988). Yet it is still very difficult to keep up with developments in this productive field of research and, despite all these research efforts, several fundamental questions concerning *Giardia* and giardiasis remain unresolved (Table 1), particularly with respect to the relationship of *Giardia* and disease, and the role of zoonotic transmission. Indeed, it is only recently that we have started to appreciate the clinical significance of *Giardia* infections in developing countries and among disadvantaged groups. *Giardia* is now recognized as one of the 10 major parasites of humans, being equal to ascariasis as a cause of death in the developing world (Warren, 1989; Meyer, 1990b). In developed countries, *Giardia* has the distinction of being the most commonly reported human intestinal parasite (Acha and Szyfres, 1987; Thompson *et al.*, 1990a; Schantz, 1991). Regrettably, however, the range of drugs available to treat giardiasis is limited and their efficacy leaves much to be desired. There is an urgent need for new antigiardial agents, yet this search is hampered by our lack of understanding of many fundamental aspects of *Giardia* biochemistry and metabolism. In addition, although the application of molecular biological techniques to research on *Giardia* has revealed new avenues of investigation, it has also given rise to many new questions about this intriguing organism concerning its phylogenetic position, reproductive behaviour and genetic diversity.

TABLE 1 Giardia *and giardiasis: outstanding questions*

Mode of attachment
Reproductive behaviour
Ploidy
Nuclear function
Mechanisms of pathogenesis
Determinants of susceptibility and virulence
Role of intracellular symbionts
Host immune mechanisms
Metabolism
Mechanisms of drug resistance
Genetic structure of populations
Phylogenetic relationships
Zoonotic potential

To review *Giardia* and giardiasis in detail would require at least an entire volume of *Advances in Parasitology*. Such treatment in depth is not warranted at this time in view of the excellent book recently edited by Meyer (1990a). Our intention here is to give an up-to-date overview of *Giardia* and giardiasis and provide an insight into the enormous wealth of literature on the subject, as well as highlight the most important recent developments and unresolved questions.

II. Morphology and Ultrastructure

A. Overall Morphology

The gut-dwelling trophozoite stage has a characteristic and distinctive morphological appearance. It is a pear-shaped binucleate organism, usually within the range of 10–15 μm in length and 6–10 μm in width, with eight flagella, a pair of distinctive median bodies and a structure unique to *Giardia*, the ventral disc (Fig. 1). The infective transmissible stage is the cyst which is usually 8–12 μm long and 7–10 μm wide and is surrounded by a fibrous proteinaceous (Erlandsen *et al.*, 1990a) wall of approximately 0.3 μm in thickness. Within the cyst are visible 2–4 nuclei (depending upon whether nuclear division has been completed), basal bodies, median bodies and structural elements of the ventral disc and flagella.

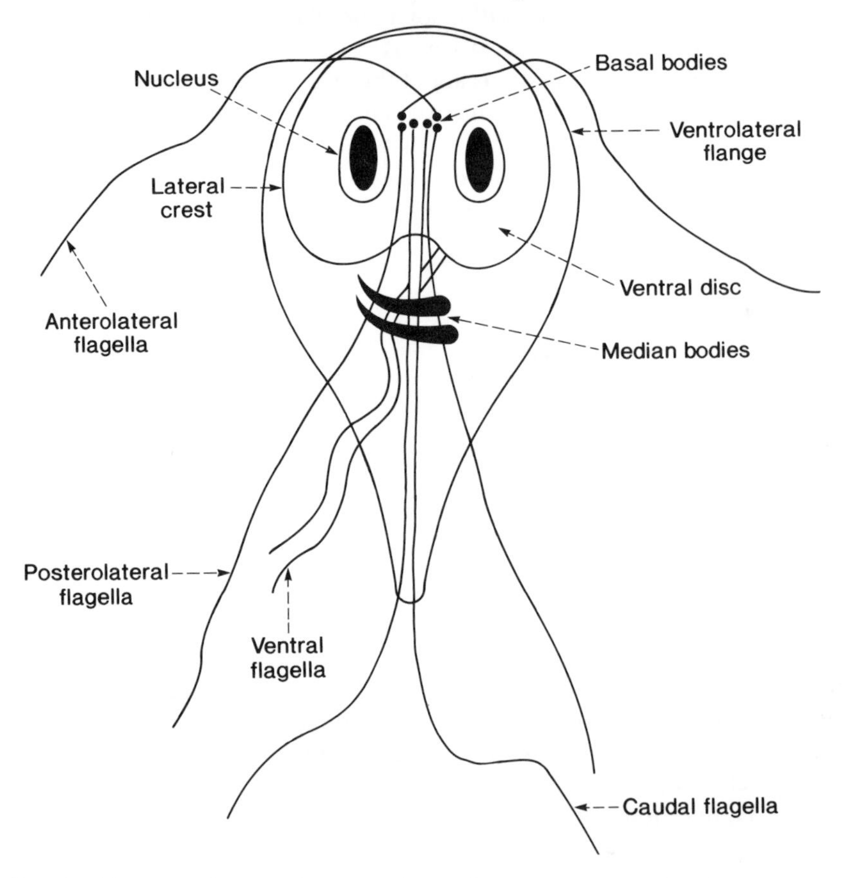

FIG. 1. Diagram showing the general morphology of the trophozoite of *Giardia duodenalis*.

B. VENTRAL DISC

Scanning electron microscopy has provided considerable information about the surface structure of *Giardia*, particularly in relation to the ventral (attachment) disc which is not readily visible by light microscopy. This concave structure comprises approximately two-thirds of the ventral surface (Figs 1, 2) and serves to distinguish *Giardia* from all other flagellates. It is a complex, relatively rigid structure, made up of several evenly spaced cytoskeletal elements (Fig. 3). Unique structures called microribbons project dorsally into the cytoplasm; these are connected by cross-bridges and attached ventrally to microtubules adjacent to the plasma membrane (Fig. 3) (Peattie *et al.*, 1989; Peattie 1990; Feely *et al.*, 1990). The projecting rim of the disc is formed by a large electron-dense fibrous structure called the lateral crest (Fig. 3) to which the external plasma membrane is tightly applied (Feely *et al.*, 1990). The ventral disc is usually surrounded by a distinct overlying cytoplasmic extension called the ventrolateral flange which encircles and forms a lateral border to the disc (Figs 1–3). The ventrolateral flange is supported internally by two filamentous plates of unknown biochemical composition. It is thought that these plates may have a contractile role associated with the motility of the flange lip (Feely *et al.*, 1990).

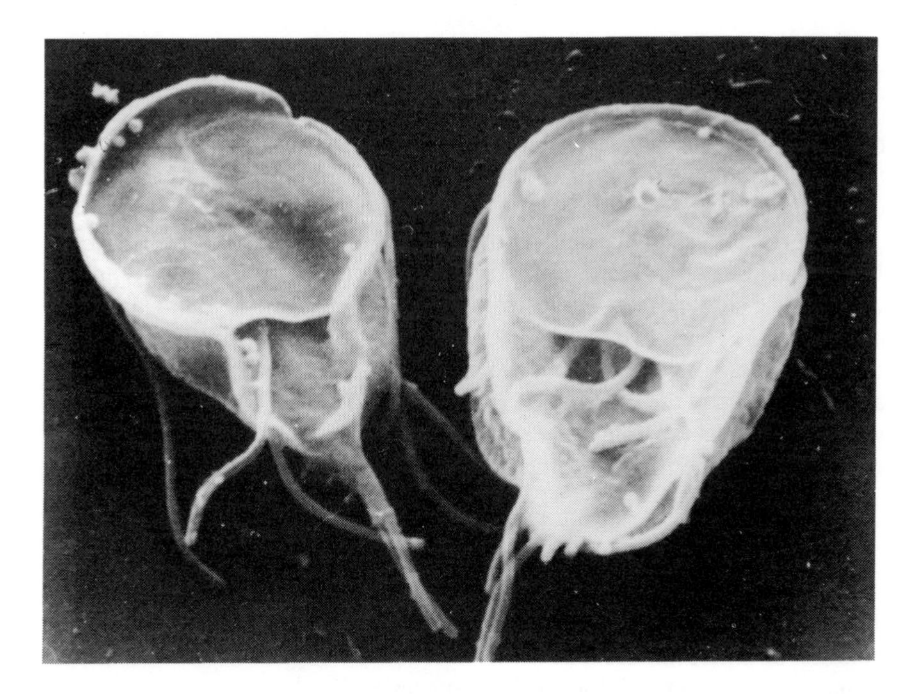

FIG. 2. Scanning electron micrograph of the ventral surface of two *Giardia duodenalis* showing the ventral disc and flagella (× 10 000). Photograph taken by Dr A. Warton.

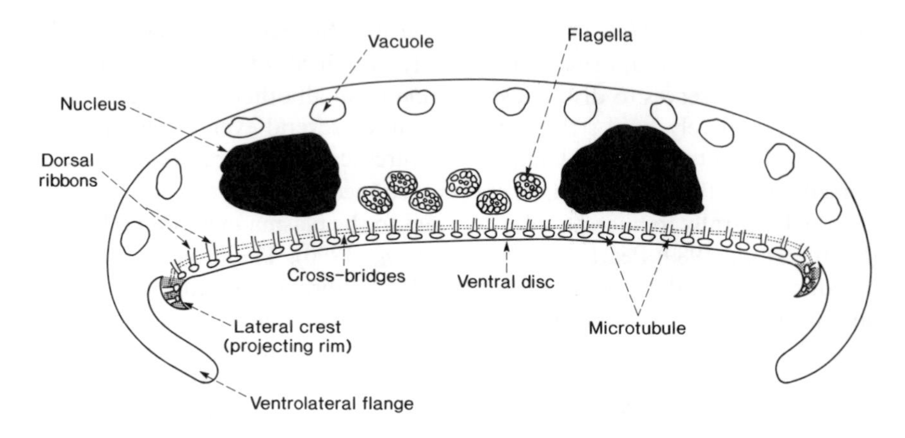

FIG. 3. Diagram showing details of a cross-section through a trophozoite of *Giardia duodenalis*.

C. CYTOSKELETON

The cytoskeleton of *Giardia* consists of four organelle systems composed of microtubules: the ventral disc; basal bodies, axonemes and flagella; median bodies; and fibrils associated with the caudal axonemes (= "body" or "funis"). Tubulin is the most widespread cytoskeletal protein in *Giardia* but, in addition, the microribbons of the ventral disc have been shown to contain unique proteins called giardins (Crossley and Holberton, 1983; Peattie *et al.*, 1989; Peattie, 1990). These *Giardia*-specific proteins are found only in the ventral disc and range in size from approximately 29 to 38 kDa. There appear to be at least two major types of giardin: the β-giardins (29 kDa) and the α-giardins (33.8 kDa), but how they interact with each other and aggregate to form the ventral disc is unclear (Peattie, 1990). Presumably, the giardins play a major role in the functioning of the ventral disc and are good candidates for specific chemotherapeutic attack (see Section VIII.C.2.e).

D. INTERNAL ORGANIZATION

The paired median bodies (Fig. 1) are unique to *Giardia* spp. They consist of bundles of microtubules but their function is unknown. A role in the formation of the ventral disc has been suggested (Feely *et al.*, 1990) but their biochemical affinities require further study.

There are few endomembranous organelles in *Giardia*. Mitochondria, peroxisomes, glycosomes and hydrogenosomes are absent, and it is only recently that evidence for the existence of a Golgi apparatus in *Giardia* has been produced (Reiner *et al.*, 1990; Gillin *et al.*, 1991). Recent interest has also focused on the peripheral vacuoles, or vesicles, of *Giardia* (Fig. 3) which may form part of an endosomal–lysosomal system in the trophozoites (Feely *et al.*, 1990; Kattenbach *et al.*, 1991). The vacuoles represent an acidic compartment containing acid phosphatase. The endoplasmic reticulum has been proposed as the site of formation and transport of this enzyme (Feely *et al.*, 1990) which probably assists in the degradation of accumulated macromolecules that have been ingested and localized in the peripheral vacuoles through an endocytic process (Kattenbach *et al.*, 1991).

Giardia is also unique in possessing two equal-sized nuclei, one on either side of the midline. Preliminary studies suggest that the two nuclei may be equivalent with respect to the amount of deoxyribonucleic acid (DNA) contained in each nucleus, ribosomal DNA sequences and transcriptional activity (Kabnick and Peattie, 1990). However, whether the two nuclei are indeed morphologically and functionally equivalent requires further investigation in view of conflicting data regarding genome size in *Giardia* (Nash *et al.*, 1985; Boothroyd *et al.*, 1987; Fan *et al.*, 1991). Similar confusion surrounds determinations of the ploidy of *Giardia* (see Section III.C.1).

E. ENDOSYMBIONTS

A variety of bacterial, mycoplasmal and viral inclusions has been demonstrated in trophozoites and cysts of many isolates of *Giardia* (reviewed by Sogayar and Gregorio, 1989; Adam, 1991). Their presence appears to be more frequent than previously recognized, but the most interesting and well characterized is the 32 nm double-stranded ribonucleic acid (*ds*RNA) virus, Giardiavirus (GLV), which specifically infects *Giardia* and has been identified in numerous isolates (Wang and Wang, 1986; De Jonckheere and Gordts, 1987). It is a small icosahedral virus comprising a non-segmented dsRNA genome of 7 kilobases and a major capsid protein of 100 kDa (Miller *et al.*, 1988a). It resembles similar viruses that infect yeast or fungi and is classified in the Totiviridae family (Tai *et al.*, 1991). GLV most probably replicates by means of a virus-encoded RNA polymerase (White and Wang, 1990). It can transfect uninfected *Giardia* isolates (De Jonckheere and Gordts, 1987; Miller *et al.*, 1988b) and is associated with decreased adherence and growth rate of trophozoites *in vitro* (Miller *et al.*, 1988a; Wang and Wang, 1991). Such adversely affected cells contain an average of 5×10^5 viral particles per cell which may be the threshold intracellular density of viral particles that inhibits growth of *Giardia* (Wang and Wang, 1991). It therefore appears that the consequence of heavy viral infection is reduced growth rate rather than cell lysis (Wang and Wang, 1991). Of 76 isolates of *Giardia* so far examined, 28 harboured the virus and the majority of virus-free isolates were found to be susceptible to GLV infection; only nine were resistant; there was no correlation between GLV infection and virulence of *Giardia*, and the virus is found in the nuclei of infected trophozoites and is eventually released into the culture medium (Wang and Wang, 1991). Feely *et al.* (1990) have raised the possibility of viral transfer between *Giardia* and the host genome. However, GLV virus appears to be unable to infect other protozoa or mammalian cell lines, which suggests there is no transfer of GLV between parasite and host during infection (Wang and Wang, 1991).

III. Taxonomy, Nomenclature and Genetics

A. Genera and Species

The characteristics of the genus *Giardia* (Fig. 1) are clearly defined and well accepted (reviewed by Kulda and Nohynkova, 1978; Meyer and Radulescu, 1979; Meyer, 1990c; Thompson *et al.*, 1990a). *Giardia* is widely recognized as the correct generic name and nowadays there is only minimal use of the synonym *Lamblia*.

Over 40 species have been described in the genus *Giardia* and their details have recently been compiled by Thompson *et al.* (1990a). The majority of these species were based on host occurrence but most authorities do not now accept the concept of rigid host specificity and consider the evaluation of Filice (1952) as being the most realistic means of species determination in the genus (reviewed by Thompson *et al.*, 1990a). This system divided *Giardia* into three morphologically distinct groups differentiated primarily on the shape of the median bodies (Table 2). The largest of these is the "*duodenalis*" group in which Filice (1952) placed over 20 formerly described species, mostly from mammals and a few from birds. Most attention has focused on this group because of its significance in human and veterinary medicine. Unfortunately, other names are used for the "*duodenalis*" group: *G. lamblia* and *G. intestinalis*. A detailed examination of the use of species names in *Giardia* (Thompson *et al.*, 1990a) clearly shows that, according to the rules of zoological nomenclature, the name *G. duodenalis* is correct. Use of the names *G. intestinalis* and *G. lamblia* is therefore not valid, but more importantly may give rise to confusion in the scientific literature (Meyer, 1985). The present review, in common with influential leaders in the field (Meyer, 1990a,d), will use the name *G. duodenalis*.

TABLE 2 *The major species groups in the genus* Giardia

Species group	*Host*	*Morphological characteristics*	*Trophozoite dimensions*
G. agilis	Amphibians	Long, narrow trophozoites with club-shaped median bodies	20–30 μm long 4–5 μm wide
G. muris	Rodents (possibly also birds and reptiles)	Rounded trophozoites with small round median bodies	9–12 μm long 5–7 μm wide
G. duodenalis	Mammals (possibly also birds and reptiles)	Pear-shaped trophozoites with claw-shaped median bodies	12–15 μm long 6–8 μm wide

In addition to the three species recognized by Filice (1952), two additional species occurring in birds have recently been proposed. *G. psittaci* was described from budgerigars (*Melopsittacus undulatus*) and is most unusual in lacking a ventrolateral flange (Erlandsen and Bemrick, 1987), which normally encircles and forms a lateral border to the ventral disc (Figs 1–3). Erlandsen *et al.* (1990b) also re-established the species *G. ardeae* from herons (*Ardea herodias*) which had previously been included in the "*muris*" group by Filice (1952). They emphasized its distinct morphology, having only a single caudal flagellum and variable median body morphology. Electrophoretic karyotyping also demonstrated differences between *G. ardeae, G. duodenalis* and *G. muris* (Campbell *et al.*, 1990). In addition, van Keulen *et al.* (1991a) concluded that *G. ardeae* was a distinct species on the basis of nucleotide sequence data of the 5.8S and large subunit ribosomal RNA (rRNA) genes when compared with isolates of *G. duodenalis*. This conclusion is supported by the results of restriction enzyme mapping of ribosomal DNA (van Keulen *et al.*, 1991b) and DNA analysis using probes and primers based on the giardin gene (Mahbubani *et al.*, 1992) which differentiated between *G. ardeae* and members of the *G. duodenalis* group. Recent cross-transmission experiments by Erlandsen *et al.* (1991) also tended to support the distinctness of both *G. psittaci* and *G. ardeae*.

B. VARIATION IN *GIARDIA*

Considerable genetic variation has been reported in *Giardia*, particularly in

G. duodenalis, and this has recently been extensively reviewed by Thompson *et al.* (1990a). Several criteria have been used in comparative studies and have revealed widespread heterogeneity (Table 3).

TABLE 3 *Differential criteria used in comparative studies on isolates of* Giardia[a]

Morphology
Host specificity and experimental cross-transmission
Growth and development *in vivo* and *in vitro*
Infectivity, virulence and pathogenicity
Sensitivity to drugs
Antigenic characteristics
Electrophoresis of proteins and enzymes
Restriction site analysis and DNA hybridization
Molecular karyotyping

[a]See Thompson *et al.* (1990a) for further information.

The nature of the differences which have been found between isolates of *G. duodenalis* will have a significant influence on the epidemiology and control of giardiasis, particularly differences in host specificity, growth and development, virulence, drug sensitivity and antigenicity. In addition, it is now clear that much of this variation has a genetic basis and the extent of this variation indicates the existence of a number of species within the *G. duodenalis* morphological group (Meloni *et al.*, 1988a, 1989, 1992; Andrews *et al.*, 1989). There is no evidence that these species are associated with different hosts, or that they are morphologically distinguishable (Thompson *et al.*, 1990a). Studies are urgently required on *G. muris, G. agilis* and putative avian species to determine whether similar levels of genetic diversity exist within these groups. The source of genetic diversity in *Giardia* is uncertain and will not be resolved until a clearer understanding of the reproductive mechanisms of this parasite is obtained.

C. MODE OF REPRODUCTION AND THE RECOGNITION OF SPECIES

1. *Ploidy*

Each nucleus in a *Giardia* trophozoite appears to be derived from the division of its corresponding parent nucleus (Filice, 1952). However, it is still not clear whether the nuclei are functionally and morphologically identical. Studies by Kabnick and Peattie (1990; see Section II.D) suggested that the nuclei are equivalent and that each nucleus is haploid. However, estimates of DNA content, genome complexity and karyotype have provided conflicting data. Using electrophoretic karyotyping procedures, the number of chromosomes has been variably estimated by different laboratories (Adam *et al.*, 1988a; Upcroft *et al.*, 1989a; Kabnick and Peattie, 1990; Fan *et al.*, 1991). Similarly, estimates of genome size and complexity have provided conflicting data with size estimates of 11 megabases (Mb), 30 Mb and 80 Mb (Nash *et al.*, 1985; Boothroyd *et al.*, 1987; Fan *et al.*, 1991). Meloni *et al.* (1988a), using isoenzyme electrophoresis, demonstrated multiple-banded enzyme patterns in a number of isolates of *G. duodenalis*. These patterns were retained in cloned cultures of the original isolates (Meloni *et al.*, 1989; Binz *et al.*, 1991), indicating that they were not produced by genetically different haploid organisms. The observed patterns could be explained by a single, functional haploid nucleus with isoenzymes specified by multiple loci, but this requires many gene duplications and subsequent mutations (Meloni *et al.*, 1988a). More likely explanations are that the isolates were either haploid with two functional nuclei or diploid with either one or both nuclei functional, and isoenzymes specified by alternative alleles at the same locus (Thompson *et al.*, 1990a). Clearly, further studies using different approaches are required to determine accurately the ploidy of *Giardia*. In this respect, direct measurement of DNA content in *Giardia* using cytometric techniques may be rewarding since such techniques have recently been successfully applied to *Sarcocystis* and *Babesia* (Mackenstedt *et al.*, 1990a,b).

2. *Asexual and/or sexual reproduction*

Giardia has always been presumed to reproduce entirely asexually by a process of binary fission with genetic diversity arising solely by mutation. Recent analyses of isoenzyme data support mutation and clonal selection as the mechanism maintaining genetic diversity (Tibayrenc *et al.*, 1990). Studies using isoenzyme and DNA analyses found that zymodeme classification of *G. duodenalis* was strongly correlated with schizodeme groupings of the same isolates (Meloni *et al.*, 1989). Such a correlation between two independent sets of genetic markers supports a clonal theory of reproduction (Tibayrenc *et al.*, 1990; Tibayrenc and Ayala, 1991). However, substantial homozygosity at enzyme loci in isolates of *G. duodenalis* has also been demonstrated, suggesting, for the first time, that a sexual phase may be present in *Giardia* (Meloni *et al.*, 1988a, 1989). Morphological studies have not provided any evidence of gamete formation, karyogamy or meiosis (Filice, 1952). However, now that isolates of *Giardia* can be routinely cloned in the laboratory (Binz *et al.*, 1991), it will be possible to carry out hybridization experiments *in vitro* and *in vivo* and analyse progeny genetically, as has been done successfully in studying genetic exchange in *Trypanosoma* (Jenni *et al.*, 1986) and *Plasmodium* (Walliker *et al.*, 1987).

Knowledge of the mode of reproduction in *Giardia* will not only assist in determining the source and maintenance of genetic diversity exhibited by this organism, but will also have a profound effect on our understanding of the epidemiology of giardiasis and attempts to control the causative agents. The clonal model of asexual reproduction, if applicable to *Giardia*, has significant implications with respect to drug development, diagnosis and treatment (Tibayrenc and Ayala, 1991). This is because of the predictability and stability inherent in such a model. However, as emphasized by Cibulskis (1988), rare or occasional bouts of sexual recombination in a normally asexual organism can have a major effect on the extent of genetic diversity. Indeed, as pointed out by Tibayrenc and Ayala (1991), the probability of identifying, within a given species, particular genetic make-ups that might be studied separately is inversely proportional to the extent of sexual reproduction.

3. *Recognition of species*

The taxonomic interpretation of genetic variation in *Giardia* is hindered by uncertainty concerning the appropriate species concept for the genus. The biological species concept (Mayr, 1940) was designed solely for sexually reproducing outbreeders, although it appears applicable in groups where sexual reproduction occurs only occasionally (White, 1978). If reproduction in *Giardia* is entirely asexual, however, the concept would not be appropriate.

The taxonomic recognition of presumed sibling species in *Giardia* requires a detailed genetic study of natural populations (Thompson *et al.*, 1990a). If sexual reproduction is frequent, then specific designation requires evidence that groups do not interbreed. This is straightforward if the groups are sympatric, but may require a phenetic approach if they are allopatric (Thompson *et al.*, 1990a). If reproduction is purely asexual, then specific designation will require evidence that a group of clones forms a lineage and that this lineage is distinct from others. This will require the study of laboratory cloned lines of *Giardia* using highly resolving molecular characterizing procedures (e.g., DNA fingerprinting, see Upcroft, 1991).

D. PHYLOGENETIC POSITION

Interest has focused on the phylogenetic affinities of *Giardia* following an analysis of its small subunit rRNA (16S-like rRNA) by Sogin *et al.* (1989).

They estimated evolutionary distances on the basis of sequence comparisons between the 16S-like rRNAs of *Giardia* and some other eukaryotes and found the distances to be greater than between *Giardia* and several prokaryotes. Sogin *et al.* (1989) concluded that *Giardia* must have separated very early in the evolution of eukaryotes so that it is further removed from its eukaryotic relatives than several of its prokaryotic neighbours. More recently, van Keulen *et al.* (1991a) similarly concluded that *Giardia* was one of the most primitive eukaryotes studied to date based on the structure of the large subunit rRNA of *G. duodenalis, G. muris* and *G. ardeae*. However, translation and transcription appear to be more similar to what is seen in other eukaryotes than in prokaryotes, although the translation apparatus of *Giardia* is quite different from that of other eukaryotes (reviewed by Adam, 1991).

In the light of the rRNA data, Kabnick and Peattie (1990) have speculated further with respect to the two putatively haploid nuclei of *Giardia*. They suggest that a single haploid nucleus eventually gave rise to a second, thus making the entire organism diploid. Two haploid nuclei could have fused to produce the single diploid nucleus characteristic of higher eukaryotes (Kabnick and Peattie, 1990). Adam (1991) and Kabnick and Peattie (1991) also pointed to the apparent lack of typical eukaryotic organelles in *Giardia*, such as mitochondria, peroxisomes and nucleoli, in support of its classification as a primitive eukaryote. Indeed, Kabnick and Peattie (1990) suggested that, since *Giardia* probably represents the earliest diverging lineage of eukaryotes, its biological tactics may therefore be transitional. However, this notion has been strongly criticized by Siddall *et al.* (1992), who considered that obligate parasites, such as *Giardia*, are more likely to have lost unnecessary organelles in their evolution. They also emphasized that the microsporidia lack mitochondria and Golgi apparatus and that this group probably diverged earlier than diplomonads such as *Giardia*, which has recently been shown to possess a Golgi apparatus (Reiner *et al.*, 1990; Gillin *et al.*, 1991).

Siddall *et al.* (1992) carried out a detailed phylogenetic analysis of members of the order Diplomonadida using a range of ultrastructural characters. Their results do not support the presumed pivotal position of *Giardia* in the evolution of the eukaryotes. Indeed, Siddall *et al.* (1992) expressed reservations about the method of distance analysis used by Sogin *et al.* (1989). They also provided evidence to suggest that the characteristic ventral disc of *Giardia* with its unique giardin cytoskeletal proteins is a derived character, probably evolving after divergence from other eukaryotes (Siddall *et al.*, 1992), thus negating the significance of this structure as a forerunner of the cytoskeleton of higher eukaryotes (Peattie *et al.*, 1989; Peattie, 1990). Siddall *et al.* (1992) also pointed to the conceptual difficulties in giving an ancestral role to a parasitic organism that is supposed to be the source of features characteristic of the hosts upon which it depends for survival.

Although many authorities have enthusiastically embraced the notion of *Giardia* being one of the most primitive eukaryotes, some caution may be prudent until more extensive comparative studies have been undertaken, particularly on a greater range of free-living protozoa. It would also be interesting to examine the rRNA of presumed close relatives to *Giardia* such as *Hexamita* and the trichomonads. It is quite clear that *Giardia* is an unusual organism with a variety of seemingly unique characteristics. However, further studies, particularly relating to the biochemistry of *Giardia*, are warranted before its true phylogenetic affinities can be resolved.

IV. Life Cycle and Laboratory Maintenance

Giardia has a simple direct life cycle (Fig. 4). Trophozoites colonize the duodenum and jejunum where they attach to the intestinal mucosa. Attachment is an essential feature of the relationship between *Giardia* and its host, and the ability to attach *in vitro* is an important indicator of viability (Meloni *et al.*, 1990; Crouch *et al.*, 1991a,b; Magne *et al.*, 1991). Trophozoites may detach intermittently but periods of detachment are likely to be of minimal duration since the parasite may be in danger of losing its position in the gut and being swept away as the result of peristalsis.

A. Attachment

Three possible mechanisms have been proposed to explain how *Giardia* attaches to intestinal epithelial cells (reviewed by Inge *et al.*, 1986): (i) attachment is mediated through suction force generated beneath the ventral disc by propulsive efforts of the ventral flagella (= hydrodynamic theory), (ii) mechanical processes related to contractile protein elements of the ventral disc and ventrolateral flange (VLF) are responsible for attachment, (iii) trophozoites of *Giardia* have surface mannose-binding lectins which bind to receptors on host epithelial cells.

Most evidence to date indicates that the ventral disc plays the major role in attachment and that the cytoskeletal elements of the disc are the major mediators in this process. This is indicated by the fact that microtubule inhibitors, including known β-tubulin antagonists, have been shown to inhibit adherence *in vitro* (Meloni *et al.*, 1990; Edlind *et al.*, 1990a; Magne *et al.*, 1991; and see section VIII.C). An important observation was that, although *Giardia* was unable to attach to the wall of culture vessels when exposed to benzimidazoles, such as albendazole, flagellar activity was not affected (Meloni *et al.*, 1990). These results support a major role for the disc cytoskeleton in attachment but do not favour the hydrodynamic theory since, on their own, functional flagella are not capable of maintaining attachment. Furthermore, the ability of *G. psittaci*, which lacks a VLF, to attach is not consistent with the hydrodynamic model (Erlandsen and Bemrick, 1987), which depends on the integrity of the VLF. It has also been observed that attachment can occur in the absence of flagellar movement (Feely *et al.*, 1982).

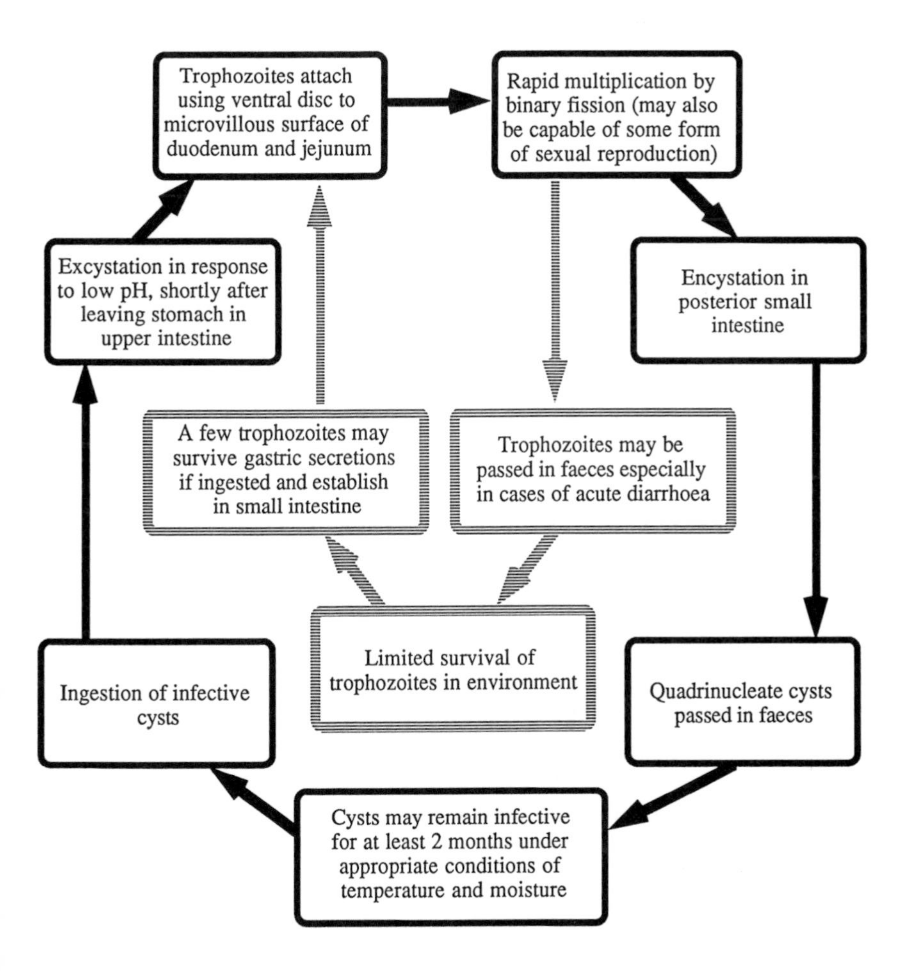

FIG. 4. Life-cycle of *Giardia*.

Although the ventral disc itself appears to be relatively rigid, the projecting rim (lateral crest; Fig. 3) around the disc is flexible (Adam, 1991) and several contractile proteins have been identified in the rim (Feely *et al.*, 1982). It is the rim which actually makes contact with the microvillous border of the small intestine, and it must be flexible in order to counter the variable topography of the epithelial surface so as to achieve uniform attachment and thus allow the concave rigid ventral disc to maintain suction pressure.

The role of lectins in attachment is uncertain since they are presumed to function by binding to receptors in the host (Farthing *et al.*, 1986a; Lev *et al.*, 1986). However, there is no evidence that such a mechanism is specific for intestinal epithelial cells nor for any particular region on the *Giardia* trophozoite. In addition, as emphasized by Adam (1991), the fact that *Giardia* attaches to glass and plastic surfaces of culture vessels does not support receptor-mediated binding. Any role for surface lectins is therefore likely to be of secondary importance.

B. GROWTH DYNAMICS

Trophozoites multiply rapidly in the small intestine of their host, although the growth rate has been shown to vary between different strains of *G. duodenalis* (Binz *et al.*, 1992) and, in addition, is likely to be influenced by host factors—particularly the immune and nutritional status of the host.

Trophozoites multiply asexually. Nuclear division is cryptomitotic (nuclear membranes retained), with the two nuclei dividing slightly out of phase (Filice, 1952; Wiesehahn *et al.*, 1984). Following nuclear division, trophozoites reproduce by binary fission, and the plane of cell division appears to be oblique (Kabnick and Peattie, 1990). As discussed above (Section III. 2.C) there is also evidence that *Giardia* may be capable of some form of sexual reproduction.

Little is known about the cell cycle kinetics of *Giardia*, and the events leading to cell division have not been clearly determined. Flow cytometry of DNA content suggests that the cell cycle of *Giardia* may be controlled in a different manner to that of mammalian cells (Hoyne *et al.*, 1989). However, more accurate information is needed, which will entail the production of synchronous cultures and the use of specific agents which can selectively inhibit phases of the cell cycle.

C. ENCYSTATION AND EXCYSTATION

The trigger for trophozoites to encyst is not completely understood but encystation takes place as trophozoites pass to the posterior regions of the small intestine. Cyst wall formation is completed within approximately 44–70 h (Reiner *et al*., 1990) and appear to be initiated by the presence of bile salts (Gillin *et al*., 1988) in the lower small intestine. Asexual reproduction occurs in the cyst with nuclear division either preceding (Filice, 1952) or following (Meyer, 1985) cyst wall formation, resulting in a cyst with four nuclei. Fission (cytokinesis) is delayed until excystation, when two binucleate trophozoites arise from each quadrinucleate cyst. It is not clear whether all cysts are immediately infective when passed since there is evidence that some undergo a maturation period of up to 7 days before becoming infective (Grant and Woo, 1978; Bingham *et al*., 1979). Cysts are resistant and can survive for at least 2 months in suitable temperature and moisture conditions (Meyer and Jarroll, 1980).

Excystation follows ingestion and takes place shortly after cysts leave the stomach. The low pH of the stomach environment appears to be the major factor which initiates the excystation process (Bingham and Meyer, 1979; Boucher and Gillin, 1990). However, excystation may occur at higher pH, possibly in response to pancreatic secretions and/or carbon dioxide (Feely *et al*., 1991). Excystation leads to rapid colonization of the small intestine, and subsequent cyst production commencing after a further 4–15 days (Swan, 1984). Trophozoites may also intermittently be passed in the faeces, particularly during acute infection, and could be a source of infection, especially in situations where direct transmission between individuals is likely to occur. Although far more labile than cysts and with a very limited survival capacity in the environment, the routine use of trophozoites to initiate experimental infections indicates that trophozoites can survive passage through the stomach and become established in the duodenum.

D. LABORATORY MAINTENANCE

1. *Cultivation* in vitro

The first successful and reproducible method of culturing *Giardia* trophozoites was devised by Karapetyan (1960, 1962). Cultures were monoxenic and required the presence of yeast for trophozoite survival. Meyer (1970) reported the first axenic cultivation of *Giardia* from the rabbit, chinchilla and cat, and later (Meyer, 1976) from a human. The majority of isolates of *Giardia* that have been successfully established *in vitro* belong to the *G. duodenalis* group. There have been only single reports of the axenic cultivation of *G. muris* (Gonzalez-Castro *et al*., 1986) and *G. ardeae* (Erlandsen *et al*., 1990b), and these require confirmation, particularly with respect to *G. muris* which has proved refractory to axenization *in vitro* in the past.

The development of a bile supplemented medium for cultivating *Giardia* (Keister, 1983) and techniques for the excystation *in vitro* of trophozoites from cysts has enabled the isolation, amplification and maintenance of a large number of different isolates in the laboratory, and has greatly facilitated investigations into trophozoite metabolism and growth requirements, biochemical, antigenic and molecular characteristics, and drug sensitivity (reviewed by Radulescu and Meyer, 1990; Thompson *et al.*, 1990a). Two recent developments will enhance the research value of *in vitro* cultivation of *Giardia*. Boucher and Gillin (1990) succeeded in inducing encystation of *Giardia in vitro* and demonstrated that cysts derived *in vitro* could excyst, thus enabling the complete life cycle to be completed in axenic culture. In addition, a simple method has recently been developed for establishing cloned lines of *G. duodenalis* from either single trophozoites or single cysts (Binz *et al.*, 1991). These two developments will allow accurate studies on the growth dynamics (Binz *et al.*, 1992), reproductive behaviour, cell cycle kinetics and competitive interactions of *Giardia* isolates to be undertaken, and will enable more meaningful research on genetic characterization of this parasite.

It should be emphasized that a certain proportion of *G. duodenalis* isolates are refractory to establishment *in vitro* and subsequent amplification (Meloni and Thompson, 1987; Thompson *et al.*, 1990a). Therefore, much of our knowledge of the biochemistry and genetics of *G. duodenalis* is based on cultivated varieties and, as emphasized by Nash *et al.* (1985), the isolates we study may be the result of laboratory-induced artificial selection. Efforts should be made to develop *in vitro* or *in vivo* procedures to amplify isolates which have not so far proved amenable to laboratory amplification. To this end, the increasing use of polymerase chain reaction (PCR) techniques and the future application of random amplified polymorphic DNA (RAPD) methods (Welsh and McClelland, 1990; Williams *et al.*, 1990; Morgan *et al.*, in press) will enable the genetic characterization of isolates for which only limited quantities of material are available.

2. *Animal models*

During the last 40 years, natural and experimental infections of *Giardia* have been studied in a variety of animals. Mice, rats and gerbils have been most commonly used as laboratory hosts, although dogs, cats and rabbits have also been utilized (reviewed by Faubert and Belosevic, 1990; Stevens, 1990; Thompson *et al.*, 1990a).

Mice have been the most widely used experimental host for both *G. muris* and *G. duodenalis* in studies on the immunobiology and pathogenesis of *Giardia* infections, as well as in drug efficacy trials and for amplifying isolates which cannot be established in axenic culture. However, certain limitations have reduced the usefulness of the murine models in studies involving isolates of *G. duodenalis*. Infections can be established only in very young mice, with most workers using suckling or weanling animals, and are usually of short duration. This age-dependent susceptibility greatly reduces the usefulness of mice for studies on immunity and pathogenesis, although mice with specific immunological defects have proved to be of some value in dissecting aspects of the immune response and for prolonging infections (den Hollander *et al.*, 1988; Gottstein and Nash, 1991). In addition, mice have a rapid gut transit time which is thought to limit their usefulness in studies on drug efficacy (Reynoldson *et al.*, 1991a,b). In this respect, the rat has been advocated as a more useful host in studies on anti-giardial agents since its gut transit time is much slower than that of mice and more closely reflects the situation in humans (Reynoldson *et al.*, 1992, and paper in preparation).

The most promising animal model is the Mongolian gerbil (*Meriones unguiculatus*), which can be used to study infections with *G. duodenalis* and *G. muris* (Faubert *et al.*, 1983). The gerbil offers a much better alternative to mice as a laboratory model for infections with *G. duodenalis* isolates. Adult gerbils are highly and reproducibly susceptible to infection following inoculation of either cysts or trophozoites and are considered to be ideal for studying host immunity and pathophysiological changes during the course of infection (Faubert and Belosevic, 1990).

V. Biochemistry and Metabolism

A. Energy and Carbohydrate Metabolism

1. *Introduction*

Given that eukaryotic protozoan parasites such as *Giardia, Entamoeba* and *Trichomonas* spp. contain no morphologically recognizable mitochondria it is not surprising that they are dependent predominantly on anaerobic catabolic pathways to maintain their energy and redox homeostasis. In *Giardia*, as in most aerotolerant anaerobes (see Section V.A.7.a on oxygen consumption), the Embden–Meyerhof pathway (EMP) constitutes the basis of energy and redox homeostasis. Thus the absence of the tricarboxylic acid (TCA) cycle from the trophozoites and cystic stages of *Giardia* (Lindmark, 1980; Lindmark and Miller, 1988) is not surprising. The absence of a functional TCA cycle obviates cytochrome-mediated electron transport and oxidative phosphorylation in *Giardia* spp. In fact no cytochrome (Lindmark, 1980; Jarroll *et al.* 1981; Lindmark and Jarroll, 1984; Paget *et al.*, 1989a) nor haem iron (Weinbach *et al.*, 1980) has thus far been detected in *Giardia* cysts or trophozoites.

2. *Embden–Meyerhof pathway*

Exogenous glucose or reserve carbohydrates are converted to phosphoenol pyruvate (PEP) and subsequently to pyruvate (Fig. 5) via a classical EMP (Lindmark, 1980; Lindmark and Miller, 1988). Two steps of this pathway are accompanied by substrate level phosphorylations. In the amitochondrial protozoa, the purpose of anaerobic coupling (see Sections V.A.4.a and b) is to maintain the intracellular free NAD/NADH ratio (the ratio between nicotinamide adenine dinucleotide (NAD) and its reduced form (NADH) —the redox ratio—) at an optimal level. This cytoplasmic redox homeostasis is vital for the uninterrupted flux though the EMP in all organisms, and *Giardia* is no exception. If this ratio falls too low glyceraldehyde-3-phosphate dehydrogenase can no longer function in the forward direction and glycolytic flux would be attenuated (Barrett, 1984). The presence of most of the key enzymes of the EMP has been demonstrated in *G. duodenalis* and *G. muris* (Lindmark, 1980; Lindmark and Miller, 1988), except for lactate dehydrogenase (LDH) and adenosine triphosphate (ATP)-dependent phosphofructokinase (PFK-1). The occurrence of pyruvate kinase (PK) in the Portland-1 strain (Lindmark, 1980), which could not be confirmed employing continuous or stopped assay techniques (Schofield *et al.*, 1991), has been substantiated by Mertens (1990). This exemplifies the conflicting information on the occurrence and the regulatory properties of the EMP and related enzymes in *Giardia*.

In all glycolytic systems investigated thus far, the major rate-controlling step is PFK-1 (Barrett and Beis, 1973). It is unfortunate that, until 1990, metabolic studies had not attempted, or had overlooked, the determination of this important enzyme in *Giardia*. PFK-1 is important in EMP flux considerations as the maximum rate of glycolysis can be determined by the specific activity of PFK, or from the combined rates of hexokinase (HK) and phosphorylase (Crabtree and Newsholme, 1972). The phosphorylase activity of *Giardia* has not been reported. The HK : phosphorylase ratio would indicate the relative importance of free hexose and stored glycogen as the major monosaccharide source for *Giardia*. The HK activity in *Giardia* (24 ± 8 (Lindmark, 1980) and $39 + 12$ nmol min^{-1} mg^{-1} protein (Schofield *et al.*, 1991)) compares well with that of the helminth *Ascaris lumbricoides* (25.25 ± 1.5 nmol min^{-1} mg^{-1} protein (Barrett and Beis, 1973)).

It has been shown that *Giardia* (Portland-1 strain) contains an active pyrophosphate : fructose-6-phosphate-1-phosphotransferase (PPi-PFK) but no detectable PFK-1 (Mertens, 1990). *Giardia* extracts catalysed the ATP-

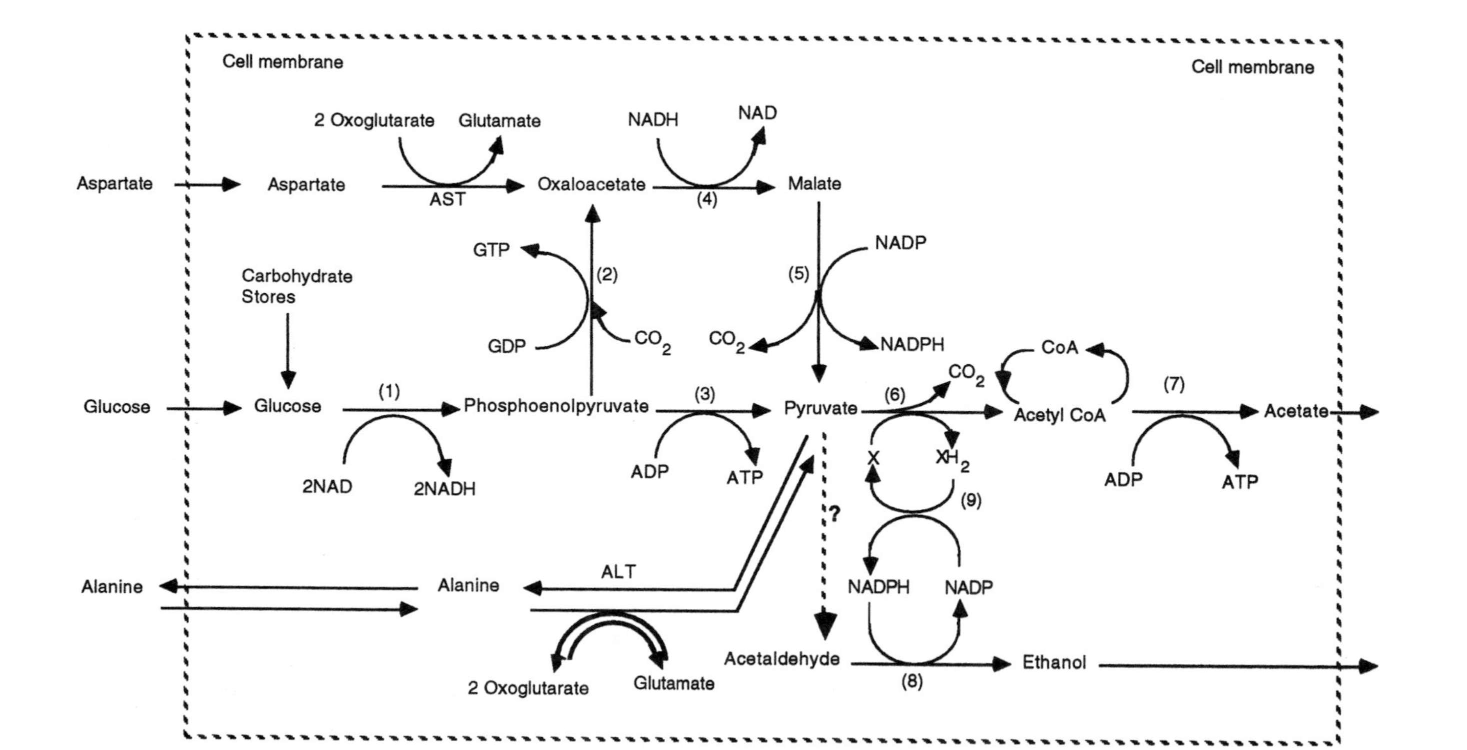

FIG. 5. Intermediary metabolic pathways operative in *Giardia*, showing the points of entry for asparate and alanine. (1) Enzymes of the Embden–Meyerhof pathway; (2) phosphoenol pyruvate carboxykinase (GDP); (3) pyruvate kinase (ADP); (4) malate dehydrogenase (NAD); (5) malate dehydrogenase (decarboxylating) (NADP); (6) pyruvate synthase (synonym, pyruvate ferredoxin oxidoreductase); (7) acetyl-CoA synthetase (ADP); (8) alcohol dehydrogenase (NADP); (9) NADPH oxidoreductase; AST: aspartate-2-oxoglutarate amino transferase; ALT: alanine-2-oxoglutarate amino transferase. Modified, with permission, from Mendis *et al.* (1992).

dependent formation of fructose-1,6,-biphosphate (Fru-1,6-P_2), namely PFK-1-like activity, at only 3% of the rate observed with PPi as phosphate donor (Mertens, 1990). It has been suggested (Mertens, 1990) that this PFK-1-like activity is insufficient to account for the glycolytic flux, estimated to be 100 nmol hexose equivalents min^{-1} mg^{-1} protein observed by Lindmark (1980).

Like *Trichomonas foetus, T. vaginalis* and *Isotricha prostoma, Giardia* (Portland-1 strain) contained no detectable levels of fructose-2,6,-biphosphate (Fru-2,6-P_2) a potent stimulator of glycolysis and negative modulator of gluconeogenesis (Mertens, 1990). Furthermore, *Giardia* had undetectable levels of phosphofructo-2-kinase or its hypothetical PPi-linked counterpart. Significantly Fru-2,6-P_2 had no modulatory effect on PPi-PFK of *Giardia* nor on the organism's PK (Mertens, 1990). The potential advantage of the presence of PPi-PFK is that it can significantly enhance the ATP yield of fermentative glycolysis using an otherwise useless metabolic byproduct (i.e. PPi). In possessing PPi-PFK *Giardia* demonstrates convergent evolution with some higher plants (Carnal and Black, 1979), at least one free-living protozoan, *Euglena gracilis*, and several important protozoan parasites (Mertens, 1990).

3. *Metabolism of phosphoenolpyruvate*

As in most other eukaryotic parasites, *Giardia* has a branched pathway ("metabolic loop") at the level of PEP (Fig. 5). Two enzymes shown by Lindmark (1980), Lindmark and Jarroll (1984) and Lindmark and Miller (1988) to be present in *Giardia* compete for the steady-state PEP pool; these are PK and phosphoenol pyruvate carboxykinase (PEPCK). PK has been measured in *Giardia* (Lindmark, 1980; Lindmark and Miller, 1988; Mertens, 1990), although its occurrence was disputed by Schofield *et al.* (1991), and the PK : PEPCK ratio of 6 : 1 has been calculated using the data of Lindmark (1980). The PK activity of *Giardia* is approximately 16-fold greater than that of *A. lumbricoides* (Barrett and Beis, 1973). The K_m (dissociation constant) values of the giardial PK and PEPCK for PEP (40 μM and 125 μM) suggest the predominant partitioning of carbon flow in the direction of pyruvate rather than oxaloacetate (OAA).

Entamoeba histolytica, which lacks PK, possesses a pyruvate phosphate dikinase (PPD) (Muller, 1988). It is possible that *Giardia* may possess a similar enzyme. If this were so, it might also explain the lack of success of Schofield *et al.* (1991) in demonstrating PK activity in *Giardia*. The role of PEPCK is carried out in *E. histolytica* by a phosphopyruvate carboxylase (PEPC) which generates PPi and OAA. Careful analysis of the CO_2-fixing enzymes in *Giardia* may resolve the apparent differences from other parasitic protozoans.

4. *Metabolism of pyruvate*

(a) Generation of acetate, ethanol and CO_2. Two distinct enzymes may catalyse the decarboxylation of pyruvate in eukaryotes: ferredoxin oxido-reductase (PFOR) (= pyruvate synthase) (molecular mass ≈240 kDa), which is considered to be the ancestral one, and the more advanced pyruvate dehydrogenase (PDH) complex (molecular mass $> 10^3$ kDa). The former is reversible while the latter is not. PFOR is a dimeric or tetrameric protein, utilizes an Fe-S protein (ferredoxin) as an electron acceptor, and has a more negative midpoint potential than pyridine nucleotides. PDH and PFOR are regarded as mutually exclusive in any given organism (Kersscher and Oesterhelt, 1982) and the latter is restricted to strictly anaerobic and nitrogen-fixing organisms and halophilic archaebacteria. PFOR has been purified from only two protozoa, i.e. *E. histolytica* and *T. vaginalis*. That of *Giardia* remains to be purified although its activity in whole cell homogenates has been negatively correlated with metronidazole resistance (Smith *et al.*, 1988) (see Section V.A.7.b).

PFOR is functional in *Giardia* (Fig. 5), as indicated by the production of acetate and CO_2 from D-glucose (Lindmark, 1980; Edwards *et al.*, 1989; Paget *et al.*, 1990). The exact mechanism of reoxidation of the giardial ferredoxin which is reduced during these processes remains unknown. The possibility that it provides reducing equivalents for ethanol production (Lindmark, 1980) is plausible since the required alcohol dehydrogenase (ADH) and/or aldehyde reductase (Schofield *et al.*, 1991) are dependent on nicotinamide adenine dinucleotide phosphate (NADP). Ethanol production was attenuated by the aldehyde reductase inhibitor valproate but not by pyrazole, an inhibitor of NAD-dependent ADH (Schofield *et al.*, 1991).

Although PFOR activity has been reported to be sensitive to normobaric oxygen (Smith *et al.*, 1988), the redox characteristics of the ferredoxin and its involvement in the above processes under hypobaric (microaerobic) conditions remains an interesting area to be investigated, especially since the reduction of metronidazole involves the giardial PFOR (Smith *et al.*, 1988). Electron paramagnetic resonance (EPR) spectroscopy would be useful in resolving these free-radical mechanisms (Docampo *et al.*, 1987).

(b) Generation of alanine. Under anaerobic conditions, alanine is one of the major end products generated by *Giardia* (Edwards *et al.*, 1989). Excretion of alanine is a common strategy employed by several helminths (Barrett, 1984; Barrett *et al.*, 1986) and some parasitic protozoa (MacKenzie *et al.*, 1983; Darling *et al.*, 1987) for maintaining a stable intracellular pH, while sustaining a high rate of flux via the EMP to produce acidic end products such as propionate lactate and acetate.

Under strictly anoxic conditions, *Giardia* produces equimolar ethanol, alanine, and CO_2 with small amounts of acetate (Edwards *et al.*, 1989; Paget *et al.*, 1990). Aerobically, alanine and ethanol production is attenuated (Paget *et al.*, 1990) and acetate and CO_2 predominate (Lindmark, 1980; Paget *et al.*, 1990). When subjected to 46 μM O_2, alanine production by *Giardia* was undetectable (Paget *et al.*, 1990). In the gut of vertebrates, where O_2 tension fluctuates between 0 and 60 μM (Paget *et al.*, 1990), the ability to produce a range of end products gives *Giardia* a high degree of flexibility in balancing its redox couples.

Alanine generation in *Giardia* may proceed via three routes: (i) direct formation from pyruvate via L-alanine dehydrogenase; (ii) via the coupling of glutamate-dehydrogenase and alanine-2-oxoglutarate transaminase (ALT) regulated by the NADP : NADPH ratio, or (iii) via ALT *per se*. (Fig. 5). ALT has been demonstrated in *Giardia* (Meloni *et al.*, 1988a; Stranden and Köhler, 1991; Mendis *et al.*, 1993) and alanine production and excretion in *Giardia* was inhibited by the ALT inhibitors L-cycloserine and carboxymethoxylamine (Edwards *et al.*, 1989). It is unlikely that alanine arises solely from internalized polypeptides or proteins as suggested by Edwards *et al.* (1989) especially since it has been shown that *Giardia* (clone PIC 10) has the capacity to take up and metabolize [U-^{14}C] alanine to $^{14}CO_2$ (see Section V.A.6)

5. *Hexose monophosphate pathway*

Radiorespirometric studies using specifically radiolabelled glucose indicate that *Giardia* trophozoites metabolize glucose at approximately equal rates (Jarroll *et al.*, 1981; Jarroll and Lindmark, 1990) via the EMP (as measured by the conversion of [3,4-^{14}C]glucose to $^{14}CO_2$) and via the HMP (as measured by the conversion of [1-^{14}C]glucose to $^{14}CO_2$ (Jarroll and Lindmark, 1990). Data also indicate that the rate of conversion of [6-^{14}C]glucose to $^{14}CO_2$ lower than that of [1-^{14}C] glucose and [3,4-^{14}C]glucose. (Jarroll *et al.*, 1981). However, the specific activities of the enzymes of the HMP have yet to be reported.

6. *Amino acids as putative energy sources*

Giardia trophozoites utilize monosaccharides, predominantly glucose, as the main energy source (Lindmark, 1980). Reducing the glucose concentration in growth media had little effect on trophozoite growth and/or end product formation (Edwards *et al.*, 1989; Schofield *et al.*, 1991). In "near zero" glucose the growth rate was reduced by 50% but the end products were similar. Substitution of 10 M fructose or ribose produced little change from the "zero-glucose" growth and end product profile (Schofield *et al.*, 1991).

Discounting CO_2, glucose utilization was sufficient to account for only 50% of the total carbon appearing in alanine, ethanol and acetate (Edwards *et al.*, 1989). This led to the suggestion that *Giardia* trophozoites may have alternative sources of energy, although the contribution by endogenous sources of glucose (i.e., stored glycogen) was not measured (Edwards *et al.*, 1989; Schofield *et al.*, 1991).

Certain helminth parasites utilize amino acids such as glutamine, and carboxylic acids such as citrate, malate and succinate, to supplement their energy requirements (Mendis *et al.*, 1986, 1987). Linstead and Cranshaw (1983) showed that *T. vaginalis* utilized arginine via the arginine-dihydrolase pathway. *Giardia* also possesses the arginine dihydrolase pathway (ADHP; Schofield *et al.*, 1990, 1992). The specific activities of the enzymes of the above pathway were greater than those reported for *T. vaginalis* and the flux through the pathway was unaffected by valinomycin (0.1 mM), nigericin (3 mM), azide (5 mM) or cyanide (1 mM), and was only marginally affected by 10 mM glucose (Schofield *et al.*, 1990, 1992).

It has been suggested that the potential ATP yield from this pathway may exceed that from glucose (Schofield *et al.*, 1992) when $^{14}CO_2$ liberation from [1-^{14} C] glucose was taken as a measure. However, conversion of [1-^{14} C] glucose to $^{14}CO_2$ is more a measure of flux through the hexose monophosphate pathway (HMP) (Lindmark, 1980; Jarroll and Lindmark, 1990) (see Section V.A.5) rather than flux through the EMP. Further carbon flux studies using [3,4-^{14}C] glucose and [6-^{14}C] glucose must be cautiously undertaken to evaluate the position of arginine as a significant energy source.

It has been shown recently that intact *Giardia* trophozoites can take up and convert L-[U-^{14}C] aspartate and, more significantly, L-[U-^{14}C] alanine to $^{14}CO_2$ (Mendis *et al.*, 1992) (Fig. 5). On account of these findings it appears that *Giardia* may be a metabolic opportunist capable of supplementing its energy requirements by scavenging amino acids prevalent in its environment. It has been suggested that the intracellular redox status of the parasite may dictate whether alanine be channelled via pyruvate into metabolic pathways to supplement its energy charge and redox homeostasis, or be actively or passively externalized (Mendis *et al.*, 1992). More detailed carbon flux studies under strictly defined conditions are necessary in order to compare the significance of these substrates such as arginine, alanine, and aspartate (Schofield *et al.*, 1991; Mendis *et al.*, 1992) as putative energy sources for *Giardia*. The possibility exists that *Giardia* may also utilize glutamine and other amino acids as bioenergetic substrates.

7. *The oxygen consumption of* Giardia

(a) Effects of exogenous substrates. Over a decade ago Weinbach *et al.* (1980) and Lindmark (1980) independently observed that intact *G.*

"lamblia" (= *G. duodenalis*) trophozoites exhibited a high affinity for oxygen. The endogenous rate of oxygen consumption (E-QO_2) was observed to be 64 (Weinbach *et al.*, 1980) and 93 nmol min^{-1} mg^{-1} protein (Lindmark, 1980) at 37°C. No Pasteur effect has been reported in *Giardia*, although this requires careful assessment under strictly defined conditions.

Several substrates, such as glucose, malate, ethanol and the electron-donor couple ascorbate/N,N,N^1,N^1-tetramethyl-*p*-phenylenediamine (TMPD), reportedly stimulated the E-QO_2 of intact *Giardia* trophozoites (Weinbach *et al.*, 1980), while pyruvate, lactate, succinate, α-ketoglutarate, isocitrate, glycerol-3-phosphate, L-glutamate, L-serine and isopropanol had no effect on the E-QO_2. In similar studies, Lindmark (1980) reported that the E-QO_2 of *Giardia* was unaffected by D-fructose, D-ribose, maltose, D-mannose, mannitol and TCA cycle intermediates such as succinate, α-ketoglutarate, citrate and/or organic acids (pyruvate, acetate and L-lactate). The stimulation of the E-QO_2 by malate observed by Weinbach *et al.* (1980) was not observed by Lindmark (1980). This discrepancy needs further investigation.

Apparent K_m values for O_2 suggest that trophozoites have a greater affinity for O_2 than cysts, indicating that cysts are more resistant to oxygen-induced toxicity. O_2 maxima have been reported in *T. vaginalis* (Yarlett *et al.*, 1987) and in several gastrointestinal helminth parasites (Paget *et al.*, 1988, 1989b). O_2 toxicity for the latter was correlated with the production of active O_2 species such as H_2O_2 and O_2. The situation in *Giardia* is bound to be similar but has not been investigated fully. The effects of temperature on the oxygen consumption (QO_2) of cysts indicated that, below 7°C, little O_2 consumption occurred. Since low temperatures also favour cyst viability (Bingham *et al.*, 1979), low temperatures would be expected to extend the half-life of cysts, thus prolonging their infective period.

(b) Effects of inhibitors on the endogenous oxygen consumption. The E-QO_2 of *Giardia* was sensitive to inhibition by iodoacetamide and quinacrine but not cyanide, rotenone, malonate, sodium arsenite, or sodium arsenate (Weinbach *et al.*, 1980; Lindmark, 1980). The uncoupler 2,4-dinitrophenol (2,4DNP) had no effect, but 2-deoxyglucose inhibited the E-QO_2 (50% inhibitory concentration = 50 mM), suggesting glucose to be the source of endogenous substrates responsible for the E-QO_2 of *Giardia* trophozoites (Weinbach *et al.*, 1980).

As one would expect, the E-QO_2 of intact trophozoites was not affected by exogenous NADH or NADPH, but upon treatment with 1% Triton X-100® the occurrence in *Giardia* of an active "DT-diaphorase" (NADPH; quinone acceptor-oxidoreductase; E.C. 1.6.99.2), which did not differentiate between NADH and NADPH, was evident (Weinbach *et al.*, 1980). The mammalian "DT-diaphorase" inhibitors rotenone and dicoumarol were effective inhibitors of NADPH-induced oxygen consumption by an 11 000 ***g*** particulate preparation of *G. "lamblia"* (= *G. duodenalis*), as were chloroquine and quinacrine (Weinbach *et al.*, 1980). However, Lindmark (1980) observed no inhibition of the E-QO_2 of *Giardia* by rotenone. These reports with respect to rotenone have significant implications on the electron transport characteristics of *Giardia*, and thus need further verification. Interestingly, the NADPH-induced QO_2 of *Giardia* was susceptible to inhibition by bathophenanthroline (0.6 mM) and 4,4,4-trifluor-1(2-naphthyl)-1,3 butanedione (0.2 mM), which are transition-metal chelators (Weinbach *et al.*, 1980). The insensitivity of the E-QO_2 of *Giardia* to cyanide is not unusual. Cyanide-insensitive QO_2 has been observed in a variety of free-living and parasitic organisms (Lloyd *et al.*, 1983; Mendis and Evans, 1984a,b; Mendis and Townson, 1985; Paget *et al.*, 1988; Mendis *et al.*, 1991). Inhibitors of azide, cyanide and antimycin A-insensitive alternative electron transport pathways such as salicyl hydroxamic acid (SHAM) and *o*-hydroxy diphenyl (Mendis and Evans, 1984a,b; Paget *et al.*, 1989b) had no effect on the giardial E-QO_2 (Paget *et al.*, 1989a), suggesting that electron transport in *Giardia* was diferent from that of other organisms. The giardial E-QO_2 was sensitive to thiol reactive agents such as *p*-chloromercuribenzoate and iodoacetamide (Paget *et al.*, 1989a). The status of quinones in the electron transport pathways of *Giardia* needs to be verified since naphthoquinones have been shown to be useful in the treatment of other protozoal diseases (Hudson *et al.*, 1985).

Most inhibitors had similar effects on cysts and trophozoites. The ATPase inhibitors ouabain and quercetin (Glick, 1970) inhibited ethanol-induced QO_2 but not E-QO_2. Chloroquine (1.5 mM) and quinacrine (1.25 mM) inhibited less than 20% of the E-QO_2 (Paget *et al.*, 1989a). These results agree with those of Weinbach *et al.* (1980). High concentration of acetate (40 mM) inhibited 22% of the E-QO_2, suggesting that acetate accumulation within trophozoites may have a negative modulatory effect on the E-QO_2 of trophozoites. Significantly, 6 mM metronidazole inhibited 55% of the E-QO_2 of trophozoites but had no effect on the E-QO_2 of cysts. This differential susceptibility suggests that viable cysts may be excreted by individuals under treatment with metronidazole (Paget *et al.*, 1989a). $NaNO_2$, which disrupts the Fe-S centre of PFOR, had very similar effects, albeit at very high (40 mM) concentrations. PFOR-mediated bioreduction of metronidazole generates toxic nitro-radicals which are cytotoxic to *Giardia*, inhibiting the E-QO_2 of trophozoites (Smith *et al.*, 1988; Paget *et al.*, 1989a). Menadione (0.3 mM), a redox cycling naphthoquinone, stimulated the E-QO_2 of *Giardia* six-fold before attenuating it (Paget *et al.*, 1989a). Menadione was effective against cysts and trophozoites, affecting their viability. The action of menadione is to generate toxic oxygen species within trophozoites via the mediation of electron transporting components intrinsic to *Giardia* trophozoites.

8. *Flavins, non-haem iron and iron sulphur compounds in* Giardia

In contrast to findings with *E. histolytica*, virtually all the flavin in *G. "lamblia"* (= *G. duodenalis*) was acid extractable, suggesting the absence of peptide bound or covalently bound flavonucleotides (Weinbach *et al.*, 1980). The involvement of flavins in electron transport processes in *Giardia* was excluded due to the lack of response to catalase (Weinbach *et al.*, 1980). However, intact trophozoites of aerotolerant anaerobes such as *Giardia* may not produce and externalize H_2O_2, which may explain the lack of response of catalase. Studies employing cell homogenates or specific subcellular fractions may yield interesting data with respect to catalase. On the contrary, the inhibitory effects of flavoantagonists such as chloroquine and quinacrine on the QO_2 of *Giardia* (Weinbach *et al.*, 1980; Paget *et al.*, 1989a) suggest otherwise. The majority of iron was non-haem and all of the sulphur was acid labile, indicating strong evidence for the occurrence of ferredoxin(s) (Weinbach *et al.*, 1980).

B. ANTIOXIDANT MECHANISMS

1. *Glutathione/thiol cycle*

In most organisms the so-called glutathione (GSH) cycle constitutes one of the primary defence mechanisms against toxic oxygen species (particularly peroxide) generated under microaerobic or aerobic conditions. It is curious that the presence of an endogenous GSH pool could not be demonstrated in *Giardia* despite a total thiol pool estimate of 56–64 mmol GSH equivalents/10^8 cells (Smith *et al.*, 1988). However a 14 000 ***g*** supernatant fraction exhibited glutathione reductase and glutathione peroxidase activity in response to exogenous glutathione (Smith *et al.*, 1988). The latter enzymic activities were thus referred to as thiol-dependent reductase (TDR) and thiol-dependent peroxidase (TDP). *E. histolytica* is the only other eukaryote reported to lack glutathione (Fahey *et al.*, 1984), glutathione reductase, glutathione peroxidase, and catalase (Murray *et al.*, 1981).

It is possible that *Giardia* may possess an alternative thiol, analogous to, and isofunctional with, glutathione in other eukaryotes. Such an alternative, trypanothione, has been identified in trypanosomes (Fairlamb *et al.*, 1985). It has been suggested that *Giardia* may "scavenge" glutathione from an exogenous source (Smith *et al.*, 1988). Irrespective of whether *Giardia* employs exogenous GSH, or possesses a unique thiol metabolism, it is significant that the specific activities of TDP and TDR show marked positive correlations with nitofuran tolerance among several stocks of *Giardia duodenalis* (Smith *et al.* 1988). This may have important implications with reference to furazolidone resistance in *Giardia*.

2. *Superoxide dismutase*

Giardia occupies a unique position among eukaryotes in that, despite its classification as an aerotolerant anaerobe, it appears to lack superoxide dismutase (SOD) (Smith *et al.*, 1988) as assayed by the method of Salin and McCord (1974). This differs from Lindmark's (1980) observations, which demonstrated the presence of SOD in the Portland-1 strain of *Giardia*. Attempts to assay SOD employing the pyrogallol-autoxidation assay (Marklund and Marklund, 1974) also yielded a negative result (Smith *et al.*, 1988). Meloni *et al.* (1988a) were also unable to demonstrate SOD activity in several human and feline isolates of *Giardia*. However, xanthine oxidase-mediated O_2 generation was attenuated by a 14 000 ***g*** supernatant fraction of *Giardia* (Smith *et al.*, 1988). This stimulation of O_2 generation is interesting and needs further verification. A more thorough investigation of the prevalence of SOD in *Giardia* spp. is imperative in order to clarify these equivocal reports.

3. *Catalase*

In contrast to SOD, the lack of catalase in *Giardia* is unequivocal (Lindmark, 1980; Smith *et al.*, 1988). Several other aerotolerant eukaryotes have also been shown to lack catalase (Lindmark, 1980).

C. LIPID METABOLISM

Giardia trophozoites are essentially incapable of synthesis *de novo* of cellular lipids. Radiolabelled glucose, threonine, acetate and glycerol were not incorporated into cellular lipids (Jarroll *et al.*, 1981). Exogenous radiolabelled cholesterol was readily taken up and assimilated into cellular sterols, as were arachidonic and palmitic acids. The latter were incorporated into the phospholipid and neutral lipid fractions (Blair and Weller, 1987). The phospholipid and neutral lipid profiles of cysts and trophozoites of *G. duodenalis* were almost identical. The neutral lipids identified include mono-, di- and triacylglycerides, sterol esters and free sterols (Jarroll *et al.*, 1981; Kaneda and Goutsu, 1988). Free sterols were the most abundant.

Of the phospholipids, phosphotidyl choline, phosphotidyl inositol, phosphotidyl serine, phosphotidyl ethanolamine, sphingomyelin and phosphotidyl glycerol have been identified (Jarroll *et al.*, 1981; Blair and Weller, 1987; Jarroll and Lindmark, 1990). Evidence suggests biliary lipids to be the source of lipids for *Giardia in vivo* (Farthing *et al.*, 1983, 1985), a view substantiated by the fact that *Giardia* (WB strain) can be cultured on serum-free media if supplemented with phosphotidyl choline, cholesterol and specific bile salts (Gillin *et al.*, 1986). Gillin *et al.* (1983) and Reiner *et al.* (1986) have shown that normal human milk is toxic to trophozoites *in vitro*. Toxic lipolytic products (probably free fatty acids) released by bile salt stimulated lipase present in milk have been shown to mediate this observed toxicity of milk (Gillin, 1987) (see also Section VI.C.1). Future studies into the mechanisms and kinetics of cholesterol and free fatty acid uptake may yield interesting chemotherapeutic targets since *Giardia* has an absolute requirement for exogenous lipids.

D. PURINE METABOLISM

Purine metabolism by *Giardia* has been reviewed recently by Jarroll and Lindmark (1990) and Adam (1991). All protozoan parasites investigated thus far lack the capacity for purine synthesis *de novo* and are reliant solely on salvage pathways for their purine nucleotide requirements (Wang, 1984). *Giardia* is no exception. Radiolabelled precursors formate and glycine were not incorporated into purine nucleotides nor were hypoxanthine, xanthine or inosine (Wang and Aldritt, 1983; Aldritt *et al.*, 1985; Aldritt and Wang, 1986). However, adenine, adenosine, guanine and guanosine incorporation was readily observed. No interconversion between the two purines has been reported. *Giardial* guanine phosphoribosyltransferase (GPRT) has been purified (Aldritt and Wang, 1986). Purine nucleoside kinase, purine nucleoside phosphotransferase, hypoxanthine phosphoribosyltransferase (HPRT) and xanthine phosphoribosyltransferase (XPRT) were essentially absent. Baum *et al.* (1989) have shown the absence of ribonucleotide reductases, suggesting the reliance of this parasite on exogenous purine deoxynucleosides. Additionally, *Giardia* can apparently convert inosine to hypoxanthine (Berens and Marr, 1986).

E. PYRIMIDINE METABOLISM

Pyrimidine metabolism in *Giardia* has been reviewed by Jarroll and Lindmark (1990) and Adam (1991). Trophozoite pyrimidine requirements are essentially met by the salvage of preformed pyrimidines. Orotate, bicarbonate, and aspartate are not incorporated into pyrimidine nucleotides (Lindmark and Jarroll, 1982; Aldritt *et al.*, 1985). Exogenous thymidine, cytidine and uridine are taken up by trophozoites via active-carrier mediated transport mechanisms and kinetic studies indicate a common carrier site for uridine and cytosine, while thymidine intake occurs via a separate carrier (Jarroll *et al.*, 1987). Uridine competitively inhibited the uptake of cytidine ($K_i = 20\ \mu M$) and vice versa ($K_i = 8\ \mu M$). Thymidine uptake was not inhibited by cytidine or uridine. The K_m values for uridine, cytidine and thymidine uptake were 0.77, 6.67 and 0.57 μM respectively (Jarroll and Lindmark, 1990). Saturation kinetics were evident for all three nucleoside transporters at concentrations exceeding 1.6 nmol ml^{-1}. Preincubation of trophozoites with iodoacetate (5 mM) or N-ethylmaleimide (2.5 mM) inhibited uridine, cytidine and thymidine uptake by 98.8, 95 and 99% respectively (Jarroll *et al.*, 1987; Jarroll and Lindmark, 1990). Dihydrofolate reductase and thymidylate synthetase activities were not detected in *Giardia* (Aldritt *et al.*, 1985). A phosphoribosyl pyrophosphate (P-Rib-PP) synthetase has recently been purified and kinetically characterized (Lee *et al.*, 1992).

F. NUCLEIC ACIDS

The characteristics of giardial DNA, RNA species and the associated genetics have been recently reviewed (Adam, 1991). Attempts to resolve the genome size and ploidy in *Giardia* have yielded rather conflicting data (see Section III.C.1).

1. *DNA and chromosomes*

The guanosine + cytidine (G + C) content of the *G. duodenalis* genome has been estimated to be 42% and 48% respectively (Nash *et al.*, 1985; Boothroyd *et al.*, 1987), whilst that of the ribosomal DNA (rDNA) gene was estimated to be 75% (Healey *et al.*, 1990). Protein coding genes thus far sequenced appear to have a G + C content ranging from 49 to 60%, while the non-coding regions appear to be rich in adenosine + thymidine (A + T).

Pulse field gel electrophoresis (PFGE) has demonstrated the presence of at least five distinct sets (bands) of chromosomes, the sizes of which varied between 1×10^6 and 4×10^6 base pairs (bp), corresponding to a total of 1.2×10^7 bp (Adam *et al.*, 1988a) and 1.1×10^7 bp (Fan *et al.*, 1991) for the five chromosomes. PFGE separation suggests that each trophozoite (i.e., two nuclei) contains 30–50 chromosomal DNA molecules (Adam *et al.*, 1988b). Field inversion gel electrophoresis (FIGE) has revealed differences in the number of chromosomes in the 650–800 kb size range (Upcroft *et al.*,

1989a). Light microscopy findings of four chromosomes, of 1–4 Mb in size per nucleus, would not explain the C_0^t analysis of 3×10^7 and 8×10^7 or the apparent presence of 30–50 chromosomal DNA molecules per trophozoite (Adam, 1991) (see Sections III.C.1 and III.C.2).

2. *Ribosomal RNA*

The ribosomal RNA (rRNA) transcripts of *G. duodenalis* are unique in that they are smaller than those of other eukaryotes and, curiously, smaller than those of eubacteria (Boothroyd *et al.*, 1987; Edlind and Chakraborty, 1987). The small (16S) subunit rRNA sequence exhibits greater similarity to those of archaebacteria than to the corresponding sequence of other eukaryotes (Edlind *et al.*, 1990b). The rDNA gene contains only 5566 bp (Healey *et al.*, 1990) and is tandemly repeated in the genome. The rDNA genes of *G. muris* and *G. ardeae* contain 7600 bp on account of a lengthy intergenic spacer region (van Keulen *et al.*, 1991a,b). Differences were also apparent in the coding regions for the 5.8S rRNA and the 3′ region of the large 28S subunit rRNA (van Keulen *et al.*, 1991a,b). Copies of rDNA genes were identified on multiple chromosomes with extensive variation between otherwise closely related isolates. Le Blancq *et al.* (1991) have shown a telomeric location for rDNA repeats and demonstrated frequent rearrangements of the rDNA repeats within cloned lines of WB strain trophozoites. Investigation of three *G. duodenalis* telomeric clones have revealed abrupt transition from rDNA to telomeric repeats, TAGGG, two from the same location at the beginning of the 28S subunit and another at the beginning of the intergenic spacer with the sequence CCCCGGA being present at each of the breakpoints (Adam *et al.*, 1991) (see also Section III.D).

3. *Translation*

Translational processes in *Giardia* have received little attention. The short leader sequences in *G. "lamblia"* (= *G. duodenalis*) messenger RNAs (mRNAs) suggest that binding of the ribosome to the mRNA and the initiation of translation may be more closely linked than in vertebrates (Adam, 1991). Comparatively short leader sequences have also been reported to occur in actin (11 bases) and ferredoxin (nine bases) genes of *E. histolytica* (Edman *et al.*, 1987; Huber *et al.*, 1988) and the ferredoxin gene (16 bases) of *T. vaginalis* (Johnson *et al.*, 1990). It is of interest to note that, while only four of 699 vertebrate mRNAs carry leader sequences less than 10 nucleotides in length (Kozak, 1987), those of *Giardia* spp. show very short untranslated regions with one to six bases. In vertebrates, mRNAs with short leader sequences are inefficiently translated (Adam, 1991). Despite differences between mRNAs from *Giardia* spp. and those of vertebrates, translation *in vitro* of *Giardia* mRNA has been achieved in a rabbit reticulocyte lysate system (Aggarwal and Nash, 1987a; Peattie *et al.*, 1989).

4. *Ribosome recognition sequences*

Prokaryotic ribosome recognition (Shine-Dalgarno) sequences (consensus UAAGGAGG) are generally located 5–10 nucleotides anterior to the initiation codon. The occurrence of Shine-Dalgarno sequences has been cited in some of the giardial ribosome recognition sequences investigated, although little locational consistency between the putative Shine-Dalgarno sequence and the initiation codon has been observed (Adam, 1991). The ribosome recognition sequences in eukaryotes include the initiation codon and the consensus sequences which have been identified in vertebrates, *Drosophila melanogaster* and *Saccharomyces cerevisiae* (Cavener, 1987; Cigan and Donahue, 1987; Kozak, 1987). Although there appears to be little similarity between the region proximal to the initiation codon in *Giardia* and those of vertebrates, the α- and β-tubulin sequences in *Giardia* essentially match the *D. melanogaster* and *S. cerevisiae* recognition consensus sequences (Adam, 1991), although this needs to be substantiated by N-terminal amino acid sequencing (Peattie *et al.*, 1989). Exacting research in this area has great potential not only in delineating the evolutionary position of *Giardia* but also in shedding light on the cardinal characteristics of translation in vertebrates.

5. *Transcription*

Giardia seems to share with other eukaryotes certain similarities in the transcriptional characteristics. A TATAA box, 9–134 bases upstream from the initiation codon, has been identified in the genomic clones that have thus far been sequenced (Kirk-Mason *et al.*, 1989; Aggarwal *et al.*, 1990; Gillin *et al.*, 1990). Whether or not these TATAA sequences are putative recognition sites with respect to RNA polymerase has yet to be investigated (Adam, 1991). The sequence AGTPuAA has been suggested to be the polyadenylation signal for *Giardia* spp. (Peattie *et al.*, 1989), whereas in vertebrates and plants the polyadenylation consensus sequence (AATAAA) is invariably located 10–30 nucleotides from the poly(A) tail (Proudfoot, 1991). In *Giardia* spp. a poly(A) tail follows a short 3′ untranslated region while a 6–19 nucletotide spacer separates the sequence AGTPu AAPy from the stop codon, which itself is separated by 7–10 bases from the beginning of the poly(A) tail. The occurrence of introns in protozoan protein coding genes has been rarely observed (Adam, 1991) and none has been identified to date in *Giardia* spp.

G. HYDROLASES

Ultrastructural studies of trophozoites and cysts of *G. muris* have demonstrated the presence of ovoid vacuoles (100–400 nm in diameter) or organelles located peripheral to the plasmalemma (Cheissin, 1964; Bockman and Winborn, 1968, see Section II.D; Fig. 3). These are known to sequester exogenous ferritin (Bockman and Winborn, 1968), the significance of which is unknown (Adam, 1991). Cytochemical localization of acid phosphatase in these ovoid vacuoles (Feely, 1985; Feely and Erlandsen, 1985; Feely and Dyer, 1987) of *G. duodenalis* trophozoites (see Section II.D) led to the suggestion that they were isofunctional with lysosomes. Differential and isopycnic centrifugation of homogenates (Lindmark and Miller, 1988) demonstrated that giardial hydrolases (Table 4) were localized in a single particle population sedimenting at 7200 ***g*** for 30 min. The particles had a buoyant density of 1.15 in sucrose and exhibited latency (Lindmark, 1988) and the hydrolase activities remained tightly bound to the organelle membrane (Lindmark, 1988).

G. duodenalis trophozoites contain two major cysteine proteases with approximate molecular masses of 40 kDa and 105 kDa which are active in the pH range of 3.8–5.8, typical of other cysteine proteases (Hare *et al.*, 1989). The characteristics and inhibitor sensitivities of these have been investigated (Hare *et al.*, 1989). The 40 kDa cysteine protease has the capacity to degrade haemoglobin and immunoglobulin (Ig)A_1 (Parenti, 1989). These observations suggest that proteases may represent virulence determinants (Parenti, 1989; Adam, 1991). Whether proteases are involved in the encystation and/or excystation processes in *Giardia* remains to be investigated.

H. CHITIN

Chitin, a linear polysaccharide composed of $\beta 1 \rightarrow 4$-linked N-acetylglucosamine (GlcNAc) units, is a major structural component of yeasts, fungi, insects and crustacea (Cabib, 1981; Ward *et al.*, 1985) and is synthesized from uridine diphosphate N-acetylglucose amine (UDP-GlcNAc) by the action of chitin synthetase $(GlcNAc)_n$-synthetase (Cabib, 1981). The occurrence of chitin in the cyst walls of *G. duodenalis* was demonstrated by lectin-binding studies (Ward *et al.*, 1985). Increased chitin synthetase activity,

TABLE 4 *Specific activities of hydrolases in* Giardia[a]

Enzyme	Activity[b]	
Acid phosphatase	(EC 3.1.3.2)	80±4(10)
Deoxyribonuclease	(EC 3.1.4.5)	58±15(8)
Ribonuclease	(EC 2.7.7.16)	42±12(10)
β-N-acetyl glucosaminidase	(EC 3.2.1.30)	ND (25)
β-galactosidase	(EC 3.2.1.23)	ND (10)
β-D-glucosidase	(EC 3.2.1.21)	ND (2)
α-D-glucosidase	(EC 3.2.1.20)	ND (14)
β-glucuronidase	(EC 3.2.1.31)	ND (14)
β-D-xylosidase	(EC 3.2.1.37)	ND (7)
Aryl sulphatase	(EC 3.1.6.1)	Present

[a] Information based on Jarroll and Lindmark (1990) and (for aryl sulphatase) Feely and Dyer (1987).
[b] Milliunits per mg of protein ± standard deviation; number of replicate assays given in parentheses. Enzyme units defined as the amount of enzyme necessary to degrade 1 mmol of substrate min^{-1} at 30°C. All assays conducted at 30°C in the presence of 0.1% Triton X-100®. ND: not detectable at 30°C or 37°C at pH 3–8.

measured by the incorporation of [^{3}H]N-acetylglucosamine into trichloroacetic acid (TCA) precipitable material, has been observed in encysting *Giardia* trophozoites (Gillin *et al.*, 1987). Tritiated-GlcNAc incorporation into TCA precipitable material was inhibited by the chitin synthetase inhibitors nikkomycin and polyoxin D, and the resultant TCA pellet was degraded by the action of affinity purified chitinase (Gillin *et al.*, 1987). Although these studies provided only indirect evidence (Jarroll and Lindmark, 1990) for the occurrence of chitin in the cyst wall of *Giardia*, the fact that nikkomycin and polyoxin D are structural analogues (Arron *et al.*, 1982) of UDP-GlcNAc may be interpreted as support for the presence of chitin in cysts. Mass spectrometric analysis has shown the occurrence of galactosamine ($GalNH_2$) (Jarroll *et al.*, 1989) and N-acetylglucosamine (Ortega-Barria *et al.*, 1990) in the outer cyst wall of *Giardia*. Although mass spectrometric evidence for the occurrence of GlcNAc in the outer cyst wall has been questioned (Adam, 1991) it confirms the presence of GlcNAc in the inner layers of *Giardia* cysts and may lend support to the occurrence of chitin (GlcNAc) in cysts. However, the absence of β-galactosidase, β-D-glucuronidase, β-N-acetylglucosaminidase (Lindmark, 1988; Jarroll and Lindmark, 1990) in lysates of trophozoites and/or cysts is curious if chitin is a component of the cysts. The lack of staining of the cyst wall with chitin specific reagents (Filice, 1952) has also been cited as evidence against the occurrence of chitin (Adam, 1991).

The identification of UDP-GlcNAc-4^1-epimerase in incipient cysts suggests the conversion of GlcNAc to GalNAc to be a key event during encystation (Jarroll *et al.*, cited by Adam, 1991). Whether enzymic polymerization of GalNAc yields a substance similar to chitin $(GalNAc)_n$, is debatable. Enzymes catalysing such a homopolymerization as well as GlcNAc-4^1-epimerase would be theoretically susceptible to inhibition by polyoxin-D and nikkomycin by virtue of the fact that the latter are structural analogues of UDP-GlcNAc. Since chitin $(GlcNAc)_n$, and possibly its galactosamine counterpart $(GalNAc)_n$, are not present in mammals or humans (Ward *et al.*, 1985), chitin metabolism represents an ideal biochemical target for the rational design of novel antigiardial agents. The involvement of $(GlcNAc)_n$ hydrolase or $(GalNAc)_n$ hydrolase during excystation requires further study.

I. STRUCTURAL PROTEINS

1. *Tubulins*

Tubulin is the most prevalent cytoskeletal protein in *Giardia* (see Sections II.B and II.C). As in most other eukaryotes, tubulin-derived microtubules carry out vital functions at the cellular level in *Giardia* (Dustin, 1984; Lacey, 1988). In *Giardia*, additional to the above functions, microtubules are vital to the structure and function of the flagellar axoneme and the ventral disc (see Sections II.B and C). The genes coding for α- and β-tubulin of *Giardia* have been sequenced (Kirk-Mason *et al.*, 1988, 1989). A 1.35 kb cDNA insert exhibited 85% isology to human β-tubulin (Kirk-Mason *et al.*, 1988). Crossley *et al.* (1986) and Torian *et al.* (1984) have shown that disc and flagellar tubulins are antigenically different. These findings seem to substantiate the differential effects observed with benzimidazoles (see Sections IV.A and VIII), which inhibited adherence with no apparent effect on flagellar function, suggesting the possibility that flagellar tubulin lacks a binding site for benzimidazoles (Edlind *et al.*, 1990a) (see Section VIII.C.2.e).

2. *Giardins*

The giardins (see also Sections II.C and VIII.C.2.e), first identified by Crossley and Holberton (1983)), are a group of specific proteins found only in the ventral disc of *Giardia* spp. with molecular masses ranging from 29 to 38 kDa (Crossley and Hoberton, 1983, 1985; Holberton *et al.*, 1988; Peattie *et al.*, 1989; Alonso and Peattie, 1992). Although 23 distinct giardins have been identified some are antigenically similar and have identical N-terminal sequences, suggesting that some may be derived by post-translational modification of the same protein (Peattie *et al.*, 1989).

Two distinct classes of giardins (α and β) have thus far been cloned and sequenced. α_1-Giardin (molecular mass 33.8 kDa) (Peattie *et al.*, 1989) is predicted to have a predominantly α-helical secondary conformation, and is very similar to α_2-giardin in its amino acid and DNA isology, which are 77 and 81% respectively (Alonso and Peattie, 1992). In contrast, β-giardin (molecular mass 29.4 kDa), exhibits no DNA or amino acid sequence similarity to α-giardin. Although predictions of secondary structure of β-giardin suggest a predominantly α-helical structure, it has a heptad repeat consistent with a coiled-coil structure similar to that found in myosin and tropomyosin (Holberton *et al.*, 1988). Published sequences of β-giardin differ only in the G + C regions and in the putative initiation site (Baker *et al.*, 1988; Aggarwal *et al.*, 1989a). The giardins are distinct from the tubulin accessory protein copolymers or microtubule-associated proteins (MAPs) identified *in vitro* (Crossley and Holberton, 1983). In *Giardia* microribbons, tubulin and giardin copolymerize in flat sheets with a structured subunit arrangement (Holberton, 1981). Although the net charge on denatured giardin is acidic, the charge characteristics of the native giardin have not been reported. Unlike MAPs, giardin occurs at high concentration (>40%) in the ribbon protein and its stoichiometry is vastly greater than that of the MAP:tubulin ratio (Crossley and Holberton, 1983); unlike MAPs, giardins aggregate in the absence of dispersing agents.

J. HEAT SHOCK PROTEINS

Exposure to 43°C repressed *Giardia* protein synthesis, and induced the synthesis of four heat shock proteins (hsps) (Lindley *et al.*, 1988), three of which fell into a high molecular mass category (100, 83 and 70 kDa) and one of low molecular mass (30 kDa). Sequence analysis of a gene whose transcript level was found to increase during the heat shock regime, was found to have little or no similarity to the HSP70 genes of other organisms (Aggarwal *et al.*, 1990). Exposure of *Giardia* to ethanol, an end product of carbohydrate metabolism, induced the production of four hsps, while exposure to oxygen following an anaerobic regime inhibited protein synthesis. The putative *Giardia* heat shock gene has been identified and its promoter sequenced (Aggarwal *et al.*, 1988). The corresponding hsp contains 142 amino acids and a TATAA box −28 to −33 upstream to the initiation site (Aggarwal *et al.*, 1988, 1990). The oxygen shock response and the reported ethanol-induced generation of hsps, or rather "metabolic shock proteins," in *Giardia* warrant further investigation.

K. CYSTEINE-RICH PROTEINS

Surface antigen analysis of *Giardia* employing iodogen and/or lactoperoxidase surface iodination techniques has revealed the occurrence of several cysteine-rich proteins (CRPs) (see Section VI.C.3). A portion of the gene coding for the elaboration of 170 kDa surface antigen (*Giardia* isolate WBa) was cloned and found to encode a CRP referred to as CRP170 (12% cysteine) (Adam *et al.*, 1988b). Radiolabelling studies have confirmed the occurrence of CRPs in other isolates as well (Aggarwal *et al.*, 1989b). Several other CRPs have since been identified and cloned: TSA417 (Gillin *et al.*, 1990), VSP1267 (Adam, 1991) and CRP72 (Adam, 1991). All CRPs have a cysteine content of 11–12%, appear to be variable within single isolates (Adam, 1991), and have highly conserved 3′ C-terminal amino acid sequences (Adam, 1991). The functions of these CRPs have yet to be elucidated. It is possible they may have high turnover rates and constitute a defensive biochemical barrier against oxidant stress.

VI. Host–Parasite Relationships

A. CLINICAL EFFECTS

The clinical signs associated with *Giardia* infections in humans vary greatly (Smith, 1985; Farthing, 1986; Wolfe, 1990; Table 5). There may be total latency, acute short-lasting diarrhoea or chronic syndromes associated with nutritional disorders, weight loss and failure to thrive.

In the majority of untreated patients, infection resolves spontaneously and symptoms usually disappear in a few weeks (Moore *et al.*, 1969; Brodsky *et al.*, 1974; Wolfe, 1990). Occasionally, symptoms may persist for reasons that are not entirely clear and certain patients, such as those with hypogammaglobulinaemia, are at increased risk of chronic clinical giardiasis associated with malabsorption syndromes (Ament *et al.*, 1973; Perlmutter *et al.*, 1985). The pronounced variability in symptomatology that is so characteristic of *Giardia* infections is the result of a myriad of factors including host immune and nutritional status, concurrent infections and heterogeneity in infectivity, and the virulence and pathogenicity of strains of *G. duodenalis*. As we learn more about the mechanisms associated with pathogenesis, our understanding of how these host and parasite factors interact will become more evident.

Children appear to be at most risk of contracting clinical giardiasis, particularly those from developing countries and from disadvantaged groups such as Australian Aborigines (Gracey, 1983; Farthing, 1986; Meloni *et al.*, 1988b; Islam, 1990; Kaminsky, 1991). In these children, *Giardia* is of particular concern in the aetiology of protein calorie malnutrition and the retardation of growth and development. Giardiasis is most common in children under 10 years of age (Dupont and Sullivan, 1986; Speelman and Ljungström, 1986; Meloni *et al.*, 1993). Severe forms of the disease affecting growth appear to occur most frequently in the second year of life (Farthing *et al.*, 1986b). The presence and frequency of diarrhoea, duration of infection and opportunities for reinfection are the essential factors affecting growth (Mason and Patterson, 1987; Brucker, 1989; Islam, 1990).

TABLE 5 *Clinical features of* Giardia *infection*

Clinical features
Asymptomatic/latent
Acute and/or chronic
Nausea
Cramps
Anorexia
Fever
Diarrhoea
Constipation
Flatulence
Heartburn
Abdominal pain
Fatigue
Headache
Abdominal distension
Foul-smelling stool
Mucus in stool
Weight loss
Malabsorption syndromes
Protein-energy malnutrition
Failure to thrive

Little attention has been given to the clinical effects of giardiasis in animals other than humans. *Giardia* is common in dogs and cats yet is rarely associated with clinical disease (Thompson, 1992a). However, Kirkpatrick (1989) concluded that giardiasis is probably an underestimated cause of malabsorption and diarrhoea in many species of immature large mammals. In dogs and cats, clinical giardiasis appears to be uncommon and is usually associated with kennel or cattery situations involving young animals. However, establishing the role of *Giardia* as a cause of disease in animals is very difficult (Thompson, 1992a). Concurrent infections are often present and, since diarrhoea is the most common symptom of giardiasis in dogs and cats, a variety of nutritional factors must also be considered. The potential role of *Giardia* as a possible cause of disease in birds was highlighted recently with a report of infection in sick and dying nestling straw-necked ibis (*Threskiornis spinicollis*) by Forshaw *et al.* (1992). Although no other obvious pathogen was present in these birds, further investigations are required to determine the role, if any, of *Giardia* in such avian disease outbreaks.

B. PATHOGENESIS

The mechanisms by which *Giardia* causes disease have been largely unknown and it is only recently that a co-ordinated picture has emerged of the factors involved in the disease process. Epithelial damage, increased epithelial cell turnover, villous shortening and disaccharidase deficiency have all been reported as manifestations of giardiasis (Yardley *et al.*, 1964; Ament and Rubin, 1972; Gillon and Ferguson, 1984; Buret *et al.*, 1990a,b).

Using rodent models for infections with both *G. muris* and *G. duodenalis*, it has been possible to study in detail these clinical manifestations during the course of infection. Villous atrophy and damage to the microvilli have been correlated with brush border enzyme deficiencies which return to normal levels of activity once infection is removed (Buret *et al.*, 1990b, 1991). Brush border injury is expressed as lowered disaccharidase activity and decreased microvillous surface area which, in turn, result in impaired digestion and malabsorption which lead to reduced growth (Buret *et al.*, 1991). In both mice infected with *G. muris*, and gerbils infected with *G. duodenalis*, the loss of brush border thickness was directly correlated with observed decreases in disaccharidase activity (Buret *et al.*, 1990b, 1991). Furthermore, these studies also revealed a correlation during the course of infection between these changes and impaired weight gain. The degree of brush border injury and loss of surface area of the microvillous border are the limiting factors for disaccharidase activity and appear to be dependent on parasite load (Buret *et al.*, 1991). Such a mechanism of pathogenesis could account for both of the major clinical features of giardiasis: diarrhoea and failure to thrive. The roles of the parasite and host in inducing these changes remain to be established. The physical damage inflicted by the adherence of *Giardia* to the gut is likely to play some part in the disease process but the role of other factors such as toxins is uncertain. Not surprisingly, proteolytic enzyme activity has been demonstrated in *Giardia* (Parenti, 1989). Two major cysteine proteases of 40 and 50 kDa have been characterized (Hare *et al.*, 1989). The smaller protease cleaves haemoglobin and immunoglobulin (Ig) A1 and has been proposed as a virulence determinant (Parenti, 1989). However, there is no evidence for the involvement of these proteases in pathogenesis in the host and thus their role, if any, remains speculative.

The characteristics of both parasite and host most probably determine the outcome of infection with *Giardia*. The fact that symptomatic giardiasis is more likely to occur in patients who are immunocompromised, malnourished, or very young is a clear indication that host factors are important. However, in other cases, there is increasing evidence that a particular strain, or isolate, of *Giardia* may have a greater potential to cause disease in a particular host (reviewed by Thompson *et al.*, 1990a). The nature and competence of immune mechanisms are likely to be central in both cases, particularly in view of mounting evidence that differing virulence between *Giardia* isolates is related to the expression of different surface antigens (Nash *et al.*, 1991).

C. IMMUNOBIOLOGY

Relatively little is known about how the host responds to the presence of *Giardia* or of the parasite's ability to survive host resistance mechanisms.

1. *Non-immunological factors*

There is ample evidence that factors in human milk have pronounced anti-giardial activity. Most prevalence surveys have shown that *Giardia* is uncommon in infants, particularly those under 6 months of age, and that this is associated with breast feeding (Islam *et al.*, 1983; Farthing *et al.*, 1986b; Shetty *et al.*, 1990; Meloni *et al.*, 1993). Gillin *et al.* (1983) demonstrated the lethal effect of human milk against *Giardia in vitro*, which appears to be due to the non-specific cytotoxic (Gillin, 1987) and/or anti-adherent (Crouch *et al.*, 1991a) actions of fatty acids in the milk.

2. *Immunological factors*

An important component of the humoral immune response in humans and other animals appears to be specifically induced IgA (Nayak *et al.*, 1987; reviewed by den Hollander *et al.*, 1988; Adam, 1991). The appearance of IgA in the intestinal secretions of mice infected with *G. muris* has been correlated with the resolution of infection (Snider *et al.*, 1988). The IgA response is directed against a protein of molecular mass of approximately 30 kDa and may inhibit attachment, thus leading to parasite removal from the gastrointestinal tract (Heyworth and Pappo, 1989). In murine giardiasis (*G. muris*), the antibody response is dependent on intact cell-mediated immunity (Carlson *et al.*, 1987; Heyworth *et al.*, 1987). Whether the situation is similar in human infections with *G. duodenalis* is not known, but T cell-dependent mechanisms are likely to be important (Gottstein *et al.*, 1991). Gottstein and

Nash (1991) demonstrated that the course of *G. duodenalis* infection is mainly mediated by functional thymus-dependent T lymphocytes, the absence of which results in chronic persistence of infection. Using a neonatal mouse model of *G. duodenalis* infection, Gottstein *et al.* (1990) have shown that the predominant parasite-specific humoral response consisted of IgM and IgG isotypes. It has been suggested (Gottstein and Nash, 1991) that such B cell mechanisms are most probably responsible for the surface antigen switch (see Section VI.C.3).

It has been demonstrated that macrophages are capable of ingesting *Giardia in vitro* (Hill and Pohl, 1990) and may function in antigen presentation (Adam, 1991). Recent studies by Crouch *et al.* (1991b) *in vitro* using human neutrophils and monocytes showed that such cells were capable of interfering with trophozoite adherence. These authors emphasized the potential vulnerability of the adherence mechanism of *Giardia* which could be targeted immunologically by phagocytic cells and affected through the activities of microbicidal enzymes and/or reactive oxygen species (Crouch *et al.*, 1991b).

3. *Antigens and antigenic variation*

A variety of antigens has been isolated and characterized from *Giardia*. Many surface and internal proteins, particularly cytoskeletal elements, are antigenic and their characteristics have been recently reviewed (den Hollander *et al.*, 1988; Adam, 1991). Comparative studies have also revealed considerable diversity in antigenic characteristics between different isolates of *Giardia*, but the method used to detect differences may influence the degree of heterogeneity reported (Thompson *et al.*, 1990a). Such variability between isolates is an important consideration in the development of diagnostic agents. However, in relation to resistance, most attention has focused on a group of surface cysteine-rich proteins which exhibit antigenic variation within isolates of *Giardia*.

Surface antigenic variation is a well-established phenomenon in *Giardia* and has been shown to occur *in vitro* and *in vivo* (Adam *et al.*, 1988b; Nash *et al.*, 1988; Nash, 1989; Gottstein *et al.*, 1990). Cloned lines of *Giardia* give rise to trophozoites possessing a different surface antigen (Nash *et al.*, 1988), but the ability to express a certain range of epitopes is characteristic of the parent isolate and thus appears to be genetically determined (Nash *et al.*, 1990a). The variant antigens, which all appear to be cysteine-rich (Aggarwal *et al.*, 1989b; Gillin *et al.*, 1990), cover the entire cell surface uniformly and only one variant appears to be present on the surface at any one time (Pimenta *et al.*, 1991).

Isolates have been shown to differ in the rate of antigenic variation (Nash *et al*., 1990b), the proportion of a population which expresses a particular antigen (Pimenta *et al*., 1991; Hopkins *et al*., in press) and the range of epitopes they are able to express (Nash *et al*., 1990a). These differences are also genetically determined (Nash *et al*., 1990a). The frequency of surface antigen switching has been estimated as once every 6–13 generations, depending on the particular isolate of *Giardia* (Nash *et al*., 1990b).

The biological relevance of antigenic variation is uncertain. Although evasion of host immunological effector mechanisms was originally thought to be a possibility, as occurs with salivarian trypanosomes, no more than one cyclical change in surface antigens has so far been demonstrated with *Giardia* infections in humans and other animals (Pimenta *et al*., 1991). However, recent studies have shown that *Giardia* isolates possessing unique antigenic variants differ in their susceptibility to the host digestive proteases, trypsin and α-chymotrypsin; thus, certain surface antigens seem to protect *Giardia* from enzyme attack (Nash *et al*., 1991).

VII. Epidemiology and Transmission

A. Faecal–Oral Transmission

Transmission of *Giardia* is predominantly by faecal–oral contamination. Levels of infection are therefore highest under conditions of poor hygiene and sanitation, particularly in tropical and subtropical environments. Under such conditions, transmission of infection can be achieved with only minute quantities of infective material. As few as 10 cysts (Rendtorff, 1954) or a single trophozoite (De Carneri *et al*., 1977) appear to be able to initiate an infection, thus explaining why ingestion of very small amounts of faecal material may be sufficient for transmission of *Giardia* (Owen, 1984).

Giardiasis is considered to constitute an urgent public health problem in many developing countries and among disadvantaged groups such as Australian Aborigines (Brucker, 1989; Islam, 1990; Thompson, 1992b). Very high prevalence rates of between 20 and 60% are common, particularly in children (Gilman *et al*., 1985; Dupont and Sullivan, 1986; Miotti *et al*., 1986; Mason and Patterson, 1987; Simango and Dindiwe, 1987; Meloni *et al*., 1988b; Brucker, 1989; Islam, 1990; Meloni *et al*., 1993). For example, in rural areas of Guatemala, all children were infected during the first 3 years of life (Mata, 1978). In such highly endemic foci of transmission, reinfection is very common (Mason and Patterson, 1987; Thompson, 1992b; Meloni *et al*., 1993). In such circumstances, although effective chemotherapy may be available, long term control requires education and improved levels of hygiene (see Section VIII).

Other predisposing factors may enhance the frequency of faecal–oral transmission. For example, conditions conducive to faecal–oral contamination are common in day care centres and high prevalence rates of *Giardia* infection have often been observed in children attending such centres, particularly in those under 3 years of age (Pickering *et al.*, 1981; Boreham and Shepherd, 1984; Polis *et al.*, 1986; Woo and Paterson, 1986).

Direct transmission between homosexual men is being increasingly recognized as significant in the epidemiology of giardiasis (Phillips *et al.*, 1981). Food-borne transmission, usually as a result of contamination during food preparation, is also a well recognized source of *Giardia* infection (Barnard and Jackson, 1984; Petersen *et al.*, 1988).

B. WATER-BORNE TRANSMISSION

Numerous cases of *Giardia* infection have been associated with contaminated water and, in the USA, giardiasis is the most frequently diagnosed water-borne disease (Levine *et al.*, 1990). Water-borne outbreaks of giardiasis have also been reported in Canada (Wallis *et al.*, 1986; Moorehead *et al.*, 1990) and Europe (Jephcott *et al.*, 1986; Neringer *et al.*, 1986). The potential for water-borne transmission has been recognized in Australia (Wade and Yapp, 1989) and in New Zealand recent surveys have identified *Giardia* in a number of water supplies (Ampofo *et al.*, 1991; Fraser and Cooke, 1991). Water-borne transmission is also a well documented cause of giardiasis in travellers who usually contracted infection from drinking local tapwater (Jokipii and Jokipii, 1974). Investigations of endemic water-borne giardiasis in the USA have usually found the contamination of water supply to have resulted from inadequate water treatment, ineffective filtration, or contamination with human sewage (Craun, 1986, 1990; Schantz, 1991). Filtration is necessary to remove *Giardia* as chlorination alone is insufficient without high concentrations of chlorine and long contact times (Schantz, 1991).

However, the fact that human infections with *Giardia* have been traced to the consumption of water from streams in rural areas well away from urban environments has implicated a number of species of wild animals, particularly beavers and muskrats, as reservoirs of infection (Davies and Hibler, 1979; Craun, 1990). However, the role of animals in water-borne transmission is controversial (see Thompson *et al.*, 1990a). Whether animals serve as the original source of contamination or amplify the numbers of the originally contaminating isolate, or both, remains to be determined (Bemrick and Erlandsen, 1988; Thompson *et al.*, 1990a).

C. ZOONOTIC TRANSMISSION

The significance and interpretation of the results obtained from experimental infections of animals with human isolates of *Giardia* have been controversial (reviewed by Thompson *et al.*, 1990a). However, there is sufficient evidence to show that at least some isolates of *Giardia* are not host-specific and that humans and a variety of other animals naturally share the parasite. Most authorities therefore regard giardiasis as a zoonosis (Acha and Szyfres, 1987; Meyer, 1990b; Schantz, 1991).

Studies of the prevalence of *Giardia* in a variety of mammals and birds underlie the fact that a potential reservoir of human infection exists. Numerous surveys of dogs and cats have been undertaken in urban and semi-urban areas of the world, and all have shown that *Giardia* is very common, with rates of infection of 7–35% in cats and 12–50% in dogs (Swan and Thompson, 1986; Baker *et al.*, 1987; Collins *et al.*, 1987; Winsland *et al.*, 1989; Castor and Lindquist, 1990; Tonks *et al.*, 1991; Thompson, 1992b; Meloni *et al.*, 1993). Some authorities have recommended that all *Giardia*-infected dogs and cats should be treated, whether or not they are clinically ill, because of the potential zoonotic risk (Kirkpatrick, 1984). If both family members and their pets are found to be infected with *Giardia*, as recently reported from Canada (Cribb and Spracklin, 1986), this must be strong circumstantial evidence of zoonotic transmission.

Very few surveys of the prevalence of *Giardia* have been undertaken in domestic ruminants. A recent study in Canada found 17.7% of sheep and 10.4% of cattle to be infected with *Giardia*, although much higher prevalence rates were recorded in lambs (35.6%) and calves (27.7%) (Buret *et al.*, 1990c). *Giardia* has also been reported to be common in sheep and cattle in Switzerland (Taminelli and Eckert, 1989). Apart from the potential zoonotic reservoir of *Giardia* in sheep and cattle, high levels of infection in young animals are likely to be clinically significant (Kirkpatrick, 1989) and any resultant effect on weight gain will inflict an important economic loss to agriculture (Buret *et al.*, 1990c).

Some authors consider that birds may be a source of zoonotic contamination of open water with *Giardia* (Bemrick, 1984). *Giardia* in a blue heron was suggested as a possible source of water-borne giardiasis in humans, but not confirmed (Georgi *et al.*, 1986), although the species of *Giardia* involved was not determined. Cross-infection experiments with the avian species *G. psittaci* and *G. ardeae* were taken to indicate that birds should not be considered as likely potential reservoirs of infection for mammalian hosts (Erlandsen *et al.*, 1991), at least in the USA. The zoonotic significance of avian outbreaks in Australia (Forshaw *et al.*, 1992) is not known as the species of *Giardia* has not yet been identified.

Although the potential for zoonotic transmission of *Giardia* is now well accepted, irrefutable evidence for such transmission outside the laboratory has been difficult to obtain. Because all isolates of *G. duodenalis* are morphologically indistinguishable, and since some are more host-specific than others (Thompson *et al.*, 1990a), it has not been possible to prove that a particular outbreak of infection in humans was contracted from an animal source, or vice versa.

The situation is compounded by the enormous genetic heterogeneity of *G. duodenalis*. Molecular characterization procedures such as enzyme electrophoresis and the use of DNA hybridization using specific probes have found some mammalian isolates to be very similar to human ones (Thompson *et al.*, 1988a,b; Meloni *et al.*, 1989). However, cat isolates of *Giardia* from Western Australia, for example, are quite different from isolates of *Giardia* infecting cats in Europe (Meloni *et al.*, 1992). As a consequence, an indication of genetic identity between isolates of *Giardia* from humans and other animals in one endemic area cannot be extrapolated to the situation in another endemic area (Thompson *et al.*, 1990a,b). The isolates of *Giardia* involved may have quite different host specificities. Comparative studies which utilize molecular characterization techniques to provide evidence of zoonotic transmission must therefore compare isolates at a local level. The most appropriate techniques to apply in such studies remain to be determined. Enzyme electrophoresis has been shown to be of value (Meloni *et al.*, 1988a, 1989), but may not provide the level of genotypic discrimination necessary for differentiating and demonstrating affinities between closely related individuals. In this respect, DNA fingerprinting may be of value (Upcroft, 1991), and should be applied to the study of *Giardia* transmission in the field. However, caution will be needed in the interpretation of results. If genetic identity of isolates from humans and other animals in the same location can be found, this is very strongly suggestive of cross-transmission (Archibald *et al.*, 1991). However, it does not necessarily follow that non-identity means that zoonotic transmission is not occurring, because particular species of host are known to be susceptible to more than one genetically distinct variant of *Giardia* (Andrews *et al.*, 1989; Meloni *et al.*, 1989; Thompson *et al.*, 1990a).

VIII. Control

A. PUBLIC HEALTH

Groups at risk of *Giardia* infection include inhabitants of underdeveloped countries or underprivileged groups within developed countries, children in day care centres, pre-school children and nursing home dwellers, and patients taking immunosuppressive drugs or those with acquired immune deficiency syndrome (AIDS).

1. *Underdeveloped countries and underprivileged communities*

Giardiasis is one of the common causes of acute or persisting diarrhoea in developing countries (Anonymous, 1987) and is a major health problem (Islam, 1990). Within these countries the incidence of giardiasis is often over four times higher than in the USA (7.4%; Quinn, 1971) or the rest of the world (7.4%, Nikolic *et al.*, 1990), and it varies considerably between regions in the same country (3.9–18.2% in Ghana; Annan *et al.*, 1986). Children are more often affected (Gracey, 1983; Farthing, 1986; Meloni *et al.*, 1988b, 1993; Islam, 1990; Kaminsky, 1991), as discussed in Section VI.A, and this is of particular concern because of the ensuing repeated exposure to potentially toxic drugs in some regions. The higher prevalence of giardiasis in children is likely to be a consequence of a greater risk of infection, as protective immunity is acquired with age and multiple exposure (Islam, 1990; Farthing, 1986).

Infection varies inversely with socio-economic status and is high in regions where water supplies are poor or non-existent and sanitation and hygiene standards are poor (Knight, 1980; Meyer, 1985; Meloni *et al.*, 1988b; Boreham *et al.*, 1990; Islam, 1990). A greater incidence of giardiasis is seen in urban than in rural communities (Hossain *et al.*, 1983; Kaminsky, 1991) and risk factors identified as important include high environmental faecal contamination; lack of potable water, education and good housing; overcrowding and high population density (Islam, 1990; Kaminsky, 1991).

Consequences of high infection rates include protein-calorie malnutrition, retardation of growth and development (Gracey, 1983; Farthing, 1986; Meloni *et al.*, 1988b, 1993), and malabsorption of other nutrients including D-xylose and fat (Mahlanabis *et al.*, 1979; Judd *et al.*, 1983), lactose (Rabassa *et al.*, 1975; Jove *et al.*, 1983), vitamin B12 (Cordingley and Crawford, 1986) and vitamin A (Mahlanabis *et al.*, 1979; Chavalittamrong *et al.*, 1980). In at least three of these conditions (malabsorption of D-xylose, fat, vitamin B12) there was also a correction of the deficit upon anti-protozoal treatment. It is difficult to attribute all the above pathological effects to *Giardia* infection since there are usually multiple infections present and the drug often used for treatment (metronidazole) is also effective against anaerobic bacterial infections. However, the reversal of observed malabsorption suggests a causative role for *Giardia*.

In terms of the effects on health it is now well documented that chronic or repetitive infection with gastrointestinal parasites, including *Giardia*, causes reduction in growth and development of children, although it is not possible to rule out effects of other intestinal pathogens, but the evidence strongly suggests that *Giardia* has a causative role in "failure to thrive" and that anti-giardial treatment is effective in reversing these effects (Gupta and Urrutia, 1982; Gracey, 1983; Cordingley and Crawford, 1986; Farthing *et al.*, 1986b).

2. *Day care, pre-school and nursing home groups*

In these situations the observed incidence of giardiasis of 17–54% is often close to that found in underdeveloped countries. Giardiasis is often identified as an epidemic either alone or in association with other intestinal pathogens (Black *et al.*, 1977; Keystone *et al.*, 1978; Pickering *et al.*, 1981; Boreham and Shepherd, 1984; Bartlett *et al.*, 1985a,b). Pickering and Engelkirk (1990) summarized the incidence ("attack rate"), age and number of centres involved in a series of well documented outbreaks, and emphasized the young age at which infection seemed most common (under 3 years). Prospective studies are few in number but indicate that *Giardia* is responsible for a significant number of outbreaks of diarrhoea in day care centres (Pickering *et al.*, 1981; Sullivan *et al.*, 1984; Bartlett *et al.*, 1985a,b).

The major factors involved in epidemics in these institutions are presence of young non-toilet-trained children, contamination of hands, communal classroom objects and lack of infection control measures (Sullivan *et al.*, 1984). A proportion of each group is asymptomatic and can act as carriers of infection to relatives and the community (Pickering *et al.*, 1981; Bartlett *et al.*, 1985a,b), together with transmission by affected children (Polis *et al.*, 1986). However, day care centre infections are prolonged (approximately 6 months) and children tolerate the infection well. Pickering and Engelkirk (1990) concluded that *Giardia* was endemic in day care centres and that treatment to eradicate the parasite from asymptomatic carriers was not practical. However, the public health problem of transmission to the community from these centres is a significant one and needs further evaluation and control through infection control measures and education.

Similar problems in public health are encountered in nursing homes and should be tackled in similar ways. A notable case from a nursing home, which also served as a day care centre for children, was reported by White *et al.* (1989), in which 88 cases of giardiasis were identified in a population of 312 in a period of 6 weeks, involving transmission by multiple routes including food-borne and person-to-person transmission. It should be remembered that *Giardia* may also be underestimated as a cause of diarrhoea, malnutrition and gastrointestinal illness in adults (Beaumont and James, 1986).

3. *Immunodepressed patients and AIDS patients*

Bromiker *et al.* (1989) reported two cases of severe giardiasis in two patients who underwent bone marrow transplantation. Both cases were successfully treated with metronidazole by standard dose regimes. Although this group comprises only a small number of patients, giardiasis represents a significant post-implantation disease risk in such cases and infection status should be determined using appropriate techniques (see Section VIII.B).

Parasitic infections in AIDS have been a focus of much attention in recent times, with most concern placed on *Pneumocystis carinii* infections of the lung (Lockwood and Weber, 1989). Similarly, *Toxoplasma gondii*, *Cryptosporidium* species and *Isospora belli* are recognized as opportunistic infective agents in AIDS, whereas there have been no reports of *Giardia* as a serious problem in AIDS patients despite a high prevalence in homosexual men and a high possibility of encountering a heavy *Giardia* load and a concomitant $CD4^{+}$T cell deficiency. One report considered *Giardia* to be an opportunistic infection in AIDS (Laughon *et al.*, 1988). Microsporidia and *Strongyloides stercoralis* have since been added to the list of opportunistic parasites (Farram and Smithyman, 1991). Janoff *et al.* (1988) reported lower levels of IgG, IgM and IgA antibodies to *Giardia* trophozoites, although they did note that infection with this parasite was of little consequence since available therapy is effective.

4. Giardia *reservoirs and control of the disease*

The potential zoonotic transmission of *Giardia* has been discussed above-(Section VII.C) and remains a possibility with implications for public health. The presence of symptomless carriers who are able to infect relatives and others in the community also remains a problem of public health. Treatment of dogs and cats with giardiasis has been recommended "because of the close contact they have with children" (Meyer and Jarroll, 1982; Acha and Szyfres, 1987). Several cases of families with affected pets and symptomless carrier relatives have been reported and highlight the need to undertake therapy in these situations in order to prevent repeated infections (Pancorbo *et al.*, 1985; Cribb and Spracklin, 1986). Conversely, other authors have reported cases from hyperendemic communities in which treatment with highly effective drugs such as tinidazole is soon followed by reinfection and doubt the value of treatment except when clinically indicated (Gilman *et al.*, 1988).

The control of endemic gastrointestinal parasitic infections may also offer a useful entry point for other primary health care activities including health education and nutrition. Intestinal parasite control programmes are regarded as highly desirable since the recipients see the beneficial effects of intervention and learn basic health care facts, and sectors of the health care services become integrated (Anonymous, 1987).

Costs of not having a control programme for giardiasis include those related to nutrition, growth and development, work and productivity, and medical care (Anonymous, 1987). Strategies used to control giardiasis are usually based on epidemiology providing basic information on transmission, the reduction of incidence rates with control of transmission and health education, and a reduction in number of infections through therapy (Anonymous, 1987).

Control measures to counter the high infection rates with *Giardia* in developing regions include those listed in Table 6 (Anonymous, 1987; Boreham, 1987; Islam, 1990). Practical approaches are mainly centred on sanitation control and education about indiscriminate defaecation and on personal hygiene concerning the handling of domestic animals (Anonymous, 1987). The central role of health education in prevention and control programmes should be adapted for each target community with education as a primary focus in each phase of a programme including preparation, implementation and follow-up involving all sectors of the community and health workers (Anonymous, 1987; Halloran *et al.*, 1989).

TABLE 6 *Desirable outcomes and practical approaches to controlling giardiasis*

Desired result	*Practical approach*
Prevention of water-borne transmission	Disinfection of town water supplies Adequate disinfection or filtration of water or boiling in endemic areas Treatment of wastes used as fertilizer
Prevention of food-borne transmission	Effective surveillance and hygiene instruction Screening of food handlers
Reduced morbidity	Individual medical treatment
Prevention and control of epidemics	Improved personal hygiene Improved water quality Quality of health education Treatment of wastes used as fertilizer
Reduction of prevalence	Non-specific hygienic approach, water quality, hand washing, sanitation, food handling instruction
Elimination of reservoirs of giardiasis	Treatment of people with infection who are asymptomatic cyst passers

B. DIAGNOSIS

1. *Traditional methods*

Diagnosis of *Giardia* by traditional methods remains a reliable indicator of giardiasis despite the intermittent nature of cyst excretion by animals and humans (Wolfe, 1990). Thus three faecal samples taken on non-consecutive days are required for reliable assessment of the presence of *Giardia* since a single sample identifies only 50–75% of positive patients (Heymans *et al.*, 1987; Goka *et al.*, 1990; Wolfe, 1990; Adam, 1991). Other methods such as duodenal fluid/biopsy aspirate or the Entero-Test® (adherence of trophozoites to a swallowed nylon line) are far more invasive and costly (Beal *et al.*, 1970; Hall *et al.*, 1988). Environmental surveillance currently employs filtration and concentration methods and microscopy using stains and fluorescent antibodies although many authorities are moving toward the use of immunological tests (see Section VIII.B.2). Comparative trials of traditional methods have been carried out but the relative effectiveness of each is still in some doubt (Rosenthal and Liebman, 1980; Goka *et al.*, 1990). Faecal examination is tedious and relies on visual identification of cysts by light microscopy which may be facilitated by concentration using zinc sulphate or formol-ether and the use of trichrome or iron haematoxylin stains (Baker *et al.*, 1987).

Recent developments using the fluorogenic dye fluorescein diacetate, which is taken up by viable cysts, and propidium iodide, which is taken up by non-viable cysts, may facilitate identification of infective material. While these tests are not 100% accurate they are rapid, can accommodate a large through-put of samples, lend themselves to sorting of viable/non-viable cysts by fluorescence-activated cell sorters, and may be useful in the field for environmental monitoring (Schupp and Erlandsen, 1987; Smith and Smith, 1989). These two dyes have also been shown to be useful quantitative markers of the killing of trophozoites, using two-colour flow cytometry (Heyworth and Pappo, 1989).

2. *Immunological techniques*

The greatest advances in detection of giardiasis have arisen from recent work on development of immunological tests for *Giardia* antigens. Serodiagnostic tests have been relatively unsuccessful, exhibiting problems related to "false positive" identification of old resolved infections, lack of sensitivity, or lack of specificity (Goka *et al.*, 1986; Nash *et al.*, 1987a; Jokipii *et al.*, 1988; McNally *et al.*, 1990).

Giardia antigen detection in faecal specimens has had greater success, leading to immunologically based test systems for detection in the clinic and in the field, and for use with water and other environmental samples (Ungar *et al.*, 1984; Green *et al.*, 1985; Sauch, 1985; Vinayak *et al.*, 1985; Rosoff and Stibbs, 1986a,b; Nash *et al.*, 1987b; Janoff *et al.*, 1989; Engelkirk and Pickering, 1990; Hopkins *et al.*, 1993). These systems have a sensitivity and specificity of 90–100% and are now available in several forms as commercial enzyme-linked immunosorbent assay (ELISA) kits. These tests are rapidly developing as more sensitive and reliable tests than traditional methods such as microscopy and more invasive diagnostic procedures (Janoff *et al.*, 1989; Dutt and Vinayak, 1990; Goldin *et al.*, 1990; Isaac-Renton, 1991). We have recently evaluated a capture ELISA which recognizes *Giardia* antigens of molecular mass 30 and 66 kDa (CELISA® Cellabs, Sydney, Australia) in cases of giardiasis from both metropolitan and remote Aboriginal communities in Western Australia. These antigens have the same molecular mass as the immunodominant antigens recognized by sera from *Giardia*-infected humans and rabbits immunized with *Giardia* cysts and trophozoites (Taylor and Wenman, 1987; Janoff *et al.*, 1988, 1989). Antigens of this size may be similar to those observed by Crossley and Holberton (1983), thought to represent the giardins of the ventral disc. We found a specificity of 91% for human samples and 95% for canine samples, with a sensitivity of 100% for human samples but only 55–64% for canine samples (Hopkins *et al.*, 1993). Clear results were obtained with ease under difficult field conditions reading visually and, later, spectrophotometrically with equivalent results.

Reiner *et al.* (1989) have shown that the most prominent specific cyst antigens are from a hetero-disperse group with molecular masses of 21–39 kDa and 66–103 kDa, are protein rather than carbohydrate, and are located in encystation specific vesicles (ESV) early in encystation but throughout the whole cyst later. Torian *et al.* (1984) had shown earlier that tubulins were likely candidates for *Giardia* antigens and Holberton and Ward (1981) have reported that major cytoskeletal components in *Giardia* are composed of tubulin and giardins. Major surface antigens with differing apparent molecular masses (82 kDa) have also been identified and may represent adhesion molecules or immune components (Einfeld and Stibbs, 1984). One of the most interesting *Giardia*-specific antigens identified is the GSA 65 antigen of *Giardia* cysts reported by Rosoff and Stibbs (1986a). This antigen may relate to those found earlier by Craft and Nelson (1982), but it appears to differ from those identified by Vinayak *et al.* (1985), which showed reactivity against trophozoite material, and by Stibbs *et al.* (1990), which showed a lack of correlation with GSA 65 in human and gerbil faeces when studied by Western immunoblotting. However, GSA 65 is stable under a variety of conditions and appears to be a glycosylated glycoprotein (Rosoff and Stibbs, 1986b). These findings have now been translated to a commercially available ELISA which demonstrated 96 and 100% sensitivity and specificity respectively, compared with 74 and 100% for microscopical investigation (Rosoff *et al.*, 1989). However, Moss *et al.* (1990) warn that coprodiagnostic assays based on monoclonal antibodies will need to include many isolates of *Giardia* because of antigen variation between strains.

3. *Alternative approaches*

Barker (1990) has reviewed the development of diagnostic techniques for parasitic infections based on DNA probes and the use of the polymerase chain reaction (PCR) to amplify target DNA, and this method has been used to detect and differentiate live from dead cysts of *Giardia* (Mahbubani *et al.*, 1991). However, sensitivity has been low and, until recently, only 1 ng or more of *Giardia* DNA (equivalent to 5×10^3 trophozoites) has been detectable (Butcher and Farthing, 1989). This technique has produced a sensitive, specific assay for *Giardia*, based on detection of the giardin gene by PCR. Further development of this approach has led to an assay which can detect a single *Giardia* cyst: a powerful tool for environmental detection (Abbaszadegan *et al.*, 1991). In addition, the PCR technique is reported to be able to distinguish between pathogenic and non-pathogenic *Giardia* strains (Mahbubani *et al.*, 1992).

C. DRUG EFFECTS AND TREATMENT

1. *Laboratory tests of drug efficacy*

The primary tests for the assessment of the killing effect of drugs on *Giardia in vitro* are those based on morphological changes which occur before death of trophozoites, histological appearance and absence of flagellar activity (Belosevic and Faubert, 1986; Rohrer *et al.*, 1986). Subculturing of *Giardia* from cultures exposed to drugs has also been used extensively, as has growth of *Giardia* trophozoites and counting techniques (Favennec *et al.*, 1992). A semi-solid agarose phase has also been used as a test system with moderate success and, although the diffusion and distribution of drugs may be restricted (Gillin and Diamond, 1981; Smith *et al.*, 1982), general assessment of drug effects in such a system correlates quite well with results from a thymidine uptake assay (Boreham *et al.*, 1984).

Greater reliability has followed the use of vital dyes which may be excluded or taken up by non-viable trophozoites. Such dyes include trypan blue, neutral red, fluorescein diacetate, acridine orange and ethidium bromide (see also Section VIII.B.1) (Hill *et al.*, 1984; Farthing and Inge, 1986;

Deguchi *et al*, 1987; Heyworth and Pappo, 1987, 1989; Magne *et al.*, 1991).

Adherence tests, based on the ability of viable *Giardia* trophozoites to adhere to culture vessel or cell surfaces, are also reliable (Favennec *et al.*, 1990; Meloni *et al.*, 1990; Magne *et al.*, 1991; Favennec *et al.*, 1992). Indeed, it has been suggested that adherence inhibition may be a more efficient mechanism for host resistance to *Giardia* than killing and that the adherence mechanism of *Giardia* is a feasible and more efficient target for therapy (Crouch *et al.*, 1991b).

Attempts at further accuracy have been made with the use of labelled markers including tritium-labelled thymidine uptake and ^{51}Cr release (Smith *et al.*, 1982; Boreham *et al.*, 1984; Aggarwal and Nash, 1986; Kanwar *et al.*, 1986). Thymidine uptake is a simple objective measure of anti-giardial activity and minimum inhibitory concentration values obtained with this and standard viability and morphology assays were indistinguishable (Inge and Farthing, 1987). However, these assays involve the use of radioactive substances and many of the viability assays or adherence tests, although only semiquantitative, provide excellent, simple measures of drug effects (Meloni *et al.*, 1990). In some situations the thymidine uptake test fails to detect cytotoxicity (Aggarwal and Nash, 1986) and it may also be possible for some drugs to alter parasite DNA, hence its requirement for thymidine, giving the appearance of *Giardia* cytotoxicity while the trophozoites remain viable. However, the test is reported as simple and reliable in the assessment of drug action *in vitro* and has been used quite successfully (Boreham *et al.*, 1984).

2. *Drug efficacy, adverse drug reactions and emerging therapy*

The several drugs commonly used for therapy of giardiasis worldwide have in general been effective. However, in those regions of the world where *Giardia* is endemic there are some drawbacks associated with long dose regimes, poor palatability and potential toxicity, especially in children who may require treatment often because of reinfection.

(a) *Quinacrine*. In a study of *Giardia* cyst and trophozoite respiration, quinacrine was shown to reduce cyst viability but not by an effect on respiration (Paget *et al.*, 1989a). This drug also kills trophozoites *in vitro*, reduces clonal growth, causes morphological changes, and reduces excystation rates after exposure *in vitro* and with cysts from treated patients (Gillin and Diamond, 1981; Namgung *et al.*, 1985).

Clinical evaluations indicate that quinacrine is effective but causes severe adverse reactions such as vomiting, intestinal disturbance, occasional ocular toxicity, headache and dizziness, and discoloration of skin and urine

(Webster, 1990). A particularly disturbing side-effect is the 0.1–1.5% incidence of toxic psychosis induced by this drug (Lindenmayer and Vargas, 1981). Quinacrine may also cause haemolysis in patients deficient in glucose 6-phosphate dehydrogenase, and has caused disulfiram-like reactions (Pratt and Fekety, 1986). Significantly, the bitter taste and vomiting may lead to treatment non-compliance (Namgung *et al.*, 1985), as does the usual requirement for a five to seven day dose regime (Davidson, 1984).

Quinacrine appears to act by binding to DNA and inhibiting nucleic acid synthesis, its selective toxicity being a function of different relative uptake into host and parasite (O'Brien *et al.*, 1966; Looker *et al.*, 1986; Pratt and Fekety, 1986; Webster, 1990). Quinacrine has also been used in the dog concurrently with metronidazole in a regime which was more successful than any of the standard antigiardial drugs alone (Zimmer and Burrington, 1986).

(b) *Furazolidone.* Furazolidone is more effective *in vitro* and *in vivo* and better tolerated than quinacrine, and induces better patient compliance (Craft *et al.*, 1981; Crouch *et al.*, 1986). Furazolidone is the preferred drug for children in many countries since it lacks the bitter taste and subsequent gastrointestinal disturbance of quinacrine. Drawbacks with its use include the volume required, the need for a 7–10 day treatment regime and adverse drug reactions, which are not infrequent. These include nausea and vomiting, diarrhoea, mild haemolysis in patients deficient in glucose-6-phosphate dehydrogenase, and disulfiram-like reactions (Dietrich and Hellerman, 1963; Klein, 1964; Chamberlain, 1976). Of particular concern are the reports of mutagenicity in bacteria and tumours produced in rats, although no such effect has yet been reported for humans (Anonymous, 1976; McCalla, 1979).

Furazolidone is thought to undergo reductive activation in parasites to compounds which damage DNA either directly or via oxygen radicals and its selective toxicity is related to its relative lack of absorption and selective activation by the parasite (McCalla, 1979; Pratt and Fekety, 1986). An alternative speculation is that furazolidone acts similarly to the 5-nitroimidazoles as an alternative electron acceptor in the anaerobic electron transport system (Crouch *et al.*, 1986), although other authors believe that the nitrofurantoin mechanism of action is different from that of the 5-nitroimidazoles (Edwards *et al.*, 1973).

(c) *Paromomycin.* This drug is a member of the large, diverse aminoglycoside group of anti-bacterial agents and exhibits anti-giardial activity *in vitro* (Gillin and Diamond, 1981; Gordts *et al.*, 1985, 1987; Boreham *et al.*, 1986; Edlind, 1989a). Clinical efficacy has also been demonstrated and the drug is administered over 10 days (Carter *et al.*, 1962; Kreutner *et al.*, 1981).

Adverse reactions to paromomycin include ulceration of the gastrointestinal tract and potential renal toxicity when the drug is given parenterally (Davidson, 1990). Selective toxicity is achieved by a lack of absorption from the gut of the host and a mode of action which relies on an unusual size and sequence of *Giardia* rRNA (Edlind, 1989a). It appears that paromomycin is an acceptable drug for treatment of giardiasis in patients for whom the other standard drugs are too toxic (e.g. in pregnancy), although there is little clinical experience with its use (Kovacs and Masur, 1986).

(d) *5-Nitroimidazoles.* Metronidazole has been repeatedly shown to be effective against *Giardia* trophozoites *in vitro* and *in vivo* and is routinely used in the therapy of giardiasis (Boreham *et al.*, 1986; Paget *et al.*, 1989a; Webster, 1990). However, cysts of *Giardia* are unaffected by the drug and it has been suggested that patients treated with metronidazole would probably pass viable cysts for several days after treatment (Paget *et al.*, 1989a). Metronidazole is well absorbed following administration and is widely distributed throughout the body (Ralph, 1983).

Selective toxicity is achieved in anaerobic organisms having electron transport proteins with low redox potentials, in which metronidazole acts as an electron sink, drawing electrons from the reduced electron transfer protein ferredoxin. The ensuing nitro group reduction of the drug is a necessary feature of its action and the change in its structure also allows the concentration gradient to be retained so that drug accumulates intracellularly (Ings *et al.*, 1974; Pratt and Fekety, 1986). *E. histolytica* and *T. vaginalis* metabolize metronidazole to N-(2-hydroxyethyl)oxamic acid and acetamide, although a drug-resistant strain of *Bacillus fragilis* does this poorly (Koch and Goldman, 1979; Chrystal *et al.*, 1980; Beaulieu *et al.*, 1981; McLafferty *et al.*, 1982). The proposed final step in the mode of action of this drug is the covalent binding of a cytotoxic intermediate compound with macromolecules including DNA within the parasite cell (LaRusso *et al.*, 1977; Knight *et al.*, 1978; Muller, 1983).

Disadvantages in the use of metronidazole include the 5-day dose regime required to eliminate *Giardia*, an unpleasant taste, gastrointestinal disturbances, occasional pancreatitis, vertigo and headache, and a disulfiram-like action (Roe, 1977; Hager *et al.*, 1980; Ralph, 1983; Friedman and Selby, 1990). Central nervous system toxicity has been seen with high doses of the drug and transient leukaemia has occurred (Kusumi *et al.*, 1980).

Although no similar effect has been seen in humans, metronidazole is mutagenic in bacteria (Rosenkranz and Speck, 1975; Lindmark and Muller, 1976) and produces lung tumours in mice and hepatic and mammary tumours in rats with prolonged high dose regimes (Legator *et al.*, 1975; Rustia and Shubik, 1979). However, mammalian mutagenicity tests have been negative (Bost, 1977), the drug has been used in pregnant women with no evidence of malformations (Peterson *et al.*, 1966), and a large retrospective study has shown no tumorigenic effect in patients receiving metronidazole (Friedman, 1980). Despite these findings, the use of metronidazole in giardiasis in regions where the disease is endemic is worrying since children in these areas often require repeated dosing because of reinfection (J. A. Reynoldson and R. C. A. Thompson, unpublished observations). An additional problem with nitroimidazoles, often neglected although important in underdeveloped regions, is the unpleasant taste of both metronidazole and tinidazole which may lead to poor patient compliance. However, there is a tasteless form of metronidazole available (benzoylmetronidazole) which has not been widely used against *Giardia* and which exhibits slower absorption, more prolonged plasma concentrations and slower elimination (Homeida *et al.*, 1986).

As a result of the success of metronidazole against anaerobic infections, many related compounds have now been produced. Relatively few of these are available for clinical use but several have been tested against *Giardia*. In contrast to metronidazole, tinidazole is 100% effective following a single oral dose (Jokipii and Jokipii, 1979, 1985; Speelman, 1985; Webster, 1990). It is interesting that several studies confirm that concentration at the site of infection is paramount rather than serum drug levels (Jokipii and Jokipii, 1979, 1982). However, tinidazole may be superior to metronidazole since plasma concentrations normally achieved with the former are approximately 1.5 times the 4-h exposure minimum lethal concentration values obtained *in vitro* (Forsgren and Wallin, 1974; Jokipii and Jokipii, 1979; Mattila *et al.*, 1983; Meloni *et al.*, 1990). Failure of therapy ascribed to possibly poor intestinal absorption of metronidazole has been reported (Kane *et al.*, 1961).

Boreham *et al.* (1986) have reported high activity of a series of nitroimidazoles including secnidazole, which has also been shown to be effective against *Giardia* with a single dose (Cimerman *et al.*, 1989, 1990). Secnidazole has a long plasma half-life (17 h, Cimerman *et al.*, 1989) and mild side-effects comprising nausea, anorexia and abdominal pain in approximately 5% of children treated with a single oral dose, with the incidence for all symptoms after treatment being lower than that before treatment (Katz *et al.*, 1989).

(e) *Emerging therapeutics—the benzimidazoles.* De Souza *et al.* (1973) noted that mebendazole was ineffective against *Giardia*, but Hutchison *et al.* (1975) recognized that the benzimidazoles could be effective anti-giardial agents when they noted that mebendazole (100 mg every 12 h for 3 days) resulted in a cure rate of 37% against *Giardia* while examining its anti-nematodal activity in patients. Meyer and Radulescu (1979) reported similar findings and there have been a number of positive clinical reports since that time of the use of mebendazole and albendazole (Zhong *et al.*, 1986; Cheng-i, 1988; Al-Waili *et al.*, 1988; Al-Waili, 1990). The claims for mebendazole have been disputed by Gascon *et al.* (1989), who administered mebendazole in divided doses over 1 day but also failed to achieve a high cure rate with the same dose regime as Al-Waili (Gascon *et al.*, 1990). Furthermore, 600 mg per day of mebendazole in divided doses was effective against the protozoan *T. hominis*, and mebendazole was effective against *T. vaginalis* (Al-Waili, 1987; Sears and O'Hare, 1988). More systematic laboratory work has now shown the potential of the benzimidazoles against *Giardia in vitro* using both mebendazole and albendazole (Meloni *et al.*, 1990; Edlind *et al.*, 1990a; Nohynkova, 1990; Reynoldson *et al.*, 1991a, 1992a). Magne *et al.* (1991) supported our earlier hypothesis that albendazole might be affecting the microtubular cytoskeleton by an interaction with tubulin (Meloni *et al.*, 1990). Support for the broader efficacy of the benzimidazoles against *Giardia* has come from studies with mebendazole, albendazole, nocodazole, oxfendazole, fenbendazole and thiabendazole, all effective against *Giardia* (Edlind *et al.*, 1990a; Reynoldson *et al.*, 1992a; Morgan *et al.*, 1993). Albendazole has now also been shown to be effective against *Giardia in vivo* (Reynoldson *et al.*, 1991b) and to be a clinically effective anti-giardial agent (Hall and Nahar, 1993; Reynoldson *et al.*, 1992b).

The advantages of the benzimidazoles in the treatment of giardiasis are their relatively poor absorption from the intestine, low levels of adverse reactions, a broad spectrum of action against nematodes, cestodes and trematodes, and the production of active metabolites by a number of these agents (Marriner and Bogan, 1980; Firth, 1983; Morris *et al.*, 1983; Saimot *et al.*, 1983; Botero, 1986). Although the benzimidazoles cause some teratogenesis and embryo toxicity in laboratory animals, mutagenicity tests show a lack of genotoxicity with albendazole and there is no evidence of long-term high dose carcinogenicity in male or female rats (Botero, 1986; Theodorides and Daly, 1989). Similar lack of toxicity has been noted with other benzimidazoles and both mebendazole and albendazole have been used at high dose rates for long periods to treat human hydatidosis (Bryceson *et al.*, 1982a,b; Morris *et al.*, 1983; Saimot *et al.*, 1983). Flubendazole should be tested more extensively against *Giardia* since it has no reported teratogenic effect, such as is observed with other benzimidazoles in rodents, and has been used in humans to treat echinococcosis (Schantz, 1982; Botero, 1986).

The potential for use of the benzimidazoles and their postulated mode of action involving interaction with the tubulins and/or giardins of the cytoskeleton of *Giardia* and other parasites have been reviewed recently (Lacey, 1988; Reynoldson *et al.*, 1991a, 1992a). The benzimidazoles appear to bind to the colchicine site on tubulin causing disruption of microtubule assembly and disassembly (Lacey, 1988). As described in Section II.C, recent studies have shown that the microtubule-based cytoskeleton and microtubule-associated proteins of protozoa represent important potential targets for chemotherapy (De Souza, 1984; Russell and Dubremetz, 1986; Seebeck *et al.*, 1988, 1990; Peattie *et al.*, 1989; MacRae and Gull, 1990; Peattie, 1990). Although the function of the ventral disc in *Giardia* remains unclear, the integral role of microtubules and the giardins in its structure, along with the presence of actin, myosin, α-actinin and tropomyosin (Feely *et al.*, 1982) suggest a crucial role for the integrity of the parasite. The contention that tubulin binding of benzimidazoles leads to their anthelmintic activity is well supported (Borgers and De Nollin, 1975; Lacey, 1985, 1988, 1990; Sangster *et al.*, 1985; Lacey and Prichard, 1986). An alternative hypothesis is that the benzimidazoles disrupt parasite bioenergetics (McCracken and Stillwell, 1991). However, additional evidence for the tubulin binding hypothesis comes from demonstrations of reduced tubulin binding in resistant worms and the fact that resistance in *Aspergillus nidulans* and *Caenorhabditis elegans* results from mutations affecting β-tubulin (Sheir-Neiss *et al.*, 1978; Lacey, 1985; Sangster *et al.*, 1985; Lacey and Prichard, 1986; Lacey and Snowdon, 1988; Woods *et al.*, 1989). With respect to *Giardia*, it is notable that flagellar activity was little affected by the benzimidazoles despite the presence of tubulin in flagella; thus it is tempting to postulate that these compounds may have greater affinity for the ventral disc tubulin or the related protein giardin, resulting in the observed detachment from the site of adherence (Edlind *et al.*, 1990a; Meloni *et al.*, 1990). Different subunit forms of tubulin are present in protozoa and the higher and lower eukaryotes, possibly explaining the selectivity for disc tubulin and for the parasite (Crossley and Holberton, 1983). Fundamental structural and antigenic differences have also been found between the tubulins and giardins of the ventral disc and flagella in *Giardia* (Clark and Holberton, 1988). Thus agents known to bind to tubulin may offer opportunities for further development of anti-tubulin anti-protozoal chemotherapy (Hamel and Lin, 1983; McGown and Fox, 1989).

(f) *Other active compounds and potential targets for therapy*. The wide range of side effects observed with the standard anti-giardial agents has led to an extensive search for less toxic compounds.

Recent data concerning the importance of arginine in *Giardia* metabolism, and also the production of alanine as an end product anaerobically (see Section V and Edwards *et al.*, 1989; Paget *et al.*, 1990; Schofield *et al.*, 1990), may provide a stimulus in the search for new drug targets related to this extensive utilization of protein or amino acids by *Giardia*.

Purine and pyrimidine uptake may also offer lethal targets since *Giardia* lacks the ability to synthesize these compounds and pyrimidine analogues inhibited pyrimidine uptake (Lindmark and Jarroll, 1982; Aldritt *et al.*, 1985; Jarroll *et al.*, 1987). Similarly, purine salvage has been shown to be essential for *Giardia* viability and this uptake pathway, unique to eukaryotic cells, also offers possibilities (Baum *et al.*, 1989). Furthermore, incorporation of free purine bases into nucleotides requires a nucleoside hydrolase step followed by a purine phosphoribosyltransferase, a potential target for chemotherapy as shown by the reduced viability of *Giardia* in the presence of guanine arabinoside and adenosine analogues (Berens and Marr, 1986; Miller *et al.*, 1987; Marr, 1991).

Numerous compounds have been tested against *Giardia* trophozoites in culture with mixed success. Crouch *et al.* (1986) reported a lack of effect of most anti-malarial agents, diaminopyrimidines, sulphonamides, acetylspiramycin, 5-fluorocytosine and rifampin while the nitroimidazoles, several tetracyclines and furazolidone were effective. The potential of the tetracyclines against *Giardia* has been further explored by Edlind (1989b), identifying increased activity with greater lipophilicity. Similarly the aminoglycoside antibiotics have been assessed with some success, as described above, by Edlind (1989a). Other anti-parasitic agents have been screened against *Giardia*, with positive results obtained for bithionol, dichlorophene and hexachlorophene (Takeuchi *et al.*, 1985).

Recent work has shown the presence of methylthioribose kinase of the methionine salvage pathway in *Giardia*, which is absent from mammalian cells. Thus analogues of methylthioribose may be useful anti-giardial agents (Riscoe *et al.*, 1988, 1991).

Calmodulin, the common primary intracellular receptor for calcium in eukaryotic cells, is present in *Giardia* and its inhibition has led to reduced growth of the parasite (Weinbach *et al.*, 1985). Similarities to mammalian calmodulin may present future problems through lack of selective toxicity (De Lourdes Munoz *et al.*, 1987). Other novel agents to be tested include the plant alkaloid berberine, which was reported as effective by Kaneda *et al.* (1991).

An alternative approach to finding new anti-giardial compounds is to use drug combinations which may have a potentiation effect. Thus dyadic combinations of azithramycin, doxycycline, mefloquine, tinidazole and furazolidone showed synergism *in vitro*, offering improvements over existing therapies (Crouch *et al.*, 1990).

3. *Benefits of chemotherapy*

Chemotherapy has become very effective as a tool for controlling giardiasis and other parasite infections in underdeveloped countries with the ability to treat individual infections promptly, minimizing spread of infection. Mass medication based on survey results in defined regions along with sanitation and educational measures can also be beneficial (Anonymous, 1987). With this role of chemotherapy in mind selection of appropriate drugs, their efficacy in relation to identified problems, and their safety are critical factors in choice of agent, particularly if drug use is contemplated in areas with high prevalence and reinfection rates where children are being exposed to drugs regularly (Anonymous, 1987). Therapy may also be an important consideration with particular isolates which have different antigenic properties and which lead to repeated infections (Aggarwal and Nash, 1987b).

It is important to recognize that anti-parasitic drug use may also have benefits in terms of growth rate, clinical and nutritional status, educational participation and performance, pubertal development, fitness and productivity in addition to providing a parasitological cure (Cole *et al.*, 1982, Pamba *et al.*, 1989; Stephenson *et al.*, 1989, 1990, 1991; Bundy, 1990; Crompton and Stephenson, 1990; Adams *et al.*, 1991).

4. *Treatment failure*

Although resistance *per se* has not been described, treatment failures occur with the most common anti-giardial therapeutics and are often treated with repeat therapy or an alternative drug (Mendelson, 1980; Davidson, 1984; Geary *et al.*, 1986). There is also evidence of differential sensitivity of strains of *Giardia* to drugs (Boreham *et al.*, 1987), although isolates from patients not cured clinically have not proven to be drug resistant *in vitro* (Gordts *et al.*, 1985, 1987; Geary *et al.*, 1986; McIntyre *et al.*, 1986). Treatment failure could therefore be related to host immune mechanisms (Doenhoff *et al.*, 1991). However, in one study, high resistance *in vitro* was shown to correlate with treatment failure in the patient from whom the strain had been isolated (Majewska *et al.*, 1991).

Mechanisms underlying tolerance of *Giardia* to metronidazole appear to include a lowered uptake and ability to reduce the drug and lowered pyruvate : ferredoxin oxidoreductase activity in both naturally "resistant" strains and in a strain selected for "resistance" (Muller, 1983; Boreham *et al.*, 1988; Smith *et al.*, 1988). However, a similar mechanism did not pertain to tinidazole, and furazolidone "resistance" took a different form, depending on efficiency of thiol cycling, a feature seen only with a 7–8 fold increase in tolerance to metronidazole (Smith *et al.*, 1988). Recent evidence suggests that metronidazole tolerance is associated with DNA changes both at chromosomal and repetitive DNA levels (Upcroft *et al.*, 1989b, 1990).

IX. Conclusions

The last 10 years have seen a most fruitful and productive period of research on *Giardia* and giardiasis. We hope that this review has highlighted the numerous developments which have taken place over this period and that it will serve as a useful resource for workers in this field. At the start of this review we listed a number of outstanding questions concerning *Giardia* and giardiasis (Table 1). It is clear from the preceding sections that we are close to answering a number of these questions.

Major advances have been made in our understanding of the mode of attachment of *Giardia*, and we can now accept one of a number of hypotheses which were put forward over the last 10 years. Significant advances have also been made concerning the pathogenesis of *Giardia* infection and in this regard the utilization of appropriate animal models has proved invaluable. Various intracellular symbionts have been demonstrated in *Giardia* and, of these, considerable progress has been made in characterizing the ubiquitous GLV (*Giardiavirus*) and its relationships with its protozoan host.

We still know remarkably little about the immunology of host responses in *Giardia* infections. In contrast, knowledge of the antigenic characteristics of *Giardia* has advanced significantly, although the advantages to the parasite of possessing variable surface antigens remain to be determined. Exciting developments have been made in studies on the biochemistry of *Giardia*, particularly in the area of energy metabolism and the organism's substrate requirements. It is to be hoped that *Giardia* biochemistry will become a growing area of interest over the next few years, since such research is an essential prerequisite in the search for new chemotherapeutic agents. Although differences in drug tolerance have been demonstrated between isolates of *Giardia*, complete resistance to currently used anti-giardial drugs does not appear to be a problem. However, there are drawbacks to the use of some of these drugs and the recent evidence of the anti-giardial activity of benzimidazoles augers well for a new approach to the chemotherapy of giardiasis. In addition, studies on the mode of action of benzimidazole drugs against *Giardia* may yield new information about the cytoskeletal proteins of the organism.

Major advances have been made in our knowledge of the nature and extent of genetic variation in *Giardia*, but more work is required to determine the epidemiological significance of this variation. In addition, we as yet know little about the origin and maintenance of the extensive genetic diversity exhibited by *Giardia*. The phylogenetic affinities of *Giardia* have been the subject of much recent discussion in the light of RNA sequence data. However, the interpretation of these data has been questioned and other phylogenetic analyses question the assumption of the primitive phylogenetic position of *Giardia*.

Few authorities would today disagree with the inclusion of *Giardia* in a list of zoonotic disease agents. However, definitive proof of zoonotic transmission in nature has still to be obtained. It is to be expected that the further application of molecular characterization procedures may provide useful data in this area.

Limited progress has been made in a number of other areas highlighted in Table 1, particularly with respect to reproductive behaviour, ploidy, nuclear function and determinants of susceptibility and virulence. The current revolution in molecular genetics will have a great bearing on the resolution of these outstanding questions and by the time of the next review on *Giardia* and giardiasis we are confident that not only will answers have been forthcoming but that important new questions will have arisen.

Acknowledgements

Our research is supported by grants from the National Health and Medical-Research Council of Australia, SmithKline Beecham, The Australian Research Council and the World Health Organization. We thank Nicolette Binz, Bruno Meloni and Richard Hopkins for critical comments on the manuscript, Sue Lyons and Luba Sanderson for typing it, and Jim Cooper and his colleagues for the artwork.

References

Abbaszadegan, M., Gerba, C.P. and Rose, J.B. (1991). Detection of *Giardia* cysts with a cDNA probe and applications to water samples. *Applied and Environmental Microbiology* **57**, 927–931.

Acha, P.N. and Szyfres, B. (1987). "Zoonoses and communicable diseases common to man and animals". Pan American Health Organization, Washington, Scientific Publication No. 503.

Adam, R.D. (1991). The biology of *Giardia* spp. *Microbiological Reviews* **55**, 706–732.

Adam, R.D., Nash, T.E. and Wellems, T.E. (1988a). The *Giardia lamblia* trophozoite contains sets of closely related chromosomes. *Nucleic Acids Research* **16**, 4555–4567.

Adam, R.D., Aggarwal, A., Lal, A.A., de la Cruz, F.P., McCuchan, T. and Nash, T.E. (1988b). Antigenic variation of a cysteine-rich protein in *Giardia lamblia*. *Journal of Experimental Medicine* **167**, 109–118.

Adam, R.D., Nash, T.E. and Wellems, T.E. (1991). Telomeric location of *Giardia* rDNA genes. *Molecular and Cell Biology* **11**, 3326–3330.

Adams, E.J., Stephenson, L.S., Latham, M.C. and Kinoti, S.N. (1991). Albendazole treatment improves growth and physical activity of Kenyan school children with hookworm, *T. trichiura* and *A. lumbricoides* infections. *American Journal of Clinical Nutrition* **53** (3), 30.

Aggarwal, A. and Nash, T.E. (1986). Lack of cellular cytotoxicity by human mononuclear cells to *Giardia*. *Journal of Immunology* **136**, 3486–3488.

Aggarwal, A. and Nash, T.E. (1987a). *Giardia lamblia*: RNA translation products. *Experimental Parasitology* **64**, 336–341.

Aggarwal, A. and Nash, T.E. (1987b). Comparison of two antigenically distinct *Giardia lamblia* isolates in gerbils. *American Journal of Tropical Medicine and Hygiene* **36**, 325–332.

Aggarwal, A., Romans, P., de la Cruz, V.F. and Nash, T.E. (1988). Conserved sequences of the hsp gene family. *In* "Advances in *Giardia* Research" (P.M. Wallis and B.R. Hammond, eds), pp. 173–175. University of Calgary Press, Calgary.

Aggarwal, A., Adam, R.D. and Nash, T.E. (1989a). Characterization of a 33-kilodalton structural protein of *Giardia lamblia* and localization to the ventral disk. *Infection and Immunity* **57**, 1305–1310.

Aggarwal, A., Merritt, J.W. and Nash, T.E. (1989b). Cysteine-rich variant surface proteins of *Giardia lamblia*. *Molecular and Biochemical Parasitology* **32**, 39–47.

Aggarwal, A., de la Cruz, V.F. and Nash, T.E. (1990). A heat shock protein gene in *Giardia lamblia* unrelated to HSP70. *Nucleic Acids Research* **18**, 3409.

Aldritt, S.M. and Wang, C.C. (1986). Purification and characterisation of guanine phosphoribosyl transferase from *Giardia lamblia*. *Journal of Biological Chemistry* **261**, 8528–8533.

Aldritt, S.M., Tien, P. and Wang, C.C. (1985). Pyrimidine salvage in *Giardia lamblia*. *Journal of Experimental Medicine* **161**, 437–445.

Alonso, R.A. and Peattie, D.A. (1992). Nucleotide sequence of a second alpha giardin gene and molecular analysis of alpha giardin genes and transcripts in *Giardia lamblia*. *Molecular and Biochemical Parasitology* **50**, 95–104.

Al-Waili, N.S. (1987). Mebendazole in *Trichomonas hominis*. *Clinical and Experimental Pharmacology and Physiology* **14**, 679–680.

Al-Waili, N.S. (1990). Mebendazole in giardial infections: inappropriate doses. *Transactions of the Royal Society of Tropical Medicine and Hygiene* **84**, 753–754.

Al-Waili, N.S., Al-Waili, B.H. and Saloom, K.Y. (1988). Therapeutic use of mebendazole in giardial infections. *Transactions of the Royal Society of Tropical Medicine and Hygiene* **82**, 438.

Ament, M.E. and Rubin, C.E. (1972). Relation of giardiasis to abnormal intestinal structure and function in gastrointestinal immunodeficiency syndromes. *Gastroenterology* **62**, 216–226.

Ament, M.E., Ochs, H.D. and Davis, S.D. (1973). Structure and function of the gastrointestinal tract in primary immunodeficiency syndromes: a study of 39 patients. *Medicine* **52**, 227–248.

Ampofo, E., Fox, E.G. and Shaw, C.P. (1991). "*Giardia* and Giardiasis in New Zealand". Environmental Health Unit, Department of Health, Wellington.

Andrews, R.H., Adams, M., Boreham, P.F.L., Mayrhofer, G. and Meloni, B.P. (1989). *Giardia intestinalis*: electrophoretic evidence for a species complex. *International Journal for Parasitology* **19**, 183–190.

Annan, A., Crompton, D.W.T., Walters, D.E. and Arnold, S.E. (1986). An investigation of the prevalence of intestinal parasites in pre-school children in Ghana. *Parasitology* **92**, 209–217.

Anonymous (1976). Treatment of giardiasis. *Medical Letter* **18**, 39.

Anonymous (1987). "Report of a WHO Expert Committee: Prevention and Control of Intestinal Parasitic Infections". World Health Organization, Geneva. Technical Report Series no. 749.

Archibald, S.C., Mitchell, R.W., Upcroft, J.A., Boreham, P.F.L. and Upcroft, P. (1991). Variation between human and animal isolates of *Giardia* as demonstrated by DNA fingerprinting. *International Journal for Parasitology* **21**, 123–124.

Arron, B., Deutch, R.M. and Mirelman, D. (1982). Chitin synthesis inhibitors prevent cyst formation by *Entamoeba* trophozoites. *Biochemical and Biophysical Research Communications* **108**, 815–821.

Baker, D.A., Holberton, D.V. and Marshall, J. (1988). Sequence of a giardin subunit cDNA from *Giardia lamblia. Nucleic Acids Research* **16**, 7177.

Baker, D.G., Strombeck, D.R. and Gerschwin, L.J. (1987). Laboratory diagnosis of *Giardia duodenalis* infection in dogs. *Journal of the American Veterinary Medical Association* **190**, 53–56.

Baker, R.H., jr. (1990). DNA probe diagnosis of parasitic infections. *Experimental Parasitology* **70**, 494–499.

Barnard, R.J. and Jackson, G.J. (1984). *Giardia lamblia.* The transfer of human infections by food. *In* "*Giardia* and Giardiasis" (S.L. Erlandsen and E.A. Meyer, eds), pp. 365–378. Plenum Press, New York.

Barrett, J. (1984). The anaerobic end-products of helminths. *Parasitology* **88**, 179–198.

Barrett, J. and Beis, I. (1973) The redox state of the free nicotinamide-adenine dinucleotide couple in the cytoplasm and mitochondria of muscle tissue from *Ascaris lumbricoides* (Nematoda). *Comparative Biochemistry and Physiology* **44A**, 331–340.

Barrett, J., Mendis, A.H.W. and Butterworth, P.E. (1986). Carbohydrate metabolism in *Brugia pahangi. International Journal for Parasitology* **16**, 465–472.

Bartlett, A.V., Moore, M., Gary, G.W., Starko, K.M., Erben, J.J. and Meredith, B.A. (1985a). Diarrheal illness among infants and toddlers in day care centers. I. Epidemiology and pathogens. *Journal of Pediatrics* **107**, 495–502.

Bartlett, A.V., Moore, M., Gary, G.W., Starko, K.M., Erben, J.J. and Meredith, B.A. (1985b). Diarrheal illness among infants and toddlers in day care centers. II. Comparison with day care homes and households. *Journal of Pediatrics* **107**, 503–509.

Baum, K.F., Berens, R.L., Marr, J.J., Harrington, J.A. and Spector, T. (1989). Purine deoxynucleoside salvage in *Giardia lamblia. Journal of Biological Chemistry* **264**, 21087–21090.

Beal, C.B., Viens, P., Grant, R.G.L. and Hughes, J.M. (1970). A new technique for sampling duodenal contents—demonstration of upper small-bowel pathogens. *American Journal of Tropical Medicine and Hygiene* **19**, 349–352.

Beaulieu, B.B., McLafferty, M.A., Koch, R.K. and Goldman, P. (1981). Metronidazole metabolism in cultures of *Entamoeba histolytica* and *Trichomonas vaginalis. Antimicrobiological Agents and Chemotherapy* **20**, 410.

Beaumont, D.M. and James, O.F.W. (1986). Unsuspected giardiasis as a cause of malnutrition and diarrhoea in the elderly. *British Medical Journal* **293**, 554–555.

Belosevic, M. and Faubert, G.M. (1986). Killing of *Giardia muris* trophozoites *in vitro* by spleen, mesenteric lymph node and peritoneal cells from susceptible and resistant mice. *Immunology* **59**, 269–275.

Bemrick, W.J. (1984). Some perspectives on the transmission of giardiasis. *In*

"*Giardia* and Giardiasis" (S.L. Erlandsen and E.A. Meyer, eds), pp. 379–400. Plenum Press, New York.

Bemrick, W.J. and Erlandsen, S.L. (1988). Giardiasis—is it really a zoonosis? *Parasitology Today* **4**, 69–71.

Berens, R.L. and Marr, J.J. (1986). Adenosine analog metabolism in *Giardia lamblia*. *Biochemical Pharmacology* **35**, 4191–4197.

Bingham, A.K. and Meyer, E.A. (1979). *Giardia* excystation can be induced *in vitro* in acidic solutions. *Nature* **277**, 301–302.

Bingham, A.K., Jarroll, E.L., Meyer, E.A. and Radulescu, S. (1979). *Giardia* sp.: physical factors of excystation *in vitro*, and excystation *vs* eosin exclusion as determinants of viability. *Experimental Parasitology* **47**, 284–291.

Binz, N., Thompson, R.C.A., Meloni, B.P. and Lymbery, A.J. (1991). A simple method for cloning *Giardia duodenalis* from cultures and faecal samples. *Journal of Parasitology* **77**, 627–631.

Binz, N., Thompson, R.C.A., Lymbery, A.J. and Hobbs, R.P. (1992). Comparative studies on the growth dynamics of two genetically distinct isolates of *Giardia duodenalis in vitro*. *International Journal for Parasitology* **22**, 195–202.

Black, R.E., Dykes, A.C., Sinclair, S.P. and Wells, J.G. (1977). Giardiasis in day care centers: evidence of person to person transmission. *Pediatrics* **60**, 486–491.

Blair, R.J. and Weller, P.F. (1987). Uptake and esterification of arachidonic acid by trophozoites of *Giardia lamblia*. *Molecular and Biochemical Parasitology* **25**, 11–18.

Bockman, D.E., and Winborn, W.B., (1968) Electron microscopic localisation of exogenous ferritin within vacuoles of *Giardia muris*. *Journal of Protozoology* **15**, 26–30.

Boothroyd, J.C., Wang, A., Campbell, D.A. and Wang, C.C. (1987). An unusually compact ribosomal DNA repeat in the protozoan *Giardia lamblia*. *Nucleic Acids Research* **15**, 4065–4084.

Boreham, P.F.L. (1987) Transmission of *Giardia* by food and water. *Food Technology in Australia* **39** (2), 61–63.

Boreham, P.F.L. and Shepherd, R.W. (1984). Giardiasis in child-care centres. *Medical Journal of Australia* **141**, 263.

Boreham, P.F.L., Philips, R.E. and Shepherd, R.W. (1984). The sensitivity of *Giardia intestinalis* to drugs *in vitro*. *Journal of Antimicrobial Chemotherapy* **14**, 449–461.

Boreham, P.F.L., Phillips, R.E. and Shepherd, R.W. (1986). The activity of drugs against *Giardia intestinalis* in neonatal mice. *Journal of Antimicrobial Chemotherapy* **18**, 393–398.

Boreham, P.F.L., Phillips, R.E. and Shepherd, R.W. (1987). Heterogeneity in the responses of clones of *Giardia intestinalis* to anti-giardial drugs. *Transactions of the Royal Society of Tropical Medicine and Hygiene* **81**, 406–407.

Boreham P.F.L., Philips, R.E. and Shepherd R.W. (1988). Altered uptake of metronidazole *in vitro* by stocks of *Giardia intestinalis* with different drug sensitivities. *Transactions of the Royal Society of Tropical Medicine and Hygiene* **82**, 104–106.

Boreham, P.F.L., Upcroft, J.A. and Upcroft, P. (1990). Changing approaches to the study of *Giardia* epidemiology: 1681–2000. *International Journal for Parasitology* **20**, 479–487.

Borgers, M. and De Nollin, S. (1975) Ultrastructural changes in *Ascaris suum* intestine after mebendazole treatment *in vivo*. *Journal of Parasitology* **60**, 110–122.

Bost, R.G. (1977). Metronidazole: mammalian mutagenicity. In "Metronidazole.

Proceedings of the International Metronidazole Conference, Montreal" (S.M. Finegold, ed.), pp. 126–131. Excerpta Medica, Amsterdam.

Botoro, D. (1986). Nematode infections of man: intestinal infections. In: "Chemotherapy of Parasitic Diseases" (W.C. Campbell and R.S. Rew, eds), pp. 267–276. Plenum Press, New York.

Boucher, S.E.M. and Gillin, F.D. (1990). Excystation of *in vitro*-derived *Giardia lamblia* cysts. *Infection and Immunity* **58**, 3516–3522.

Brodsky, R.E., Spencer, H.D., jr and Schultz, M.G. (1974). Giardiasis in American travelers to the Soviet Union. *Journal of Infectious Diseases* **130**, 319–323.

Bromiker, R., Korman, S.H., Jor, R., Hardan, I., Naparstek, E., Cohen, P., Ben-Shahar, M., and Engelhard, D. (1989). Severe giardiasis in two patients undergoing bone marrow transplantation. *Bone Marrow Transplant* **4**: 701–703.

Brucker, G. (1989). Prevalence of amoebiasis and giardiasis in severe intestinal disorders in intertropical countries. *In* "Secnidazole: a New Approach in 5-Nitroimidazole Therapy" (N. Katz and A.T. Willis, eds), pp. 3–11. Excerpta Medica, Amsterdam.

Bryceson, A.D.M., Cowie, A.G.A., MacLeod, C., White, S., Edwards, D., Smyth, J.D. and McManus, D.P. (1982a). Experience with mebendazole in the treatment of inoperable hydatid disease in England. *Transactions of the Royal Society of Tropical Medicine and Hygiene* **76**, 510–518.

Bryceson, A.D.M., Woestenborghs, R., Michiels, M. and van den Bossche, H. (1982b). Bioavailability and tolerability of mebendazole in patients with inoperable hydatid disease. *Transactions of the Royal Society of Tropical Medicine and Hygiene* **76**, 563–564.

Bundy, D.A.P. (1990). New initiatives in the control of helminths. *Transactions of the Royal Society of Tropical Medicine and Hygiene* **84**, 467–468.

Buret, A., Gall, D.G., Nation, P.N. and Olson, M.E. (1990a). Intestinal protozoa and epithelial cell kinetics, structure and function. *Parasitology Today* **6**, 375–380.

Buret, A., Gall, D.G., and Olson, M.E. (1990b). Effects of murine giardiasis on growth, intestinal morphology and disaccharidase activity. *Journal of Parasitology* **76**, 403–409.

Buret, A., den Hollander, N., Wallis, P.M., Befus, D. and Olson, M.E. (1990c). Zoonotic potential of giardiasis in domestic ruminants. *Journal of Infectious Diseases* **162**, 231–237.

Buret, A., Gall, D.G. and Olson, M.E. (1991). Growth activities of enzymes in the small intestine, and ultrastructure of microvillous border in gerbils infected with *Giardia duodenalis*. *Parasitology Research* **77**, 109–114.

Butcher, P.D. and Farthing, M.J.G. (1989). DNA probes for the faecal diagnosis of *Giardia lamblia* infections in man. *Biochemical Society Transactions* **17**, 363–364.

Cabib, E. (1981) Chitin: structure, metabolism and regulation of biosynthesis. In "Encyclopedia of Plant Physiology" (W. Tanner and F.A. Lowens, eds), Vol. 13B, pp. 395–415. Springer, New York.

Campbell, S.R., van Keulen, H., Erlandsen, S.L., Senturia, J.B. and Jarroll, E.L. (1990). *Giardia* sp.: comparison of electrophoretic karyotypes. *Experimental Parasitology* **71**, 470–482.

Carlson, J.R., Heyworth, M.F. and Owen, R.L. (1987). T-lymphocyte subsets in nude mice with *Giardia muris* infection. *Thymus* **9**, 189–196.

Carnal, N.W. and Black, C.C. (1979). Pyrophosphate-dependent 6-phosphofructokinase, a new glycolytic enzyme in pineapple leaves. *Biochemical and Biophysical Research Communications* **86**, 20–26.

Carter, C.H., Bayles, A. and Thompson, P.E. (1962). Effects of paromomycin sulfate in man against *Entamoeba histolytica* and other intestinal protozoa. *American Journal of Tropical Medicine and Hygiene* **11**, 448–451.

Castor, S.B. and Lindquist, K.B. (1990). Canine giardiasis in Sweden: no evidence of infectivity to man. *Transactions of the Royal Society of Tropical Medicine and Hygiene* **84**, 249–250.

Cavener, D.R. (1987). Comparison of the consenus sequence flanking translational start sites in *Drosophila* and vertebrates. *Nucleic Acids Research* **15**, 1353–1361.

Chamberlain, R.E. (1976). Chemotherapeutic properties of prominent nitrofurans. *Journal of Antimicrobial Chemotherapy* **2**, 325.

Chavalittamrong, B., Sunlornpoch, V. and Siddhikal, C. (1980). Vitamin A concentration in children with giardiasis. *Southeast Asian Journal of Tropical Medicine and Public Health* **11**, 245–249.

Cheissin, E.M. (1964). Ultrastructure of *Lamblia duodenalis*. *Journal of Parasitology* **15**, 26–30.

Cheng-i, W. (1988). Parasitic diarrhoeas in China. *Parasitology Today* **4**, 284–286.

Chrystal, E.J.T., Koch, R.L., McLafferty, M.A. and Goldman, P. (1980). Relationship between metronidazole metabolism and bactericidal activity. *Antimicrobial Agents and Chemotherapy* **18**, 566.

Cibulskis, R.E. (1988). Origins and organization of genetic diversity in natural populations of *Trypanosoma brucei*. *Parasitology* **96**, 303–322.

Čigan, A.M. and Donahue, T.F. (1987). Sequence and structural features associated with translational initiator regions in yeast—a review. *Gene* **59**, 1–18.

Cimerman, B., Boruchovski, H., Cury, F.M. Bichued, L.M. and Ieiri, A. (1989). A comparative study of secnidazole and metronidazole in the treatment of giardiasis. *In* "Secnidazole: a New Approach in 5-Nitroimidazole Therapy" (N. Katz and A.T. Willis, eds), pp. 28–34. Excerpta Medica, Amsterdam.

Cimerman, B., Katz, N., Zingano, A.G., Zingano, R.G. and Rocha, R.S. (1990). Treatment of intestinal giardiasis, with a single does of secnidazole. *Bulletin de la Société Française de Parasitologie* **8**, 423.

Clark, J.T. and Holberton, D.V. (1988). Triton-labile antigens in flagella isolated from *Giardia lamblia*. *Parasitology Research* **74**, 415–423.

Cole, T.J., Sohair, I., Salem, A.S., Hafez, O.M. Galal, and Massoud, A. (1982). Plasma albumin, parasitic infection and pubertal development in Egyptian boys. *Transactions of the Royal Society of Tropical Medicine and Hygiene* **76**, 17–20.

Collins, G.H., Pope, S.E., Griffin, D.L., Walker, J. and Connor, G. (1987). Diagnosis and prevalence of *Giardia* spp. in dogs and cats. *Australian Veterinary Journal* **64**, 89–90.

Cordingley, F.T. and Crawford, G.P.M. (1986). *Giardia* infection causes vitamin B_{12} deficiency. *Australian and New Zealand Journal of Medicine* **16**, 78–79.

Crabtree, B. and Newsholme, E.A. (1972). The activities of phosphorylase, hexokinase, phosphofructokinase, lactate dehydrogenase, and glycerol-3-phosphate dehydrogenase in muscles from vertebrates and invertebrates. *Biochemical Journal* **126**, 49–58.

Craft, J.C. and Nelson, J.D. (1982). Diagnosis of giardiasis by counterimmunoelectrophoresis of feces. *Journal of Infectious Diseases* **145**, 499–504.

Craft, J.C., Murphy, T. and Nelson, J.D. (1981). Furazolidone and quinacrine. *American Journal of Diseases of Children* **135**, 164–166.

Craun, G.F. (1986). Water-borne giardiasis in the United States 1965–1984. *Lancet* **ii**, 513–514.

Craun, G.F. (1990). Water-borne giardiasis. *In* "Giardiasis" (E.A. Meyer, ed.), pp. 267–293. Elsevier, Amsterdam.

Cribb, A.E. and Spracklin, D. (1986). Giardiasis in a home. *Canadian Veterinary Journal* **27**, 169.

Crompton, D.W.T. and Stephenson, L.S. (1990). Hookworm infection, nutritional status and productivity. *In* "Hookworm Disease: Current Status and New Directions" (G.A. Schad and K.S. Warren, eds), pp. 231–264. Taylor and Francis, London.

Crossley, R. and Holberton, D.V. (1983). Characterization of the proteins from the cytoskeleton of *Giardia lamblia. Journal of Cell Science* **59**, 81–103.

Crossley, R. and Holberton, D.V. (1985). Assembly of 2.5 nm filaments from giardin, a protein associated with cytoskeletal microtubules in *Giardia. Journal of Cell Science* **78**, 205–231.

Crossley, R., Marshall, J., Clark, J.T. and Holberton, D.V. (1986). Immunocytochemical differentiation of microtubules in the cytoskeleton of *Giardia lamblia* using monoclonal antibodies to alpha-tubulin and polyclonal antibodies to associated low molecular weight proteins. *Journal of Cell Science* **80**, 233–252.

Crouch, A.A., Seow, W.K. and Thong, Y.H. (1986). Effect of twenty-three chemotherapeutic agents on the adherence and growth of *Giardia lamblia in vitro. Transactions of the Royal Society of Tropical Medicine and Hygiene* **80**, 893–896.

Crouch, A.A., Seow, W.K., Whitman, L.M. and Thong, Y.H. (1990). Sensitivity *in vitro* of *Giardia intestinalis* to dyadic combinations of azithromycin, doxycycline, mefloquine, tinidazole and furazolidone. *Transactions of the Royal Society of Tropical Medicine and Hygiene* **84**, 246–248.

Crouch, A.A., Seow, W.K., Whitman, L.M. and Thong, Y.H. (1991a). Effect of human milk and infant milk formulae on adherence of *Giardia intestinalis. Transactions of the Royal Society of Tropical Medicine and Hygiene* **85**, 617–619.

Crouch, A.A., Seow, W.K., Whitman, L.M., Smith, S.E. and Thong, Y.H. (1991b). Inhibition of adherence of *Giardia intestinalis* by human neutrophils and monocytes. *Transactions of the Royal Society of Tropical Medicine and Hygiene* **85**, 375–379.

Darling, T.N., Davis, D.G., London, R.E. and Blum, J.J. (1987). Products of *Leishmania braziliensis* glucose catabolism: release of D-lactate and glycerol under anaerobic conditions. *Proceedings of the National Academy of Sciences of the USA* **84**, 7129–7133.

Davidson, R.A. (1984). Issues in clinical parasitology: the treatment of giardiasis. *American Journal of Gastroenterology* **79**, 256–261.

Davidson, R.A. (1990). "Treatment of Giardiasis: the North American Perspective", pp. 325–334. Elsevier, Amsterdam.

Davies, R.B. and Hibler, C.P. (1979). Animal reservoirs and cross-species transmission of *Giardia. In* "Waterborne Transmission of Giardiasis" (W. Jakubowski and J.C. Hoff, eds), pp. 104–126. Environmental Protection Agency, Cincinnati, USA.

De Carneri, I., Trane, F. and Mandelli, V. (1977). *Giardia muris*: oral infection with one trophozoite and generation time in mice. *Transactions of the Royal Society of Tropical Medicine and Hygiene* **71**, 438.

Deguchi, M., Gillin, F.D. and Gigli, I. (1987). Mechanism of killing of *Giardia lamblia* trophozoites by complement. *Journal of Clinical Investigation* **79**, 1296–1302.

De Jonckheere, J.F. and Gordts, B. (1987). Occurrence and transfection of a *Giardia*

virus. *Molecular and Biochemical Parasitology* **23**, 85–89.
De Lourdes Munoz, M., Weinbach, E.C., Wieber, S.C., Claggett, C.E. and Levenbook, L. (1987). *Giardia lamblia*: detection and characterization of calmodulin. *Experimental Parasitology* **63**, 42–48.
Den Hollander, N., Riley, D. and Befus, D. (1988). Immunology of giardiasis. *Parasitology Today* **4**, 124–131.
De Souza, W. (1984). Cell biology of *Trypanosoma cruzi*. *International Reviews in Cytology* **86**, 197–283.
De Souza, D.W., de Souza, M.S. and Neves, J. (1973). Therapeutic action of mebendazole (R17,635), in multiple parasitised patients. Preliminary results. *Revista do Instituto de Medicina Tropical de São Paulo* **15**, 30–33.
Dietrich, R.A. and Hellerman, L. (1963). Diphosphopyridine nucleotide-linked aldehyde dehydrogenase. II. Inhibitors. *Journal of Biological Chemistry* **238**, 1683.
Docampo, R., Moreno, S.N.J. and Mason, R.P. (1987). Free radical intermediates in the reaction of pyruvate : ferredoxin oxidoreductase in *T. foetus*. *Journal of Biological Chemistry* **262**, 12417–12420.
Doenhoff, M.J., Modha, J., Lambertucci, J.R. and McLaren, D.J. (1991). The immune dependence of chemotherapy. *Parasitology Today* **7**, 16–18.
Dupont, H.L. and Sullivan, P.S. (1986). Giardiasis: the clinical spectrum, diagnosis and therapy. *Pediatric Infectious Disease* **5**, 131–138.
Dustin, P. (1984). "Microtubules". Springer, Berlin.
Dutt, P. and Vinayak, V.K. (1990). Enzyme linked immunosorbent assay for coprodiagnosis of giardiasis and characterisation of *Giardia lamblia* specific stool antigen. *Bulletin de la Société Française de Parasitologie* **8**, 1008.
Edlind, T.D. (1989a). Susceptibility of *Giardia lamblia* to aminoglycoside protein synthesis inhibitors: correlation with rRNA structure. *Antimicrobial Agents and Chemotherapy* **33**, 484–488.
Edlind, T.D. (1989b). Tetracyclines as antiparasitic agents: lipophilic derivatives are highly active against *Giardia lamblia in vitro*. *Antimicrobial Agents and Chemotherapy* **33**, 2144–2145.
Edlind, T.D. and Chakraborty, P.R. (1987). Unusual ribosomal RNA of the intestinal parasite *Giardia lamblia*. *Nucleic Acids Research* **15**, 7889–7901.
Edlind, T.D., Hang, T.L. and Chakraborty, P.R. (1990a). Activity of anthelmintic benzimidazoles against *Giardia lamblia in vitro*. *Journal of Infectious Diseases* **162**, 1408–1411.
Edlind, T.D., Sharetzsky, D. and Cha, M.E. (1990b). Ribosomal RNA of the primitive eukaryote *Giardia lamblia*: large subunit domain I and potential processing signals. *Gene* **96**, 289–293.
Edman, U., Meza, I. and Agabian, N. (1987). Genomic and cDNA actin sequences from a virulent strain of *Entamoeba histolytica*. *Proceedings of the National Academy of Sciences of the USA* **84**, 3024–3028.
Edwards, D.I., Dye, M. and Carne, H. (1973). The selective toxicity of antimicrobial nitroheterocyclic drugs. *Journal of General Microbiology* **76**, 135–145.
Edwards, M.R., Gilroy, F.V., Jimenez, B.M. and O'Sullivan, W.J. (1989). Alanine is a major end product of metabolism by *Giardia lamblia*: a proton nuclear magnetic resonance study. *Molecular and Biochemical Parasitology* **37**, 19–26.
Einfeld, D.A. and Stibbs, H.H. (1984). Indentification and characterization of a major surface antigen of *Giardia lamblia*. *Infection and Immunity* **46**, 377–383.
Engelkirk, P.G. and Pickering, L.K. (1990). Detection of *Giardia* by immunologic methods. *In* "Giardiasis" (E.A. Meyer, ed.), pp. 187–198. Elsevier, Amsterdam.

Erlandsen, S.L. and Bemrick, W.J. (1987). SEM evidence for a new species, *Giardia psittaci*. *Journal of Parasitology* **73**, 623–629.

Erlandsen, S.L. and Meyer, E.A. (1984). "*Giardia* and Giardiasis". Plenum Press, New York.

Erlandsen, S.L., Bemrick, W.J., Schupp, D.E., Shields, J.M., Jarroll, E.L., Sauch, J.F. and Pawlay, J.B. (1990a). High resolution immunogold localization of *Giardia* cyst wall antigens using field emission SEM with secondary and backscatter electron imaging. *Journal of Histochemistry and Cytochemistry* **38**, 625–632.

Erlandsen, S.L., Bemrick, W.J., Wells, C.L., Feely, D.E., Knudsen, L., Campbell, S.R., van Keulen, H. and Jarroll, E.L. (1990b). Axenic culture and characterization of *Giardia ardeae* from the great blue heron (*Ardea herodias*). *Journal of Parasitology* **76**, 717–724.

Erlandsen, S.L., Bemrick, W.J. and Jakubowski, W. (1991). Cross-species transmission of avian and mammalian *Giardia* spp.: inoculation of chicks, ducklings, budgerigars, Mongolian gerbils and neonatal mice with *Giardia ardeae, Giardia duodenalis* (*lamblia*), *Giardia psittaci* and *Giardia muris*. *International Journal of Environmental Health Research* **1**, 144–152.

Fahey, R.C., Newton, G.L., Arricks, B., Overdank-Bogart, T. and Aley, S.B. (1984). *Entamoeba histolytica*: an eukaryote without glutathione metabolism. *Science* **224**, 70–72.

Fairlamb, A.H., Blackburn, P., Ulrich, P., Chait, B.T. and Cerami, A. (1985). Tryptathione: a novel bis(glutathionyl)spermidine cofactor for glutathione reductase in trypanosomatids. *Science* **227**, 1484–1487.

Fan, J.B., Korman, S.H., Cantor, C.R. and Smith, C.L. (1991). *Giardia lamblia*: haploid genome size determined by pulsed field gel electrophoresis is less than 12 Mb. *Nucleic Acids Research* **19**, 1905–1908.

Farram, E. and Smithyman, A.M. (1991). Opportunistic infections in AIDS and their diagnosis. *Australian Journal of Biotechnology* **5**, 37–43.

Farthing, M.J.G. (1986). Clinical impact of giardiasis. *In* "Interactions of Parasitic Diseases and Nutrition" (C. Chagas and G.T. Keusch, eds), pp. 185–202. Pontificia Academia Scientiarum, Città del Vaticano.

Farthing, M.J.G. and Inge, P.M.G. (1986). Antigiardial activity of the bile salt-like antibiotic sodium fusidate. *Journal of Antimicrobial Chemotherapy* **17**, 165–171.

Farthing, M.J.G., Varon, S.R. and Keusch, G.T. (1983). Mammalian bile promotes the growth of *Giardia lamblia* in axenic culture. *Transactions of the Royal Society of Tropical Medicine and Hygiene* **77**, 467–469.

Farthing, M.J.G., Keusch, G.T. and Carey, M.C. (1985). Effects of bile and bile salts on growth and membrane lipid uptake by *Giardia lamblia*: possible implications for pathogenesis of intestinal diseases. *Journal of Clinical Investigation* **76**, 1727–1732.

Farthing, M.J.G., Pereira, M.E.A. and Keusch, G.T. (1986a). Description and characterisation of a surface lectin from *Giardia lamblia*. *Infection and Immunity* **51**, 661–667.

Farthing, M., Mata, L., Urrutia, J. and Kronmal, R. (1986b). Natural history of *Giardia* infection of infants and children in rural Guatemala and its impact on physical growth. *American Journal of Clinical Nutrition* **43**, 395–405.

Faubert, G.M. and Belosevic, M. (1990). Animal models for *Giardia duodenalis* type organisms. *In* "Giardiasis" (E.A. Meyer, ed.), pp. 77–90. Elsevier, Amsterdam.

Faubert, G.M., Belosevic, M., Walker, T.S., MacLean, J.D. and Meerovitch, E. (1983). Comparative studies on the pattern of infection with *Giardia* spp. in

Mongolian gerbils. *Journal of Parasitology* **69**, 802–805.
Favennec, L., Chochillon, C., Meillet, D., Magne, D., Savel, J., Raichvarg, D., and Gobert, J.G. (1990). Adherence and multiplication of *Giardia intestinalis* on human enterocyte-like differentiated cells *in vitro*. *Parasitology Research* **76**, 581–584.
Favennec, L., Chochillon, C., Magne, D., Meillet, D., Raichvarg, D., Savel, J. and Gobert, J.G. (1992). A new screening assay for antigiardial compounds: effects of various drugs on the adherence of *Giardia duodenalis* to Caco2 cells. *Parasitology Research* **78**, 80–81.
Feely, D.E. (1985). Acid hydrolase cytochemistry of *Giardia*. *Microecological Therapy* **15**, 149–156.
Feely, D.E. and Dyer, J.K. (1987). Localization of acid phosphatase activity in *Giardia lamblia* and *Giardia muris* trophozoites. *Journal of Protozoology* **34**, 80–83.
Feely, D.E. and Erlandsen, S.L. (1985). Morphology of *Giardia agilis*: observation by scanning electron microscopy and interference reflexion microscopy. *Journal of Protozoology* **32**, 691–693.
Feely, D.E., Schollmeyer, J.V. and Erlandsen, S.L. (1982). *Giardia*: distribution of contractile proteins in the attachment organelle. *Experimental Parasitology* **53**, 145–154.
Feely, D.E., Holberton, D.V. and Erlandsen, S.L. (1990). The biology of *Giardia*. *In* "Giardiasis" (E.A. Meyer, ed.), pp. 11–49. Elsevier, Amsterdam.
Feely, D.E., Gardner, M.D. and Hardin, E.L. (1991). Excystation of *Giardia muris* induced by a phosphate-bicarbonate medium: localisation of acid phosphatase. *Journal of Parasitology* **77**, 441–448.
Filice, F.P. (1952). Studies on the cytology and life history of a *Giardia* from the laboratory rat. *University of California Publication in Zoology* **57**, 53–146.
Firth, M. (1983). "Albendazole in Helminthiasis". Royal Society of Medicine, Academic Press/Grune & Stratton, London, Royal Society of Medicine International Congress and Symposium Series, no. 57.
Forsgren, A. and Wallin, J. (1974). Tinidazole—a new preparation for *Trichomonas vaginalis* infections. *British Journal of Venereal Disease* **50**, 146.
Forshaw, D., Palmer, D.G., Halse, S.A., Hopkins, R.M. and Thompson, R.C.A. (1992). *Giardia* infection in straw necked ibis (*Threskiornis spinicollis*). *Veterinary Record*, **131**, 267–268.
Fraser, G.G. and Cooke, K.R. (1991). Endemic giardiasis and municipal water supply. *American Journal of Public Health* **81**, 760–762.
Friedman, G.D. (1980). Cancer after metronidazole. *New England Journal of Medicine* **302**, 519.
Friedman. G.D. and Selby, J.V. (1990). How often does metronidazole induce pancreatitis? *Gastroenterology* **98**, 1702–1703.
Gascon, J., Moreno, A., Valls, M.E., Miró, J.M. and Corachán, M. (1989). Failure of mebendazole treatment in *Giardia lamblia* infection. *Transactions of the Royal Society of Tropical Medicine and Hygiene* **83**, 647.
Gascon, J., Abós, R., Valls, M.E. and Corachán, M. (1990). Mebendazole and metronidazole in giardial infections. *Transactions of the Royal Society of Tropical Medicine and Hygiene* **84**, 694.
Geary, T.G., Edgar, S.A. and Jensen, J.B. (1986). Drug resistance in protozoa. *In* "Chemotherapy of Parasitic Diseases" (W.L. Campbell and R.E.W. Campbell, eds), pp. 209–235. Plenum Press, New York.

Georgi, M.E., Carlisle, M.S. and Smiley, L.E. (1986). Giardiasis in a great blue heron (*Ardea herodias*) in New York State: another potential source of water-borne giardiasis. *American Journal of Epidemiology* **123**, 916–917.

Gillin, F.D. (1987). *Giardia lamblia*: the role of conjugated and unconjugated bile salts in killing by human milk. *Experimental Parasitology* **63**, 74–83.

Gillin, F.D. and Diamond, L.S. (1981). Inhibition of clonal growth of *Giardia lamblia* and *Entamoeba histolytica* by metronidazole, quinacrine, and other antimicrobial agents. *Journal of Antimicrobial Chemotherapy* **8**, 305–316.

Gillin, F.D., Reiner, D.S. and Wang, C.C. (1983). Human milk kills parasitic protozoa. *Science* **221**, 1290–1292.

Gillin, F.D., Gault, M.J., Hofmann, A.F., Gurantz, D. and Sauch, J.F. (1986). Biliary lipids support serum-free growth of *Giardia lamblia. Infection and Immunity* **53**, 641–645.

Gillin, F.D., Reiner, D.S., Gault, M.J., Douglas, H., Das, S., Wunderlich, A. and Sauch, J.F. (1987). Encystation and expression of cyst antigens by *Giardia lamblia in vitro. Science* **235**, 1040–1043.

Gillin, F.D., Reiner, D.S. and Boucher, S.E. (1988). Small intestinal factors promote encystation of *Giardia lamblia in vitro. Infection and Immunity* **56**, 705–707.

Gillin, F.D., Hagblom, P., Harwood, J. and Aley, S.B. (1990). Isolation and expression of the gene for a major surface protein of *Giardia lamblia. Proceedings of the National Academy of Sciences of the USA* **87**, 4463–4467.

Gillin, F.D., Reiner, D.S. and McCaffery, M. (1991). Organelles of protein transport in *Giardia lamblia. Parasitology Today* **7**, 113–116.

Gillon, J. and Ferguson, A. (1984). Changes in the small intestinal mucosa in giardiasis. *In* "*Giardia* and Giardiasis" (S.L. Erlandsen and E.A. Meyer, eds), pp. 163–183. Plenum Press, New York.

Gilman, R.H., Brown, K.H., Visvesvara, G.S., Mondal, G., Greenberg, B., Sack, R.B., Brandt, F. and Khan, M.U. (1985). Epidemiology and serology of *Giardia lamblia* in a developing country: Bangladesh. *Transactions of the Royal Society of Tropical Medicine and Hygiene* **79**, 469–473.

Gilman, R.H., Marquis, G.S., Miranda, E., Vistegui, M., Miranda, E. and Montinez, H. (1988). Rapid reinfection by *Giardia lamblia* after treatment in a hyperendemic Third World community. *Lancet* **ii**, 343–345.

Glick, N.B. (1970). Inhibitors of transport reactions. *In* "Metabolic Inhibitors: a Complete Treatise" (R.M. Hochster, M. Kates and J.A. Quastel, eds), Vol. 3, pp. 8–17. Academic Press, London.

Goka, A.K.J., Rolston, D.D.K., Mathan, V.I. and Farthing, M.J.G. (1986). Diagnosis of giardiasis by specific IgM antibody enzyme-linked immunosorbent assay. *Lancet* **ii**, 184–186.

Goka, A.K.J., Rolston, D.D.K., Mathan, V.I. and Farthing, M.J.G. (1990). The relative merits of faecal and duodenal juice microscopy in the diagnosis of giardiasis. *Transactions of the Royal Society of Tropical Medicine and Hygiene* **84**, 66–67.

Goldin, A.J., Apt, W., Aguilera, X., Zulantay, I., Warhurst, D.C. and Miles, M.A. (1990). Efficient diagnosis of giardiasis among nursery and primary school children in Santiago, Chile by capture ELISA for the detection of fecal *Giardia* antigens. *American Journal of Tropical Medicine and Hygiene* **42**, 538–545.

Gonzalez-Castro, J., Bermejo-Vicedo, M.T. and Palacios-Gonzalez, F. (1986). Desenquistamiento y cultivo de *Giardia muris. Revista Iberica Parasitologia* **46**, 21–25.

Gordts, B., Hemelhof, W., Asselman, C. and Butzler, J.P. (1985). *In vitro* susceptibilities of 25 *Giardia lamblia* isolates of human origin to six commonly used antiprotozoal agents. *Antimicrobial Agents and Chemotherapy* **28**, 378–380.

Gordts, B., Kasprzak, W., Majewska, A.C. and Butzler, J.P. (1987). *In vitro* activity of antiprotozoal drugs against *Giardia intestinalis* of human origin. *Antimicrobial Agents and Chemotherapy* **31**, 672–673.

Gottstein, B. and Nash, T.E. (1991). Antigenic variation in *Giardia lamblia*: infection of congenitally athymic nude and *scid* mice. *Parasite Immunology* **13**, 649–659.

Gottstein, B., Harriman, G.R., Conrad, J.T. and Nash, T.E. (1990). Antigenic variation in *Giardia lamblia*: cellular and humoral immune response in a mouse model. *Parasite Immunology* **12**, 659–673.

Gottstein, B., Stocks, N.I., Shearer, G.M. and Nash, T.E. (1991). Human cellular immune response to *Giardia lamblia*. *Infection* **19**, 421–426.

Gracey, M. (1983). Enteric disease in young Australian Aborigines. *Australian and New Zealand Journal of Medicine* **3**, 576–579.

Grant, D.R. and Woo, P.T.K. (1978). Comparative studies of *Giardia* spp. in small mammals in southern Ontario. II. Host specificity and infectivity of stored cysts. *Canadian Journal of Zoology* **56**, 1360–1366.

Green, E.L., Miles, M.A. and Warhurst, D.C. (1985). Immunodiagnostic detection of *Giardia* antigen in faeces by a rapid visual enzyme-linked immunosorbent assay. *Lancet* **ii**, 691–695.

Gupta, M.C. and Urrutia, J.J. (1982). Effect of periodic antiascaris and antigiardial treatment on nutritional status of preschool children. *American Journal of Clinical Nutrition* **36**, 79–86.

Hager, W.D., Brown, S.T., Kraus, S.T., Kleris, G.S., Perkins, G.J. and Henderson, M. (1980) Metronidazole for vaginal trichomoniasis. *Journal of the American Medical Association* **244**, 1219.

Hall, A. and Nahar, Q. (1993). Albendazole as a treatment for infections with *Giardia duodenalis* in children in Bangladesh. *Transactions of the Royal Society of Tropical Medicine and Hygiene* **87**, 84–86.

Hall, E.J., Rutgers, H.C. and Batt, R.M. (1988). Evaluation of the peroral string test in the diagnosis of canine giardiasis. *Journal of Small Animal Practice* **29**, 177–183.

Halloran, M.E., Bundy, D.A.P. and Pollitt, E. (1989). Infectious disease and the Unesco basic education initiative. *Parasitology Today* **5**, 359–362.

Hamel, E. and Lin, C.M. (1983). Interactions of combretastatin, a new plant derived antimitotic agent, with tubulin. *Biochemical Pharmacology* **32**, 3864–3867.

Hare, D.F., Jarroll, E.L. and Lindmark, D.G. (1989). *Giardia lamblia*: characterization of proteinase activity in trophozoites. *Experimental Parasitology* **68**, 168–175.

Healey, A., Mitchell, R., Upcroft, J.A., Boreham, P.F.L. and Upcroft, P. (1990). Complete nucleotide sequence of the ribosomal RNA tandem repeat unit from *Giardia intestinalis*. *Nucleic Acids Research* **18**, 4006.

Heymans, H.S.A., Aronson, D.C. and van Hooft, M.A.J. (1987). Giardiasis in childhood: an unnecessarily expensive diagnosis. *European Journal of Pediatrics* **146**, 401–403.

Heyworth, M.F. and Pappo, J. (1987). Use of fluorescent dyes to assess viability of *Giardia muris* trophozoites. *Gastroenterology* **92**, 1435 [abstract].

Heyworth, M.F. and Pappo, J. (1989). Use of two-colour flow cytometry to assess killing of *Giardia muris* trophozoites by antibody and complement. *Parasitology* **99**, 199–203.

Heyworth, M.F. Carlson, J.R. and Ermak, T.H. (1987). Clearance of *Giardia muris* infection requires helper/inducer T lymphocytes. *Journal of Experimental Medicine* **165**, 1743–1748.

Hill, D.R. and Pohl, R. (1990). Ingestion of *Giardia lamblia* trophozoites by murine Peyer's patch macrophages. *Infection and Immunity* **58**, 3202–3207.

Hill, D.R., Burge, J.J. and Pearson, R.D. (1984). Susceptibility of *Giardia lamblia* trophozoites to the lethal effect of human serum. *Journal of Immunology* **132**, 2046–2052.

Holberton, D.V. (1981). Arrangement of subunits in microribbons from *Giardia*. *Journal of Cell Science* **47**, 167–185.

Holberton, D.V. and Ward, A.P. (1981). Isolation of the cytoskeleton from *Giardia*. Tubulin and a low molecular weight protein associated with microribbon structures. *Journal of Cell Science* **47**, 139–166.

Holberton, D., Baker, D.A. and Marshall, J. (1988). Segmented alpha-helical coiled-coil structure of the protein giardin from the *Giardia* cytoskeleton. *Journal of Molecular Biology* **204**, 789–795.

Homeida, M.A., Daneshmend, T.K., Ali, H.M. and Kaye, C.M. (1986). Metronidazole metabolism following oral benzoylmetronidazole suspension in children with giardiasis. *Journal of Antimicrobial Chemotherapy* **18**, 213–219.

Hopkins, R.M., Deplazes, P., Meloni, B.P., Reynoldson, J.A. and Thompson, R.C.A. (1993b). A field and laboratory evaluation of a commercial ELISA for the detection of *Giardia* copro-antigens in humans and dogs. *Transactions of the Royal Society of Tropical Medicine and Hygiene* **87**, 39–41.

Hopkins, R.M., Thompson, R.C.A., Hobbs, R.P., Lymbery, A.J., Villa, N. and Smithyman, T.M. (in press). Differences in antigen expression within and between 10 isolates of *Giardia duodenalis*. *Acta Tropica*.

Hossain, M.M., Ljungström, I., Glass, R.I., Lundin, L., Stoll, B.J. and Huldt, G. (1983). Amoebiasis and giardiasis in Bangladesh: parasitological and serological studies. *Transactions of the Royal Society of Tropical Medicine and Hygiene* **77**, 552–554.

Hoyne, G.F., Boreham, P.F.L., Parsons, P.G., Ward, C. and Biggs, B. (1989). The effect of drugs on the cell cycle of *Giardia intestinalis*. *Parasitology* **99**, 333–339.

Huber, M., Garfinkel, L., Gitler, C., Mirelman, D., Revel, M. and Rozenblatt, S. (1988). Nucleotide sequence analysis of an *Entamoeba histolytica* ferredoxin gene. *Molecular and Biochemical Parasitology* **31**, 27–34.

Hudson, A.T., Randall, A.W., Fry, M., Ginger, C.D., Hill, B., Latter, V.S., McHardy, N. and Williams, R.B. (1985). Novel anti-malarial hydroxynaphthoquinones with potent broad spectrum anti-protozoal activity. *Parasitology* **90**, 45–55.

Hutchison, J.G.P., Johnston, N.M., Plevey, M.V.P. and Thongkhiew, C.A. (1975). Clinical trial of mebendazole: a broad-spectrum anthelmintic. *British Medical Journal* **ii**, 309–310.

Inge, P.M.G. and Farthing, M.J.G. (1987). A radiometric assay for antigiardial drugs. *Transactions of the Royal Society of Tropical Medicine and Hygiene* **81**, 345–347.

Inge, R.M.J., McFadzean, J.A. and Ormerod, W.E. (1974). The mode of action of metronidazole in *Trichomonas vaginalis* and other microorganisms. *Biochemical Pharmacology* **23**, 1421–1429.

Inge, P.M.G., Edson, C.M. and Farthing, M.J.G. (1986). Attachment of *Giardia lamblia* to rat intestinal epithelial cells. *Gut* **29**, 795–801.

Isaac-Renton, J.L. (1991). Immunological methods of diagnosis in giardiasis: an overview. *Annals of Clinical and Laboratory Science* **21**, 116–122.

Islam, A. (1990). Giardiasis in developing countries. *In* "Giardiasis" (E.A. Meyer, ed.), pp. 235–266. Elsevier, Amsterdam.

Islam, A. Stoll, B.J., Ljungström, I., Biswas, J., Nazrul, H. and Huldt, G. (1983). *Giardia lamblia* infections in a cohort of Bangladeshi mothers and infants followed for one year. *Journal of Pediatrics* **103**, 996–1000.

Janoff, E.N., Smith, P.D. and Blaser, M.J. (1988). Acute antibody responses to *Giardia lamblia* are depressed in patients with AIDS. *Journal of Infectious Diseases* **157**, 798–803.

Janoff, E.N., Craft, J.C., Pickering, L.K., Novotny, T., Blaser, M.J., Knisley, C.V. and Reller, L.B. (1989). Diagnosis of *Giardia lamblia* infections by detection of parasite-specific antigens. *Journal of Clinical Microbiology* **27**, 431–435.

Jarroll, E.L. and Lindmark, D.G. (1990). *Giardia* metabolism. *In* "Giardiasis" (E.A. Meyer, ed.), pp. 61–76. Elsevier, Amsterdam.

Jarroll, E.L., Muller, P.J., Meyer, E.A. and Morse, S.A. (1981). Lipid and carbohydrate metabolism of *Giardia lamblia. Molecular and Biochemical Parasitology* **2**, 187–196.

Jarroll, E.L., Hammond, M.M. and Lindmark, D.G. (1987). *Giardia lamblia*: uptake of pyrimidine nucleosides. *Experimental Parasitology* **63**, 152–156.

Jarroll, E.L., Manning, P., Lindmark, D.G., Coggins, J.R. and Erlandsen, S.L. (1989). *Giardia* cyst wall specific carbohydrate: evidence for the presence of galactosamine. *Molecular and Biochemical Parasitology* **32**, 121–132.

Jenni, L., Marti, S., Schweizer, J., Betschart, B., Le Page, R.W.F., Wells, J.M., Tait, A., Paindavoines, P., Pays, E. and Steinert, M. (1986). Hybrid formation between African trypanosomes during cyclical transmission. *Nature* **322**, 173–175.

Jephcott, A.E., Begg, N.T. and Baker, I.A. (1986). Outbreak of giardiasis associated with mains water in the United Kingdom. *Lancet* **i**, 730–732.

Johnson, P.J., d'Oliveira, C.E., Gorrell, T.E. and Muller, M. (1990). Molecular analysis of the hydrogenosomal ferredoxin of the anaerobic protist *Trichomonas vaginalis. Proceedings of the National Academy of Sciences of the USA* **87**, 6097–6101.

Jokipii, L. and Jokipii, A.M.M. (1974). Giardiasis in travellers: a prospective study. *Journal of Infectious Diseases* **130**, 295–299.

Jokipii, L. and Jokipii, A.M.M. (1979). Single-dose metronidazole and tinidazole as therapy for giardiasis: success rates, side effects, and drug absorption and elimination. *Journal of Infectious Diseases* **140**, 984–988.

Jokipii, L and Jokipii, A.M.M. (1982). Treatment of giardiasis: comparative evaluation of ornidazole and tinidazole as a single oral dose. *Gastroenterology* **83**, 399–404.

Jokipii, L. and Jokipii, A.M.M. (1985). Comparative evaluation of the 2-methyl-5-nitroimidazole compounds dimetridazole, metronidazole, secnidazole, ornidazole, tinidazole, carnidazole, and panidazole against *Bacteroides fragilis* and other bacteria of the *Bacteroides fragilis* group. *Antimicrobial Agents and Chemotherapy* **28**, 561–564.

Jokipii, L., Miettinen, A. and Jokipii, A.M.M. (1988). Antibodies to cysts of *Giardia lamblia* in primary giardiasis and in the absence of giardiasis. *Journal of Clinical Microbiology* **26**, 121–125.

Jove, S., Fangundes-Neto, U., Wehba, J., Machado, N.L. and Silvia Patricio, F.R. (1983). Giardiasis in childhood and its effect on the small intestine. *Journal of Pediatric Nutrition* **2**, 472–477.

Judd, R., Deckelbaum, R.J., Weizman, Z., Granot, E., Ron, N. and Okon, E. (1983). Giardiasis in childhood: poor clinical and histological correlations. *Israel Journal of Medical Science* **19**, 818–823.

Kabnick, K.S. and Peattie, D.A. (1990). *In situ* analyses reveal that the two nuclei of *Giardia lamblia* are equivalent. *Journal of Cell Science* **95**, 353–360.

Kabnick, K.S. and Peattie, D.A. (1991). *Giardia* as a missing link between prokaryotes and eukaryotes. *American Scientist* **79**, 34–43.

Kaminsky, R.G. (1991). Parasitism and diarrhoea in children from two rural communities and marginal barrio in Honduras. *Transactions of the Royal Society of Tropical Medicine and Hygiene* **85**, 70–73.

Kane, P.O., McFadzean, J.A. and Squires, S. (1961). Absorption and excretion of metronidazole: II. Studies on primary failures. *British Journal of Venereal Disease* **37**, 276–277.

Kaneda, Y. and Goutsu, T. (1988). Lipid analysis of *Giardia lamblia* and its culture medium. *Annals of Tropical Medicine and Parasitology* **82**, 83–90.

Kaneda, Y., Torii, M., Tanaka, T. and Aikawa, M. (1991) *In vitro* effects of berberine sulphate on the growth and structure of *Entamoeba histolytica, Giardia lamblia* and *Trichomonas vaginalis*. *Annals of Tropical Medicine and Parasitology* **85**, 417–425.

Kanwar, S.S., Ganguly, N.K., Walia, B.N.S. and Mahajan, R.C. (1986). Direct and antibody dependent cell mediated cytotoxicity against *Giardia lamblia* by colonic and intestinal lymphoid cells in mice. *Gut* **27**, 73–77.

Karapetyan, A.E. (1960). Methods of *Lamblia* cultivation. *Tsitologiia* **2**, 379–384.

Karapetyan, A.E. (1962). *In vitro* cultivation of *Giardia duodenalis*. *Journal of Parasitology* **48**, 337–340.

Kattenbach, W.M., Pimenta, P.F.P., de Souza, W. and Pinto da Silva, P. (1991). *Giardia duodenalis*: a freeze-fracture, fracture-flip and cytochemistry study. *Parasitology Research* **77**, 651–658.

Katz, N., Cimerman, B., Zingano, A.G., Zingano, R.G. and Rocha, R.S. (1989). A clinical trial using secnidazole suspension in children infected with *Giardia lamblia*. *In* "Secnidazole: a New Approach in 5-Nitroimidazole Therapy" (N. Katz and A.T. Willis, eds), pp. 35–40. Excerpta Medica, Amsterdam.

Keister, D.B. (1983). Axenic culture of *Giardia lamblia* in TYI-S-33 medium supplemented with bile. *Transactions of the Royal Society of Tropical Medicine and Hygiene* **77**, 487–488.

Kerscher, L. and Oesterhelt, D. (1982). Pyruvate: ferredoxin oxidoreductase—new findings on an ancient enzyme. *Trends in Biochemical Science* **7**, 371–374.

Keystone, J.S., Krajden, S. and Warren, M.R. (1978). Person-to-person transmission of *Giardia lamblia* in day-care nurseries. *Canadian Medical Journal* **119**, 241–248.

Kirk-Mason, K.E., Turner, M.J. and Charkraborty, P.R. (1988). Cloning and sequence of beta tubulin cDNA from *Giardia lamblia*. *Nucleic Acids Research* **16**, 2733.

Kirk-Mason, K.E., Turner, M.J. and Charkraborty, P.R. (1989). Evidence for unusually short tubullin mRNA leaders and characterization of tubulin genes in *Giardia lamblia*. *Molecular and Biochemical Parasitology* **36**, 87–100.

Kirpatrick, C.E. (1984). Enteric protozoal infections. *In* "Clinical Microbiology and Infectious Diseases of the Dog and Cat" (C.E. Greene, ed.), pp. 806–823. W.B. Saunders, Philadelphia.

Kirkpatrick, C.E. (1989). Giardiasis in large animals. *Compendium for Continuing Education for Practicing Veterinarians* **11**, 80–84.

Klein, H. (1964). Alkohol und Medikamente: I. Durch Medikamente verusachte

Alkoholunvertraglichleit und verstarkte Alkoholwirkung. *Forschritte der Medizin* **82**, 169.

Knight, R. (1980) Epidemiology and transmission of giardiasis. *Transactions of the Royal Society of Tropical Medicine and Hygiene* **74**, 433–436.

Knight, R.C., Skolimowski, I.M. and Edwards, D.I. (1978). The interaction of reduced metronidazole with DNA. *Biochemical Pharmacology* **27**, 2089.

Koch, R.L. and Goldman, P. (1979). The anaerobic metabolism of metronidazole forms N-(2-hydroxyethyl)oxamic acid. *Journal of Pharmacology and Experimental Therapeutics* **208**, 406.

Kovacs, J.A. and Masur, H. (1986). Protozoan infections of man: other infections. *In* "Chemotherapy of Parasitic Diseases" (W.C. Campbell and R.S. Rew, eds), pp. 139–158. Plenum Press, New York.

Kozak, M. (1987). An analysis of 5′-noncoding sequences from 699 vertebrate messenger RNAs. *Nucleic Acids Research* **15**, 8125–8148.

Kreutner, A.K., Del Bene, V.E. and Amstey, M.S. (1981). Giardiasis in pregnancy. *American Journal of Obsterics and Gynecology* **140**, 895–901.

Kulda, J. and Nohynkova, E. (1978). Flagellates of the human intestine and of intestines of other species. *In* "Parasitic Protozoa" (J.P. Kreier, ed.), Vol. 2, pp. 2–139. Academic Press, New York.

Kusumi, R.K. Plouffe, J.F., Wyatt, R.H. and Fass, R.J. (1980). Central nervous system toxicity associated with metronidazole therapy. *Annals of International Medicine* **93**, 59.

Lacey, E. (1985). The biochemistry of anthelmintic resistance. *In* "Resistance in Nematodes to Anthelmintic Drugs" (N. Anderson and P. Waller, eds), pp. 69–78. CSIRO, Glebe, NSW.

Lacey, E. (1988). The role of the cytoskeletal protein, tubulin, in the mode of action and mechanism of drug resistance to benzimidazoles. *International Journal for Parasitology* **18**, 855–936.

Lacey, E. (1990). Mode of action of benzimidazoles. *Parasitology Today*, **6**, 112–115.

Lacey, E. and Prichard, R.K. (1986). Interactions of benzimidazoles (BZ) with tubulin from BZ-sensitive and BZ-resistant isolates of *Haemonchus contortus*. *Molecular and Biochemical Parasitology* **19**, 171–181.

Lacey, E. and Snowdon, K.L. (1988). A routine diagnostic assay for the detection of benzimidazole resistance in parasitic nematodes using tritiated benzimidazole carbamates. *Veterinary Parasitology* **27**, 309–324.

LaRusso, N.F., Tomaz, M., Muller, M. and Lipman, R. (1977). Interaction of metronidazole with nucleic acids *in vitro*. *Molecular Pharmacology* **13**, 872.

Laughon, B.E., Druckman, D.A., Vernon, A., Quinn, T.C., Polk. B.F., Modlin, J.F., Yolken, R.H. and Bartlett, J.G. (1988). Prevalence of enteric pathogens in homosexual men with and without acquired immunodeficiency syndrome. *Gastroenterology* **94**, 984–993.

Le Blancq, S.M., Korman, S.H. and Van der Ploeg, L.H.T. (1991). Frequent rearrangements of rRNA-encoding chromosomes in *Giardia lamblia*. *Nucleic Acids Research* **19**, 4405–4412.

Lee, C.S., Takashi, A. and O'Sullivan, W.J. (1992). Studies on phosphoribosyl pyrophosphate synthetase from *Giardia intestinalis*. *International Journal for Parasitology* **22**, 129–133.

Legator, M.S., Connor, T.H. and Stoeckel, M. (1975). Detection of mutagenic activity of metronidazole and nitridazole in body fluid of humans and mice. *Science* **188**, 1118.

Lev, B., Ward, H., Keusch, G.T. and Pereira, M.E.A. (1986). Lectin activation in *Giardia lamblia* by host protease: a novel host-parasite interaction. *Science* **232**, 71–73.

Levine, W.C., Stephenson, W.T. and Craun, G.F. (1990). Water-borne disease outbreaks, 1986–1988. *Morbidity and Mortality Weekly Report* **39**, 1–13.

Lindenmayer, J.P. and Vargas, P. (1981). Toxic psychosis following use of quinacrine. *Journal of Clinical Psychiatry* **42**, 162.

Lindley, T.A., Chakraborty, P.R. and Edlind, T.D. (1988). Heat shock and stress response in *Giardia lamblia*. *Molecular and Biochemical Parasitology* **28**, 135–144.

Lindmark, D.G. (1980). Energy metabolism of the anaerobic protozoon *Giardia lamblia*. *Molecular and Biochemical Parasitology* **1**, 1–12.

Lindmark, D.G. (1988). *Giardia lamblia*: localization of hydrolase activities in lysosome-like organelles of trophozoites. *Experimental Parasitology* **65**, 141–147.

Lindmark, D.G. and Jarroll, E.L. (1982). Pyrimidine metabolism in *Giardia lamblia* trophozoites. *Molecular and Biochemical Parasitology* **5**, 291–296.

Lindmark, D.G. and Jarroll E.L. (1984). The metabolism of trophozoites. *In* "*Giardia* and Giardiasis" (S.L. Erlansden and E.A. Meyer, eds), pp. 65–80. Plenum Press, New York.

Lindmark, D.G., and Miller, J.J. (1988). Enzyme activities of *Giardia lamblia* and *Giardia muris* trophozoites and cysts. *In* "Advances in *Giardia* Research" (P. Wallis and B. Hammond, eds), pp.187–189. University of Calgary Press, Calgary.

Lindmark, D.G. and Muller, M. (1976). Antitrichomonal action, mutagenicity, and reduction of metronidazole and other nitroimidazoles. *Antimicrobiological Agents and Chemotherapy* **10**, 476.

Linstead, D. and Cranshaw, M.R. (1983). The pathway of arginine catabolism in the parasitic flagellate *Trichomonas vaginalis*. *Molecular and Biochemical Parasitology* **8**, 241–252.

Lloyd, D., Mellor, H. and Williams, J.L. (1983). Oxygen affinity of the respiratory chain of *Acanthamoeba castellanii*. *Biochemical Journal* **214**, 47–51.

Lockwood, D.N.J. and Weber, J.N. (1989). Parasite infections in AIDS. *Parasitology Today* **5**, 310–316.

Looker, D.L., Marr, J.J. and Stotish, R.L. (1986). Modes of action of antiprotozoal agents. *In* "Chemotherapy of Parasitic Diseases" (W.C. Campbell and R.S. Rew, eds), pp.193–207. Plenum Press, New York.

Mackenstedt, U., Gauer, M., Mehlhorn, H., Schein, E. and Hauschild, S. (1990a). Sexual cycle of *Babesia divergens* confirmed by DNA measurements. *Parasitology Research* **76**, 199–206.

Mackenstedt, U., Wagner, D., Heydorn, A.O. and Mehlhorn, H. (1990b). DNA measurements and ploidy determination of different stages in the life cycle of *Sarcocystis muris*. *Parasitology Research* **76**, 662–668.

MacKenzie, N.E., Hall, J.E., Flynn, I.W. and Scott, A.I. (1983) ^{13}C nuclear magnetic resonance studies of anaerobic glycolysis in *Trypanosoma brucei*. *Biosciences Report* **3**, 141–151.

MacRae, T.H. and Gull, K. (1990). Purification and assembly *in vitro* of tubulin from *Trypanosoma brucei brucei*. *Biochemical Journal* **265**, 87–93.

Magne, D., Favennec, L., Chochillon, C., Gorenflot, A., Meillet, D., Kapel, N., Raichvarg, D., Savel, J. and Gobert, J.G. (1991). Role of cytoskeleton and surface lectins in *Giardia duodenalis* attachment to Caco2 cells. *Parasitology Research* **77**, 659–662.

Mahbubani, M.H., Bej, A.K., Perlin, M., Schaefer, F.W., Jakubowski, W. and Atlas,

R.M. (1991). Detection of *Giardia* cysts by using the polymerase chain reaction and distinguishing live from dead cysts. *Applied and Environmental Microbiology* **57**, 3456–3461.

Mahbubani, M.G., Bej, A.K., Perlin, M.G., Schaefer, F.W., Jakubowski, W. and Atlas, R.M. (1992). Differentiation of *Giardia duodenalis* from other *Giardia* spp. by using polymerase chain reaction and gene probes. *Journal of Clinical Microbiology* **30**, 74–78.

Mahlanabis, D., Simpson, T.W., Chakraborty, M.L., Ganguli, C., Bhattarcharjuee, A.K. and Mukherjee, K.L. (1979). Malabsorption of water miscible vitamin A in children with giardiasis and ascariasis. *American Journal of Clinical Nutrition* **32**, 313–318.

Majewska, A.C., Kasprzak, W., DeJonckheere, J.F. and Kaczmarek, E. (1991). Heterogeneity in the sensitivity of stocks and clones of *Giardia* to metronidazole and ornidazole. *Transactions of the Royal Society of Tropical Medicine and Hygiene* **85**, 67–69.

Marklund, S. and Marklund, G. (1974). Involvement of the superoxide anion radical in antoxidation of pyrogallol and a convenient assay for superoxide dismutase. *European Journal of Biochemistry* **47**, 469–474.

Marr, J.J. (1991). Purine metabolism in parasitic protozoa and its relationship to chemotherapy. *In* "Biochemical Protozoology" (G. Coombs and M. North, eds), pp. 524–536. Taylor and Francis, London.

Marriner, S. and Bogan, J.A. (1980). Pharmacokinetics of albendazole in sheep. *American Journal of Veterinary Research* **41**, 1126–1129.

Mason, P. and Patterson, B. (1987). Epidemiology of *Giardia lamblia* infection in children: cross-sectional and longitudinal studies in urban and rural communities in Zimbabwe. *American Journal of Tropical Medicine and Hygiene* **37**, 277–282.

Mata, L.J. (1978). "The Children of Santa Maria Cauqu: a Prospective Field Studyof Health and Growth". Massachusetts Institute of Technology Press, Cambridge, Massachussetts.

Mattila, T., Mannisto, P.T., Mantyla, R., Nykanen, S. and Lamminsivu, U. (1983). Comparative pharmacokinetics of metronidazole and tinidazole as influenced by administration route. *Antimicrobial Agents and Chemotherapy* **23**, 721–725.

Mayr, E. (1940). Speciation phenomena in birds. *American Naturalist* **74**, 249–278.

McCalla, D.R. (1979). Nitrofurans. *In* "Antibiotics" (F.E. Hahn, ed.), Vol. 1, pp. 176–213. Springer, New York.

McCracken, R.O. and Stillwell, W.H. (1991). A possible mode of action for benzimidazole anthelmintics. *International Journal for Parasitology* **21**, 99–104.

McGown, A.T. and Fox, B.W. (1989). Interaction of the novel agent amphethinile with tubulin. *British Journal of Cancer* **59**, 865–868.

McIntyre, P., Boreham, P.F.L., Phillips, R.E. and Shepherd, R.W. (1986). Chemotherapy in giardiasis: clinical responses and *in vitro* drug sensitivity of human isolates in axenic culture. *Journal of Pediatrics* **108**, 1005–1010.

McLafferty, M.A., Koch, R.L. and Goldman, P. (1982). Interaction of metronidazole with resistant and susceptible *Bacteroides fragilis*. *Antimicrobiological Agents and Chemotherapy* **21**, 131.

McNally, P.R., Herrera, J.L., Brewer, T.G., Visvesvara, G.S., MacRae, T.H. and Gull, K. (1990) Purification and assembly *in vitro* of tubulin from *Trypanosoma brucei brucei*. *Biochemistry Journal* **265**, 87–93.

Meloni, B.P. and Thompson, R.C.A. (1987). Comparative studies on the axenic *in vitro* cultivation of *Giardia* of human and canine origin: evidence for intraspecific variation. *Transactions of the Royal Society of Tropical Medicine and Hygiene* **81**, 637–640.

Meloni, B.P., Lymbery, A.J. and Thompson, R.C.A. (1988a). Isoenzyme electrophoresis of 30 isolates of *Giardia* from humans and felines. *American Journal of Tropical Medicine and Hygiene* **38**, 65–73.

Meloni, B.P., Lymbery, A.J., Thompson, R.C.A. and Gracey, M. (1988b). High prevalence of *Giardia lamblia* in children from a WA Aboriginal community. *Medical Journal of Australia* **149**, 715.

Meloni, B.P., Lymbery, A.J. and Thompson, R.C.A. (1989). Characterisation of *Giardia* isolates using a non-radiolabelled DNA probe, and correlation with the results of isoenzyme analysis. *American Journal of Tropical Medicine and Hygiene* **40**, 629–637.

Meloni, B.P., Thompson, R.C.A., Reynoldson, J.A. and Seville, P. (1990). Albendazole: a more effective antigiardial agent *in vitro* than metronidazole or tinidazole. *Transactions of the Royal Society of Tropical Medicine and Hygiene* **84**, 375–379.

Meloni, B.P., Thompson, R.C.A., Stranden, A.M., Kohler, P. and Eckert, J. (1992). Critical comparison of *Giardia duodenalis* from Australia and Switzerland using isoenzyme electrophoresis. *Acta Tropica* **50**, 115–124.

Meloni, B.P., Thompson, R.C.A., Hopkins, R.M., Reynoldson, J.A. and Gracey, M. (1993). The prevalence of *Giardia* and other intestinal parasites in children, dogs and cats from Aboriginal communities in the Kimberley. *Medical Journal of Australia* **158**, 157–159.

Mendelson, R.M. (1980). The treatment of giardiasis. *Transactions of the Royal Society of Tropical Medicine and Hygiene* **74**, 438–439.

Mendis, A.H.W. and Evans, A.A.F. (1984a). Substrates respired by mitochondrial fractions of two isolates of the nematode *Aphelenchus avenae* and the effects of electron transport inhibitors. *Comparative Biochemistry and Physiology* **78B**, 373–378.

Mendis, A.H.W. and Evans, A.A.F. (1984b). The effect of respiratory inhibitors and ethylene dibromide (EDB) on the respiratory oxygen uptake of three isolates of *Aphelenchus avenae* under aerobic and anaerobic regimes. *Nematologica* **30**, 457–451.

Mendis, A.H.W. and Townson, S. (1985). Evidence for the occurrence of respiratory electron transport in adult *Brugia pahangi* and *Dipetalonema viteae*. *Molecular and Biochemical Parasitology* **14**, 337–354.

Mendis, A.H.W., O'Dowd, A.B., Selwood, D. and Hudson, A.T. (1986). The incorporation of radiolabelled bicarbonate, malate and succinate by *D. viteae in vitro*. *Tropenmedizin und Parasitologie* **37**, 85.

Mendis, A.H.W., Selwood, D. and Hudson, A.T. (1987). Electron transport inhibitors as potential macrofilaricides. *In* "Proceedings of the Upjohn/WHO Onchocerciasis Symposium April 1986" (G.A. Conder and J.F. Williams, eds), pp. 67–82. Upjohn Company Press, Kalamazoo, Michigan.

Mendis, A.H.W., Armson, A. and Grubb, W.B. (1991). The response of *Strongyloides ratti* infective (L3) larvae to substrates and inhibitors of respiratory electron transport. *International Journal for Parasitology* **21**, 965–968.

Mendis, A.H.W., Thompson, R.C.A., Reynoldson, J.A., Armson, A., Meloni, B.P. and Gunsberg, S. (1992). The uptake and conversion of [U^{14}C] aspartate and [U^{14}C] alanine to $^{14}CO_2$ by *Giardia duodenalis*. *Comparative Biochemistry and Physiology* **102B**, 235–239.

Mertens, E. (1990). Occurrence of pyrophosphate: fructose 6-phosphate-1-phosphotransferase in *Giardia lamblia* trophozoites. *Molecular and Biochemical Parasitology* **40**, 147–150.

Meyer, E.A. (1970) Isolation and axenic cultivation of *Giardia* trophozoites from the rabbit, chinchilla and cat. *Experimental Parasitology* **27**, 179–183.

Meyer, E.A. (1976). *Giardia lamblia*: isolation and axenic cultivation. *Experimental Parasitology* **39**, 101–105.

Meyer, E.A. (1985). The epidemiology of giardiasis. *Parasitology Today* **1**, 101–105.

Meyer, E.A., ed. (1990a). "Giardiasis". Elsevier, Amsterdam.

Meyer, E.A. (1990b). Preface. *In* "Giardiasis" (E.A. Meyer, ed.). Elsevier, Amsterdam.

Meyer, E.A. (1990c). Taxonomy and nomenclature. *In* "Giardiasis" (E.A. Meyer, ed.), pp, 51–60. Elsevier, Amsterdam.

Meyer, E.A. (1990d). Introduction. *In* "Giardiasis" (E.A. Meyer, ed.), pp. 1–9. Elsevier, Amsterdam.

Meyer, E.A. and Jarroll, E.J. (1980). Giardiasis. *American Journal of Epidemiology* **111**, 1–12.

Meyer, E.A. and Jarroll E.L. (1982) Giardiasis. *In* "CRC Handbook Series in Zoonoses" (L. Jacobs and P. Arambulo), Section C, Vol. 1, pp. 25–40. CRC Press, Boca Raton, Florida.

Meyer, E.A. and Radulescu, S. (1979). *Giardia* and giardiasis. *Advances in Parasitology* **17**, 1–48.

Miller, R.L., Nelson, D.J., LaFon, S.W., Miller, W.H. and Krenitsky, T.A. (1987). Antigiardial activity of guanine arabinoside. *Biochemical Pharmacology* **36**, 2519–2525.

Miller, R.L., Wang, A.L. and Wang, C.C. (1988a). Purification and characterisation of the *Giardia lamblia* double-stranded RNA virus. *Molecular and Biochemical Parasitology* **28**, 189–196.

Miller, R.L., Wang, A.L. and Wang, C.C. (1988b). Identification of *Giardia lamblia* isolates susceptible and resistant to infection by the double-stranded RNA virus. *Experimental Parasitology* **66**, 118–123.

Miotti, P.G., Gilman, R.H., Santosham, M., Ryder, R.W. and Yolken, T.H. (1986). Age-related rate of seropositivity and antibody to *Giardia lamblia* in four diverse populations. *Journal of Clinical Microbiology* **24**, 972–975.

Moore, G.T., Cross, W.M., McGuire, D., Mollohan, C.S., Gleason, N.N., Healy, G.R. and Newton L.H. (1969). Epidemic giardiasis at a ski resort. *New England Journal of Medicine* **281**, 402–407.

Moorehead, P., Guasparini, R., Donovan, C.A., Mathias, R.G., Cottle, R. and Baytalan, G. (1990). Giardiasis outbreak from a chlorinated community water supply. *Canadian Journal of Public Health* **81**, 358–362.

Morgan, U.M., Reynoldson, J.A. and Thompson, R.C.A. (1993). Activities of several benzimidazoles and tubulin inhibitors against *Giardia* spp. *in vitro*. *Antimicrobial Agents and Chemotherapy* **37**, 328–331.

Morgan, U.M., Constantine, C.C., Greene, W.K. and Thompson, R.C.A. (in press). RAPD (random amplified polymorphic DNA) analysis of *Giardia* DNA and correlation with isoenzyme data. *Transactions of the Royal Society of Tropical Medicine and Hygiene*.

Morris, D.L, Dykes, P.K., Dickson, B., Marriner, S.E., Bogan, J.A. and Burrows, F.G. (1983). Albendazole in hydatid disease. *British Medical Journal* **286**, 103–104.

Moss, D.M., Mathews, H.M., Visvesvara, G.S., Dickerson, J.W. and Walker, E.M.

(1990). Antigenic variation of *Giardia lamblia* in the feces of mongolian gerbils. *Journal of Clinical Microbiology* **28**, 254–257.
Muller, M. (1983). Mode of action of metronidazole on anaerobic bacteria and protozoa. *Surgery* (St Louis) **93**, 165–171.
Muller, M. (1988). Energy metabolism of protozoa without mitochondria. *Annual Reviews on Microbiology* **42**, 465–488.
Murray, H.W., Aley, S.B., Scott, W.A. (1981). Susceptibility of *Entamoeba histolytica* to oxygen intermediates. *Molecular and Biochemical Parasitology* **3**, 381–391.
Namgung, R., Ryu, J.-S., Lee, K.-T. and Soh, C.-T. (1985). The effect of metronidazole and quinacrine on the morphology and the excystation of *Giardia lamblia*. *Yonsei Reports in Tropical Medicine* **16**, 28–44.
Nash, T.E. (1989). Antigenic variation in *Giardia lamblia*. *Experimental Parasitology* **68**, 238–241.
Nash, T.E., McCutchan, T., Keister, D., Dame, J.B., Conrad, J.D. and Gillin, F.D. (1985). Restriction-endonuclease analysis of DNA from 15 *Giardia* isolates obtained from humans and animals. *Journal of Infectious Diseases* **152**, 64–73.
Nash, T.E., Herrington, D.A., Losonsky, G.A. and Levine, M.M. (1987a). Experimental human infections with *Giardia lamblia*. *Journal of Infectious Diseases* **156**, 974–984.
Nash, T.E., Herrington, D.A. and Levine, M.M. (1987b). Usefulness of an enzyme-linked immunosorbent assay for detection of *Giardia* antigen in feces. *Journal of Clinical Microbiology* **25**, 1169–1171.
Nash, T.E., Aggarwal, A., Adam, R.D., Conrad, J.T. and Merritt, J.W. (1988). Antigenic variation in *Giardia lamblia*. *Journal of Immunology* **141**, 636–641.
Nash, T.E., Conrad, J.T. and Merritt, J.W. (1990a). Variant specific epitopes of *Giardia lamblia*. *Molecular and Biochemical Parasitology* **42**, 125–132.
Nash, T.E., Banks, S.M., Allin, D.W., Merritt, J.W. and Conrad, J.T. (1990b). Frequency of variant antigens in *Giardia lamblia*. *Experimental Parasitology* **71**, 415–442.
Nash, T.E., Merritt, J.W. and Conrad, J.T. (1991). Isolate and epitope variability in susceptibility of *Giardia lamblia* to intestinal proteases. *Infection and Immunity* **59**, 1334–1340.
Nayak, N., Ganguly, N.K., Walia, B.N.S., Wahi, V., Kanway, S.S. and Mahajan, R.C. (1987). Specific secretory IgA in the milk of *Giardia lamblia*-infected and uninfected women. *Journal of Infectious Diseases* **155**, 724–727.
Neringer, R., Andersson, Y. and Baker, I.A. (1986). A water-borne outbreak of giardiasis in Sweden. *Scandinavian Journal of Infectious Disease* **19**, 85–90.
Nikolic, A., Petrovic, Z. and Radovic, M. (1990). Prevalence of *Giardia lamblia* in school children in Serbia, Yugoslavia. *Epidemiology of Directly Transmitted Parasites* **S6.D 25**, 755.
Nohynkova, E. (1990). The effect of albendazole on *Giardia intestinalis in vitro*. *Bulletin de la Société Française de Parasitologie* **8**, 199.
O'Brien, R.L., Olenick, J.G. and Hahn, F.E. (1966). Reactions of quinine, chloroquine, and quinacrine with DNA and their effects on the DNA and RNA polymerase reactions. *Proceedings of the National Academy of Sciences of the USA* **55**, 1511.
Ortega-Barria, E., Ward, H.D., Evans, J.E. and Pereira, M.E.A. (1990). N-acetyl-D-glucosamine is present in cysts and trophozoites of *Giardia lamblia* and serves as a receptor for wheatgerm agglutinin. *Molecular and Biochemical Parasitology* **43**, 151–166.

Owen, R.L. (1984). Direct faecal–oral transmission of giardiasis. *In* "*Giardia* and Giardiasis" (S.L. Erlandsen and E.A. Meyer, eds), pp. 329–338. Plenum Press, New York.

Paget, T.A., Fry, M. and Lloyd, D. (1988). The O_2 dependence of respiration in the parasitic nematode *Ascaridia galli. Biochemical Journal* **256**, 880–889.

Paget, T.A., Jarroll, E.L., Manning, P., Lindmark, D.G. and Lloyd, D. (1989a). Respiration in the cysts and trophozoites of *Giardia muris. Journal of General Microbiology* **135**, 145–154.

Paget, T.A., Mendis, A.H.W. and Lloyd, D. (1989b). Respiration in the filarial nematode *Brugia pahangi. International Journal for Parasitology* **19**, 337–343.

Paget, T.A., Raynor, M.H., Shipp, D.W.E. and Lloyd, D. (1990). *Giardia lamblia* produces alanine anaerobically but not in the presence of oxygen. *Molecular and Biochemical Parasitology* **42**, 63–68.

Pamba, H.O., Bwibo, N.O., Chunge, C.N. and Estambale, B.B.A. (1989). A study of the efficacy and safety of albendazole (Zentel) in the treatment of intestinal helminthiasis in Kenyan children less than 2 years of age. *East African Medical Journal* **66**, 197–202.

Pancorbo, J.M.C., Munoz, M.T.G. and Badia, J.L.S. (1985). Giardiasis: treatment of carriers. *Lancet* **ii**, 984.

Parenti, D.M. (1989). Characterization of a thiol proteinase in *Giardia lamblia. Journal of Infectious Diseases* **160**, 1076–1080.

Peattie, D.A. (1990). The giardins of *Giardia lamblia*: genes and proteins with promise. *Parasitology Today* **6**, 52–56.

Peattie, D.A., Alonso, R.A., Hein, A. and Caulfield, J.P. (1989). Ultrastructural localization of giardins to the edges of disk microribbons of *Giardia lamblia* and the nucleotide and deduced protein sequence of alpha giardin. *Journal of Cell Biology* **109**, 2323–2335.

Perlmutter, D.H., Leichtner, A.M., Goldman, H. and Winter, H.S. (1985). Chronic diarrhea associated with hypogammaglobulinemia and enteropathy in infants and children. *Digestive Disease Science* **30**, 1149–1155.

Peterson, W.F., Stauch, J.E. and Ryder, C.D. (1966). Metronidazole in pregnancy. *Americal Journal of Obstetrics and Gynecology* **94**, 343.

Petersen, L.R., Carter, M.L. and Hadler, J.L. (1988). A food-borne outbreak of *Giardia lamblia. Journal of Infectious Diseases* **157**, 846–848.

Phillips, S.C., Mildvan, D., Williams, D.C., Gelb, A.M. and White, M.C. (1981). Sexual transmission of enteric protozoa and helminths in a venereal-disease clinic population. *New England Journal of Medicine* **305**, 603–606.

Pickering, L.K. and Engelkirk, P.G. (1990). *Giardia* among children in day care. *In* "Giardiasis" (E.A. Meyer, ed.), pp. 235–266. Elsevier, Amsterdam.

Pickering, L.K., Evans, D.G., Du Pont, H.L., Vollet, J.J. and Evans, D.J. (1981). Diarrhea caused by *Shigella*, rotavirus, and *Giardia* in day care centers: prospective study. *Journal of Pediatrics* **99**, 51–56.

Pimenta, P.F.P., Pinto da Silva, P. and Nash, T. (1991). Variant surface antigens of *Giardia lamblia* are associated with the presence of a thick cell coat: thin section and label fracture immunocytochemistry survey. *Infection and Immunity* **59**, 3989–3996.

Polis, M.A., Tuazon, C.U., Alling, D.W. and Talmanis, E. (1986). Transmission of *Giardia lamblia* from a day care center to the community. *American Journal of Public Health* **76**, 1142–1144.

Pratt, W.B. and Fekety, R. (1986). "The Antimicrobial Drugs." Oxford University Press, New York.

Proudfoot, N. (1991). Poly(A) signals. *Cell* **64**, 671–674.
Quinn, R.W. (1971). The epidemiology of intestinal parasites of importance in the United States. *Southern Medical Bulletin* **59**, 29–30.
Rabassa, E.B., Arbelo, T.F., Guillot, C.C. and Gonzales, E.S. (1975). Malabsorcion por *Giardia lamblia*. *Review of Cuban Pediatrics* **47**, 247–263.
Radulescu, S. and Meyer, E.A. (1990). *In vitro* cultivation of *Giardia* trophozoites. *In* "Giardiasis" (E.A. Meyer, ed.), pp. 99–110. Elsevier, Amsterdam.
Ralph, E.D. (1983). Clinical pharmacokinetics of metronidazole. *Clinical Pharmacokinetics* **8**, 43–62.
Reiner, D.S., Wang, C. and Gillin, F.D. (1986). Human milk kills *Giardia lamblia* by generating toxic lipolytic products. *Journal of Infectious Diseases* **154**, 825–832.
Reiner, D.S., Douglas, H. and Gillin, F.D. (1989). Identification and localization of cyst-specific antigens of *Giardia lamblia*. *Infection and Immunity* **57**, 963–968.
Reiner, D.S., McCafferty, M. and Gillin, F.D. (1990). Sorting of cyst wall proteins to a regulated secretory pathway during differentiation of the primitive eukaryote, *Giardia lamblia*. *European Journal of Cell Biology* **53**, 142–153.
Rendtorff, R. (1954). The experimental transmission of human intestinal protozoan parasites. II. *Giardia lamblia* cysts given in capsules. *American Journal of Hygiene* **59**, 209–220.
Reynoldson, J.A., Thompson, R.C.A. and Meloni, B.P. (1991a). *In vivo* efficacy of albendazole against *Giardia duodenalis* in mice. *Parasitology Research* **77**, 325–328.
Reynoldson, J.A., Thompson, R.C.A. and Meloni, B.P. (1991b). The mode of action of benzimidazoles against *Giardia* and their chemotherapeutic potential against *Giardia* and other parasitic protozoa. *In* "Biochemical Protozoology" (G.H. Coombs and M.J. North, eds), pp. 587–593. Taylor and Francis, London.
Reynoldson, J.A., Thompson, R.C.A. and Meloni, B.P. (1992a). The potential and possible mode of action of the benzimidazoles against *Giardia* and other protozoa. *Journal of Pharmaceutical Medicine* **2**, 35–50.
Reynoldson, J.A., Thompson, R.C.A. and Horton, R.J. (1992b). Albendazole as a future antigiardial agent. *Parasitology Today* **8**, 412–413.
Riscoe, M.K., Ferro, A.J. and Fitchen, J.H. (1988). Analogs of 5-methylthioribose, a novel class of antiprotozoal agents. *Antimicrobiological Agents and Chemotherapy* **32**, 1904–1906.
Riscoe, M.K., Tower, P.A., Peyton, D.H., Ferro, A.J. and Fitchen, J.H. (1991). Methionine recycling as a target for antiprotozoal drug development. *In* "Biochemical Protozoology" (G. Coombs and M. North, eds), pp. 450–457. Taylor and Francis, London.
Roe, F.J.C. (1977). Metronidazole: a review of uses and toxicity. *Journal of Antimicrobial Chemotherapy* **3**, 205.
Rohrer, L., Winterhalter, K.H., Eckert, J. and Kohler, P. (1986). Killing of *Giardia lamblia* by human milk is mediated by unsaturated fatty acids. *Antimicrobial Agents and Chemotherapy* **30**, 254–257.
Rosenkranz, H.S. and Speck, W.T. (1975). Mutagenicity of metronidazole: activation by mammalian liver microsomes. *Biochemistry and Biophysical Research Communications* **66**, 520.
Rosenthal, P. and Liebman, W.M. (1980). Comparative study of stool examinations, duodenal aspirations and pediatric Entero-Test for the diagnosis of giardiasis in children. *Journal of Pediatrics* **96**, 278–279.
Rosoff, J.D. and Stibbs, H.H. (1986a). Isolation and identification of a *Giardia lamblia*-specific stool antigen (GSA 65) useful in coprodiagnosis of giardiasis.

Journal of Clinical Microbiology **23**, 905–910.

Rosoff, J.D. and Stibbs, H.H. (1986b). Physical and chemical characterization of a *Giardia lamblia*-specific antigen useful in the coprodiagnosis of giardiasis. *Journal of Clinical Microbiology* **24**, 1079–1083.

Rosoff, J.D., Sanders, C.A., Sonnad, S.S., DeLay, P.R., Hadley, W.K., Vincenzi, F.F., Yajko, D.M. and O'Hanley, P.D. (1989). Stool diagnosis of giardiasis using a commercially available enzyme immunoassay to detect *Giardia*-specific antigen 65 (GSA 65). *Journal of Clinical Microbiology* **27**, 1997–2002.

Russell, D.G. and Dubremetz, J.F. (1986). Microtubular cytoskeletons of parasitic protozoa. *Parasitology Today* **2**, 177–179.

Rustia, M. and Shubik, P. (1979). Experimental induction of hepatomas, mammary tumors and other tumors with metronidazole in non-inbred Sas:MRC(WI)BR rats. *Journal of the National Cancer Institute* **63**, 863.

Saimot, A.G., Meulemans, A., Cremieux, A.C., Giovanangeli, M.D., Hay, J.M., Delaitre, B. and Coulaud, J. (1983). Albendazole as a potential treatment for human hydatidosis. *Lancet* **ii**, 652–656.

Salin, M.L. and McCord, J.M. (1974). Superoxide dismutases in polymorphonuclear leukocytes. *Journal of Clinical Investigation* **54**, 1005–1009.

Sangster, N.C., Prichard, R.K. and Lacey, E. (1985). Tubulin and benzimidazole-resistance in *Trichostrongylus colubriformis* (Nematoda). *Journal of Parasitology* **71**, 644–651.

Sauch, J.F. (1985). Use of immunofluorescence and phase-contrast microscopy for detection and identification of *Giardia* cysts in water samples. *Applied Environmental Microbiology* **50**, 1434–1438.

Schantz, P.M. (1982). Echinococcosis. *In* "CRC Handbook Series in Zoonoses" (L. Jacobs and P. Arambulo, section eds), Section C, Vol. 1, pp. 231–237. CRC Press, Boca Raton, Florida.

Schantz, P.M. (1991). Parasitic zoonoses in perspective. *International Journal for Parasitology* **21**, 161–170.

Schofield, P.J., Costello, M., Edwards, M.R. and O'Sullivan, W.J. (1990). The arginine dihydrolase pathway is present in *Giardia intestinalis*. *International Journal for Parasitology* **20**, 697–699.

Schofield, P.J., Edwards, M.R. and Kranz, P. (1991). Glucose metabolism in *Giardia intestinalis*. *Molecular and Biochemical Parasitology* **45**, 39–48.

Schofield, P.J., Edwards, M.R., Mathews, J. and Wilson, J.R. (1992). The pathway of arginine catabolism in *Giardia intestinalis*. *Molecular and Biochemical Parasitology* **51**, 29–36.

Schupp, D.G. and Erlandsen, S.L. (1987). A new method to determine *Giardia* cyst viability: correlation of fluorescein diacetate and propidium iodide staining with animal infectivity. *Applied and Environmental Microbiology* **53**, 704–707.

Seebeck, T., Schneide, A., Schloepp, K. and Hemphill, A. (1988). Cytoskeleton of *Trypanosoma brucei*—the beauty of simplicity. *Protoplasma* **144**, 188–194.

Seebeck, T., Hemphill, A. and Lawson, D. (1990). The cytoskeleton of trypanosomes. *Parasitology Today* **6**, 49–52.

Sears, S.D. and O'Hare, J. (1988). *In vitro* susceptibility of *Trichomonas vaginalis* to 50 antimicrobial agents. *Antimicrobial Agents and Chemotherapy* **32**, 144–146.

Shetty, N., Narasimha, M., Raghuveer, T.S., Elliott, E., Farthing, M.J.G. and Macaden, R. (1990). Intestinal amoebiasis and giardiasis in southern Indian infants and children. *Transactions of the Royal Society of Tropical Medicine and Hygiene* **84**, 382–384.

Sheir-Neiss, G., Lai, M.H. and Morris, N.R. (1978). Identification of a gene for β-tubulin in *Aspergillus nidulans*. *Cell* **15,** 639–647.

Siddall, M.E., Hong, H. and Desser, S.S. (1992). Phylogenetic analysis of the Diplomonadida (Wenyon, 1926) Brugerolle, 1975: evidence for heterochrony in protozoa and against *Giardia lamblia* as a "missing link". *Journal of Protozoology* **39**, 361–367.

Simango, C. and Dindiwe, J. (1987). The aetiology of diarrhoea in a farming community in Zimbabwe. *Transactions of the Royal Society of Tropical Medicine and Hygiene* **81,** 552–553.

Smith, A.L. and Smith, H.V. (1989). A comparison of fluorescein diacetate and propidium iodide staining and *in vitro* excystation for determining *Giardia intestinalis* cyst viability. *Parasitology* **99,** 329–331.

Smith, N.C., Bryant, C. and Boreham, P.F.L. (1988). Possible roles for pyruvate-ferredoxin oxido-reductase and thiol-dependent peroxidase and reductase activities in resistance to nitroheterocyclic drugs in *Giardia intestinalis*. *International Journal for Parasitology* **18,** 991–997.

Smith, P.D. (1985). Pathophysiology and immunology of giardiasis. *Annual Reviews of Medicine* **36,** 295–307.

Smith, P.D., Gillin, F.D., Spira, W.M. and Nash, T.E. (1982). Chronic giardiasis: studies on drug sensitivity, toxin production and host immune response. *Gastroenterology* **83,** 797–803.

Snider, D.P., Skea, D. and Underdown, B.J. (1988). Chronic giardiasis in B-cell-deficient mice expressing the *xid* gene. *Infection and Immunity* **56,** 2838–2842.

Sogayar, M.I.L. and Gregorio, E.A. (1989). Uptake of bacteria by trophozoites of *Giardia duodenalis* (Say). *Annals of Tropical Medicine and Parasitology* **83,** 63–66.

Sogin, M.L., Gunderson, J.H., Elwood, H.J., Alonso, R.A. and Peattie, D.A. (1989). Phylogenetic meaning of the kingdom concept: an unusual ribosomal RNA from *Giardia lamblia*. *Science* **243,** 75–77.

Speelman, P. (1985). Single-dose tinidazole for the treatment of giardiasis. *Antimicrobial Agents and Chemotherapy* **27,** 227–229.

Speelman, P. and Ljungström, I. (1986). Protozoal enteric infections among expatriates in Bangladesh. *American Journal of Tropical Medicine and Hygiene* **35,** 1140–1145.

Stephenson, L.S., Latham, M.C., Kurz, K.M., Kinoti, S.N. and Brigham, H. (1989). Treatment with a single dose of albendazole improves growth of Kenyan school children with hookworm, *Trichuris trichiura,* and *Ascaris lumbricoides* infections. *American Journal of Tropical Medicine and Hygiene* **41,** 78–87.

Stephenson, L.S., Latham, M.C., Kinoti, S.N., Kurz, K.M. and Brigham, H. (1990). Improvements in physical fitness of Kenyan schoolboys infected with hookworm, *Trichuris trichiura* and *Ascaris lumbricoides* following a single dose of albendazole. *Transactions of the Royal Society of Tropical Medicine and Hygiene* **84,** 277–282.

Stephenson, L.S., Latham, M.C., Adams, E. Kinoti, S.N. and Pertet, A. (1991). Albendazole treatment improves physical fitness, growth and appetite of Kenyan school children with hookworm. *FASEB Journal* **5,** 1081.

Stevens, D.P. (1990). Animal model of *Giardia muris* in the mouse. *In* "Giardiasis" (E.A. Meyer, ed.), pp. 91–97. Elsevier, Amsterdam.

Stibbs, H.H., Samadpour, M. and Ongerth, J.E. (1990). Identification of *Giardia lamblia*-specific antigens in infected human and gerbil feces by Western immunoblotting. *Journal of Clinical Microbiology* **28,** 2340–2346.

Stranden, A.M. and Köhler, P. (1991). Swiss *Giardia* isolates of different host origin show great similarities in their metabolism. *Parasitology Research* **77,** 455–

457.

Sullivan, P., Woodward, W.E., Pickering, D.G. and DuPont, H.L. (1984). Longitudinal study of diarrheal disease in day care centers. *American Journal of Public Health* **74,** 987–991.

Swan, J.M. (1984). "Giardiasis in Dogs and Cats in Western Australia". Honours Thesis, Murdoch University, Western Australia.

Swan, J.M. and Thompson, R.C.A. (1986). The prevalence of *Giardia* in dogs and cats in Perth, Western Australia. *Australian Veterinary Journal* **63,** 110–112.

Tai, J.-H., Wang, A.L., Ong, S.-J., Lai, K.-S., Lo, C. and Wang, C.C. (1991). The course of *Giardiavirus* infection in the *Giardia lamblia* trophozoites. *Experimental Parasitology* **73,** 413–423.

Takeuchi, T., Kobayashi, S., Tanabe, M. and Fujiwara, T. (1985). *In vitro* inhibition of *Giardia lamblia* and *Trichomonas vaginalis* growth by bithionol, dichlorophene, and hexachlorophene. *Antimicrobial Agents and Chemotherapy* **27,** 65–70.

Taminelli, V. and Eckert, J. (1989). Haufigkeit und geographische Verbreitung des *Giardia* Befalls bei Wiederkauern in der Schweiz. *Schweizerische Archiv für Tierheilkunde* **131,** 251–258.

Taylor, G.D. and Wenman, W.M. (1987) Human immune responses to *Giardia lamblia* infection. *Journal of Infectious Diseases* **155,** 137–140.

Theodorides, V.J. and Daly, I.H. (1989). Human safety studies with albendazole. *Proceedings of the World Association for the Advancement of Veterinary Parasitology* **13,** S 8–510.

Thompson, R.C.A. (1992a). Giardiasis. *In* "Zoonoses". Proceedings 194 Postgraduate Committee in Veterinary Science, University of Sydney, pp. 88–91.

Thompson, R.C.A. (1992b). Parasitic zoonoses—problems created by people, not animals. *International Journal for Parasitology*, **22**, 555–561.

Thompson, R.C.A., Meloni, B.P. and Lymbery, A.J. (1988a). Humans and cats havegenetically-identical forms of *Giardia*: evidence of a zoonotic relationship. *Medical Journal of Australia* **148**, 207–209.

Thompson, R.C.A., Lymbery, A.J. and Meloni, B.P. (1988b). Giardiasis: a zoonosis in Australia? *Parasitology Today* **4**, 201.

Thompson, R.C.A., Lymbery, A.J. and Meloni, B.P. (1990a). Genetic variation in *Giardia* Kunstler, 1882: taxonomic and epidemiological significance. *Protozoological Abstracts* **14**, 1–28.

Thompson, R.C.A., Lymbery, A.J., Meloni, B.P. and Binz, N. (1990b). The zoonotic transmission of *Giardia* species. *Veterinary Record* **126**, 513–514.

Tibayrenc, M. and Ayala, F.J. (1991). Towards a population genetics of microorganisms: the clonal theory of parasitic protozoa. *Parasitology Today* **7**, 228–232.

Tibayrenc, M., Kjellberg, F. and Ayala, F.J. (1990). A clonal theory of parasitic protozoa: the population structures of *Entamoeba*, *Giardia*, *Leishmania*, *Naegleria*, *Plasmodium*, *Trichomonas* and *Trypanosoma* and their medical and taxonomical consequences. *Proceedings of the National Academy of Sciences of the USA* **87**, 2414–2418.

Tonks, M.C., Brown, T.J. and Ionas, G. (1991). *Giardia* infection of cats and dogs in New Zealand. *New Zealand Veterinary Journal* **39**, 33–34.

Torian, B.E., Barnes, R.C., Stephens, R.S. and Stibbs, H.H. (1984). Tubulin and high-molecular-weight polypeptides as *Giardia lamblia* antigens. *Infection and Immunity* **46**, 152–158.

Ungar, B.L.P., Yolken, R.H., Nash, T.E. and Quinn, T.C. (1984). Enzyme-linked immunosorbent assay for the detection of *Giardia lamblia* in fecal specimens. *Journal of Infectious Diseases* **149**, 90–97.

Upcroft, P. (1991). DNA fingerprinting of the human intestinal parasite *Giardia intestinalis* with hypervariable minisatellite sequences. *In* "DNA Fingerprinting: Approaches and Applications" (T. Burke, G. Dolf, A.J. Jeffries and R. Wolff, eds), pp. 70–83. Birkhauser, Basel.

Upcroft, J., Boreham, P.F.L. and Upcroft, P. (1989a). Geographic variation in *Giardia* karyotypes. *International Journal for Parasitology* **19**, 519–527.

Upcroft, J.A., Boreham P.F.L. and Upcroft, P. (1989b). Molecular mechanisms of drug resistance in *Giardia intestinalis. In* "Immunological and Molecular Basis of Pathogenesis in Parasitic Diseases" (R.C. Ko, ed.), pp. 141–153. University of Hong Kong, Hong Kong.

Upcroft, J.A., Upcroft, P. and Boreham P.F.L. (1990). Drug resistance in *Giardia intestinalis. International Journal for Parasitology* **20**, 489–496.

Van Keulen, H., Horvat, S., Erlandsen, S.L. and Jarroll, E.L. (1991a). Nucleotide sequence of the 5.8S and large subunit rRNA genes and the internal transcribed spacer and part of the external spacer from *Giardia ardeae. Nucleic Acids Research* **19**, 6050.

Van Keulen, H., Campbell, S.R., Erlandsen, S.L. and Jarroll, E.L. (1991b). Cloning and restriction enzyme mapping of ribosomal DNA of *Giardia duodenalis*, *Giardia ardeae* and *Giardia muris. Molecular and Biochemical Parasitology* **46**, 275–284.

Vinayak, V.K., Kum, K., Chandna, R., Vinkateswarlu, K. and Mehta, S. (1985). Detection of *Giardia lamblia* antigen in the feces by counterimmunoelectrophoresis. *Pediatric Infectious Diseases* **4**, 383–386.

Wade, A. and Yapp, G. (1989). "*Giardia*: An Emerging Issue in Water Management". Centre for Continuing Education, Australian National University, Canberra.

Walliker, D., Quakyi, I.A., Wellems, T.E., McCutchan, T.F., Szarfman, A., London, W.T., Corcoran, L.M., Burkot, T.R. and Carter, R. (1987). Genetic analysis of the human malaria parasite *Plasmodium falciparum. Science* **236**, 1661–1666.

Wallis, P.M. and Hammond, B.R., eds (1988). "Advances in *Giardia* Research". Kananaskis Centre for Environmental Research, University of Calgary, Alberta, Canada.

Wallis, P.M., Zammuto, R.M. and Buchanan-Mappin, J.M. (1986). Cysts of *Giardia* spp. in mammals and surface waters in southwestern Alberta. *Journal of Wildlife Diseases* **22**, 115–118.

Wang, C.C. (1984). Parasite enzymes as potential targets for antiparasitic chemotherapy. *Journal of Medical Chemistry* **27**, 1–9.

Wang, C.C. and Aldritt, S. (1983). Purine salvage networks in *Giardia lamblia. Journal of Experimental Medicine* **158**, 1703–1712.

Wang, A.L. and Wang, C.C. (1986). Discovery of a specific double-stranded RNA virus in *Giardia lamblia. Molecular and Biochemical Parasitology* **21**, 269–276.

Wang, A.L. and Wang, C.C. (1991). Viruses of parasitic protozoa. *Parasitology Today* **7**, 76–80.

Ward, H.D., Alroy, J., Lev, B.I., Keusch, G.T. and Pereira, M.E.A. (1985). Identification of chitin as a structural component of *Giardia* cysts. *Infection and Immunity* **49**, 629–634.

Warren, K.S. (1989). Selective primary health care and parasitic diseases. *In* "New Strategies in Parasitology" (K.P.W.J. McAdam, ed.), pp. 217–231. Churchill Livingstone, Edinburgh.

Webster, L.T. (1990). Drugs used in the chemotherapy of protozoal infections. *In* "Goodman and Gilman's Pharmacological Basis of Therapeutics", 8th edn. (A.G. Gilman, T.W. Rall, A.S. Nies and P. Taylor, eds), pp. 978–1007. Pergamon Press, New York.

Weinbach, E.C., Claggett, C.E., Keister, D.B. and Diamond, L.S. (1980). Respiratory metabolism of *Giardia lamblia*. *Journal of Parasitology* **66**, 347–350.

Weinbach, E.C., Costa, J.L. and Wieber, S.C. (1985). Antidepressant drugs suppress growth of the human pathogenic protozoan *Giardia lamblia*. *Research Communications in Chemical Pathology and Pharmacology* **47**, 145–148.

Welsh, J. and McClelland, M. (1990). Fingerprinting genomes using PCR with arbitrary primers. *Nucleic Acids Research* **18**, 7213–7218.

White, K.E., Hedberg, C.W., Edmonson, L.M., Jones, D.B.W., Osterholm, M.T. and MacDonald, K.L. (1989). An outbreak of *Giardiasis* in a nursing home with evidence of multiple modes of transmission. *Journal of Infectious Diseases* **160**, 298–304.

White, M.J.D. (1978). "Modes of Speciation". Freeman, San Francisco.

White, T.C. and Wang, C.C. (1990). RNA dependent RNA polymerase activity associated with the double-stranded RNA virus of *Giardia lamblia*. *Nucleic Acids Research* **18**, 553–559.

Wiesehahn, G.P., Jarroll, E.L., Lindmark, D.G., Meyer, E.A. and Hallick, L.M. (1984). *Giardia lamblia*: autoradiographic analysis of nuclear replication. *Experimental Parasitology* **58**, 94–100.

Williams, J.G.K., Kubelik, A.R., Livak, K.J., Rafalski, J.A. and Tingey, S.V. (1990). DNA polymorphisms amplified by arbitrary primers are useful as genetic markers. *Nucleic Acids Research* **18**, 6531–6535.

Winsland, J.K.D., Nimmo, S., Butcher, P.D. and Farthing, M.J.G. (1989). Prevalence of *Giardia* in dogs and cats in the United Kingdom: survey of an Essex veterinary clinic. *Transactions of the Royal Society of Tropical Medicine and Hygiene* **83**, 791–792.

Wolfe, M.S. (1990). Clinical symptoms and diagnosis by traditional methods. *In* "Giardiasis" (E.A. Meyer, ed.), pp. 175–185. Elsevier, Amsterdam.

Woo, P.T.K. and Paterson, W.B. (1986). *Giardia lamblia* in children in day-care centres in southern Ontario, Canada, and susceptibility of animals to *Giardia lamblia*. *Transactions of the Royal Society of Tropical Medicine and Hygiene* **80**, 56–59.

Woods, R.A., Malone, K.M.B., Spence, A.M., Sigurdson, W.J. and Byard, E.H. (1989). The genetics, ultrastructure, and tubulin polypeptides of mebendazole resistant mutants of *Caenorhabditis elegans*. *Canadian Journal of Zoology* **67**, 2422–2431.

Yardley, J.H., Takano, J. and Hendrix, T.R. (1964). Epithelial and other mucosal lesions of the jejunum in giardiasis: jejunal biopsy studies. *Bulletin of Johns Hopkins Hospital* **115**, 389–406.

Yarlett, N., Rowlands, C.C., Yarlett, N.C., Evans, J.C. and Lloyd, D. (1987). Reduction of niridazole by metronidazole-resistant and susceptible strains of *Trichomonas vaginalis*. *Parasitology* **94**, 93–99.

Zhong, H., Cao, W., Rossignol, J.F., Feng, M., Hu, R., Gan, S.B. and Tan, W. (1986). Albendazole in nematode, cestode, trematode and protozoan (*Giardia*) infections. *Chinese Medical Journal* (English edition) **99**, 912–915.

Zimmer, J.F. and Burrington, D.B. (1986) Comparison of four protocols for the treatment of canine giardiasis. *Journal of Small Animal Hospital Association* **22**: 168–172.

Immunology of Leishmaniasis

F. Y. LIEW AND C. A. O'DONNELL

Department of Immunology, University of Glasgow, Western Infirmary, Glasgow, G11 6NT, UK

ADVANCES IN PARASITOLOGY VOL. 32
ISBN 0-12-031732-X

I. Introduction

Leishmaniasis is caused by species of the intracellular protozoan parasite belonging to the genus *Leishmania*. There are three main categories of leishmaniasis: cutaneous leishmaniasis (oriental sore), mucocutaneous leishmaniasis (espundia) and visceral leishmaniasis (kala-azar). Leishmaniasis is found in most parts of the world (Table 1) and has an incidence of 400 000 new cases per year. Currently, the world-wide prevalence of leishmaniasis is estimated to be in the region of 12 million cases (Modabber, 1987).

Most forms of leishmaniasis are zoonotic, and humans are infected only secondarily. Animal reservoirs of species pathogenic to man include dogs, rodents and sloths. The parasites are transmitted by the female sandfly, of the genera *Phlebotomus* ("Old World") or *Lutzomyia* ("New World") to mammals (Fig. 1), where they invade and multiply within macrophages of the skin, mucous membranes and viscera (Behin and Louis, 1984). Flagellated promastigotes develop in the gut of the sandfly and in cell-free cultures. Transformation into the amastigote stage occurs within the mammalian macrophage.

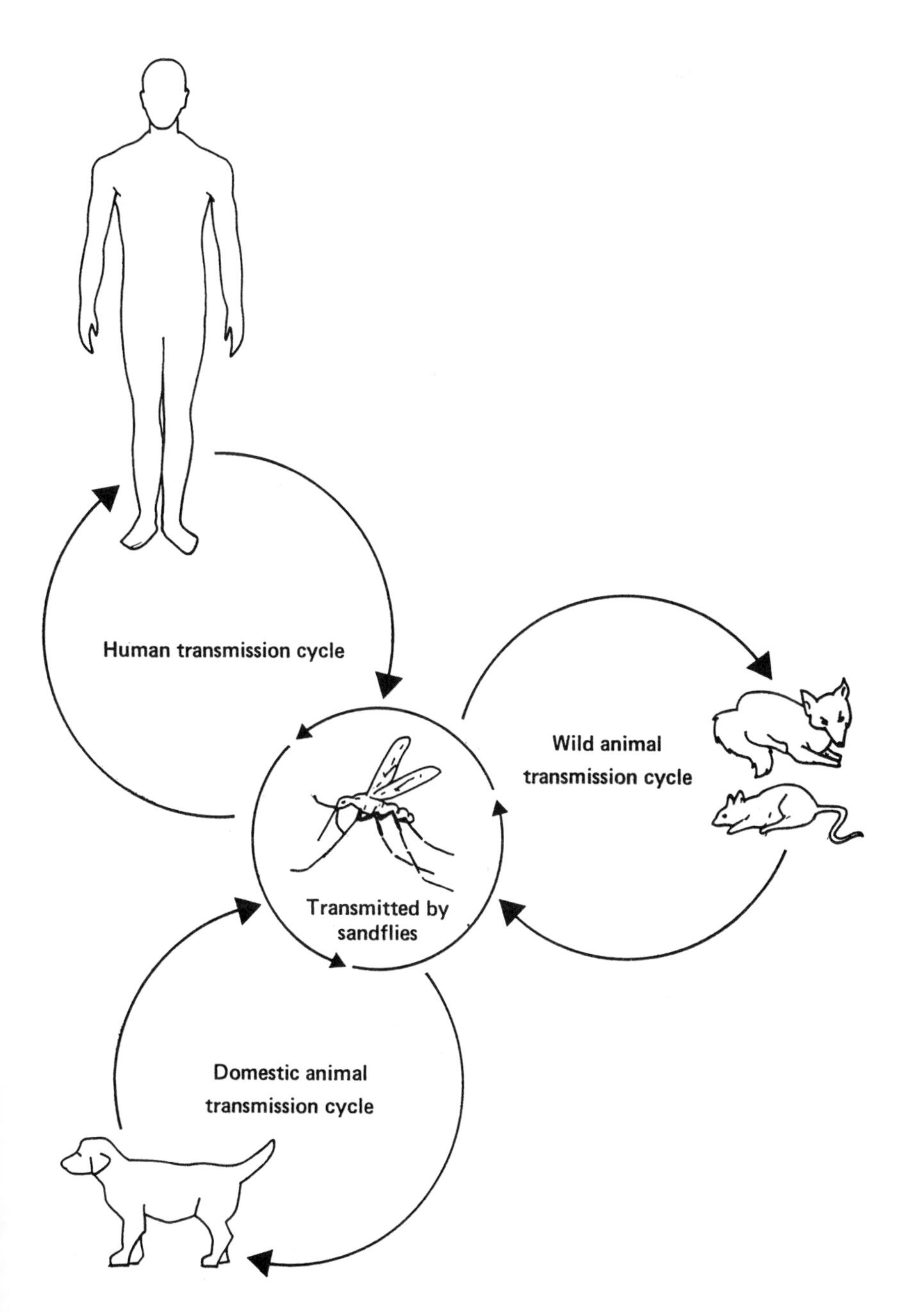

FIG. 1. A schematic representation of the reservoirs of *Leishmania* infection in humans and animals.

TABLE 1 *Major species of* Leishmania *causing disease in humans*

Species	*Region*[a]	*Disease*[b]
L. major	Old World	CL
L. tropica	Old World	CL
L. aethiopica	Old World	DCL
L. mexicana	New World	ACL/DCL
L. amazonensis	New World	ACL/MCL
L. brasiliensis	New World	ACL/MCL
L. donovani	Old World	VL
L. chagasi	New World	VL

[a]"Old World" includes southern Europe, Africa, Middle East, India and south-east Asia; "New World" includes Central and South America.
[b]CL, cutaneous leishmaniasis; DCL, diffuse cutaneous leishmaniasis; ACL, American cutaneous leishmaniasis; MCL, mucocutaneous leishmaniasis; VL, visceral leishmaniasis.

A. THE DISEASE

In humans, infection with *Leishmania* results in a spectrum of disease dependent upon the species involved (Table 1) and the efficiency of the host's immune response to the parasite (Turk and Bryceson, 1971). Cutaneous leishmaniasis, caused by *L. major* and *L. tropica,* is characterized by a skin ulcer which heals spontaneously, leaving an unsightly scar. Diffuse cutaneous leishmaniasis, caused by *L. aethiopica* and *L. mexicana,* causes widespread thickening of the skin with lesions, resembling those of lepromatous leprosy, which do not heal spontaneously. Infection with *L. braziliensis* results in mucocutaneous leishmaniasis due to metastasis of organisms to mucosal sites from a primary cutaneous lesion established much earlier. Metastatic spread may occur to the oronasal and pharyngeal mucosa, causing highly disfiguring leprosy-like tissue destruction and swelling. Visceral leishmaniasis or kala-azar is caused by *L. donovani* or *L. chagasi.* Common symptoms include fever, malaise, weight loss, coughing and diarrhoea accompanied by anaemia, skin darkening and hepatosplenomegaly. Visceral leishmaniasis is fatal if untreated due to the failure of the host to mount an effective protective immune response.

Diagnosis of active leishmaniasis is based primarily on the demonstration of the parasite in biopsies. A skin test using a killed whole parasite preparation (leishmanin) is used as a presumptive test, but is not fully specific. Serological tests, e.g., complement fixation tests, are neither specific nor sensitive. The more sophisticated techniques of immunofluorescence and counter-current electrophoresis require specialized equipment and are thus unsuitable for field use. Some new test systems, including the use of monoclonal antibodies, enzyme-linked immunosorbent assays (ELISAs), indium-based slide precipitation and deoxyribonucleic acid (DNA) probes, are now being used, with various degrees of success.

Primary drug treatment of leishmaniasis is based on antimony compounds, notably the pentavalent antimonial compounds sodium stibogluconate and N-methylglucamine antimonate (reviewed by Bryceson, 1987). However, these must be given in daily intramuscular doses for several weeks, have unpleasant side effects and are not very effective against cutaneous leishmaniasis. Vaccination may be the best answer to the control of leishmaniasis (Alexander, 1989; Liew, 1989a) as individuals who have recovered from the disease remain refractory to further infection. Therefore, vaccination is feasible in principle. However, so far, the only immunization strategy against leishmaniasis used with any success in humans has been restricted to the cutaneous disease. This is based on convalescent immunity following controlled induction of a lesion, in an aesthetically acceptable site, with viable *L. major* (Greenblatt, 1980). This process, called leishmanization, is effective but fraught with complications. Although it may be acceptable in highly endemic areas as a last resort for the control of cutaneous leishmaniasis, it is unlikely to be of general use. Therefore, there is a great need to develop a safe and effective vaccine against all forms of leishmaniasis.

Before a safe and effective vaccine against leishmaniasis can be developed, the involvement of the immune response in leishmanial infection must be fully understood. The aim of this review is to summarize the understanding of the role of the immune response to *Leishmania*, with particular reference to the cutaneous disease caused by *L. major*.

II. Genetic Regulation of Leishmaniasis

A number of excellent reviews on the genetic regulation of leishmaniasis exist (Howard, 1985; Blackwell and Alexander, 1986; Bradley, 1987; Blackwell, 1988; Blackwell *et al.*, 1991; Alexander and Russell, 1992). Therefore, we will only summarize the present knowledge, with particular reference to *L. major*.

A. Genetic Regulation in the Mouse

The use of inbred and congeneic mouse strains has greatly advanced our understanding of the genetic control of leishmaniasis. Several different host genes control the infection of distinct species of *Leishmania*. Two stages of genetic regulation have been identified—primary innate susceptibility and subsequent acquired immunity. This division is particularly clear in visceral leishmaniasis (caused by *L. donovani*), where innate susceptibility is controlled by the *Lsh* gene (Bradley, 1977), whilst acquired immunity is controlled by at least three additional genes, *Rld-1* (linked to the major histocompatibility loci, H-2) H-11 and Ir-2. In cutaneous leishmaniasis, the distinction between innate and acquired resistance is less clear, but the susceptibility of inbred mice to *L. major* infection is controlled by *Scl-1*, *Scl-2* (Blackwell *et al.*, 1985a) and H-11 linked genes (Blackwell *et al.*, 1985b).

1. Lsh

Inbred mouse strains infected with *L. donovani* were shown to separate into two groups as determined by the number of parasites in the liver by day 15 after infection—those with an increase < eight-fold (resistant) and those with an increase > 80-fold (susceptible) (Bradley, 1977). This was found to be under the control of a single gene, *Lsh*, which maps to mouse chromosome 1 (Bradley *et al.*, 1979). *Lsh* is thought to be identical to the genes designated *Ity* and *Bcg* (Plant *et al.*, 1982; Blackwell, 1989). These genes control the response to *Salmonella typhimurium* and *Mycobacterium* species respectively (Plant and Glynn, 1979; Brown *et al.*, 1982; Skamene *et al.*, 1982). It has also been suggested that this gene may be the same as *Idd-5*, one of the major genes regulating type 1 insulin-dependent diabetes (Cornall *et al.*, 1991).

Lsh gene expression occurs 2–3 days after infection, both *in vitro* and *in vivo* (Crocker *et al.*, 1984, 1987), and results in an inhibition of parasite multiplication. Although expression is T cell independent (Bonventre and Nickol, 1984), it is associated with up-regulation of major histocompatibility complex (MHC) class II expression, an increase in responsiveness to lipopolysaccharide (LPS) and interferon γ (IFN-γ) and increased production of tumour necrosis factor (TNF) and interleukin (IL) 1 (Zwilling *et al.*, 1987; Kaye *et al.*, 1988; Blackwell *et al.*, 1989; Kaye and Blackwell, 1989). Furthermore, bone-marrow-derived macrophages from Lsh^r mice produce increased levels of nitric oxide (NO) compared with macrophages from Lsh^s mice in response to IFN-γ and LPS (Blackwell *et al.*, 1991; Roach *et al.*, 1991).

Although *Lsh* gene control is important in *L. donovani* infection, it has no involvement in the genetic control of *L. major* infection as both Lsh^r and Lsh^s strains were found to be equally susceptible to infection with *L. major* (Mock *et al.*, 1985a).

2. *H-2 linked gene* (Rld-1)

When infected susceptible, homozygous recessive (*Lsh*^{s}) mouse strains were monitored for longer periods, over the immune stage of disease, three different disease profiles were observed, defined by the parasite load in the liver (Blackwell *et al.*, 1980; Ulczak and Blackwell, 1983). These were dependent on the H-2 haplotype of the strain used and were defined as early cure (H-2^s, H-2^r), cure (H-2^b) and non-cure (H-2^d, H-2^q, H-2^f). The genes controlling these responses map to I-E and a subregion to the left, presumably I-A, on the murine MHC (Blackwell, 1983). However, the disease profile obtained is dependent on the size of the infecting inoculum and the effect can be reversed by altering the infective dose (Ulczak and Blackwell, 1983). Furthermore, both the genetically determined cure phenotype and the manipulated cure response in the non-cure phenotype appear to be mediated by $CD4^+$ T cells (Blackwell and Ulczak, 1984; Ulczak *et al.*, 1989).

In contrast to *L. donovani*, the H-2 linked gene has a less pronounced effect on cutaneous *L. major* infection, although differences in the severity of lesion development can be observed (Howard *et al.*, 1980a; Mitchell *et al.*, 1981a).

3. *H-11 linked gene*

Studies with the B10.129 (10M) congenic strain of mouse, which carries alternative alleles at H-11, revealed that the normal cure phenotype of the B10 mouse had reverted to a non-cure phenotype (DeTolla *et al.*, 1980; Blackwell *et al.*, 1985b). However, unlike *Lsh* or *Rld-1*, expression of the H-11 linked gene has a similar effect on both visceral *L. donovani* and cutaneous *L. major* infection (Blackwell *et al.*, 1985b). As yet, this is the only gene known to affect susceptibility to both *L. donovani* and *L. major*.

4. *Ir-2 linked gene*

The B10.LP-H-3^b congeneic strain, which carries the alternative Ir-2^b allele on chromosome 2, has been found to revert from a cure to early cure phenotype on infection with *L. donovani* (DeTolla *et al.*, 1980).

5. Scl-1

Unlike the situation with *L. donovani* infection, it is more difficult to separate the genetic regulation involved in the innate and immune response to *L. major* infection. However, different inbred mouse strains do exhibit different patterns of lesion development following subcutaneous injection with *L.*

major (Mock *et al.*, 1985a; Bradley, 1987). When different strains were examined for their resistance or susceptibility to *L. major* infection, it was found that the BALB/c mouse was extremely susceptible to disease, even with minimal infecting doses (Howard *et al.*, 1980a). This extreme susceptibility was controlled by a single autosomal gene not linked to H-2 (Howard *et al.*, 1980a; DeTolla *et al.*, 1981), which could be transferred by donor haematopoietic cells in mouse radiation chimaeras (Howard *et al.*, 1980b). The susceptibility to cutaneous leishmaniasis (*Scl*) gene was provisionally mapped to chromosome 8 (Blackwell *et al.*, 1984). However, an alternative location on mouse chromosome 11 has also been suggested (Mock *et al.*, 1985b). Expression of *Scl-1* results in a primary macrophage defect, allowing increased parasite multiplication *in vitro* in skin macrophages from susceptible mice and differences in the ability of these macrophages to process and present antigen (Gorczynski and MacRae, 1982). This may be due to the reduced level of MHC class II antigens expressed on infected BALB/c macrophages (Handman *et al.*, 1979; Gorczynski and MacRae, 1982). Susceptible BALB/c mice develop parasite-specific suppression of cell-mediated immunity within 4 weeks of infection (Howard *et al.*, 1980c).

6. Scl–2

Scl–2 is associated with the phenotype of DBA/2 mice resistant to infection with *L. mexicana* (Blackwell and Alexander, 1986; Roberts *et al.*, 1990). This gene maps to chromosome 4.

B. GENETIC REGULATION IN HUMANS

As yet, little is known about the effect of genetic variation on leishmaniasis in humans. However, both visceral and cutaneous infections can exhibit a wide clinical spectrum in humans (Bryceson, 1970; Pampiglione *et al.*, 1974). In mucocutaneous leishmaniasis, caused by *L. braziliensis*, signs of genetic variation have been recorded. The severity of mucocutaneous leishmaniasis in co-existing populations of American Indians and Negroes, who had retained genetic integrity in a remote area of Bolivia, was assessed (Walton and Valverde, 1979). The Negro population had rapidly growing mucocutaneous lesions, resulting in severe facial damage, and strong delayed-type hypersensitivity (DTH) responses to leishmanial antigen. In contrast, the American Indian population had chronic, slow-growing, non-destructive lesions and depressed DTH responses. However, identification of human homologues for the murine genes controlling resistance and susceptibility is required, as is familial analysis of the disease.

III. Immune Response to *Leishmania*—a Historical Perspective

A. The Immune Response in Humans

The development of long-term immunity following recovery from cutaneous infection with *Leishmania* has long been recognized (reviewed by Turk and Bryceson, 1971; Mauël and Behin, 1987). In both endemic areas and in experimental volunteers, recovery from an initial infection allowed to run its course resulted in long-lasting protection from reinfection with either *L. major* or *L. mexicana* (Biagi, 1953; Adler and Gunders, 1964; Guirges, 1971). Similar results have been obtained in visceral leishmaniasis following infection with *L. donovani* (Manson-Bahr, 1961). Although different forms of the disease result in a spectrum of immune responses, both cell-mediated (CMI) and humoral immune responses can be induced (Howard, 1985; Mauël and Behin, 1987).

1. *Cutaneous leishmaniasis*

This is characterized by a strong CMI response, detectable both *in vivo* and *in vitro*. The *in vivo* response is detectable as a strong DTH skin reaction to whole killed parasite preparations. This response is often detectable before healing occurs and, once positive, remains so for many years (Guirges, 1971). *In vitro*, peripheral blood lymphocytes from actively infected or recovered individuals undergo blast transformation and proliferation in the presence of leishmanial antigen (Tremonti and Walton, 1970; Neva *et al.*, 1979; Wyler *et al.*, 1979; Green, M. S. *et al.*, 1983) and inhibition of macrophage migration has also been used to demonstrate CMI (Zeledón *et al.*, 1977). Humoral responses have been more difficult to detect (Wyler *et al.*, 1979; Green, M. S. *et al.*, 1983), but specific antibody has been detected in most patients with cutaneous leishmaniasis and the titre is directly linked to the severity of the disease (Behforouz *et al.*, 1976; Menzel and Bienzle, 1978; Roffi *et al.*, 1980).

2. *Mucocutaneous leishmaniasis*

Both CMI and humoral responses have been demonstrated in patients infected with *L. braziliensis* (Walton *et al.*, 1972; Anthony *et al.*, 1980; Ridley *et al.*, 1980). However, it is still unclear how the parasite can metastasize from a primary cutaneous lesion to the mucosa in the presence of a cell-mediated response.

3. *Visceral leishmaniasis*

Infection with *L. donovani* is associated with impaired CMI but a marked humoral response. The absence of CMI is characterized by the absence of a DTH response during the primary infection (Turk and Bryceson, 1971; Rezai *et al.*, 1978; Wyler *et al.*, 1979; Haldar *et al.*, 1983). Peripheral T cell responses are also impaired, both to parasite antigens (Wyler *et al.*, 1979; Haldar *et al.*, 1983) and to mitogens (Ghose *et al.*, 1979). However, the failure of the CMI response may be due to an inability to control the infection because, after successful chemotherapy, both the skin test response and *in vitro* lymphocyte proliferation responses become positive (Manson-Bahr, 1961). In contrast to the CMI response, there is a marked humoral response in visceral leishmaniasis including raised non-specific immunoglobulin (Ig) M and IgG levels (Rezai *et al.*, 1978; Ghose *et al.*, 1980) due to polyclonal B cell activation (Ghose *et al.*, 1980; Galvao-Castro *et al.*, 1984). High titres of specific anti-*L. donovani* antibody can also be detected (Duxbury and Sadun, 1964; Rezai *et al.*, 1978).

Although the immune response to leishmaniasis in humans is well characterized, its generation and control are not well understood. Therefore, in order to manipulate and examine the immune response, animal models of infection have been developed. We will first review the early work on animal models of cutaneous leishmaniasis.

B. THE IMMUNE RESPONSE IN THE GUINEA-PIG

The earliest animal model developed was *L. enriettii* infection of guinea-pigs. Inoculation of guinea-pigs with *L. enriettii* resulted in the development of a self-healing cutaneous lesion (Paraense, 1953). This was accompanied by the development of a DTH reaction within 1–2 weeks of infection and, following recovery, resistance to reinfection (Glazunova, 1965; Bryceson *et al.*, 1970). However, antibody could not be detected in these animals (Bryceson *et al.*, 1970). The T cell responses could also be assayed *in vitro* by measuring blast transformation in response to parasite antigens, macrophage migration inhibition and the production of lymphokines (Bryceson *et al.*, 1970; Blewett *et al.*, 1971; Gorczynski, 1983).

The requirement for a functionally intact immune system in recovery from cutaneous leishmaniasis was first demonstrated in the guinea-pig model. Suppression of the cell-mediated immune response by treatment with anti-lymphocyte serum, cyclophosphamide or X-irradiation resulted in non-healing disease and cutaneous dissemination of the parasite (Bryceson and Turk, 1971; Lemma and Yau, 1973; Belehu *et al.*, 1976).

However, the guinea-pig model has now been superseded by inbred mouse strains. These exhibit a range of disease severities, depending on the strain used, and are easier to manipulate immunologically, particularly as so much is known about the immunology and genetics of the mouse.

C. THE IMMUNE RESPONSE IN THE MOUSE

Intradermal inoculation with *L. major* results in the development of a skin lesion. The size and duration of these lesions were dependent on the infecting dose (Preston and Dumonde, 1976a; Preston *et al.*, 1978). However, as described in Section II.A.5, the genetic predisposition of the strain used was also critical to the eventual outcome of the infection, with some strains, e.g. CBA, highly resistant to *L. major* and others, e.g. BALB/c, extremely susceptible (Howard *et al.*, 1980a).

The fundamental importance of CMI in resistance to leishmaniasis has also been demonstrated in murine models. Resistant strains of mice, rendered relatively T cell deficient by thymectomy followed by irradiation and reconstitution with syngeneic bone marrow, were less able to control *L. major* infection and had delayed healing (Preston *et al.*, 1972). Similar results were obtained with *L. donovani* infection (Skov and Twohy, 1974a,b). Exacerbation of disease was even more marked in T cell deficient athymic nude (*nu/nu*) mice, where even the normally highly resistant CBA and C57BL/6 strains were unable to control *L. major* infection, resulting in progressive dissemination of the parasite (Handman *et al.*, 1979; Mitchell *et al.*, 1980). However, reconstitution with as few as 10^6 syngeneic T cells restored normal resistance in these mice and these cells appeared to be Lyt-1^+2^- (Mitchell *et al.*, 1980, 1981b). Protective immunity against both *L. major* and *L. donovani*, as a result of recovery from infection or prophylactic immunization, could be adoptively transferred to syngeneic recipients (Preston and Dumonde, 1976a; Rezai *et al.*, 1980). This adoptive transfer of immunity also required Lyt-1^+2^- T cells, but not B cells (Liew *et al.*, 1982, 1984). Finally, although treatment of resistant C3H mice with anti-μ antibody rendered them defective in antibody production and susceptible to *L. major* infection, lesion progression could be arrested and the disease outcome reversed by the adoptive transfer of T cells alone from normal C3H donors without any restoration of antibody formation (Scott *et al.*, 1986).

Therefore, these early findings demonstrated the importance of CMI in resistance to leishmaniasis. Work carried out over the past decade has examined the relative roles of different T cell subsets, macrophages and lymphokines in the host response to leishmanial infection.

IV. Immune Response to *Leishmania* Infection

A. Humoral Immunity

Anti-leishmanial antibodies have been shown *in vitro* to lyse promastigotes in the presence of complement (Pearson and Steigbigel, 1980; Mosser and Edelson, 1984), to promote phagocytosis (Herman, 1980), and to induce surface patching and capping on both promastigotes and amastigotes (Dwyer, 1976). However, there is little evidence for a corresponding role *in vivo* for antibody in determining the outcome of leishmanial infection. Although some monoclonal antibodies (Anderson *et al.*, 1983) or their Fab fragments (Handman and Mitchell, 1985), when mixed in large quantities with the parasites before infection, reduced the infectivity of promastigotes, they had no effect on the disease development if injected separately from the promastigotes.

In fact, the available evidence argues strongly against a protective role for antibody in controlling leishmaniasis. The outcome of disease in different inbred mouse strains is not associated with the titre or isotype of the antibody response (Olobo *et al.*, 1980; Rezai *et al.*, 1980). In addition, the relatively late appearance of a low level of antibody during healing is unlikely to be effective against *Leishmania*, an intracellular parasite. Mice genetically selected for low antibody response (Biozzi AB/L, selection 1) were highly resistant to *L. major* infection and developed small self-healing lesions despite the low level of antibody produced (Hale and Howard, 1981). Furthermore, prolonged administration of large amounts of hyperimmune serum or antibody fractions from donor mice protectively immunized against *L. major* failed to influence the incidence or outcome of infection in highly susceptible BALB/c mice (Howard *et al.*, 1984). Mice protectively immunized with irradiated promastigotes did develop higher antibody titres than infected mice. However, splenectomy before immunization led to a substantial decrease in the antibody response without impairing the level of protection (Howard *et al.*, 1984). Finally, treatment of mice from birth with anti-μ antibody profoundly affected the outcome of *L. major* infection (Sacks *et al.*, 1984; Scott *et al.*, 1986). When genetically susceptible BALB/c mice were depleted of B cells, they developed enhanced resistance to *L. major* infection (Sacks *et al.*, 1984). Conversely, B cell depletion in a genetically resistant strain led to an enhanced infection, as indicated by larger lesions than in control mice and an inability to heal the infection (Scott *et al.*, 1986). However, in each case, the effect was not reversed by the transfer of specific antibody (F. Y. Liew, unpublished observation) and was more likely to be due to the depletion of the antigen-presenting function of the B cells or the depletion of cytokines necessary for the development of the T cell response.

Therefore, it appears that antibody does not play a central role in the *in vivo* control of cutaneous leishmaniasis. In contrast to antibody, evidence for the role of various aspects of CMI is compelling. This will be discussed in the following sections.

B. NATURAL KILLER CELLS

Little work has been done on the possible role of NK cells in leishmaniasis. When skin cell biopsies from patients with American cutaneous leishmaniasis were examined, a substantial proportion of the cells (38%) expressed an NK cell phenotype (Ridel *et al.*, 1988), although no functional study was carried out. The involvement of NK cells has been examined in experimental systems using C57BL/6 *bg/bg* mice which are profoundly deficient in NK cell activity. These mice were shown to be modestly less capable of eliminating *L. donovani* infection than the phenotypically normal controls. However, there was no difference seen in the course of *L. major* infection (Kirkpatrick and Farrell, 1982, 1984). Therefore, although NK cells may be involved in the response to visceral leishmaniasis, they appear not to be involved in immunity to cutaneous leishmaniasis.

C. γ/δ T CELLS

In American cutaneous leishmaniasis (*L. braziliensis* infection), 20% of $CD3^+$ T cells in the skin lesions were $\gamma\delta$-TCR^+ cells (Modlin *et al.*, 1989). This contrasted with only 5% $\gamma\delta^+$ cells in the peripheral blood of these patients, in DTH lesions or in mucocutaneous lesions. Analysis of these cells within microanatomical regions of the lesions revealed their limited diversity and it may be that specific $\gamma\delta$ T cells are clonally selected by antigen in the lesions and expanded within a microanatomical region (Uyemura *et al.*, 1992). Activated $\gamma\delta$ T cells were also demonstrated in the peripheral blood of patients with visceral leishmaniasis (Raziuddin *et al.*, 1992), secreting elevated levels of cytokines involved in B cell growth and differentiation, which may contribute to the polyclonal B cell activation associated with visceral leishmaniasis. However, preliminary work in the mouse indicated that $\gamma\delta$ T cells are not involved in the pathogenesis of cutaneous leishmaniasis (Lohoff *et al.*, 1991; Titus *et al.*, 1993). $\gamma\delta$ T cells were not detected in the lymph nodes of either resistant or susceptible mice 7 weeks after infection with *L. major* (Lohoff *et al.*, 1991). In addition, mice injected subcutaneously in the footpad with a pan-anti-γδ-TCR antibody 24 h before infection in the same site with *L. major* developed similar lesions to untreated infected controls (Titus *et al.*, 1993). Therefore, the possible role of γδ T cells in leishmaniasis, particularly cutaneous leishmaniasis, requires further investigation.

D. CD8$^+$ T CELLS

A role for CD8$^+$ cytotoxic T cells in the control of cutaneous leishmaniasis has been suggested. Initial evidence came from the guinea-pig model of *L. enriettii* infection (Bray and Bryceson, 1968). Recent evidence has also implicated CD8$^+$ T cells in the response to *L. major* infection. Three weeks after infection, at the onset of lesion healing, resistant CBA mice were found to have three times as many parasite-specific CD8$^+$ T cells in the lymph-nodes draining the cutaneous lesions as did susceptible BALB/c mice (Milon *et al.*, 1986; Titus *et al.*, 1987). The possible role of these cells in infection was further elucidated using an anti-CD8 monoclonal antibody (MAb). Administration of anti-CD8 MAb exacerbated cutaneous *L. major* lesions in both resistant and susceptible strains of mice (Titus *et al.*, 1987) and resulted in substantially greater numbers of parasites in the lesions compared to untreated, infected controls. However, even though the CD8$^+$ cells were severely depleted, the genetically resistant strain still exhibited lesion healing. Deletion of CD8$^+$ T cells in mice protected by sublethal irradiation or by intravenous immunization also led to enhanced disease (Farrell *et al.*, 1989; Hill *et al.*, 1989). It may be that in certain experimental models in which BALB/c mice are rendered resistant to infection, e.g. by thymectomy, lethal irradiation and reconstitution with syngeneic bone marrow cells or by injection with anti-CD4 MAb, removal of suppressive CD4$^+$ T cells allows the expansion of a protective CD8$^+$ T cell population (Hill, 1991; Müller *et al.*, 1991). These CD8$^+$ cells can transfer *Leishmania*-specific DTH responses. However, a clear need for CD8$^+$ T cells has not been consistently demonstrated. In nude mice of both resistant and susceptible strains, the adoptive transfer of CD8$^+$ T cells had no effect, whereas CD4$^+$ T cells completely restored immunity to infection with *L. major* (Moll *et al.*, 1988). In visceral leishmaniasis, the ability of nude mice to control *L. donovani* infection was achieved only by the adoptive transfer of both CD4$^+$ and CD8$^+$ T cells (Stern *et al.*, 1988).

It has proven to be difficult to demonstrate killing of *Leishmania*-infected macrophages *in vitro* by cytotoxic lymphocytes (CTL) in conventional 4 h ^{51}chromium-release assays (Coutinho *et al.*, 1984; Smith *et al.*, 1991). However, in an indirect assay system using a *Plasmodium* Circumsporozoite protein-specific CD8$^+$ cytotoxic T cell clone, it was demonstrated that extended incubation of the $CD8^+$ cells with infected macrophages resulted in macrophage lysis and parasite killing (Smith *et al.*, 1991). This required an incubation period of 24–72 h; less than 24 h led to macrophage lysis but the released amastigotes remained viable. This killing appeared to be mediated, at least in part, by IFN-γ as addition of an anti-IFN-γ antibody partly abrogated the leishmanicidal effect.

$CD8^+$ CTL have traditionally been associated with resistance to viral infections, but there is now increasing evidence that these cells may also have an important role in immunity to intracellular microbes, including *Listeria* and *Mycobacterium* (Chiplunkar *et al.*, 1986; De Libero and Kaufmann, 1986; Kaufmann, 1988; Brunt *et al.*, 1990). This may be due to direct cytotoxicity or through cytokine production. $CD8^+$ T cells produce IFN-γ (Prystowsky *et al.*, 1982; Kelso and Glasebrook, 1984; Chiplunkar *et al.*, 1986; De Libero and Kaufmann, 1986; Fong and Mosmann, 1990) as well as tumour necrosis factor (Fong and Mosmann, 1990). Both these lymphokines are known to be important in activating macrophages to kill *Leishmania* (see Section VIII). Therefore, $CD4^+$ and $CD8^+$ T cells may interact to kill *Leishmania*-infected macrophages, both by the production of lymphokines and by direct lysis. In this regard, direct lysis of infected target cells by $CD8^+$ T cells has been demonstrated (J. A. Louis, personal communication). Therefore, it is likely that a greater role for $CD8^+$ T cells in the control of leishmaniasis will become apparent.

E. $CD4^+$ T CELLS

Although a role for $CD8^+$ T cells in leishmaniasis appears to be likely, the important cell in mediating protective immunity in leishmaniasis is undoubtedly the $CD4^+$ T cell. As described in Section III.C, T cell deficient mice were unable to control *Leishmania* infection (Preston *et al.*, 1972; Skov and Twohy, 1974a; Handman *et al.*, 1979; Mitchell *et al.*, 1980), while reconstitution with $CD4^+$ T cells restored resistance to infection (Mitchell *et al.*, 1980, 1981b; Moll *et al.*, 1988). $CD4^+$ T cells from recovered or immunized mice could also transfer protective immunity to syngeneic recipients (Preston and Dumonde, 1976a; Rezai *et al.*, 1980; Liew *et al.*, 1982, 1984). However, in BALB/c mice a different picture began to emerge.

BALB/c mice are exceptionally susceptible to infection with *L. major*, developing a uniformly fatal disseminating disease, even with a minimal infecting dose (see Section II.A.5). However, the failure of BALB/c mice to control *L. major* infection is not due to an intrinsic inability to mount an immune response. Various treatments of BALB/c mice before infection rendered these mice resistant to *L. major* infection, including sublethal whole body γ-irradiation (Howard *et al.*, 1981) and injection with anti-CD4 MAb (Titus *et al.*, 1985a) or cyclosporin A (Behforouz *et al.*, 1986; Solbach *et al.*, 1986). The recovered mice developed a classical DTH response to *Leishmania* antigen and antigen-specific culture supernatants could activate normal resident peritoneal macrophages to kill intracellular *L. major* amastigotes (Liew and Dhaliwal, 1987). This macrophage-activating potential was due, in part, to IFN-γ. In addition, T cells from the recovered mice could adoptively transfer resistance to otherwise highly susceptible BALB/c recipients. These cells were shown to be Lyt-$1^{+}2^{-}$, $CD4^{+}$ T cells (Liew *et al.*, 1984; Liew and Dhaliwal, 1987). Reconstitution of BALB/c nude mice with Lyt-$1^{+}2^{-}$, $CD4^{+}$ cells from syngeneic euthymic donors also resulted in resistance to infection with *Leishmania* (Mitchell *et al.*, 1981b).

Resistance to *L. major* infection in susceptible BALB/c mice could also be induced by repeated intravenous (i.v.) or intraperitoneal (i.p.) immunization with lethally irradiated, heat-killed or sonicated promastigotes (Howard *et al.*, 1982). The protective immunity which developed lasted for more than 150 days, extended to *L. mexicana*, *L. amazonensis* and *L. panamensis*, and was not stage specific. This immunity was transferable with splenic or lymph node T cells into normal or sublethally irradiated syngeneic recipients (Liew *et al.*, 1984). The protective T cells also belonged to the Lyt-$1^{+}2^{-}$, $CD4^{+}$ subset (Liew *et al.*, 1984) and produced IFN-γ when stimulated with leishmanial antigen *in vitro* (Liew *et al.*, 1987). However, unlike T cells from recovered mice, the protective cells from mice immunized i.v. failed to mediate DTH (Liew *et al.*, 1984). In addition, these cells could suppress both the induction and expression of specific DTH induced with killed parasites injected intradermally (Dhaliwal *et al.*, 1985). These cells were devoid of any demonstrable cytotoxic activity (Liew *et al.*, 1984), and thus appeared to be distinguishable from T cells mediating either DTH or cytotoxicity.

Therefore, Lyt-$1^{+}2^{-}$, $CD4^{+}$ T cells appeared to be essential in the protective immune response to cutaneous leishmaniasis (Table 2). However, T cells with the same phenotype can also mediate suppression of protective immunity.

V. Suppression of Protective Immunity

A. Suppression in Leishmaniasis

Immune suppression is evident in both clinical and experimental leishmaniasis. Patients with visceral leishmaniasis fail to develop leishmanial-specific skin reactions or proliferative T cell responses (Rezai *et al.*, 1978; Ho *et al.*,

TABLE 2. *Subsets of $CD4^+$ T cells induced during infection or immunization with* Leishmania major *in mice.*[a] *From Liew, F. Y. (1989b)*

Cell source	Protection	Counter-protection	Help in antibody synthesis	Delayed-type hyper-sensitivity	Lymphokine secretion[e]			
					IL-1	IL-3	IL-4	IFN-γ
Cured infection	+	−	+	+[b]	+	−	−	+
Intravenous immunization	+	−	+	−[c]	+	−	−	+
Progressive infection	−	+	+	−[c]	±	+	+	−
Subcutaneous immunization	−	+	+	+[d]	±	+	+	−

[a] +, a Positive; ±, modest but significant priming; −, negative.
[b] Classical tuberculin reaction.
[c] Suppressing delayed-type hypersensitivity.
[d] Jones–Mote hypersensitivity.
[e] IL, interleukin; IFN-γ, interferon γ.

1983), but these reactions were fully restored following successful chemotherapy (Carvalho *et al.*, 1981; Haldar *et al.*, 1983). This non-specific suppression extends to IL-2 production (Petersen *et al.*, 1984; Cillari *et al.*, 1988). In experimental models, a reduced T cell response to phytohaemagglutinin, associated with impaired IL-2 production, was observed in BALB/c mice infected with *L. donovani* (Reiner and Finke, 1983). These effects were also observed in susceptible BALB/c mice infected with *L. major* and were attributed to a population of macrophage-like adherent suppressor cells (Scott and Farrell, 1981; Cillari *et al.*, 1986). The mechanism of this macrophage-mediated suppression remains obscure.

The development, in highly susceptible BALB/c mice, of fatal disseminating disease was not due to an intrinsic inability of these mice to develop effective $CD4^+$ T cells against *Leishmania*. BALB/c mice could be rendered resistant to *L. major* infection by a number of treatments, including prior sublethal whole body γ-irradiation (Howard *et al.*, 1981) or injection with anti-CD4 MAb (Titus *et al.*, 1985a) (see Section IV.E). However, the protective effect of these treatments could be abrogated by the injection of T cells from normal syngeneic mice or, even better, T cells from mice with progressive *L. major* infection (Howard *et al.*, 1981). The disease-promoting cells from mice with progressive disease, like the protective T cells, expressed the Lyt-1^+2^-, $CD4^+$ phenotype (Liew *et al.*, 1982). They did not secrete IFN-γ in response to leishmanial antigens and were unable to mediate classical DTH (Table 2), but could suppress the expression of DTH by T cells from recovered mice (Liew and Dhaliwal, 1987). In addition, a cloned T cell line derived from these cells was able to suppress T cell proliferation, as well as the induction of DTH to *L. major*, *in vitro* (Liew, 1983).

L. major-specific $CD4^+$ T cell lines were also generated which, upon adoptive transfer to syngeneic recipients, led to the exacerbation of disease (Titus *et al.*, 1984a). This disease exacerbation was seen in both genetically susceptible BALB/c mice (Titus *et al.*, 1984a) and in genetically resistant CBA/T6 mice (Titus *et al.*, 1985b). Disease exacerbation was characterized by increased lesion size and by greatly increased parasite numbers in the lesions (Titus *et al.*, 1985b).

Injection of killed promastigotes by the subcutaneous (s.c.), intradermal or intramuscular routes, with or without adjuvant, also led to disease exacerbation if given before infection (Liew *et al.*, 1985a). This contrasted with the development of protective immunity if the i.v. or i.p. routes were used (Howard *et al.*, 1982). Splenic and lymph node T cells from mice immunized s.c. with killed promastigotes could also inhibit the prophylactic effect of i.v. immunization (Liew *et al.*, 1985b). This inhibition was achieved with a single s.c. injection, although rather less potently than with four injections, and was even effective against four repeated weekly i.v. immunizations (Liew *et al.*, 1985a). Again, these blocking T cells were Lyt-1^+2^-, $CD4^+$ cells (Liew *et al.*, 1985b). However, cells from mice immunized s.c. mediated a Jones–Mote (12–15 h) type of DTH rather than the classical tuberculin-type (24–48 h) DTH (Table 2).

Thus it became apparent that, in murine cutaneous leishmaniasis, the disseminated form of the disease observed in BALB/c mice was due to the generation of T cells which suppressed the activation of a protective cell-mediated response. However, these suppressor cells were Lyt-1^+2^-, $CD4^+$, I-J^- T cells (Liew *et al.*, 1982; Liew, 1983) which did not fit the conventional definition of T suppressor cells as Lyt-1^-2^+, $CD8^+$, I-J^+ cells (Tada and Okumura, 1979).

B. CONCEPT OF SUPPRESSOR T CELLS

The concept of T cells specifically involved in suppressing immune responses was first developed following the demonstration that T cell-deprived mice reconstituted with syngeneic T cells produced higher antibody responses than normal, control mice (Gershon and Kondo, 1970). The concept of suppressor T cells involved in down-regulating immune responses grew through the 1970s (reviewed by Golub, 1981; Green D. R. *et al.*, 1983; Dorf and Benacerraf, 1984; Asherson *et al.*, 1986). Suppressor T cells were identified as Lyt-1^-2^+ T cells (Feldmann *et al.*, 1975; Cantor *et al.*, 1976) and found to express a unique MHC restriction element, I-J (Murphy *et al.*, 1976; Tada *et al.*, 1976). The phenomenon of suppression was demonstrated in many experimental systems (reviewed by Tada and Okumura, 1979), including contact sensitivity (Battisto and Bloom, 1966; Asherson and Zembala, 1974), DTH (Liew, 1977; Bach *et al.*, 1978) and disease (Arnon, 1981; Rose *et al.*, 1981; Doughty and Phillips, 1982). However, in order to explain many of the results, a complex suppressor cascade was postulated involving two or three distinct suppressor T cell populations connected by non-specific suppressor factors (Dorf and Bennacerraf, 1984; Asherson *et al.*, 1986). These cascades involved both antigen specific and non-specific interactions and idiotype–anti-idiotype regulation, implicating suppressor cells, presenting cells and soluble suppressor factors. Then, from the mid-1980s, the idea of a distinct suppressor T cell or cells began to be discredited. The reasons included the sheer scale of the suppressor network and the problem that none of the many soluble suppressive factors had been characterized molecularly (reviewed by Bloom *et al.*, 1992). In addition, molecular analysis of antigen-specific suppressor T cell hybridomas revealed that more than half had deleted or unrearranged T cell receptors and were unable to produce a functional receptor protein (Hedrick *et al.*, 1985).

However, the most serious obstacle concerned the location of the putative MHC restriction element I-J expressed on suppressor T cells and factors. The I-J molecule was thought to map to the murine MHC between Eα and Eβ but, when this area was sequenced, there was not enough DNA to encode the I-J polypeptide (Kronenberg *et al.*, 1983; Murphy, 1987) and, as yet, the I-J molecule has still not been mapped or sequenced. Thus, although the phenomenon of immune suppression does exist, the existence of specific Lyt-1^-2^+, I-J^+ T cells is now generally thought to be unlikely.

As described above, the phenomenon of immune suppression has always been apparent in leishmaniasis. The profound impairment of a protective immune response observed in BALB/c mice was due to the generation of a population of Lyt-1^+2^- suppressor T cells (Howard *et al.*, 1980c; Liew *et al.*, 1982; Liew, 1983). Although this suppressor cell phenotype did not agree with the conventional $CD8^+$ suppressor phenotype, $CD4^+$ suppressor cells had been demonstrated in other systems, particularly in the suppression of DTH. Lyt-1^+2^- $CD4^+$ T cells were shown to suppress the DTH response to a number of antigens, including horse and sheep red blood cells (Ramshaw *et al.*, 1976; Liew, 1977) and influenza virus haemagglutinin (Liew and Russell, 1980). Therefore, it became apparent that $CD4^+$ suppressor cells could exist.

By the mid-1980s it was apparent that a number of $CD4^+$ T cell subsets were involved in the regulation of the immune response to *L. major*, both in the protection and exacerbation of disease (see Table 2). However, it needed the finding that murine $CD4^+$ T cell clones could be subdivided into at least two functional subsets on the basis of lymphokine secretion (Mosmann *et al.*, 1986; Mosmann and Coffman, 1987) to provide a rational basis for the understanding of the cellular factors involved in protection/resistance and exacerbation/susceptibility to *L. major* infection. This will be further discussed in Section VI.

VI. T Cell Subsets in Leishmaniasis

A. HETEROGENEITY OF $CD4^+$ T CELLS

Evidence had long existed that the immune response could be divided into humoral and cell-mediated immunity and that these two responses did not always appear in parallel. Animals pre-treated with antigen in saline failed to develop DTH, a typical cell-mediated response, to a subsequent challenge of the same antigen in adjuvant, but did exhibit an antigen-specific antibody response (Battisto and Miller, 1962; Asherson and Stone, 1965; Dvorak *et al.*, 1965; Asherson, 1966; Borel *et al.*, 1966; Crowle and Hu, 1966; Loewi *et al.*, 1966; Borel and David, 1970; Crowle and Hu, 1970). This phenomenon was called "immune deviation" (Asherson and Stone, 1965), "preimmunization tolerance" (Loewi *et al.*, 1966) or "split tolerance" (Crowle and Hu, 1966). The reverse could also be demonstrated, with animals rendered antibody tolerant to tuberculin but maintaining normal development of DTH (Janicki *et al.*, 1970). Later, an inverse relationship between humoral immunity and CMI became evident (Parish, 1971). This phenomenon extends to high and low zone antibody tolerance (Parish and Liew, 1972), in that DTH was detectable only at those antigen concentrations where antibody was undetectable (see Fig. 2). It later became clear that the T cells helping antibody synthesis and those mediating DTH were distinct (Liew and Parish, 1974; Silver and Benacerraf, 1974). The heterogeneity of Lyt-1^+2^-, $CD4^+$ T cells was further sustained by the demonstration that antigen-specific helper T cells and non-specific T cells could be segregated *in vitro* by limiting dilution (Marrack and Kappler, 1975), and that two populations of helper T cells were required to provide optimal help in antibody responses (Janeway, 1975).

More evidence to support $CD4^+$ T cell heterogeneity came from the study of the immune response to *Leishmania*. As mentioned earlier, highly susceptible BALB/c mice were able to control and recover from leishmanial infection if they were treated with a sublethal dose of γ-irradiation or with an anti-CD4 MAb before infection (Howard *et al.*, 1981; Titus *et al.*, 1985a). This acquired resistance could be readily reversed by transferring to the irradiated recipients T cells from donors with progressive disease (Howard *et al.*, 1981). Both the protective T cells and the disease exacerbating cells were $CD4^+$ T cells (Mitchell *et al.*, 1980; Howard *et al.*, 1981; Liew *et al.*, 1982). $CD4^+$ T cells which could transfer protective immunity were also obtained from mice immunized i.v. with killed promastigotes (Liew *et al.*, 1984). However, the s.c. injection of killed parasites resulted in $CD4^+$ T cells which could exacerbate disease (Liew *et al.*, 1985a,b). Therefore, functionally separate subpopulations of $CD4^+$ T cells were induced in BALB/c mice during infection or immunization with *L. major* (Table 2). However, it was not clear then if these were all in fact the same cell population or if different cell populations were involved. Two main hypotheses developed to explain the results concerning the role of $CD4^+$ T cells in cutaneous leishmaniasis (reviewed by Liew, 1987, 1989b; Müller *et al.*, 1989). The evidence for each hypothesis is summarized briefly below.

1. *Outcome of infection is due to quantitative differences in $CD4^+$ cell numbers*

As the T cell phenotype involved in both protection and exacerbation was the Lyt-1^+2^-, $CD4^+$ T cell, it was argued that disease protection or exacerbation was a reflection of quantitative differences in the number of

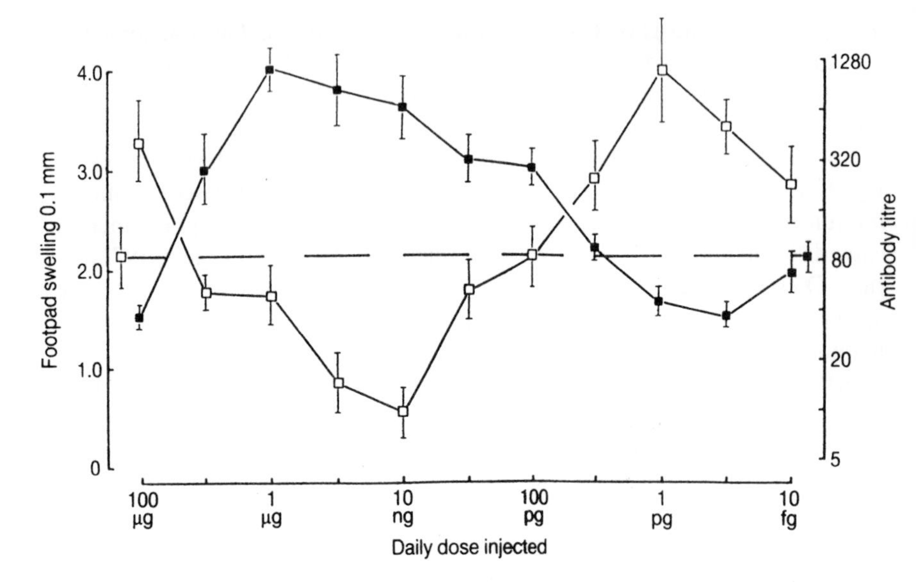

FIG. 2. Inverse relationship between humoral and cell-mediated immunity. Groups of seven rats (strain W, Wistar) were injected intraperitoneally daily for 27 days with doses of a CNBr digest of flagellin (*Salmonella adelaide*, strain SW 1338) in 10-fold dilution steps from 100 µg to 10 fg. Immunized and unimmunized controls were challenged on day 28 with 100 µg flagellin in saline in the right hind footpad. The antibody response (■) represents the mean of the serum antibody titres (reciprocal of dilution) 7, 14, 21 and 28 days after challenge determined by haemagglutination using sheep red blood cells sensitized with polymerized flagellin. Delayed-type hypersensitivity (DTH) (□) was elicited in the left hind footpad with 0.5 µg flagellin in saline 28 days after the flagellin challenge. DTH, expressed as 24 h footpad swelling, was determined by the difference in the footpad thickness between left and right hind footpads. The broken line represents the antibody and DTH responses of control rats injected with 100 µg flagellin in saline, only, on day 28. Vertical bars represent standard errors of the means. (From Parish and Liew, 1972.)

Leishmania-specific $CD4^+$ T cells generated. Thus, susceptibility in BALB/c mice was caused by excessive activation of $CD4^+$ T cells (Milon *et al*., 1986) and the reduction of these cells with anti-CD4 MAb *in vivo* resulted in disease containment (Titus *et al*., 1985a). This hypothesis argued that small numbers of specific $CD4^+$ T cells were host-protective, whereas excessive numbers of such cells were detrimental to the host (Titus *et al*., 1985a). This was consistent with the earlier findings that the transfer of 10^6 $CD4^+$ cells from normal syngeneic BALB/c mice into *nu/nu* BALB/c mice led to protection of the recipients, whereas transfer of 10^8 cells exacerbated the disease (Mitchell *et al*., 1981b, 1982). However, evidence to support the alternative hypothesis was accumulating.

2. *Outcome of infection is due to the interaction of distinct CD4$^+$ T cell subsets*

This hypothesis is supported by the following experimental evidence.

(a) *Cell titration experiments.* T cells freshly isolated from mice immunized s.c. with killed promastigotes either inhibited protective immunization ($> 10^7$ cells/recipient) or had no effect ($< 10^7$ cells/recipient). No protection was observed at any dose (Liew *et al.*, 1987). Conversely, freshly isolated T cells from mice given prophylactic i.v. immunization were either protective ($> 10^7$ cells/recipient) or were ineffective ($< 10^7$ cells/recipient). No exacerbation of disease was observed at any dose. In mixed transfer experiments, increasing numbers of T cells from s.c. immunized donors progressively inhibited the protective effect of T cells from i.v. immunized donors. These results argued against a quantitative effect of protective and counter-protective T cells and supported the hypothesis that protection and disease promotion were mediated by functionally distinct T cell subsets.

(b) *Delayed-type hypersensitivity.* T cells from protected, recovered BALB/c mice mediated the classical tuberculin-type DTH, whereas T cells from mice immunized s.c. with killed promastigotes expressed a transient Jones-Mote DTH reaction (De Rossell *et al.*, 1987; Liew and Dhaliwal, 1987). T cells from mice with progressive disease or from mice protectively immunized i.v. both failed to mediate DTH and also suppressed parasite-specific DTH (Howard *et al.*, 1981; Dhaliwal *et al.*, 1985). Therefore, DTH may be an additional criterion to differentiate protective and disease-promoting CD4$^+$ T cells.

(c) *IFN-γ/IL-4 production.* Spleen or lymph node cells from resistant, recovered mice or mice immunized i.v. produced IFN-γ when stimulated *in vitro* with leishmanial antigens. In contrast, cells from susceptible mice, mice with progressive disease or from mice immunized s.c. produced little or no IFN-γ when similarly stimulated *in vitro* (Sadick *et al.*, 1986; Liew and Dhaliwal, 1987). Additional evidence for distinct subpopulations of CD4$^+$ T cells was obtained when messenger ribonucleic acid (mRNA) was extracted from T cells of resistant C57BL/6 and susceptible BALB/c mice, both healed and non-healed, and analysed using probes for IFN-γ and IL-4 (Locksley *et al.*, 1987; Sadick *et al.*, 1987; Heinzel *et al.*, 1989). In both the lymph nodes and spleen, healed BALB/c mice contained 50 to 100-fold greater amounts of IFN-γ mRNA than were detectable in the non-healed BALB/c mice. These levels were comparable with those detected in the resistant C57BL/6 mice. However, when the level of IL-4 mRNA was assessed, susceptible BALB/c mice were found to have 50-fold greater amounts of IL-4 mRNA than the resistant mice. This again suggested that different $CD4^+$ T cell subsets were involved in the response to cutaneous leishmaniasis, both in healed and non-healed mice and in resistant and susceptible strains.

However, confirmation of the heterogeneity of $CD4^+$ T cells came with the finding that murine $CD4^+$ T cell clones could be divided into two subsets on the basis of lymphokine production (Mosmann and Coffman, 1987).

B. Th1/Th2, CD4 T CELL SUBSETS

The first report on Th1/Th2 subsets was in 1986 when it was shown that a panel of murine $CD4^+$ T cell clones could be divided into two groups on the basis of their lymphokine secretion in response to specific antigen (Mosmann *et al.*, 1986). Th1 clones were found to secrete IL-2 and IFN-γ, whereas Th2 clones secreted IL-4. Further characterization of lymphokine secretion using bioassays and detection of mRNA extended the list of lymphokines secreted: Th1 clones secrete IL-2, IFN-γ and lymphotoxin while Th2 clones secrete IL-4, IL-5 and IL-6 (Cherwinski *et al.*, 1987). IL-3, granulocyte-macrophage colony stimulating factor (GM-CSF) and TNF are secreted by both types of T cell clone. Another cytokine, IL-10 or cytokine synthesis inhibitory factor (CSIF), has been described (Fiorentino *et al.*, 1989) and is secreted by Th2 clones. The current cytokine secretion phenotypes of murine T cells are summarized in Table 3.

The characteristics and functions of Th1/Th2 cell subsets have been the subject of many reviews (Mosmann and Coffman, 1987, 1989; Coffman *et al.*, 1988, 1991a; Finkelman *et al.*, 1990; Coffman and Mosmann, 1991; Mosmann and Moore, 1991). Th1 cells mediate a DTH reaction when injected with specific antigen into the footpads of mice (Cher and Mosmann, 1987), while Th2 cells cannot. However, it is unclear if the inability of Th2 cells to mediate DTH is due to a deficiency in the production of the essential mediators *in vivo* or if they are not activated by antigen-presenting cells in the footpad. Both Th1 and Th2 cells can provide help for antibody production; however, Th2 cells are much more efficient (Coffman *et al.*, 1988; Mosmann and Coffman, 1989; Finkelman *et al.*, 1990), providing both antigen-specific (Kim *et al.*, 1985; Killar *et al.*, 1987; Stevens *et al.*, 1988) and non-specific (Boom *et al.*, 1988; Coffman *et al.*, 1988) help *in vitro*. Th1 cells provide help in antigen-specific secondary responses with primed B cell populations (Giedlin *et al.*, 1986). However, a major difference in the help provided by Th1 and Th2 cells for antibody production is their ability to stimulate different antibody isotypes. Th1 clones induced substantially more IgG2a, whereas Th2 clones induced production of IgE and IgG1 (Coffman *et al.*, 1988; Stevens *et al.*, 1988). This selective induction of the antibody isotype appears to be due to the cytokines secreted by each cell subset. IL-4, from Th2 cells, enhances production of IgE by increasing the frequency of B cell isotype switching (Lebman and Coffman, 1988; Savelkoul *et al.*, 1988). IgE production could also be inhibited by the administration of either IFN-γ or anti-IL-4 both *in vitro* (Coffman and Carty, 1986; Coffman *et al.*, 1986) and *in vivo* (Finkelman *et al.*, 1986, 1988). In contrast, IFN-γ could switch the isotype from IgE to IgG2a (Stavnezer *et al.*, 1988; Stevens *et al.*, 1988). Therefore, IFN-γ and IL-4, the products of Th1 and Th2 cells respectively, can reciprocally regulate B cell immunoglobulin production (Snapper and Paul, 1987).

TABLE 3. *Characteristics of murine Th1 and Th2 cells*

	Th1	*Th2*
Lymphokines[a]		
IL-2	++	−
IFN-γ	++	−
LT	++	−
GM-CSF	++	+
TNF-α	++	+
IL-3	++	++
IL-4	−	++
IL-5	−	++
IL-6	−	++
IL-10	−	++
B cell help[b]		
IgM, IgG1, IgA	+	++
IgG2a	++	+
IgE	−	++
Delayed-type hypersensitivity	+	−

[a] IFN-γ, interferon γ; IL, interleukin; GM-CSF, granulocyte macrophage-colony stimulating factor; LT, lymphotoxin; TNF-α, tumour necrosis factor.
[b] Ig, immunoglobulin.

Immunoglobulin production is not the only function which is reciprocally regulated by Th1 and Th2 cells. They also regulate each other's induction and proliferation. IFN-γ inhibits the proliferation of Th2 cells to IL-2 or IL-4, but has no effect on Th1 cells (Fernandez-Botran *et al.*, 1988; Gajewski and Fitch, 1988). Th2 cells secrete IL-10 which inhibits the synthesis of cytokines by Th1 cells and indirectly reduces Th1 cell proliferation (Fiorentino *et al.*, 1989; Magilavy *et al.*, 1989; Mosmann and Moore, 1991). Th1 and Th2 cells also appear to be affected by the type of antigen presenting cell in that Th1 cells are preferentially stimulated by macrophages, whereas Th2 cells are preferentially stimulated by B cells (Gajewski and Fitch, 1991; Gajewski *et al.*, 1991).

Although many murine $CD4^+$ T cell clones fit the Th1/Th2 pattern, there are examples of T cell clones which do not do so. This is particularly apparent with clones which are examined early in culture *in vitro*. These clones secreted a mixture of Th1 and Th2 cytokines (Prystowsky *et al.*, 1982; Kelso and Glasebrook, 1984; Firestein *et al.*, 1989; Street *et al.*, 1990) and were termed Th0 cells. Earlier studies indicated that human T cell clones did not appear to fit the murine $CD4^+$ clone pattern either, with clones secreting a mixture of cytokines (Maggi *et al.*, 1988; Paliard *et al.*, 1988; Umetsu *et al.*, 1988; Bacchetta *et al.*, 1990; de Vries *et al.*, 1991). However, most of these earlier human clones examined were derived from alloreactive or mitogen-stimulated human peripheral blood. Human Th1- or Th2-like $CD4^+$ T cell clones can be obtained from the tissues or peripheral blood of patients with different disease states (de Vries *et al.*, 1991; Romagnani, 1991). Thus $CD4^+$ T cell clones from patients with autoimmune thyroid disease produce IFN-γ but little or no IL-4, while $CD4^+$ clones from patients with vernal conjunctivitis produce IL-4 but not IFN-γ (Romagnani, 1990; Maggi *et al.*, 1991). Most allergen-specific T cell clones from atopic individuals produce IL-4 and IL-5, but little IFN-γ, on antigen stimulation *in vitro* (Wierenga *et al.*, 1990; Parronchi *et al.*, 1991). Analysis of a large panel of T cell clones specific for purified protein derivative (PPD) of *M. tuberculosis* or *Toxocara canis* excretory–secretory (TES) antigens established from the same healthy individuals revealed that most PPD-specific clones secreted IL-2 and IFN-γ, but not IL-4 or IL-5. In contrast, most TES-specific clones secreted IL-4 and IL-5, but not IL-2 or IFN-γ (Del Prete *et al.*, 1991). The cytokine phenotypes of these clones remained stable for the course of the 6-month study period. All the clones could provide help for antibody production, but only the Th2-like clones supplied help for IgE synthesis. A study of the cytokine profiles in tuberculoid and lepromatous leprosy lesions revealed that the tuberculoid lesions contained mRNA for IL-2 and IFN-γ but not for IL-4, while the lepromatous lesions contained mRNA for IL-4, IL-5 and IL-10 but not for IFN-γ (Yamamura *et al.*, 1991). Human $CD4^+$ T cell clones which mediated DTH produced high levels of IFN-γ but almost no IL-4, whereas $CD4^+$ clones which mediated B cell help produced predominantly IL-4 (Salgame *et al.*, 1991). T cells cloned from the synovial membrane of patients with rheumatoid arthritis also resembled Th1 cells, secreting high levels of IFN-γ but little or no IL-4 following stimulation *in vitro* (Miltenburg *et al.*, 1992). Therefore, a Th1/Th2 cytokine pattern can also be detected in humans.

Much of the understanding of how Th1 and Th2 cells interact and regulate each other has come from the study of *L. major* infection in mice. We will now discuss the evidence which shows that resistance and healing are associated with a Th1 response, while susceptibility and progressive disease is correlated with a Th2 response.

C. CD4$^+$ T CELL SUBSETS IN *L. MAJOR* INFECTION

As discussed in Section VI.A, evidence was accumulating that the differential immune response observed in resistant and susceptible mice and in healed and non-healed mice was due to the activation of different CD4$^+$ T cell subsets (Liew, 1987, 1989b). The relationship of these CD4$^+$ T cells to the Th1/Th2 subsets was also defined. This has been extensively reviewed (Liew, 1989b, 1990, 1992; Müller *et al.*, 1989; Scott, 1990, 1991a; Coffman *et al.*, 1991a,b; Locksley and Scott, 1991; Locksley *et al.*, 1991; Titus *et al.*, 1993), but will be summarized here.

As discussed previously, cells from resistant C57BL/6 and healed BALB/c mice contained predominantly IFN-γ and IL-4 mRNA, whereas cells from susceptible BALB/c mice contained mainly IL-4 and IL-10 mRNA (Locksley *et al.*, 1987; Sadick *et al.*, 1987; Heinzel *et al.*, 1989, 1991). Evidence for the reciprocal role of IFN-γ and IL-4 also came from the study of these lymphokines *in vitro*. Culture supernatant from spleen cells of BALB/c mice recovered, or protected, from infection could activate macrophages to kill intracellular amastigotes (Liew and Dhaliwal, 1987). In contrast, spleen cells from BALB/c mice with progressive disease, when stimulated *in vitro*, produced factors which inhibited this macrophage activation (Liew *et al.*, 1989). The macrophage activating factor was found to be IFN-γ and the inhibiting factors were IL-3 and IL-4 (Liew and Dhaliwal, 1987; Liew *et al.*, 1989). The macrophage activating activity of IFN-γ could be completely abrogated by the IL-3 and IL-4 (Liew *et al.*, 1989). Similar reciprocal activity of IFN-γ and IL-4 was obtained with human monocytes infected with *L. donovani* (Lehn *et al.*, 1989; Zwingenberger *et al.*, 1990). However, IL-4 had to be added to the cultures before IFN-γ for inhibition of parasite killing to occur (Liew *et al.*, 1991a). If IL-4 was added at the same time, no inhibitory effect was seen and, if it was added later, a synergistic effect with IFN-γ was observed (Liew *et al.*, 1991a). This synergistic effect of IFN-γ and IL-4 on macrophage activation was observed with macrophages from both resistant and susceptible strains of mice (Bogdan *et al.*, 1991).

Confirmation of the correlation of Th subsets and disease outcome was obtained from studies with Th1 and Th2 cell lines specific for *L. major* (Scott *et al.*, 1988). Transfer of a Th1 cell line, secreting IL-2 and IFN-γ, protected

BALB/c mice against *L. major* infection. In contrast, transfer of a Th2 line secreting IL-4 resulted in an exacerbated infection. The same results were obtained when severe combined immunodeficient (SCID) mice, which are highly susceptible to *L. major* infection, were reconstituted with either a Th1 or Th2 cell line (Holaday *et al.*, 1991). SCID mice also developed a healing response and resistance to reinfection when reconstituted with BALB/c spleen cells (Coffman *et al.*, 1991b; Varkila *et al.*, 1993). The development of this Th1 response by transferring BALB/c T cells required the presence of IFN-γ during the initial infection period and the healing response was associated with the Th1-like responses of DTH and the production of IFN-γ but not IL-4 following antigen restimulation *in vitro*. This is consistent with the finding that BALB/c mice inoculated with parasites plus IFN-γ produced higher levels of IFN-γ, and had significantly smaller lesions, than mice inoculated with parasites alone (Scott, 1991b). Thus, high levels of IFN-γ early in infection favour a Th1 response, whereas low levels promote a Th2 response.

The role of Th1 and Th2 cells in the protection and exacerbation of cutaneous leishmaniasis is fairly clear cut, although one group has reported that the transfer of an IFN-γ secreting cell line resulted in exacerbated infection (Titus *et al.*, 1984a, 1991). However, Th1 and Th2 populations have not yet been clearly demonstrated in murine *L. donovani* infection (Kaye *et al.*, 1991; Murray *et al.*, 1992). Therefore, the involvement of Th1 and Th2 cells may not be absolute and it is conceivable that other, as yet undefined, $CD4^+$T cell subsets may also be involved.

As discussed above, IFN-γ and IL-4 had a reciprocal effect on the activation of macrophages to kill *Leishmania* amastigotes (Lehn *et al.*, 1989; Liew *et al.*, 1989). The role of IFN-γ and IL-4, as well as other cytokines, in the regulation of *Leishmania* infection will now be discussed.

VII. Cytokines in Leishmaniasis

A. Interferon γ and Interleukin 4

In vitro, IFN-γ was required to activate macrophages to kill effectively *Leishmania* species (Mauël *et al.*, 1978; Nacy *et al.*, 1981, 1985; Murray *et al.*, 1983, 1987; Nathan, 1983; Titus *et al.*, 1984b; Liew and Dhaliwal, 1987; Pham and Mauël, 1987; Belosevic *et al.*, 1988, 1989). *In vivo*, the injection of resistant mice with anti-IFN-γ antibody resulted in non-healing and, ultimately, death (Belosevic *et al.*, 1989). This was achieved with even a single injection of anti-IFN-γ, but it had to be administered within 1 week of infection. When lymph node cells from these treated mice were restimulated *in vitro*, the predominant lymphokines secreted were now IL-4 and IL-5, not IFN-γ (Scott, 1991b). Inoculation of susceptible BALB/c mice with IFN-γ and *L. major* induced lymph node cells from these mice to secrete IFN-γ, a Th1 response, but did not reverse the progress of disease (Sadick *et al.*, 1990; Scott, 1991b). Conversely, susceptible BALB/c mice could be made resistant to infection by the administration *in vivo* of an anti-IL-4 MAb at the time of infection or within 1 week after (Sadick *et al.*, 1990; Coffman *et al.*, 1991a). These mice had the characteristics of a Th1-mediated healing response, with high levels of IFN-γ, low levels of IL-4, DTH and resistance to reinfection (Chatelain *et al.*, 1992). However, the development of this Th1-type response in susceptible mice treated with anti-IL-4 required the presence of IFN-γ, as co-administration of an anti-IFN-γ antibody prevented the conversion of BALB/c mice into healers (Coffman *et al.*, 1991a; Locksley *et al.*, 1991).

The administration of IL-4 itself to resistant mice at the time of infection had little effect on the course of infection (Sadick *et al.*, 1991). In contrast, the injection of IL-4 into the lesions of genetically susceptible BALB/c mice early in infection resulted in an exacerbated infection (Lezama-Davila *et al.*, 1992). This effect was observed with both *L. major* and *L. mexicana* infection, although the timing of IL-4 treatment was important as injection of IL-4 after lesion development inhibited parasite growth and lesion size (Lezama-Davila *et al.*, 1992). Indeed, IL-4 can act as a second signal with IFN-γ to induce macrophage leishmanicidal activity (Belosevic *et al.*, 1988). Direct evidence for the importance of IL-4 in the development of exacerbative disease has been obtained with IL-4 transgenic mice (Leal *et al.*, 1993). Resistant mice were rendered highly susceptible to *L. major* infection by the expression of transgenic IL-4 (Fig. 3). Spleen cells from infected transgenic mice produced significantly higher levels of IL-4 but lower levels of IFN-γ when stimulated *in vitro* with specific antigen compared with infected normal controls. IL-4 transgenic mice also had higher levels of IgE. Thus, it appears that the pervasive and continuous presence of additional IL-4 from birth is necessary to render the mice susceptible to cutaneous leishmaniasis.

It is clear that resistance to cutaneous leishmanial infection and healing in mice is associated with IFN-γ and a Th1 cell response, while susceptibility and inexorable disease is related to IL-4 and a Th2-type response. This lymphokine pattern is apparent very early, 3 days after infection (Scott, 1991b). This may explain why administration of anti-cytokine antibodies must be given within 1 week of infection to have any effect. The action of these lymphokines appears to be to regulate reciprocally the development of the T cell response. IFN-γ can mediate this by inhibiting the proliferation of Th2 cells (Gajewski and Fitch, 1988; Fernandez-Botran *et al.*, 1988). IL-4 does not affect the proliferation of Th1 cells. However, IFN-γ and IL-4 directly affect the effector function involved in the destruction of *Leishmania* species (see Section VIII).

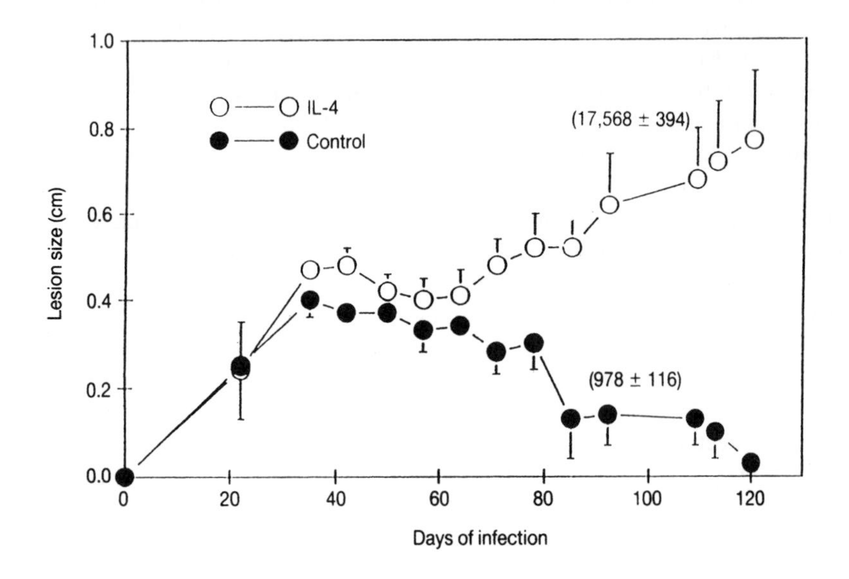

FIG. 3. Disease development in 129/Sv mice (control) and IL-4 transgenic mice infected with *L. major*. Mice were infected in the rump with 2×10^7 promastigotes and the size of the lesions measured at regular intervals. Experiments were terminated on day 115 as the lesions in the transgenic mice had reached the legally permitted size. Figures in parentheses represent relative parasite loads in the infected tissues (counts/min $\pm$ 1 standard error of the mean per 1.2×10^6 cells). Vertical bars represent standard errors of the means. (Leal *et al.*, 1993.)

B. INTERLEUKIN IL-10

IL-10 is secreted by Th2 cells and inhibits the synthesis of cytokines by Th1 cells, particularly IFN-γ (Fiorentino *et al.*, 1989). Although IL-10 does not directly inhibit the proliferation of Th1 cells, it probably has an indirect effect by inhibiting the synthesis of cytokines necessary for Th1 proliferation. IL-10 acts to inhibit cytokine synthesis only when antigen presenting cells are present (Fiorentino *et al.*, 1989) and it impairs the ability of antigen-presenting cells to stimulate cytokine synthesis by Th1 clones (Fiorentino *et al.*, 1991). IL-10 strongly reduces antigen-specific human T cell proliferation by diminishing the antigen-presenting capacity of monocytes by down-regulating class II MHC expression (de Waal Malefyt *et al.*, 1991a). This represents another pathway through which Th2 cells can regulate the function of Th1 cells (Mosmann and Moore, 1991). However, IL-10 can be produced by other cell types as well, including mast cell lines (Moore *et al.*, 1990), monocytes (de Waal Malefyt *et al.*, 1991b), macrophages and B cells (Howard and O'Garra, 1992).

A role for IL-10 in leishmaniasis is indicated by the detection of IL-10 mRNA in $CD4^+$ T cells of infected, susceptible BALB/c mice (Heinzel *et al.*, 1991). IL-10 was not detected in cells from resistant C57BL/6 mice. The presence of IL-10 was further evidence of the role of Th2 cells in exacerbated disease. Although it is likely that IL-10 functions to block the development of a Th1 response in BALB/c mice, administration of anti-IL-10 antibody at the time of infection did not alter the disease progression or alter the Th response obtained (Coffman *et al.*, 1991a), although IL-4 production was inhibited. However, IL-10 may be more important in inhibiting the production of nitric oxide (Cunha *et al.*, 1992), which appears to be the major effector mechanism in leishmaniasis. Indeed, recent evidence from other parasite systems has demonstrated that IL-10 can block the microbicidal activity of IFN-γ (Gazzinelli *et al.*, 1992; Silva *et al.*, 1992). In some cases, this is by down-regulating the production of toxic nitric oxide metabolites induced by IFN-γ (Gazzinelli *et al.*, 1992).

C. TUMOUR NECROSIS FACTOR α

Tumour necrosis factor α (TNF-α) also plays an influential role in the control of cutaneous infection in the murine model (Titus *et al.*, 1989; Liew *et al.*, 1990a,b,c). Lymph node cells from resistant mice produced large amounts of TNF-α following stimulation *in vitro*, while cells from susceptible mice produced little or no TNF-α (Titus *et al.*, 1989). However, one group was unable to demonstrate such a difference in TNF-α production between susceptible and resistant strains (Moll *et al.*, 1990). Treatment of both resistant and susceptible mice with TNF-α resulted in smaller lesions containing fewer parasites than were detected in untreated control mice (Titus *et al.*, 1989; Liew *et al.*, 1990a). Conversely, treatment of mice with an anti-TNF-α antibody resulted in exacerbated disease.

TNF-α appears to exert its leishmanicidal activity by activating macrophages, rather than by acting directly on the parasite (Liew *et al.*, 1990a). Incubation of macrophages with TNF-α, in the presence of bacterial LPS, resulted in leishmanicidal activity (Liew *et al.*, 1990a,b). Incubation of the macrophages with TNF-α plus IFN-γ resulted in increased killing of amastigotes (Bogdan *et al.*, 1990; Liew *et al.*, 1990c). However, TNF-α, in the absence of either LPS or IFN-γ, could not induce leishmanicidal activity (Bogdan *et al.*, 1990). This increased leishmanicidal activity, in the presence of TNF-α and IFN-γ or LPS, appears to be mediated by the induction of nitric oxide (Liew *et al.*, 1990c; see Section VIII). Interestingly, one group has reported a synergistic effect of IFN-γ and IL-4 in the induction of parasite killing by macrophages (Bogdan *et al.*, 1991). This now appears to be due to the ability of these cytokines to induce the endogenous production of TNF-α, which in turn interacts with the IFN-γ, resulting in leishmanicidal activity through the production of nitric oxide (Stenger *et al.*, 1991). Finally, another group has proposed that membrane-associated TNF-α on the surface of $CD4^+$ T cells may be important in sending activation signals to infected macrophages, resulting in parasite killing (Sypek and Wyler, 1991).

D. OTHER CYTOKINES

1. *Interleukin-1*

Macrophages from genetically susceptible mice were found to secrete large amounts of IL-1 in response to infection with *L. major*, while those from resistant mice produced significantly lower levels of IL-1 (Cillari *et al.*, 1989). The level of IL-1 production correlated with the degree of parasitization. This production of IL-1 in susceptible mice may be important in the generation of the Th2-type response observed as IL-1 acts as a co-factor in the activation of Th2 cells (Kurt-Jones *et al.*, 1987; Lichtman *et al.*, 1988). The high production of IL-1 in cutaneous leishmaniasis is in contrast to visceral leishmaniasis where infection of mice with *L. donovani* resulted in suppression of the IL-1 response (Reiner, 1987). These findings reflect the distinct pathology and immune responses caused by the two species of *Leishmania*.

2. *Interleukin-2*

Lymphoid cells from BALB/c mice infected with *L. major* produced significantly lower levels of IL-2 following stimulation *in vitro* compared to uninfected controls (Cillari *et al.*, 1986). This depressed IL-2 response was not seen with cells from protected BALB/c mice or from resistant mice (Cillari *et al.*, 1986; Solbach *et al.*, 1987). Suppression of IL-2 production appeared to be mediated by macrophages, possibly by the release of prostaglandins (Cillari *et al.*, 1986), and has also been observed in *L. donovani* infection (Reiner and Finke, 1983).

IL-2 mRNA was detected at a comparable level in the spleen and lymph nodes of resistant, healer and non-healer mice during *L. major* infection (Locksley *et al.*, 1987), but much of this appears to be derived from B cells (Heinzel *et al.*, 1991). $CD4^+$ T cells from resistant and healed mice also contained detectable levels of IL-2 mRNA, but not $CD4^+$ T cells from susceptible non-healer mice (Heinzel *et al.*, 1991). The role of B cell-derived IL-2 in susceptible mice is unclear; however, it may be involved in the proliferation and differentiation of Th2 cells *in vivo* (Fernandez-Botran *et al.*, 1988; Le Gros *et al.*, 1990; Swain *et al.*, 1990) and in the production of IL-4 (Powers *et al.*, 1988; Ben-Sasson *et al.*, 1990). This may help to explain some conflicting findings regarding the effect of B cell depletion by anti-IgM treatment in susceptible and resistant mice (Sacks *et al.*, 1984; Scott *et al.*, 1986). Depletion of B cells from birth rendered genetically susceptible BALB/c mice resistant to *L. major* infection (Sacks *et al.*, 1984), possibly by inhibiting the development of Th2 cells. Conversely, when resistant mice were depleted of B cells, they became susceptible to infection (Scott *et al.*, 1986), and this could also be due to a lack of IL-2, resulting in decreased Th1 proliferation and development.

The involvement of IL-2 in the development of a Th2 response in susceptible BALB/c mice was further supported by the finding that prolonged treatment of infected BALB/c mice with anti-IL-2 MAb resulted in a healing response (F. P. Heinzel and R. M. Locksley, personal communication). This healing was associated with decreased parasite numbers in the lesion and with increased IFN-γ and decreased IL-4 mRNA expression. Thus, IL-2 does appear to be necessary for the differentiation of Th2 cells in susceptible mice. In contrast, anti-IL-2 treatment had no effect on the development of a Th1 response in resistant mice.

IL-2 can also act as a second signal with IFN-γ to induce macrophage leishmanicidal activity (Belosevic *et al.*, 1988). However, administration of IL-2 itself to susceptible mice infected with *L. major* had no effect on the course of infection (Lezama-Davila *et al.*, 1992). In contrast, IL-2 did significantly enhance parasite growth and lesion development in mice infected with *L. mexicana* (Mazingue *et al.*, 1989; Lezama-Davila *et al.*, 1992). These findings again illustrate the distinct pathology and immune response caused by different species of *Leishmania*.

3. *Interleukin-3*

IL-3 is produced by susceptible mice during *L. major* infection, but not by resistant mice or by γ-irradiated recovered mice (Lelchuk *et al.*, 1988). The level of production of IL-3 correlated with disease progression. Mice immunized by the s.c. route, which leads to exacerbated disease, produced higher levels of IL-3 than mice protectively immunized by the i.v.

route (Lelchuk *et al.*, 1989). Administration of recombinant IL-3 also led to disease exacerbation (Feng *et al.*, 1988). Although the mechanism by which IL-3 exacerbates disease progression is unclear, IL-3, in combination with IL-4, can abrogate the macrophage activating capacity of IFN-γ (Liew *et al.*, 1989). It has also been suggested that IL-3 can increase the number of immature monocytes in the developing lesion and that such cells may be "safe targets" in which the parasite can replicate (Mirkovich *et al.*, 1986).

4. *Migration inhibitory factor*

Migration inhibitory factor (MIF) can activate human macrophages to inhibit the growth of, and kill, *L. donovani* (Pozzi and Weiser, 1991; Weiser *et al.*, 1991). This inhibitory activity is enhanced by LPS and, unlike IFN-γ activation, is not inhibited by IL-4 (Weiser *et al.*, 1991). $CD4^+$ T cells from genetically resistant mice secrete MIF when restimulated with specific antigen *in vitro* (Titus *et al.*, 1993). Finally, MIF may be an important regulatory molecule in the induction of nitric oxide, the effector mechanism in the destruction of *Leishmania* species (see Section VIII). MIF activates murine macrophages to express nitric oxide synthase and to produce high levels of nitric oxide *in vitro* (Cunha *et al.*, 1993).

5. *Transforming growth factor β*

Transforming growth factor β (TGF-β) can inhibit the intracellular killing of *L. major* by activated macrophages (Nacy *et al.*, 1991; Nelson *et al.*, 1991), probably by inhibiting the production of nitric oxide by IFN-γ treated macrophages (Ding *et al.*, 1990; Nelson *et al.*, 1991). TGF-β can also down-regulate the induction of nitric oxide synthesis in macrophages activated by MIF (Cunha *et al.*, 1993). Therefore, TGF-β appears to have a major role in the down-regulation of the protective responses involved in murine cutaneous leishmaniasis.

6. *Granulocyte macrophage-colony stimulating factor*

Granulocyte macrophage-colony stimulating factor (GM-CSF) can activate murine macrophages to kill intracellular *L. donovani* (Weiser *et al.*, 1987; Ho *et al.*, 1992). However, its effect is less clear cut in *L. major* infection, where it has been shown to be beneficial (Handman and Burgess, 1979; Solbach *et al.*, 1987), detrimental (Greil *et al.*, 1988), or to have no effect

(Corcoran *et al.*, 1988). Results with *L. mexicana* demonstrate that macrophages from genetically susceptible mice produce higher levels of GM-CSF *in vitro* than cells from resistant mice (Soares and Barcinski, 1992). This may be partly due to the fact that GM-CSF can act as a growth factor for *L. mexicana* promastigotes *in vitro* (Charlab *et al.*, 1990).

Although the role of some cytokines in leishmaniasis is fairly well defined, that of others is not well established. In addition, the cellular sources of some cytokines are not known, e.g., MIF and TGF-β, while other cytokines are secreted by cells other than T cells. Indeed, the requirement for IFN-γ or IL-4 at the initiation of infection (Sadick *et al.*, 1991; Scott, 1991a,b) suggests that other cell types are almost certainly involved as lymphocytes generally produce cytokines on restimulation, which may take days to develop (Weinberg *et al.*, 1990). Thus, factors other than Th1 and Th2 cells probably also influence the outcome of leishmaniasis. However, regardless of the cells and cytokines involved, the final effector mechanism appears to be similar.

VIII. The Effector Mechanism in Leishmaniasis

When resting macrophages are infected with a virulent strain of *Leishmania*, the parasites survive and subsequently replicate in the macrophages until the cells are lysed. In contrast, when the macrophages are activated *in vitro* with IFN-γ, in the presence of low levels of LPS (10 ng/ml), the parasites are killed. It was assumed for a long time that reactive oxygen intermediates (ROI) such as superoxide and hydrogen peroxide, which are products of the macrophage respiratory burst, were the major killing mechanism (Hughes, 1988). Indeed, a number of groups reported evidence that ROIs were involved in macrophage leishmanicidal activity (Murray, 1981, 1982; Haidaris and Bonventre, 1982; Pearson *et al.*, 1982; Murray and Cartelli, 1983; Buchmüller-Rouiller and Mauël, 1986). Hydrogen peroxide (H_2O_2) was shown to be leishmanicidal, with promastigotes more susceptible than amastigotes (Murray, 1981, 1982; Pearson *et al.*, 1983). This correlated with the ability of amastigotes to remove greater amounts of H_2O_2 in a phagocyte-free system than promastigotes (Channon and Blackwell, 1985a,b). However, a macrophage cell line (IC-21) deficient in the respiratory burst was as efficient at killing *Leishmania* amastigotes and promastigotes as normal peritoneal macrophages when activated by lymphokines (Scott *et al.*, 1985). In contrast to unactivated peritoneal macrophages, unactivated IC-21 cells were ineffective in killing promastigotes. Therefore, although the killing of log-phase promastigotes did require oxidative products, the killing of intracellular amastigotes was by a non-oxidative mechanism. Studies from several laboratories have demonstrated that this non-oxidative mechanism is nitric oxide, which appears to be the principal effector mechanism in this, and other related, systems.

A. NITRIC OXIDE

Nitric oxide (NO) is involved in a variety of biological activities, including endothelium-related vascular relaxation, platelet aggregation, neurotransmission, and macrophage killing of tumour cells (see Fig. 4). The importance of NO in various biological systems has resulted in a number of excellent reviews (e.g. by Marletta, 1989; James and Hibbs, 1990; Drapier, 1991; Green *et al.*, 1991; Liew and Cox, 1991; Moncada *et al.*, 1991; Nathan and Hibbs, 1991).

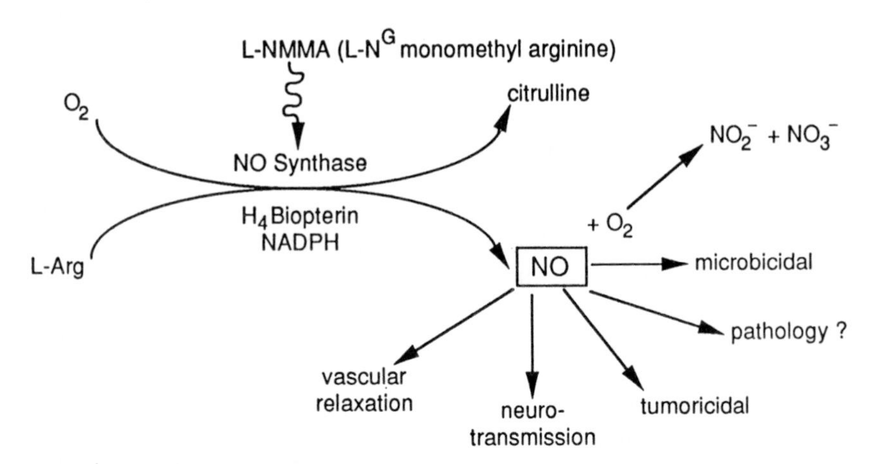

FIG. 4. The arginine : nitric oxide pathway.

Nitric oxide is derived from a reaction with the terminal guanidino nitrogen of the amino acid L-arginine and molecular oxygen, resulting in the formation of L-citrulline and NO. This reaction is catalysed by the enzyme nitric oxide synthase (NOS). The resulting NO is very unstable, having a half-life of 3–15 s. However, NO reacts with itself, water and oxygen to generate another radical, nitrogen dioxide (NO_2) and, finally, the stable end-products nitrite (NO_2^-) and nitrate (NO_3^-). The products generated are collectively known as reactive nitrogen intermediates (RNI) and the reaction can be specifically inhibited by arginine analogues such as L-N^G-monomethyl-arginine (L-NMMA).

The enzyme NOS, which catalyses the reaction, exists in at least two forms

(Moncada *et al.*, 1991; Werner-Felmayer *et al.*, 1991), a constitutive and an inducible enzyme. The constitutive enzyme, typically found in endothelial cells, platelets and the brain, is dependent on the presence of calcium/calmodulin and NADPH (B-nicotinamide adenine dinucleotide, reduced form). The second, inducible form of NOS is detected in murine macrophages (Marletta *et al.*, 1988; Stuehr *et al.*, 1989), as well as rat Kupffer cells and hepatocytes (Billiar *et al.*, 1989; Curran *et al.*, 1989), rat mast cells (Salvemini *et al.*, 1990) and rat neutrophils (McCall *et al.*, 1989, 1991). The induction of this enzyme requires protein synthesis (Marletta *et al.*, 1988); thus a lag phase of several hours is required before NO is generated. This inducible enzyme is calcium and calmodulin independent, but requires at least NADPH, flavine adenine dinucleotide and tetrahydrobiopterin as co-factors (Kwon *et al.*, 1989; Stuehr *et al.*, 1990). The constitutive NOS of the rat cerebellum and the inducible NOS of the macrophage have been cloned and sequenced (Bredt *et al.*, 1991; Lyons *et al.*, 1992; Xie *et al.*, 1992); however, other forms of these enzymes may also exist (Werner-Felmayer *et al.*, 1991).

The present discussion will concentrate on the role of NO in the immune system as a microbicidal molecule. Murine macrophages or macrophage cell lines were shown to produce substantial amounts of nitrite and nitrate in response to IFN-γ, either alone or in combination with TNF-α, TNF-β or LPS (Iyengar *et al.*, 1987; Stuehr and Marletta, 1987a,b; Ding *et al.*, 1988; Drapier *et al.*, 1988). Macrophages activated by cytokines developed potent tumouricidal activity, but this required a second signal supplied by LPS (Hibbs *et al.*, 1977; Weinberg *et al.*, 1978). This nonspecific tumour cell killing activity was mediated by NO derived from L-arginine (Hibbs *et al.*, 1987a,b, 1988). Since similar macrophages could be derived from mice infected with various intracellular pathogens, the NO effector pathway may be expected to inhibit the multiplication of such pathogens. Thus, macrophages activated with IFN-γ plus LPS had a powerful cytostatic effect on the fungal pathogen *Cryptococcus neoformans* due to the production of RNI (Granger *et al.*, 1986, 1990). Macrophage cytotoxicity against the protozoan parasite *Toxoplasma gondii* and the schistosomula of *Schistosoma mansoni* was also shown to involve the production of NO from L-arginine (James and Glaven, 1989; Adams *et al.*, 1990). Its involvement in leishmaniasis was extensively studied (see the reviews by Green *et al.*, 1991; Liew and Cox, 1991; Mauël *et al.*, 1991a; Nacy *et al.*, 1991.)

B. ROLE OF NITRIC OXIDE IN LEISHMANIASIS

Macrophages activated *in vitro* by IFN-γ or TNF-α, in the presence of LPS, or by a combination of both cytokines, developed potent leishmanicidal activity. This was directly correlated with the production of NO (Green *et al.*, 1990a,b; Liew *et al.*, 1990a–d; Mauël *et al.*, 1991b; Roach *et al.*, 1991). NO production and leishmanicidal activity could be inhibited by L-NMMA in a dose-dependent manner, but not by the D-enantiomer, D-NMMA. These results suggest that NO is necessary, and may be sufficient, to account for the killing of *Leishmania* amastigotes. Furthermore, it has been shown that NO can kill *L. major* promastigotes *in vitro* in a cell-free system (Liew *et al.*, 1990b).

The role of NO was also demonstrated *in vivo* (Liew *et al.*, 1990b). Mice injected with L-NMMA developed significantly larger lesions and parasite loads four orders of magnitude higher compared with controls injected with D-NMMA or phosphate-buffered saline. Thus, the resistance to *L. major* infection correlated with the ability of macrophages to produce NO. Macrophages from the resistant strains expressed significantly higher levels of NOS and produced larger amounts of NO compared to those from the susceptible strains when activated *in vitro* (Liew *et al.*, 1991b). Furthermore, bone marrow-derived macrophages from *Lsh*^{r} mice produced increased levels of NO compared with macrophages from *Lsh*^{s} mice in response to IFN-γ and LPS (Blackwell *et al.*, 1991; Roach *et al.*, 1991). The genetic control of resistance to a pathogen is complex and not fully understood. It is likely that the synthesis of NO is the final effector mechanism in this cascade of events determining the outcome of an infection.

An additional level of regulation in the development of leishmaniasis is also apparent. Macrophages from genetically resistant CBA mice were preincubated with IL-4 before activation with IFN-γ and LPS, followed by infection with *L. major*. The leishmanicidal activity of these cells was significantly inhibited and this correlated with the inhibition of NOS activity (Liew *et al.*, 1991a). As IL-4 is secreted by the Th2 cells associated with exacerbated disease, this may be a mechanism for the disease-promoting Th2 cells to down-regulate the host-protective effect of Th1 cells in leishmaniasis.

The mechanism by which macrophages are regulated by cytokines in the production of NO is intriguing. Macrophages exposed to IFN-γ produced low levels of NO. However, a second signal (e.g., bacterial LPS or muramyl dipeptide from the cell wall of mycobacteria) acting synergistically with IFN-γ results in the production of significantly larger amounts of NO by the macrophages (Stuehr and Marletta, 1985, 1987a,b; Iyengar *et al.*, 1987; Ding *et al.*, 1988; Drapier *et al.*, 1988). These increased NO levels were associated with the development of potent leishmanicidal activity (Green *et al.*, 1990a; Liew *et al.*, 1990d). IFN-γ could also act synergistically with TNF-α and TNF-β in the production of NO (Ding *et al.*, 1988; Drapier *et al.*, 1988). TNF-α activates peritoneal macrophages to kill *L. major in vitro* and this correlates with NO_2^- levels (Liew *et al.*, 1990a,c). The synergistic effect of IFN-γ and TNF-α results in high levels of NO_2^- and enhanced leishmanicidal activity (Liew *et al.*, 1990d). However, macrophages exposed to IFN-γ and infected with *L. major*, in the absence of LPS or TNF-α, also produced high levels of NO_2^- and killed the parasites (Green *et al.*, 1990a,b). This is thought to be due to the production of endogenous TNF-α by the macrophages in response to *L. major* infection (Green *et al.*, 1990b), which then interacts with the IFN-γ. This effect has been demonstrated both in the elimination of *L. major* amastigotes (Bogdan *et al.*, 1990) and *S. mansoni* schistosomula (Esparza *et al.*, 1987). In addition to IFN-γ and TNF-α, MIF also induces macrophage tumouricidal (Nathan *et al.*, 1971, 1973; Churchill *et al.*, 1975) and leishmanicidal (Weiser *et al.*, 1991) activity. MIF activates macrophages to express high levels of NOS and to produce large amounts of NO *in vitro* (Cunha *et al.*, 1993). MIF can act synergistically with IFN-γ in this activity and replaces the requirement for LPS.

In contrast to the positive effects of IFN-γ, TNF-α and MIF, other cytokines are inhibitors of NO synthesis. IL-4 can inhibit the expression of NOS and NO production and the leishmanicidal activity of IFN-γ activated macrophages, provided the cells were pretreated with IL-4 (Liew *et al.*, 1991a). Addition of IL-4 after IFN-γ frequently led to enhanced NO synthesis and higher levels of leishmanicidal activity. This was probably due to the production of endogenous TNF-α (Stenger *et al.*, 1991). Macrophage deactivating factor (MDF) and TGF-β1, β2 and β3 can also inhibit NO synthesis in activated macrophages (Ding *et al.*, 1990; Nacy *et al.*, 1991; Nelson *et al.*, 1991). This may explain the ability of MDF to suppress macrophage activity against *Leishmania* and *Toxoplasma* (Szuro-Sudol *et al.*, 1983). In addition, TGF-β2 can block the induction of NOS by IL-1β or TNF-α in rat mesangial cells (Pfeilschifter and Vosbeck, 1991). Finally, IL-10 down-regulates the induction of NOS by IFN-γ in murine macrophages (Cunha *et al.*, 1992). IL-10 inhibits the anti-microbial effect of activated macrophages against *Trypanosoma cruzi*, *Tox. gondii* and *S. mansoni* by down-regulating the production of NO (Gazzinelli *et al.*, 1992; Silva *et al.*, 1992). TGF-β also inhibits the antimicrobial effect of macrophages in *Tryp. cruzi* infection, although the effect on NO production was not examined in this system (Silva *et al.*, 1991).

Thus, the following conclusion may be drawn (Fig. 5). IFN-γ, TNF-α, MIF, and possibly other cytokines, occupying their respective receptors on the macrophage, send a series of signals which, together with those of co-stimulators (e.g. LPS and other bacterial components), lead to the induction of NOS. Other cytokines, such as IL-3, IL-4, TGF-β and IL-10 also send a series of signals which act in the opposite direction, inhibiting the expression of NOS. This phenomenon is consistent with the sophisticated regulatory mechanism characteristic of important biological systems. The check-and-balance of the two opposing pathways would ensure that this non-specific cytolytic effector is not overplayed, leading to possible pathology (Liew and Cox, 1991). The high degree of degeneracy observed should ensure that the total breakdown of either system is extremely remote.

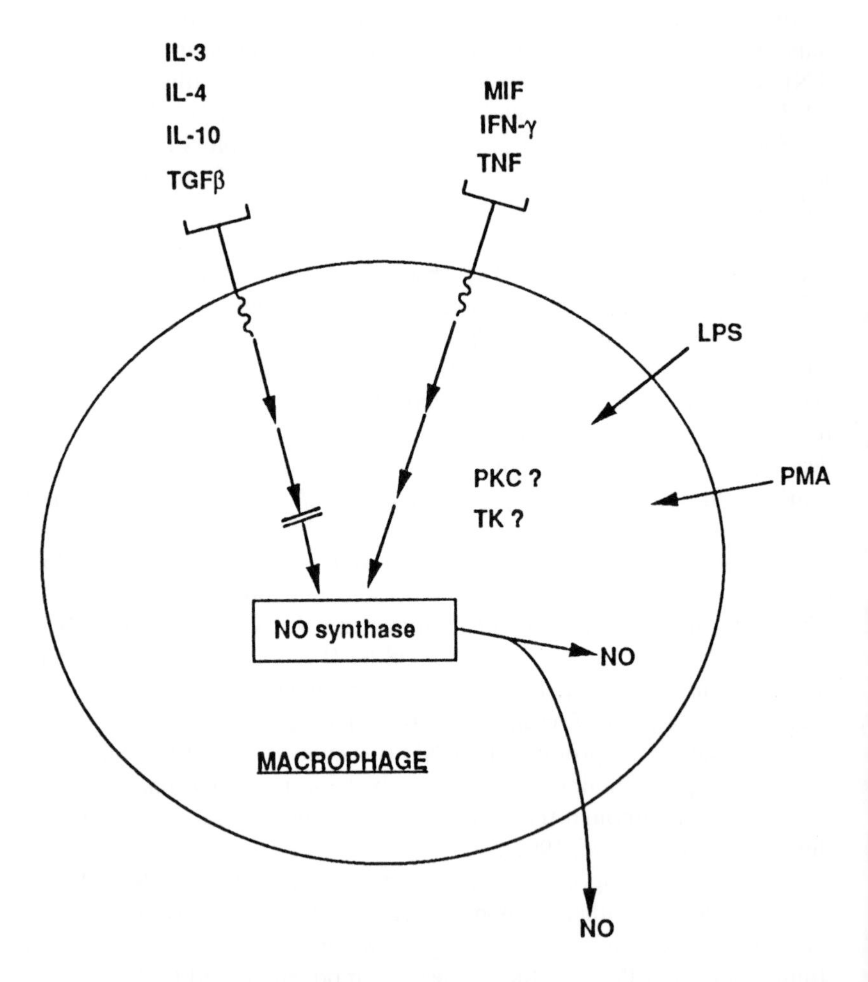

FIG. 5. The regulation of nitric oxide (NO) synthesis in macrophages by two opposing sets of cytokines; IFN-γ, interferon γ; IL, interleukin; LPS, lipopolysaccharide; MIF, migration inhibitory factor; PMA, phorbol myristate acetate; PKC, protein kinase C; TGFβ, transforming growth factor β.

Although the signalling pathway leading to the activation of NO synthesis is at present unclear, it appears that several mechanisms may be involved. IFN-γ induced leishmanicidal activity and NO_2^- production can be mimicked by exposure of the macrophages to the calcium ionophore A23187 in the presence of LPS (Buchmüller-Rouiller and Mauël, 1991; Buchmüller-Rouiller *et al.*, 1992). Therefore, it appears that some molecular events involved in the signal transduction signal are calcium-dependent. As yet, it is unclear which signal transduction pathways are involved in the activation of NOS. Conflicting evidence exists as to which protein kinases are involved in the phosphorylation of the constitutive rat cerebellar NOS. Protein kinase C does appear to be involved in the phosphorylation of the inducible macrophage NOS, as down-regulation of protein kinase C inhibits NO production (Severn *et al.*, 1993). However, the signalling cascade involved in the induction of both the constitutive and inducible NOS remains unclear.

The mechanism by which NO kills cells and parasites has yet to be clarified. NO can covalently react with intracellular iron (Salerno *et al.*, 1976), thus reacting with Fe-S prosthetic groups of susceptible enzymes, e.g. aconitase and complex I and II of the mitochondrial electron transport chain. This results in the formation of iron–nitrosyl complexes, the inactivation and degradation of these enzymes (Granger and Lehninger, 1982; Drapier and Hibbs, 1986; Lancaster and Hibbs, 1990; Pellat *et al.*, 1990) and the cessation of DNA replication (Lepoivre *et al.*, 1991). This loss of enzyme activity correlates directly with the loss of intracellular iron (Hibbs *et al.*, 1984, 1988; Drapier and Hibbs, 1986). Indeed, these are all effects observed when tumour cell targets are exposed to activated macrophages and the effect can be mimicked by exposure to NO gas (Hibbs *et al.*, 1988; Stuehr and Nathan, 1989). A role for iron loss was further supported by the finding that addition of excess iron, or the reducing agent sodium dithionite, to a lymphokine-activated macrophage culture inhibited the killing of schistosomal larvae *in vitro* (James and Glaven, 1989).

It is still unclear, however, whether NO is the final effector mechanism in all cases or if other molecules, not yet identified, exert the final damage. Nitric oxide could rapidly react with the superoxide anion radical O_2^- to form the stable peroxynitrite anion $ONOO^-$ (Beckman *et al.*, 1990). Once protonated, $ONOO^-$ decayed rapidly and appeared to form the reactive hydroxyl radical HO^- and the stable free radical nitric oxide NO_2^-. Hydroxyl radicals are highly reactive, combining with almost all molecules found in living cells with rate constants of between 10^9 and 10^{10} $\text{M}^{-1}\text{s}^{-1}$ (Anbar and Neta, 1967). Of interest, although the pathways leading to the formation of ROI and RNI are separate, IFN-γ was capable of inducing both NO_2^- and H_2O_2 release from murine peritoneal macrophages (Ding *et al.*, 1988). Evidence supporting a role for the hydroxyl radical has been obtained in a cell-free system using the chemical 3-morpholinosydnonimine N-ethylcarbamide (SIN-1) (Hogg *et al.*, 1992). SIN-1, a sydnonimine which generates both NO and superoxide simultaneously, auto-oxidized to generate peroxynitrite and, finally, hydroxyl radicals. Thus, NO may act synergistically with superoxide to form an even more potent effector molecule. Indirect evidence for the influence of reactive oxygen metabolites on NO synthesis also exists. The NO synthesis and leishmanicidal activity of macrophages activated by IFN-γ was significantly enhanced by superoxide dismutase and inhibited by catalase in a dose- and time-dependent manner (Li *et al.*, 1992), and oxygen radicals appeared to mediate their effect by respectively reducing or oxidizing the co-factor, tetrahydrobiopterin, required for NO synthesis.

So far, the production of NO by human macrophages has not been convincingly demonstrated *in vitro*. However, the capacity for endogenous nitrate production has been reported for humans (Green *et al.*, 1981) and the constitutive NOS has been identified in human platelets (Radomski *et al.*, 1990) and neutrophils (Schmidt *et al.*, 1989; Wright *et al.*, 1989). High levels of nitrite accumulation in the medium of human macrophages infected with *M. avium* and treated with GM-CSF and/or TNF-α have been reported (Denis, 1991). It may be that the infection with *M. avium* provides an additional factor resulting in the production of NO. However, data are accumulating to suggest that NO is involved in a number of human pathological conditions, including septic shock (Petros *et al.*, 1991) and alcoholic liver failure (Midgley *et al.*, 1991; Hunt and Goldin, 1992). The role of nitric oxide in clinical conditions has been reviewed (Editorial, 1991; Vallance and Moncada, 1991). The involvement of NO in various experimental models of clinical conditions has also been established, including diabetes (Kolb *et al.*, 1991; Kröncke *et al.*, 1991; Lukic *et al.*, 1991) and graft-versus-host disease (Garside *et al.*, 1992). The involvement of NO in other parasitic diseases will be discussed in Section XI.B.

IX. Vaccination Strategies in Leishmaniasis

Much work over the years has focused on the development of a vaccine for leishmaniasis (reviewed by Manson-Bahr, 1963; Preston and Dumonde, 1976b; Greenblatt, 1980, 1988; Gunders, 1987; Alexander, 1989; Liew, 1989a; Modabber, 1990). The fact that individuals who have recovered from leishmaniasis are refractory to further infection indicates that vaccination is feasible in principle.

A. EXISTING VACCINES

1. *Live vaccine*

"Leishmanization" is the ancient practice of inoculation using live parasites from an active human lesion to produce a self-healing lesion in an inconspicuous site. This method was described as early as 1911, when children in Baghdad were inoculated against "Oriental sore" (Wenyon, 1911; Marzinowsky and Schurenkowa, 1924). Later, live promastigotes from cultures *in vitro* were used, particularly in Israel and the USSR (reviewed by Gunders, 1987). This method was also used in Iran to inoculate soldiers during the Iran–Iraq war, around 1980. Although "leishmanization" can induce resistance in at least 70% of individuals treated, it can have serious problems. Apart from the technical difficulties of using live parasites for mass vaccination, there are clinical questions involved in the use of a live vaccine. "Leishmanization" can produce a large, long-lasting lesion which, in some cases, takes 3–5 years to heal (Modabber, 1990) and can also result in a depressed immune response to other vaccines (Gunders, 1987). Thus, "leishmanization" is now discontinued in Israel and rarely used anywhere else.

2. *Killed vaccines*

More recent work in Brazil concentrated on the use of a killed vaccine (Mayrink *et al.*, 1979, 1985; Antunes *et al.*, 1986). These studies used a vaccine of whole and sonicated promastigotes from five strains of *Leishmania* administered intramuscularly. Although positive skin tests were obtained and there appeared to be no side-effect, the disease disappeared from the area, making an evaluation of long-term protection impossible. However, no exacerbated disease was observed, contrasting with experimental results in which s.c. or intramuscular injection of antigen resulted in exacerbated infection (Liew *et al.*, 1985a,b; Liew, 1989b). Other groups in Iran and Venezuela are using a killed *Leishmania* vaccine with bacillus Calmette–Guérin (BCG) for additional immunotherapy (Convit *et al.*, 1987). These trials are based on similar work using killed *L. mexicana* and BCG, in which a similar rate of cure was reached as with chemotherapy, but fewer severe side-effects were seen with the killed vaccine (Convit *et al.*, 1989). However, large-scale trials are needed to produce consistent results.

B. EXPERIMENTAL VACCINES

Initial experiments used sonicated promastigotes or crude antigen–antibody complexes (Preston and Dumonde, 1976a; Handman *et al.*, 1977). Although these induced some protection against *L. major* in genetically resistant strains of mice, they had little effect in susceptible strains. Later work, using irradiated promastigotes or avirulent mutant clones of *L. major* promastigotes, did protect susceptible BALB/c mice against infection (Howard *et al.*, 1982, 1984; Liew *et al.*, 1984; Mitchell *et al.*, 1984; Titus *et al.*, 1988; McGurn *et al.*, 1990). However, the route of administration was critical. Only i.v. and i.p. immunization induced immunity, while s.c. injection resulted in exacerbated disease (Liew *et al.*, 1985a,b; McGurn *et al.*, 1990). Therefore, it was vital to ensure that any candidate vaccine would not induce exacerbated disease. One approach was to develop single, purified antigen preparations from promastigotes. A number of membrane antigens have been investigated, particularly LPG and two surface glycoproteins, gp63 and gp46.

1. *Lipophosphoglycan*

LPG is found on all *Leishmania* species and is the most abundant surface component (reviewed by Handman *et al.*, 1987; Turco, 1988). LPG is expressed in distinct forms on amastigotes and promastigotes (Glaser *et al.*, 1991) and consists of phosphorylated saccharides linked, by a carbohydrate core, to a unique lipid anchor (see Handman *et al.*, 1987; Turco, 1988). LPG can enhance the survival of the parasite within the macrophage. This may be due to a number of mechanisms, including scavenging free oxygen radicals (Chan *et al.*, 1989), inhibition of monocyte and neutrophil activity (Frankenburg *et al.*, 1990), and inhibition of protein kinase C activity (McNeely and Turco, 1987; McNeely *et al.*, 1989) and signal transduction in macrophages (Descoteaux *et al.*, 1991). LPG is highly immunogenic (Rosen *et al.*, 1988; McConville and Bacic, 1989) and immunization of susceptible mice with purified *L. major* LPG (Handman and Mitchell, 1985) or resistant and susceptible mice with *L. mexicana* LPG (Russell and Alexander, 1988) resulted in complete or partial protection. However, the presence of the lipid moiety appears to be essential for protection, as injection of lipid-free LPG, i.e. carbohydrate only, resulted in exacerbated disease after infection (Mitchell and Handman, 1986). It is not clear if LPG itself contains any T cell epitopes or if they are expressed on a tightly associated protein component of LPG (Jardim *et al.*, 1991; Russo *et al.*, 1992). One group has suggested that human T cells responding to LPG were in fact responding to protein contamination because proteinase-treated LPG was unable to stimulate a T cell response (Mendonça *et al.*, 1991). Therefore, more work is required to ascertain the usefulness of LPG as a candidate vaccine.

2. *Glycoprotein 63*

Gp63 is a major surface glycoprotein expressed on promastigotes and amastigotes of all leishmanial species tested (Bouvier *et al.*, 1987; Chaudhuri *et al.*, 1989; Medina-Acosta *et al.*, 1989; Frommel *et al.*, 1990). It is a metallo-proteinase with a conserved zinc-binding region (Etges *et al.*, 1986; Bouvier *et al.*, 1989; Chaudhuri *et al.*, 1989) and is implicated as a receptor for infecting mammalian cells (Russell and Wilhelm, 1986). The *L. major* gp63 gene has been cloned and expressed and its amino acid sequence described (Button and McMaster, 1988; Miller *et al.*, 1990). Anti-gp63 antibodies have been detected in the sera of patients with leishmaniasis (Colomer-Gould *et al.*, 1985; Reed *et al.*, 1987a) and gp63 can elicit a T cell response in patients with leishmaniasis (Jaffe *et al.*, 1990; Mendonça *et al.*, 1991; Russo *et al.*, 1991). Immunoaffinity purified gp63 reconstituted into liposomes produced protective immunity in both resistant and susceptible strains of mice (Russell and Alexander, 1988), although the protection in susceptible BALB/c mice was less complete. Other groups had difficulty in inducing protective immunity, particularly when using gp63 expressed as a fusion protein (Handman *et al.*, 1990) or when gp63-liposome preparations were injected s.c. (Kahl *et al.*, 1989). However, significant protection was achieved if the gp63-liposome preparation was administered i.v. (Kahl *et al.*, 1989). Finally, gp63 transformed into an attenuated *Salmonella* construct and administered orally protected resistant CBA mice against *L. major* infection (Yang *et al.*, 1990; see below).

3. *Glycoprotein gp/M-2*

Gp46 is a membrane glycoprotein detected on *L. amazonensis* (Kahl and McMahon-Pratt, 1987; Lohman *et al.*, 1990) and immunization with affinity purified gp46 produced significant protection in mice against *L. amazonensis* infection (Champsi and McMahon-Pratt, 1988). The gene encoding gp46 has been cloned and sequenced (Lohman *et al.*, 1990). Gp46 has recently been cloned into an attenuated *Vaccinia* virus mutant and this construct induced significant protection in susceptible BALB/c mice against a modest dose of *L. amazonensis* (McMahon-Pratt *et al.*, 1993).

As the degree of protection obtained with a single defined antigen is often less than that obtained with killed, whole promastigotes, it is likely that additional antigens will be required to achieve solid immunity. Other recently identified *Leishmania* antigens include gp10/20 (Rodrigues *et al.*, 1986), its 17 kDa precursor molecule (Rodrigues *et al.*, 1988), the PSA-2 glycoprotein complex of *L. major* (Murray *et al.*, 1989), and gp42 (Burns *et al.*, 1991). Three other *L. major* surface antigens of 60, 55 and 53 kDa, which induce significant protection in genetically resistant and susceptible mice, have also been identified (D. M. Yang and F. Y. Liew, unpublished observations). The genes encoding these antigens are being cloned and sequenced.

As demonstrated with killed whole parasites, the route of vaccine administration is critical. Intravenous administration of antigen in liposomes resulted in protection against *L. mexicana* (Russell and Alexander, 1988) and *L. major* (Kahl *et al.*, 1989) in both resistant and susceptible mice. Subcutaneous injection was generally less protective (Russell and Alexander, 1988), non-protective (Kahl *et al.*, 1989), or exacerbative (Rodrigues *et al.*, 1987). Therefore, as well as determining the best protective antigens and type of adjuvant to use, the route of immunization is also critical, with the i.v. route superior to the others. However, i.v. immunization could pose a formidable logistical obstacle to a mass vaccination programme. Hence, there is growing interest in the development of an oral vaccine which, apart from convenience of administration, also mimics the i.v. route of antigen delivery.

C. ORAL VACCINES

There are three types of heterologous carriers in use for experimental vaccines; *Vaccinia*, BCG and *Salmonella*. The *Vaccinia* system is unlikely to gain general approval by licensing authorities because of its rare but serious side-effects, unless a new generation of mutants, which are sufficiently attenuated but still suitable to function as carriers, can be found. So far, these have not been forthcoming. The BCG system is highly promising because of its in-built adjuvanticity (Aldovini and Young, 1991; Stover *et al.*, 1991), but the organism grows relatively slowly and it still requires parenteral administration. Furthermore, BCG, which is refractory to antibiotics, could develop into a systemic infection in immunodeficient individuals. However, the *Salmonella* system appears not to have any of these problems. The typhoid vaccines undergoing clincal trials so far seem to have no adverse effect given in relatively low doses, the organisms grow readily *in vitro* and, importantly, they can be administered orally. This system could, therefore, be the vaccine of the future. In addition, the ideal means of delivery for mass vaccination is by the oral route and the auxotrophic *Salmonella* mutants appear to be particularly suitable (reviewed by Winther and Dougan, 1984; Charles and Dougan, 1990).

The auxotrophic mutants of *S. typhi* (Hosieth and Stocker, 1981) are candidate oral vaccines of considerable promise and are currently undergoing clinical trials against typhoid. These mutants have deletions of the *Aro* and *Pur* genes. The *AroA*$^-$ mutants cannot synthesize aromatic amino acids and are therefore dependent on an exogenous source of these amino acids. One such mutant of *S. typhimurium*, SL3261, is highly attenuated and capable of only limited growth *in vivo* (Hosieth and Stocker, 1981). It can, however, penetrate the mucosal surface of the intestinal tract and reach the lymphoid organs, inducing both secretory and systemic immune responses and protective immunity against highly virulent wild-type organisms. This, and other attenuated *Salmonella* mutants, have been used experimentally as carriers for heterologous antigens from non-*Salmonella* species for vaccine development (Brown *et al.*, 1987; Poirier *et al.*, 1988; Sadoff *et al.*, 1988).

S. typhimurium SL3261 has recently been used as the heterologous carrier for a leishmanial antigen. The gene for *L. major* gp63 was transformed into SL3261 and the resulting SL3261-gp63 construct, which stably expressed the gp63 antigen *in vitro*, was used to immunize resistant CBA mice by the oral route (Yang *et al.*, 1990). Spleen cells from mice inoculated with the construct developed antibody and proliferative T cell responses to *L. major*, although they did not express detectable DTH reactivity. The activated cells were mainly $CD4^+$ and secreted IL-2 and IFN-γ, but not IL-4. The orally immunized mice developed significant resistance against a challenge infection with *L. major*. These results therefore demonstrated the feasibility of oral vaccination against leishmaniasis and showed that the oral route of antigen delivery, via the heterologous carrier, may preferentially induce the Th1 subset of $CD4^+$ T cells. However, as with other routes of immunization, it has not been possible to protect BALB/c mice against infection using this construct. This may be due to a number of factors, including the need for other antigens and for particular cytokines, e.g. TNF-α, IL-2 and IFN-γ. The human IL-1β gene has now been successfully expressed in the *Salmonella* SL3261 $AroA^-$ mutant, and BALB/c mice inoculated orally or i.v. with this construct produced a significant antibody response against the human IL-1β and were protected from mortality resulting from γ irradiation (Carrier *et al.*, 1992), indicating that the IL-1β gene was adequately delivered and expressed in a biologically active form *in vivo*. Other cytokine genes, e.g. TNF-α, either alone or in combination with *L. major* antigens, may be used to enhance the immunogenicity of orally delivered vaccines.

D. INDUCTION OF Th1 VS. Th2 CELLS

Since Th1 cells are host-protective, whereas Th2 cells are disease-exacerbative, in the murine *L. major* system (represented schematically in Fig. 6), it would be essential that an anti-leishmanial vaccine should preferentially induce Th1 cells and avoid the induction of Th2 cells. The possibility of identifying Th1 and Th2 "epitopes" was therefore explored.

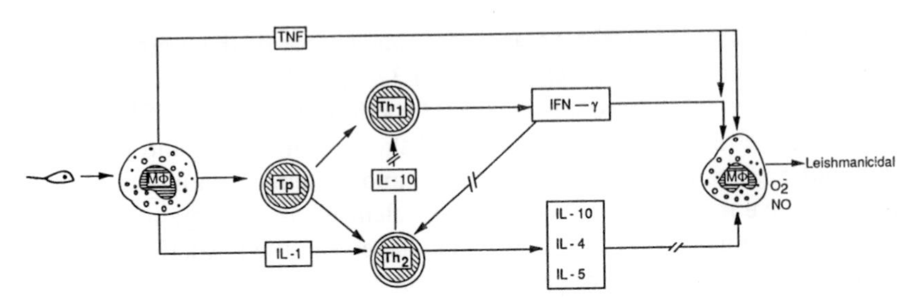

FIG. 6. Schematic representation of the induction and regulation of thymocyte helper (Th)1 and Th2 cells in murine cutaneous leishmaniasis. Macrophages (Mφ) infected with *Leishmania* can activate precursor cells (Tp) to differentiate to Th1 or Th2 cells. They can also produce interleukin (IL)-1, which potentiates the expansion of Th2 cells, and tumour necrosis factor (TNF) which can act synergistically with interferon γ (IFN-γ) or act directly to activate Mφ. Interruptions of arrows represent blockings; O_2^-, superoxide; NO, nitric oxide. (From Liew, 1993.)

Two subfractions (fractions 1 and 9) isolated from a soluble extract of *L. major* promastigotes had been shown to stimulate T cells *in vitro* (Scott *et al.*, 1987), although only fraction 9 was able to protect mice against a challenge infection. T cell lines were then generated recognizing either the protective fraction (line 9) or non-protective fraction (lines 1 and 9.2) (Scott *et al.*, 1988). Following adoptive transfer into syngeneic recipients, line 9, recognizing the protective fraction, was able to protect mice against *L. major* infection, while lines 1 and 9.2 exacerbated the infection. The lymphokine patterns of these lines were also examined. Line 9 secreted IL-2 and IFN-γ, whereas lines 1 and 9.2 secreted IL-4 and IL-5 (Scott *et al.*, 1988). Therefore, protective Th1 and exacerbative Th2 cells could be generated by different fractions of the solubilized parasite.

Wallis and McMaster (1987) had identified two cDNA clones encoding proteins containing regions of tandemly repeated peptides of 14 and 10 amino acids, respectively. Regions of repetitive peptides are a characteristic of many malaria proteins and this feature has been implicated in immune evasion (Anders, 1986). The immunogenicity of these tandemly repeated peptides was investigated (Liew *et al.*, 1990e). Using synthetic peptides corresponding to these tandemly repeating regions, an epitope was identified which can preferentially induce the disease exacerbating Th2 cells in susceptible BALB/c mice. Lymph node cells from BALB/c mice immunized s.c. with p183, an octamer of the repeating 10-mer peptide, proliferated *in vitro* against both the peptide and a soluble extract of *L. major*. These proliferative cells were $CD4^+$, H-2^d restricted and secreted high levels of IL-4 but only low levels of IL-2 or IFN-γ. T cells from BALB/c mice with progressive disease, but not healed mice, recognized this epitope. Mice injected s.c. with p183 developed exacerbated disease when infected with *L. major* and, more dramatically, the protective effect of i.v. immunization could be abrogated by prior s.c. injection with p183. However, p183 injected i.v. provoked little or no immune response in any strains of mice tested. These results indicated that there are epitopes which could preferentially induce Th2 cells but, again, other factors such as the route of immunization are also critical.

A number of epitopes which preferentially induce Th1 cells was also identified. This was done by synthesizing a series of overlapping peptides covering more than 75% of the amino acid sequence of gp63 from *L. major* (Yang *et al.*, 1991), and 11 T cell epitopes in CBA and BALB/c mice were identified. Two peptides (p146–171 and p467–482) were able to activate T cells which also recognized epitopes expressed by antigen-presenting cells infected with promastigotes. These T cells were mainly $CD4^+$, producing IL-2 and IFN-γ, but not IL-4, upon antigen stimulation *in vitro*. These two peptides could also induce a classical DTH response in CBA mice. Furthermore, CBA mice immunized with a mixture of the two peptides in *Corynebacterium parvum* or entrapped in liposomes induced significant resistance against *L. major* infection, but only when the route of immunization was i.v. The s.c. route did not induce significant protection or disease exacerbation.

Similar results were obtained by Jardim *et al.* (1990), who identified three $CD4^+$ T cell epitopes of gp63. Although all three peptides stimulated the production of IL-3, only one, PT3, stimulated the production of IL-2. None of the peptides stimulated IL-4. In addition, PT3 inoculated s.c. with adjuvant protected BALB/c mice from subsequent infection with *L. major*. Thus, PT3 appears to induce the proliferation of a Th1 subset.

Both peptides 146–171 (Yang *et al.*, 1991) and PT3 (Jardim *et al.*, 1990) cover a highly conserved region of gp63 believed to contain the zinc-binding region (Bouvier *et al.*, 1989; Chaudhuri *et al.*, 1989). The minimum length of the immunogenic peptide in this region is a heptapeptide, although optimal immunogenicity was conferred by a decapeptide containing the consensus zinc-binding residues, recognized by the T cell receptor, and two adjacent residues recognized by the class II MHC molecule (Yang *et al.*, 1993).

Therefore, it is possible preferentially to induce Th1 or Th2 cells *in vivo*, depending on the antigen and adjuvant used and the route of immunization. However, a general rule governing these conditions remains to be defined. It should also be noted that the immune response to other *Leishmania* species, particularly *L. donovani*, may not be the same as that to *L. major* (Kaye *et al.*, 1991).

X. Clinical Symptoms and Diagnosis of Leishmaniasis

A. CLINICAL SYMPTOMS

As discussed in Section III.A, infection with different species of *Leishmania* results in a spectrum of immune responses. Both cell-mediated and humoral responses can be detected (reviews include those by Turk and Bryceson, 1971; Howard, 1985; Griffiths, 1987; Mauël and Behin, 1987; Rees and Kager, 1987; Walton, 1987).

1. *Cutaneous and mucocutaneous leishmaniasis*

Infection with *L. major* or *L. tropica* results in the development of a marked cell-mediated immune response. This is detectable *in vivo* as a delayed-type skin reaction to killed promastigotes. A positive DTH response is detected within a few days to several months, and is generally positive many months before the lesions heal. However, the DTH response can then remain positive for many years (Guirges, 1971). Peripheral blood lymphocytes from infected or recovered patients proliferate in response to stimulation *in vitro* with leishmanial antigen (Tremonti and Walton, 1970; Neva *et al.*, 1979; Wyler *et al.*, 1979; Green M. S., *et al.*, 1983) and this response can occur at any stage of the clinical infection. Macrophage migration is also inhibited in cutaneous leishmaniasis (Zeledón *et al.*, 1977). The production of IL-1 *in vitro* by human monocytes infected with *L. tropica* also appears to be inhibited (Crawford *et al.*, 1985). Specific antibody has been detected in most patients with cutaneous leishmaniasis and the titre is directly linked to the severity of disease (Behforouz *et al.*, 1976; Roffi *et al.*, 1980).

The uncomplicated form of cutaneous leishmaniasis found in South and Central America after infection with *L. amazonensis*, *L. braziliensis* or *L. mexicana* has the same characteristics as those found in "Old World" cutaneous leishmaniasis, although loss of skin reactivity in over half the treated cases of American cutaneous leishmaniasis has been reported (Mayrink *et al.*, 1976). As well as the detection of a DTH reaction, individuals infected with *L. mexicana* developed detectable levels of *Leishmania*-specific IgE antibody (Lynch *et al.*, 1982). However, there appeared to be an inverse correlation between the IgE levels and DTH reactivity, possibly reflecting the differential activation of different Th cell subsets (see Section VI.B).

The development of mucocutaneous leishmaniasis sometime seen with *L. braziliensis* or *L. amazonensis* infection is also associated with the development of DTH and antibody (Walton *et al.*, 1972; Ridley *et al.*, 1980). Comparison of the proliferative responses *in vitro* of cells from patients with localized, diffuse or mucocutaneous disease showed that the highest responses were obtained from the mucocutaneous patients (Castes *et al.*, 1984). Peripheral blood lymphocytes from patients with cutaneous or mucocutaneous leishmaniasis responded *in vitro* to purified or recombinant gp63 from *L. amazonensis* (Russo *et al.*, 1991). Antibody is not detected in all patients and an inverse correlation between the level of antibody and the degree of mucosal involvement has been suggested (Ridley *et al.*, 1980).

2. *Visceral leishmaniasis*

Infection with *L. donovani* is associated with a profoundly impaired cellular response, but a marked humoral response. Visceral leishmaniasis patients fail to respond to *L. donovani* antigen in terms of DTH (Rezai *et al.*, 1978; Haldar *et al.*, 1983; Ho *et al.*, 1983), lymphocyte proliferation in response to both mitogens (Ghose *et al.*, 1979; Cillari *et al.*, 1988) and parasite antigens (Wyler *et al.*, 1979; Carvalho *et al.*, 1981; Haldar *et al.*, 1983), and IL-2 and IFN-γ production *in vitro* (Carvalho *et al.*, 1985a; Cillari *et al.*, 1988, 1991a). However, these immune responses are restored after successful chemotherapy (Manson-Bahr, 1959; Wyler, 1982; Haldar *et al.*, 1983, Carvalho *et al.*, 1985b; Cillari *et al.*, 1988). Peripheral blood lymphocytes from patients recovered from *L. chagasi* infection also proliferated in response to specific antigen, in particular to two glycoproteins of 30 and 42 kDa (Reed *et al.*, 1990a).

Although IL-2 and IFN-γ production was depressed in patients with visceral leishmaniasis (Carvalho *et al.*, 1985a; Cillari *et al.*, 1988, 1991a), IL-4 was detected in almost all patient sera (Zwingenberger *et al.*, 1990). This helps to explain the elevated levels of IgE detected in the patients' sera (Peleman *et al.*, 1989), and may indicate a role for a Th2-type response in visceral leishmaniasis. Elevated levels of TNF-α were also detected in patients with visceral leishmaniasis, but dropped rapidly after successful treatment (Barral-Netto *et al.*, 1991). However, normal human monocytes infected with *L. donovani* were unable to produce IL-1 or TNF-α (Reiner *et al.*, 1990). In contrast to the impaired cellular response, both non-specific and specific antibody levels were raised (Duxbury and Sadun, 1964; Rezai *et al.*, 1978; Ghose *et al.*, 1980), possibly due to polyclonal B cell activation (Ghose *et al.*, 1980; Galvao-Castro *et al.*, 1984).

Patients with active visceral leishmaniasis possess a population of adherent cells capable of suppressing the production of IL-2 by autologous peripheral blood lymphocytes after recovery (Cillari *et al.*, 1988). However, these adherent cells inhibit only mitogen-induced IL-2 production, whereas antigen-specific IL-2 production is inhibited by $CD4^+$, UCHL-1^- suppressor-inducer T cells (Cillari *et al.*, 1991a). In addition, the $CD4^+$, UCHL-1^-

helper-inducer T cell population, responsible for the production of IL-2 and IFN-γ, is also reduced in these patients (Cillari *et al.*, 1991a). As yet, it is unclear if these cells are analogous to the Th1 population found in murine cutaneous leishmaniasis. The discovery of two separate populations of cells, the actions of which appear to be mutually exclusive, may help to explain the earlier observation that the antigen-specific unresponsiveness of Indian kala-azar patients was not reversible by the depletion of $CD8^+$ T cells (Sacks *et al.*, 1987), because the unresponsiveness observed in these patients may also have been mediated by $CD4^+$, $UCHL\text{-}1^-$ T cells.

Visceral leishmaniasis has been detected in a number of patients with human immunodeficiency virus (HIV) infection (reviewed by Peters *et al.*, 1990). These patients are less likely to respond to treatment and there is a higher frequency of relapse. Furthermore, *Leishmania* may, by stimulating the activation of $CD4^+$ T cells and macrophages latently infected with HIV, accelerate the progression of the immunodeficiency (Pinching, 1988). As the HIV pandemic spreads, it is almost certain that more cases of both cutaneous and visceral leishmaniasis in HIV patients will be detected. This adds extra impetus to the study of clinical leishmaniasis, which undoubtedly lags behind that of the animal work. Important areas of research include the possible role of Th1- and Th2-type responses in clinical leishmaniasis and the feasibility of oral vaccination.

Finally, infection with *L. amazonensis* can result in a spectrum of illness, including cutaneous, mucocutaneous or visceral leishmaniasis (Barral *et al.*, 1991). *L. major* and *L. mexicana* have been isolated from cases of visceral leishmaniasis (Schnur *et al.*, 1985; Barral *et al.*, 1986) and both *L. donovani* and *L. chagasi* have been associated with cutaneous lesions (Hoogstraal and Heyneman, 1969; Abdalla, 1982). Thus, the use of more sensitive detection techniques, particularly in identifying the parasite species, may reveal a more heterogeneous pattern of symptoms associated with each parasite than was previously realized. Therefore, more accurate diagnosis of leishmaniasis is essential.

B. DIAGNOSIS

The initial diagnostic method generally involves demonstrating the presence of parasites either by direct smear, biopsy, culture or animal inoculation. However, the precise identification of the *Leishmania* species relies on skin tests and serological assays.

1. *Skin testing*

The skin test, also known as the leishmanin test or Montenegro test

(Montenegro, 1926), is an important tool in the diagnosis of cutaneous and mucocutaneous leishmaniasis due to the strong cellular reactivity observed in these diseases. In visceral leishmaniasis, cellular responses are detected only after recovery from infection, therefore its use is purely epidemiological. The antigen commonly used is a crude, heterogeneous preparation of promastigotes suspended in phenolized saline (leishmanin). This is injected intradermally and the response read 48–72 h later. However, a positive response does not necessarily indicate active infection, but may indicate only previous exposure to *Leishmania*. This is particularly likely in an endemic area. The use of a crude parasite preparation also results in considerable cross-reactivity with different *Leishmania* species and even with other parasites, e.g. trypanosomes (Dostrovsky and Sagher, 1946; Koufman *et al.*, 1978). As yet, the use of defined antigens which may result in species- or strain-specific skin tests has not been developed.

2. *Serological tests*

Serodiagnosis is unreliable in cutaneous leishmaniasis because of the low serum antibody titres, and up to 30% of parasitologically known cases are not detected (Edrissian and Darabian, 1979; El Safi and Evans, 1989). Serodiagnostic tests are more successful in the detection of visceral leishmaniasis because of the high anti-leishmanial antibody titres present.

A number of serodiagnostic tests is used, including indirect fluorescent antibody tests (IFAT) (Duxbury and Sadun, 1964; Badaro *et al.*, 1983), ELISAs (Hommel *et al.*, 1978; Anthony *et al.*, 1980), and the direct agglutination test (DAT) (Allain and Kagan, 1975; Harith *et al.*, 1986). In these assays, a crude preparation of whole, killed parasites is used. As the antibody levels are high in visceral leishmaniasis, the serum can be tested at high dilution, reducing the level of cross-reactivity. However, as this is not possible with cutaneous leishmaniasis, a radioimmunoassay has been developed which appears to be more sensitive (Rosen *et al.*, 1986). This assay uses the carbohydrate–lipid fraction of the parasite as antigen, which appears to reduce the problem of cross-reactivity with other *Leishmania* species and with other genera. A similar approach has been used in the detection of visceral leishmaniasis, for which serodiagnostic assays using membrane-associated antigens of both *L. donovani* (see Blaxter *et al.*, 1988; Jaffe and Zalis, 1988) and *L. chagasi* (see Reed *et al.*, 1987a, 1990b) have been reported. An assay has also been developed using species-specific *L. donovani* MAbs (Jaffe and McMahon-Pratt, 1987).

Although these techniques are in routine use, the instability of the reagents and the need for sophisticated equipment renders them unsuitable for use in the field. In addition, there is still difficulty in identifying

Leishmania species and subspecies. A new technique using DNA probes is an attractive alternative to both skin testing and serodiagnosis.

3. *DNA probes*

This technique has been extensively reviewed (Sacks *et al*., 1993) and will be only briefly summarized here. The target of the DNA probes is the kinetoplast DNA (kDNA) minicircle, which represents 10–25% of the total cell DNA and is repeated 10 000–100 000 times per kinetoplast. The minicircles contain highly repeated DNA sequences which are unique to each *Leishmania* species. Thus, the kDNA minicircle isolated from *L. braziliensis* does not have significant sequence homology with the kDNA of *L. mexicana* (see Wirth and McMahon-Pratt, 1982). A DNA probe could distinguish the two *Leishmania* species when a lesion sample was applied directly to nitrocellulose (Wirth *et al*., 1986; Rogers *et al*., 1988). This method has now been used to diagnose both visceral and cutaneous leishmaniasis (Lopes and Wirth, 1986; Wirth *et al*., 1986; Smith *et al*., 1989). DNA probes can also be used for hybridization *in situ*, when lesion material applied to a microscope slide is probed with the kDNA probe, allowing visualization of the parasite's presence (Van *et al*., 1987). Detection of *Leishmania* species in human biopsies requires only a few hundred parasites, enabling the diagnosis of cutaneous leishmaniasis. However, in mucocutaneous or visceral disease, when the parasite number can be low, the polymerase chain reaction may be used to amplify the DNA before applying the DNA probe (Rodgers *et al*., 1990). The development of non-radioactive methods will facilitate the use of this technique in developing countries and in the field.

The development of quick, simple techniques which can accurately identify *Leishmania* species from clinical samples will aid the identification of species associated with particular clinical symptoms and determine the prevalence of each species within endemic areas. This will also be of use in epidemiological studies on both the human and animal reservoirs of *Leishmania*.

XI. The *Leishmania* Model—its Impact on Other Parasite Systems

The study of the immune response against *L. major* in the murine model has had major implications in the study of the immune response to other parasitic infections. Some of these will be discussed here.

A. IMPORTANCE OF THE Th1/Th2 RESPONSE

When the existence of two distinct subsets of Th cells in a panel of murine T cell clones was initially described (Mosmann *et al.*, 1986), their relevance to the immune response against an infectious agent was not immediately clear. However, there was already evidence for different T cell populations controlling resistance and susceptibility to, as well as healing and non-healing of, cutaneous leishmaniasis in inbred mouse strains (reviewed by Liew 1987, 1989b). When the lymphokine profiles of infected animals were examined, both *in vitro* and *in vivo*, the resistant and healed mice exhibited a Th1-type pattern (IL-2 and IFN-γ), whereas the susceptible and non-healing mice exibited a Th2-type pattern (IL-4 and IL-10) (Liew and Dhaliwal, 1987; Locksley *et al.*, 1987; Sadick *et al.*, 1987; Heinzel *et al.*, 1989, 1991; Liew *et al.*, 1989). This reciprocal expression of IFN-γ and IL-4 was also demonstrated in visceral leishmaniasis (Lehn *et al.*, 1989; Zwingenberger *et al.*, 1990). Additional evidence was obtained from experiments transferring Th1 or Th2 cell lines into syngeneic recipients, with the Th1 line protecting recipients and the Th2 line exacerbating the infection (Scott *et al.*, 1988; Holaday *et al.*, 1991). Thus, the role of Th1 and Th2 cells in the differential outcome of *L. major* infection was quickly established. These findings stimulated research into the role of Th1 and Th2 cells in other parasitic infections, with interesting results.

As parasite infections often result in chronic disease, the demonstration of a polarized immune response in terms of Th1 or Th2 cells was relatively easy. However, the biological diversity of parasites and the variety of life cycles observed has resulted in different immune effector mechanisms, e.g., macrophage activation for macrophage dwelling protozoa like *Leishmania*; $CD4^+$ and $CD8^+$ T cells for tissue dwelling intracellular protozoa like *Toxoplasma*; IgE, eosinophils and mast cells for large gut-dwelling helminths like *Nippostrongylus*. Obviously, this also alters the type of regulatory Th cell response observed (reviewed by Else and Grencis, 1991a; Finkelman *et al.*, 1991; King and Nutman, 1991; Sher and Coffman, 1992). The role for Th1 and Th2 cells in a number of other parasitic infections is briefly discussed in the following sections.

1. Plasmodium *spp.*

T cell-mediated immunity is important in the immune response against malaria. In murine models of malaria, $CD4^+$ T cells are involved in the development of immunity and in clearing the infection (Cavacini *et al.*, 1986; Kumar *et al.*, 1989; Langhorne *et al.*, 1990). The adoptive transfer of a malaria-specific Th1 clone resulted in the development of immunity to *P.*

chabaudi adami in athymic nude mice (Brake *et al.*, 1988). It appears likely that Th1 cells are effective against the blood stage by the secretion of cytokines, most likely TNF and IFN-γ, leading to the production of toxic oxygen and nitrogen radicals which kill the parasite (Allison and Eugui, 1983; Clark and Hunt, 1983; Ockenhous *et al.*, 1984; Playfair, 1990; Mellouk *et al.*, 1991; Nüssler *et al.*, 1991). However, although this Th1-like response is protective in the blood-borne stage of the infection, in cerebral malaria it may be associated with disease exacerbation (cited by Sher and Coffman, 1992). In humans, this often lethal consequence of *P. falciparum* infection is associated with high levels of plasma TNF (Grau *et al.*, 1989a; Kwiatkowski, 1990). In the murine model, the pathology can be prevented or reduced by treatment with antibodies against the cytokines TNF, IFN-γ, IL-3 and GM-CSF (Grau *et al.*, 1987, 1988, 1989b). Therefore, although Thl-like cells are involved in immunity to malaria, the situation is complicated by the apparent involvement of these cells in the pathology of malaria as well.

2. Trypanosoma *spp.*

Trypanosomes, haemoflagellate protozoans, infect a wide variety of host tissues and infection is accompanied by non-specific immunosuppression. However, the immunosuppression and death observed in *T. cruzi* infection of mice was reversed by treatment *in vivo* with IFN-γ (Reed, 1988). This was due, at least partly, to the activation of macrophages. IFN-γ, TNF and GM-CSF also activated macrophages *in vitro* to kill *Tryp. cruzi* (Plata *et al.*, 1984; Wirth *et al.*, 1985; De Titto *et al.*, 1986; Reed *et al.*, 1987b). In contrast, TGF-β could block the inhibitory effect of IFN-γ on *Tryp. cruzi* both *in vitro* and *in vivo* (Silva *et al.*, 1991). Endogenous IFN-γ, a Thl cytokine, appears to be important in controlling acute infection (Silva *et al.*, 1992). However, although spleen cells from susceptible mice could produce biologically active IFN-γ *in vitro*, it was apparently ineffective *in vivo*. Susceptible B6 mice have increased amounts of IL-10 mRNA and biologically active IL-10 during the acute phase of *Tryp. cruzi* infection compared to resistant mice (Silva *et al.*, 1992). IL-10 partly, though not completely, blocks IFN-γ mediated inhibition of *Tryp. cruzi* in murine macrophages. Therefore, the production of IL-10, a Th2 cytokine, can reverse the protective effect of IFN-γ, a Thl cytokine, in experimental *Tryp. cruzi* infection. However, unlike *L. major* infection in which the reciprocal production of Thl and Th2 cytokines normally occurs, susceptible mice with *Tryp. cruzi* infection produce both sets of cytokines. The regulation of the synthesis and function of these cytokines remains obscure.

3. Toxoplasma gondii

Tox. gondii is also an intracellular, tissue-dwelling protozoan which penetrates the gut wall and forms tissue cysts. Although $CD8^+$ T cells have a major role in resistance to *Tox. gondii*, $CD4^+$ T cells can augment this immunity (Suzuki and Remington, 1988; Gazzinelli *et al.*, 1991). Again, IFN-γ can activate macrophages to kill *Tox. gondii* (Adams *et al.*, 1990), although this could be produced by Thl, $CD4^+$ T cells or $CD8^+$ T cells. However, the Th2-derived cytokine, IL-10, can inhibit the IFN-γ mediated macrophage killing of *Tox. gondii* (Gazzinelli *et al.*, 1992).

4. *Parasitic helminths*

Parasitic helminths are large, multicellular organisms which live extracellularly in the host. The host immune response to helminth infections is characterized by elevated IgE levels, eosinophilia and mastocytosis (Finkelman *et al.*, 1991). These responses are stimulated by Th2-derived cytokines—IgE by IL-4 (Finkelman *et al.*, 1988; Urban *et al.*, 1991), eosinophilia by IL-5 (Coffman *et al.*, 1989), and mastocytosis by a combination of IL-3, IL-4 and IL-10 (Madden *et al.*, 1991; Thompson-Snipes *et al.*, 1991). However, this immune response may not always be protective and, in some infections, may promote parasite survival (Finkelman *et al.*, 1991). The immune response to some helminth infections is discussed below.

(a) *Schistosomes. Schistosoma mansoni* is a parasitic trematode causing schistosomiasis. The disease is progressive and caused by the inflammatory response to egg deposition in the liver. In murine schistosomiasis, infection results in a strong Th2-type response with high levels of IL-4, IL-5 and IL-10 but little or no IL-2 or IFN-γ (Sher *et al.*, 1990, 1991; Street *et al.*, 1990; Pearce *et al.*, 1991), although this response appears to be a feature of chronic infection and is not apparent during the initial host–parasite encounter (Finkelman *et al.*, 1991). In contrast, vaccination with irradiated cercariae induced protective immunity, correlating with the elevated production of the Thl cytokines, IFN-γ and IL-2 (Pearce *et al.*, 1991). The production of eggs by the parasite appears to be the major stimulus for the Th2 cytokine production observed (Grzych *et al.*, 1991; Pearce *et al.*, 1991). The production of a Th2 response correlated with the down-regulation of the Thl response in infected mice (Pearce *et al.*, 1991). This down-regulation correlated with the production of IL-10 (Sher *et al.*, 1991) and TGF-β (Czaja *et al.*, 1989). IL-10 in particular is known to down-regulate Thl responses (Fiorentino *et al.*, 1989). Therefore, it is clear that the development of a Th2 response in mice infected with *S. mansoni* correlates with the production of at least one cytokine, IL-10, which can down-regulate the Thl response. A central role for TNF-α in the granuloma formation was evident in *S. mansoni* infection (Amiri *et al.*, 1992). Injection of SCID mice, which normally do not form granulomas in response to eggs trapped in the liver, with recombinant TNF-α resulted in the formation of granulomas and the stimulation of egg production by the adult schistosomes. Thus, schistosomes may be able to use the host immune response to enhance their own survival.

(b) Trichuris muris. *Trich. muris* is a nematode parasite of mice which dwells in the large intestine and is closely related to *Trich. trichiura*, which causes trichuriasis in man (Roach *et al.*, 1988). As with other parasitic infections, different inbred mouse strains exhibit resistance or susceptibility to infection and this is associated with distinct cytokine secretion phenotypes (reviewed by Else and Grencis, 1991a). In resistant strains, the expulsion of *Trich. muris* is associated with $CD4^+$ T cells which secrete IL-4, IL-5 and IL-9 but not IFN-γ, characteristic of a Th2 response. Conversely, in murine strains unable to expel the parasite, T cells produce mainly IFN-γ, characteristic of a Thl response. Expulsion of the parasite is associated with increased levels of IgE, eosinophilia and intestinal mastocytosis (Else and Grencis, 1991b).

This polarization of the immune response to a Th2 response during worm expulsion has also been reported for other intestinal nematode parasites, including *Nippostrongylus brasiliensis, Heligmosomoides polygyrus* and *Tri*chinella spiralis (see Mosmann and Coffman, 1989; Pond *et al.*, 1989; Street *et al.*, 1990; Finkelman *et al.*, 1991; Grencis *et al.*, 1991). Again, production of these cytokines was associated with increased IgE, eosinophilia and mastocytosis.

Thus, results from studies on *L. major*, demonstrating that Th1 cells were protective and Th2 cells exacerbative, have stimulated work with other parasite systems showing that polarized T cell responses are important for the final outcome of a variety of infections.

B. NITRIC OXIDE AS THE EFFECTOR MECHANISM

Some of the earliest demonstrations of a role for NO in the killing of a parasite were with *L. major*, both *in vitro* (Green *et al.*, 1990a,b; Liew *et al.*, 1990b,c; Mauël *et al.*, 1991b) and *in vivo* (Liew *et al.*, 1990b). It is now recognized that NO is involved in killing a number of other parasites (see the reviews by James and Hibbs, 1990; Green *et al.*, 1991; Liew and Cox, 1991; Nathan and Hibbs, 1991), including *S. mansoni* schistosomula (James and

Glaven, 1989), *Tox. gondii* (see Adams *et al.*, 1990), *Tryp. musculi* (see Vincendeau and Daulouede, 1991), *Tryp. cruzi* (see Munoz-Fernandez *et al.*, 1992), *P. berghei* (see Mellouk *et al.*, 1991) and *P. yoelii* (see Nüssler *et al.*, 1991). It has been suggested that, as well as having a protective role, NO may be involved in the pathology associated with malaria (Clark *et al.*, 1991). Thus, the high levels of plasma TNF associated with cerebral malaria may induce the production of NO from the endothelium, which could then diffuse into the central nervous system leading to vasodilatation of the cerebral blood vessels, resulting in increased intracranial pressure and the symptoms associated with cerebral malaria.

C. PREFERENTIAL INDUCTION OF Th1/Th2 CELLS

As discussed in Section IX.B, the protection obtained by immunization with killed or avirulent parasites depended on the route of immunization used (Howard *et al.*, 1982, 1984; Liew *et al.*, 1984, 1985a,b; Mitchell *et al.*, 1984; Titus *et al.*, 1988; McGurn *et al.*, 1990). Although intravenous or intraperitoneal administration protected susceptible mice against challenge with *L. major*, subcutaneous immunization generally resulted in exacerbated disease. Intravenous immunization generated IFN-γ, characteristic of a Th1-mediated response, whereas subcutaneous immunization generated IL-4, characteristic of a Th2-mediated response (Liew *et al.*, 1985a,b, 1987, 1989; Liew and Dhaliwal, 1987). Work by many groups has demonstrated that IFN-γ secreted by Th1 cells is host-protective, whereas IL-4 produced by Th2 cells is disease-promoting (see Section VII.A). This finding has important implications for vaccine development, both in *Leishmania* and in other systems. Using *L. major* membrane proteins, T cell epitopes can be selected which preferentially induce Th1 (Scott *et al.*, 1988; Jardim *et al.*, 1990; Yang *et al.*, 1991) or Th2 (Scott *et al.*, 1988; Liew *et al.*, 1990e) cells. Therefore, a careful assessment of the immune response induced by individual proteins or peptides is required to ensure that exacerbative immune responses are avoided.

IL-4 can inhibit NO synthesis and the leishmanicidal activity of IFN-γ activated macrophages (Liew *et al.*, 1991a). As IL-10, another Th2-derived cytokine, also inhibits the production of NO (Cunha *et al.*, 1992; Gazzinelli *et al.*, 1992), this may represent a novel *in vivo* pathway by which Th2 cells can regulate the activity of Th1 cells. This pathway requires further investigation both in *L. major* and in other parasitic infections in which a role for Th1 and Th2 cells and NO has been demonstrated.

Other factors may also play a role in the susceptibility of mice to cutaneous leishmaniasis. One of these appears to be ageing (Cillari *et al.*,

1992). Ageing is associated with a decline in T cell function (reviewed by Hirokawa, 1988) and an increase in susceptibility to infection with intracellular organisms, such as *Tox. gondii* and *Listeria monocytogenes* (see Gardner and Remington, 1977; Patel, 1981). This has been extended to experimental leishmaniasis with the demonstration that susceptibility to *L. major* infection in BALB/c mice increased with age (Cillari *et al.*, 1992). This was associated with reduced production of IFN-γ and IL-2, but increased IL-4 production. However, the thymic hormone thymopentin increased production of IFN-γ and IL-2 and decreased the production of IL-4, resulting in a significant increase in resistance to *L. major* infection in aged BALB/c mice (Cillari *et al.*, 1992). Thus, a thymic hormone can differentially affect the activity of Th1 and Th2 cells and increase the resistance to infection, which may have considerable implications for future therapy in age-related diseases.

Similar effects have been demonstrated with the immunopotentiating drug isoprinosine (Cillari *et al.*, 1991b). BALB/c mice treated orally with isoprinosine before infection with *L. major* exhibited an increase in IFN-γ, but a decrease in IL-3 and IL-4, secretion and a significantly delayed onset of disease. This may also have been due to differential effects on Th1 and Th2 activation and may be of relevance in other infectious diseases.

D. IMMUNOTHERAPY

Immunopotentiators have been used to modify the course of *Leishmania* infection by non-specifically activating macrophages. These agents include BCG (Fortier *et al.*, 1987) and BCG plus killed parasites (Convit *et al.*, 1987, 1989). However, cytokines alone also stimulate macrophage leishmanicidal activity. These include the Th1-derived cytokine IFN-γ (Nacy *et al.*, 1981, 1985; Murray *et al.*, 1983, 1987; Nathan, 1983; Liew and Dhaliwal, 1987; Belosevic *et al.*, 1988, 1989) and TNF-α (Liew *et al.*, 1990a). TNF-α can prevent the disease enhancement seen after s.c. immunization, even with a peptide, p183, known to stimulate Th2 cells (see Section IX.D), and induces protective immunity instead (Liew *et al.*, 1991c). However, the TNF-α had to be present in the immunizing inoculum.

Studies using recombinant IL-2 to treat patients with lepromatous leprosy demonstrated an enhancement of their cell-mediated immune response as measured by a reduction in the total body burden of *M. leprae* (Kaplan *et al.*, 1991). However, this had to be administered intradermally. As discussed in Section XI.C, oral administration of immunotherapeutic agents or vaccines is preferable. At least one cytokine, IL-1β, has been shown to be biologically active when administered orally in an attenuated *Salmonella* carrier (Carrier *et al.*, 1992). The effect of incorporating other cytokine genes into this carrier as a means of enhancing the host immune response against challenge with *L. major* seems feasible. This will have major implications on the possible development of immunotherapy and vaccination against cutaneous leishmaniasis and other diseases.

XII. Conclusions

There have been impressive advances in the understanding of the immune mechanisms involved in leishmanial infection during the past decade, making leishmaniasis arguably the infectious disease best understood immunologically. In addition, studies of murine leishmaniasis have also contributed substantially to the unravelling of some of the fundamental principles of immunology. This achievement was greatly helped by the availability of the experimental murine models which are susceptible to most strains of *Leishmania* pathogenic to humans. However, these models are imperfect and a number of major unanswered questions in leishmaniasis remains an important challenge to workers in this field.

(i) It is unclear whether the reciprocal role of Th1 and Th2 cells found in the murine model of cutaneous leishmaniasis is also applicable to clinical leishmaniasis. Although patients with progressive visceral leishmaniasis have a higher ratio of $UCHL\text{-}1^-$: $UCHL\text{-}1^+$ cells, and this ratio is reversed in the same patients following successful chemotherapy (Cillari *et al.*, 1991a), it remains to be established whether this is generally applicable to other clinical leishmaniases with more extensive geographical distributions. This is particularly important, since the murine visceral leishmanial model does not appear to follow the Th1/Th2 rule (Kaye *et al.*, 1991). However, it is arguable that the murine *L. donovani* model differs substantially from the disease pattern of clinical visceral leishmaniasis. A human-SCID mouse model, together with extensive clinical analyses of a large patient population from different endemic areas, will be needed to clarify this important question.

(ii) What is the genetic basis of the susceptibility to leishmanial infection and the phenotypic expression of the susceptible–resistant genes? Although a number of genetic factors has been identified for the murine model, the search for the susceptible genes (*Lsh* and others) in the mouse and in humans continues. Identification of these genes and their phenotypic expression would contribute substantially to the understanding of the mechanism of antigen presentation leading to the preferential induction of T cell subsets.

(iii) How do some of the parasites survive in the hostile host environment and what are the factors contributing to the virulence of some, but not other, parasites in the same macrophages? Important progress has been made in this area and this has been extensively reviewed (Alexander and Russell, 1992; Sacks *et al.*, 1993). However, our understanding of the mechanism of parasite entry into macrophages and the contribution of various proteinases to the survival of the intracellular parasites remains incomplete. Detailed knowledge in this area would have important implications in the design of effective chemo- and immunotherapy.

(iv) What is the optimal vaccine for clinical leishmaniasis? Several experimental vaccines against cutaneous leishmaniasis in the murine model appear promising. These include the oral *Salmonella*-gp63 vaccine (Yang *et al.*, 1990) and the *Vaccinia*-gp46 vaccine (McMahon-Pratt *et al.*, 1993). However, these prototype vaccines remain to be optimized and their clinical relevance to vaccination against visceral leishmaniasis, the most important clinical form of leishmaniasis, is still unclear.

Cellular immunology has served the study of leishmaniasis extremely well in the past decade. It will continue to play an influential role in addressing the biological questions mentioned above. However, major advances in this field in the future will come from the effective deployment of some of the newer technologies, such as biochemical analysis of the signal transduction pathways, gene targeting and molecular vaccine delivery. The next decade should witness the application of the immunological knowledge and the new technology in the control of this important parasitic disease.

Acknowledgements

We thank all our colleagues for helpful discussions and for the supply of unpublished material. We particularly thank Professor J. M. Blackwell, Professor E. Cillari, Professor F. E. G. Cox, Dr R. L. Coffman, Dr F. Cunha, Dr Y. Li, Dr D. Sacks, Dr A. Severn, Dr R. Titus, Dr D. Yang, L. M. C. C. Leal and D. W. Moss. C. A. O'Donnell is supported by the Wellcome Trust.

References

Abdalla, R. E. (1982). Parasites in Sudanese cutaneous and mucosal leishmaniasis. *Annals of Tropical Medicine and Parasitology* **76**, 299–307.

Adams, L. B., Hibbs, J. B., jr, Taintor, R. R. and Krahenbuhl, J. L. (1990). Microbiostatic effect of murine-activated macrophages for *Toxoplasma gondii*: role for synthesis of inorganic nitrogen oxides from L-arginine. *Journal of Immunology* **144**, 2725–2729.

Adler, S. and Gunders, A. E. (1964). Immunity to *Leishmania mexicana* following spontaneous recovery from Oriental sore. *Transactions of the Royal Society of Tropical Medicine and Hygiene* **58**, 274–277.
Aldovini, A. and Young, R. A. (1991). Humoral and cell-mediated immune responses to live recombinant BCG-HIV vaccines. *Nature* **351**, 479–482.
Alexander, J. (1989). Vaccination and immunological control of leishmaniasis. *In* "Leishmaniasis. The Current Status and New Strategies for Control" (D. T. Hart, ed.), pp. 839–843. NATO ASI Series, Plenum Press, New York.
Alexander, J. and Russell, D. J. (1992). The interaction of *Leishmania* species with macrophages. *Advances in Parasitology* **31**, 175–254.
Allain, D. S. and Kagan, I. G. (1975). A direct agglutination test for leishmaniasis. *American Journal of Tropical Medicine and Hygiene* **24**, 232–236.
Allison, A. C. and Eugui, E. M. (1983). The role of cell-mediated immune responses in resistance to malaria with special reference to oxidant stress. *Annual Review of Immunology* **1**, 361–392.
Amiri, P., Locksley, R. M., Parslow, T. G., Sadick, M. D., Rector, E., Ritter, D. and McKerrow, J. H. (1992). Tumour necrosis factor α restores granulomas and induces egg-laying in schistosome-infected SCID mice. *Nature* **356**, 604–607.
Anbar, M. and Neta, P. (1967). A compilation of specific bimolecular rate constants for the reactions of hydrated electrons, hydrogen atoms and hydroxyl radicals with inorganic and organic compounds in aqueous solution. *International Journal of Applied Radiation and Isotopes* **18**, 493–523.
Anders, R. F. (1986). Multiple cross-reactivities amongst antigens of *Plasmodium falciparum* impair the development of protective immunity against malaria. *Parasite Immunology* **8**, 529–539.
Anderson, S., David, J. R. and McMahon-Pratt, D. (1983). *In vivo* protection against *Leishmania mexicana* mediated by monoclonal antibodies. *Journal of Immunology* **131**, 1616–1618.
Anthony, R. L., Christensen, H. A. and Johnson, C. M. (1980). Micro enzyme-linked immunosorbent assay (ELISA) for the serodiagnosis of New World leishmaniasis. *American Journal of Tropical Medicine and Hygiene* **29**, 190–194.
Antunes, C. M. F., Mayrink, W., Magalhaes, P. A., Costa, C. A., Melo, M. N., Dias, M., Michalick, M. S. M., Williams, P., Oliveira-Lima, A., Vieira, J. B. F. and Schettini, A. P. M. (1986). Controlled field trials of a vaccine against New World cutaneous leishmaniasis. *International Journal of Epidemiology* **15**, 572–580.
Arnon, R. (1981). Experimental allergic encephalomyelitis—susceptibility and suppression. *Immunological Reviews* **55**, 5–30.
Asherson, G. L. (1966). Selective and specific inhibition of 24-hour skin reactions in the guinea pig. II. The mechanism of immune deviation. *Immunology* **10**, 179–186.
Asherson, G. L. and Stone, S. H. (1965). Selective and specific inhibition of 24-hour skin reactions in the guinea pig. I. Immune deviation: description of the phenomenon and the effect of splenectomy. *Immunology* **9**, 205–217.
Asherson, G. L. and Zembala, M. (1974). Suppression of contact sensitivity by T cells in the mouse. I. Demonstration that suppressor cells act on the effector stage of contact sensitivity, and their induction following *in vitro* exposure to antigen. *Proceedings of the Royal Society B* **187**, 329–348.
Asherson, G. L., Colizzi, V. and Zembala, M. (1986). An overview of T-suppressor cell circuits. *Annual Review of Immunology* **4**, 37–68.
Bacchetta, R., De Waal Malefyt, R., Yssel, H., Abrams, J., de Vries, J. E., Spits, H. and Roncarolo, M. G. (1990). Host-reactive $CD4^+$ and $CD8^+$ T cell clones

isolated from a human chimera produce IL–5, IL–2, IFN-gamma and granulocyte/macrophage-colony-stimulating factor but not IL–4. *Journal of Immunology* **144**, 902–908.

Bach, B. A., Sherman, L., Benacerraf, B. and Greene, M. I. (1978). Mechanisms of regulation of cell-mediated immunity. II. Induction and suppression of delayed-type hypersensitivity to azobenzenearsonate-coupled syngeneic cells. *Journal of Immunology* **121**, 1460–1467.

Badaro, R., Reed, S. G. and Carvalho, E. M. (1983). Immunofluorescent antibody test in American visceral leishmaniasis: sensitivity and specificity of different morphological forms of two *Leishmania* species. *American Journal of Tropical Medicine and Hygiene* **32**, 480–484.

Barral, A., Carvalho, E. M., Badaro, R. and Barral-Netto, M. (1986). Suppression of lymphocyte proliferative responses by sera from patients with American visceral leishmaniasis. *American Journal of Tropical Medicine and Hygiene* **35**, 735–742.

Barral, A., Pedral-Sampaio, D., Grimaldi, G., jr, Momen, H., McMahon-Pratt, D., Ribeiro De Jesus, A., Almeida, R., Badaro, R., Barral-Netto, M., Carvalho, E. M. and Johnson, W. D., jr (1991). Leishmaniasis in Bahia, Brazil: evidence that *Leishmania amazonensis* produces a wide spectrum of clinical disease. *American Journal of Tropical Medicine and Hygiene* **44**, 536–546.

Barral-Netto, M., Badaro, R., Barral, A., Almeida, R. P., Santos, S. B., Badaro, F., Sampaio, D. P., Carvalho, E. M., Falcoff, E. and Falcoff, R. (1991). Tumor necrosis factor (cachectin) in human visceral leishmaniasis. *Journal of Infectious Diseases* **163**, 853–857.

Battisto, J. R. and Bloom, B. R. (1966). Dual immunological unresponsiveness induced by cell membrane coupled hapten or antigen. *Nature* **212**, 156–157.

Battisto, J. R. and Miller, J. (1962). Immunological unresponsiveness produced in adult guinea pigs by parenteral introduction of minute quantities of hapten or protein antigen. *Proceedings of the Society of Experimental Biology and Medicine* **111**, 111–115.

Beckman, J. S., Beckman, T. W., Chen, J., Marshall, P. A. and Freeman, B. A. (1990). Apparent hydroxyl radical production by peroxynitrite: implications for endothelial injury from nitric oxide and superoxide. *Proceedings of the National Academy of Sciences of the USA* **87**, 1620–1624.

Behforouz, N., Rezai, H. R. and Gettner, S. (1976). Application of immunofluorescence to detection of antibody in *Leishmania* infections. *Annals of Tropical Medicine and Parasitology* **70**, 293–301.

Behforouz, N. C., Wenger, C. D. and Mathison, B. A. (1986). Prophylactic treatment of BALB/c mice with cyclosporine A and its analog B-5-49 enhances resistance to *Leishmania major*. *Journal of Immunology* **136**, 3067–3075.

Behin, R. and Louis, J. A. (1984). Immune response to *Leishmania*. *In* "Critical Reviews in Tropical Medicine" (R.K. Chandra, ed.), Vol. 2, pp. 141–188. Plenum Press, New York.

Belehu, A., Poulter, L. W. and Turk, J. L. (1976). Modification of cutaneous leishmaniasis in the guinea-pig by cyclophosphamide. *Clinical and Experimental Immunology* **24**, 125–132.

Belosevic, M., Davis, C. E., Meltzer, M. S. and Nacy, C. A. (1988). Regulation of activated macrophage antimicrobial activities. Identification of lymphokines that co-operate with IFN-gamma for induction of resistance to infection. *Journal of Immunology* **141**, 890–896.

Belosevic, M., Finbloom, D. S., van der Meide, P. H., Slayter, M. V. and Nacy, C. A.

(1989). Administration of monoclonal anti-IFN-gamma antibodies *in vivo* abrogates natural resistance of C3H/HeN mice to infection with *Leishmania major*. *Journal of Immunology* **143**, 266–274.

Ben-Sasson, S. Z., Le Gros, G., Conrad, D. H., Finkelman, F. D. and Paul, W. E. (1990). IL-4 production by T cells from naive donors. IL-2 is required for IL-4 production. *Journal of Immunology* **145**, 1127–1136.

Biagi, F. (1953). Sintesis de 70 historias clinicas de leishmaniasis tegumentaria de Mexico (ulcera de los chicleros). *Medicina México* **33**, 385–395.

Billiar, T. R., Curran, R. D., Stuehr, D. J., West, M. A., Bentz, B. G. and Simmons, R. L. (1989). An L-arginine-dependent mechanism mediates Kupffer cell inhibition of hepatocyte protein synthesis *in vitro*. *Journal of Experimental Medicine* **169**, 1467–1472.

Blackwell, J. M. (1983). *Leishmania donovani* infection in heterozygous and recombinant H-2 haplotype mice. *Immunogenetics* **18**, 101–109.

Blackwell, J. M. (1988). Protozoal infections. *In* "Genetics of Resistance to Bacterial and Parasitic Infection" (D. M. Wakelin and J. M. Blackwell, eds), pp. 103–152. Taylor and Francis, London.

Blackwell, J. M. (1989). The macrophage resistance gene *Lsh/Ity/Bcg*. *Research in Immunology* **140**, 767–769.

Blackwell, J. M. and Alexander, J. (1986). Different host genes recognise and control infection with taxonomically distinct *Leishmania* species. *In* "Proceedings of an International Symposium on Taxonomy and Physiology of *Leishmania*" (J. A. Rioux, ed.), pp. 211–219. Louis-Jean Imprimerie, Montpellier.

Blackwell, J. M. and Ulczak, O. M. (1984). Immunoregulation of genetically controlled acquired responses to *Leishmania donovani* infection in mice: demonstration and characterization of suppressor T cells in noncure mice. *Infection and Immunity* **44**, 97–102.

Blackwell, J. M., Freeman, J. and Bradley, D. J. (1980). Influence of H-2 complex on acquired resistance to *Leishmania donovani* infection in mice. *Nature* **283**, 72–74.

Blackwell, J. M., Howard, J. G., Liew, F. Y. and Hale, C. (1984). Mapping of the gene controlling susceptibility to cutaneous leishmaniasis. *Mouse News Letter* **70**, 86.

Blackwell, J. M., Roberts, M. B. and Alexander, J. (1985a). Response of BALB/c mice to leishmanial infection. *Current Topics in Microbiology and Immunology* **122**, 97–106.

Blackwell, J. M., Hale, C., Roberts, M. B., Ulczak, O. M., Liew, F. Y. and Howard, J. G. (1985b). An H-II-linked gene has a parallel effect on *Leishmania major* and *Leishmania donovani* infections in mice. *Immunogenetics* **21**, 385–395.

Blackwell, J. M., Roach, T. I. A., Kiderlen, A. and Kaye, P. M. (1989). Role of *Lsh* in regulating macrophage priming/activation. *Research in Immunology* **140**, 798–805.

Blackwell, J. M., Roach, T. I. A., Atkinson, S. E., Ajioka, J. W., Barton, C. H. and Shaw, M.-A. (1991). Genetic regulation of macrophage priming/activation: the *Lsh* gene story. *Immunology Letters* **30**, 241–248.

Blaxter, M. L., Miles, M. A. and Kelly, J. M. (1988). Specific diagnosis of visceral leishmaniasis using a *Leishmania donovani* antigen identified by expression cloning. *Molecular and Biochemical Parasitology* **30**, 259–270.

Blewett, T. M., Kadivar, D. M. H. and Soulsby, E. J. L. (1971). Cutaneous leishmaniasis in the guinea pig. Delayed-type hypersensitivity, lymphocyte stimulation, and inhibition of macrophage migration. *American Journal of Tropical Medicine and Hygiene* **20**, 546–551.

Bloom, B. R., Salgame, P. and Diamond, B. (1992). Revisiting and revising suppressor T cells. *Immunology Today* **13**, 131–136.

Bogdan, C., Moll, H., Solbach, W. and Röllinghoff, M. (1990). Tumor necrosis factor-α in combination with interferon-gamma, but not with interleukin-4 activates murine macrophages for elimination of *Leishmania major* amastigotes. *European Journal of Immunology* **20**, 1131–1135.

Bogdan, C., Stenger, S., Röllinghoff, M. and Solbach, W. (1991). Cytokine interaction in experimental cutaneous leishmaniasis. Interleukin 4 synergises with interferon-gamma to activate murine macrophages for killing of *Leishmania major* amastigotes. *European Journal of Immunology* **21**, 327–333.

Bonventre, P. F. and Nickol, A. D. (1984). *Leishmania donovani* infection in athymic mice derived from parental strains of the susceptible (Lsh^s) or resistant (Lsh^r) phenotype. *Journal of Leukocyte Biology* **36**, 651–658.

Boom, W. H., Liano, D. and Abbas, A. K. (1988). Heterogeneity of helper/inducer T lymphocytes. II. Effects of interleukin 4- and interleukin 2-producing T cell clones on resting B lymphocytes. *Journal of Experimental Medicine* **167**, 1350–1363.

Borel, Y. and David, J. R. (1970). *In vitro* studies of the suppression of delayed hypersensitivity by the induction of partial tolerance. *Journal of Experimental Medicine* **131**, 603–610.

Borel, Y., Fauconnet, M. and Miescher, P. A. (1966). Selective suppression of delayed hypersensitivity by the induction of immunological tolerance. *Journal of Experimental Medicine* **123**, 585–598.

Bouvier, J., Etges, R. and Bordier, C. (1987). Identification of the promastigote surface protease in seven species of *Leishmania*. *Molecular and Biochemical Parasitology* **24**, 73–79.

Bouvier, J., Bordier, C., Vogel, H., Reichelt, R. and Etges, R. (1989). Characterisation of the promastigote surface protease of *Leishmania* as a membrane-bound zinc endopeptidase. *Molecular and Biochemical Parasitology* **37**, 235–246.

Bradley, D. J. (1977). Regulation of *Leishmania* populations within the host. II. Genetic control of acute susceptibility of mice to *Leishmania donovani* infection. *Clinical and Experimental Immunology* **30**, 130–140.

Bradley, D. J. (1987). Genetics of susceptibility and resistance in the vertebrate host. *In* "The Leishmaniases in Biology and Medicine" (W. Peters and R. Killick-Kendrick, eds), Vol. 2, pp. 551–581. Academic Press, London.

Bradley, D. J., Taylor, B. A., Blackwell, J., Evans, E. P. and Freeman, J. (1979). Regulation of *Leishmania* populations within the host. III. Mapping of the locus controlling susceptibility to visceral leishmaniasis in the mouse. *Clinical and Experimental Immunology* **37**, 7–14.

Brake, D. A., Long, C. A. and Weidanz, W. P. (1988). Adoptive protection against *Plasmodium chabaudi adami* malaria in athymic nude mice by a cloned T cell line. *Journal of Immunology* **140**, 1989–1993.

Bray, R. S. and Bryceson, A. D. M. (1968). Cutaneous leishmaniasis of the guinea-pig. Action of sensitized lymphocytes on infected macrophages. *Lancet* **ii**, 898–899.

Bredt, D. S., Hwang, P. M., Glatt, C. E., Lowenstein, C., Reed, R. R. and Snyder, S. H. (1991). Cloned and expressed nitric oxide synthase structurally resembles cytochrome P-450 reductase. *Nature* **351**, 714–718.

Brown, A., Hormaeche, C. E., Demarco de Hormaeche, N. A. R., Winter, M. D., Dougan, G., Maskell, D. J. and Stocker, B. A. D. (1987). An attenuated AroA *Salmonella typhimurium* vaccine elicits humoral and cellular immunity to cloned β-galactosidase in mice. *Journal of Infectious Diseases* **155**, 86–92.

Brown, I. N., Glynn, A. A. and Plant, J. E. (1982). Inbred mouse strain resistance to *Mycobacterium lepraemurium* follows the *Ity/Lsh* pattern. *Immunology* **47**, 149–156.

Brunt, L. M., Portnoy, D. A. and Unanue, E. R. (1990). Presentation of *Listeria monocytogenes* to $CD8^+$ T cells requires secretion of hemolysin and intracellular bacterial growth. *Journal of Immunology* **145**, 3540–3546.

Bryceson, A. D. M. (1970). Diffuse cutaneous leishmaniasis in Ethiopia. III. Immunological studies. *Transactions of the Royal Society of Tropical Medicine and Hygiene* **64**, 380–387.

Bryceson, A. D. M. (1987). Therapy in man. *In* "The Leishmaniases in Biology and Medicine" (W. Peters and R. Killick-Kendrick, eds), Vol. 2, pp. 848–907. Academic Press, London.

Bryceson, A. D. M. and Turk, J. L. (1971). The effect of prolonged treatment with antilymphocyte serum on the course of infections with BCG and *Leishmania enriettii* in the guinea-pig. *Journal of Pathology* **104**, 153–165.

Bryceson, A. D. M., Bray, R. S., Wolstencroft, R. A. and Dumonde, D. C. (1970). Immunity in cutaneous leishmaniasis of the guinea-pig. *Clinical and Experimental Immunology* **7**, 301–341.

Buchmüller-Rouiller, Y. and Mauël, J. (1986). Correlation between enhanced oxidative metabolism and leishmanicidal activity in activated macrophages from healer and nonhealer mouse strains. *Journal of Immunology* **136**, 3884–3890.

Buchmüller-Rouiller, Y. and Mauël, J. (1991). Macrophage activation for intracellular killing as induced by calcium ionophore: correlation with biological and biochemical events. *Journal of Immunology* **146**, 217–223.

Buchmüller-Rouiller, Y., Corradin, S. B. and Mauël, J. (1992). Macrophage activation for intracellular killing as induced by calcium ionophore: Dependency on L-arginine-derived nitrogen oxidation products. *Biochemical Journal* **284**, 387–392.

Burns, J. M., jr, Scott, J. M., Carvalho, E. M., Russo, D. M., March, C. J., Van Ness, K. P. and Reed, S. G. (1991). Characterization of a membrane antigen of *Leishmania amazonensis* that stimulates human immune responses. *Journal of Immunology* **146**, 742–748.

Button, L. L. and McMaster, W. R. (1988). Molecular cloning of the major surface antigen of *Leishmania*. *Journal of Experimental Medicine* **167**, 724–729.

Cantor, H., Shen, F. W. and Boyse, E. A. (1976). Separation of helper T cells from suppressor T cells expressing different Ly components. II. Activation by antigen: after immunization, antigen-specific suppressor and helper activities are mediated by distinct T-cell subclasses. *Journal of Experimental Medicine* **143**, 1391–1401.

Carrier, M. J., Chatfield, S. N., Dougan, G., Nowicka, U. T. A., O'Callaghan, D. and Liew, F. Y. (1992). Expression of human interleukin-1β in *Salmonella typhimurium*: a model system for the delivery of recombinant therapeutic protein *in vivo*. *Journal of Immunology* **148**, 1176–1181.

Carvalho, E. M., Teixeira, R. and Johnson, W. D. (1981). Cell-mediated immunity in American visceral leishmaniasis: reversible immunosuppression during acute infection. *Infection and Immunity* **33**, 498–502.

Carvalho, E. M., Badaro, R., Reed, S., Jones, T. C. and Johnson, W. D. (1985a). Absence of gamma-interferon and interleukin-2 production during active visceral leishmaniasis. *Journal of Clinical Investigation* **76**, 2066–2069.

Carvalho, E. M., Johnson, W. D., Barreto, E., Marsden, P. D., Costa, J. L. M., Reed, S. and Rocha, H. (1985b). Cell mediated immunity in American cutaneous and mucosal leishmaniasis. *Journal of Immunology* **135**, 4144–4148.

Castes, M., Agnelli, A. and Rondon, A. J. (1984). Mechanisms associated with immunoregulation in human American cutaneous leishmaniasis. *Clinical and Experimental Immunology* **57**, 279–286.

Cavacini, L. A., Long, C. A. and Weidanz, W. P. (1986). T-cell immunity in murine malaria: adoptive transfer of resistance to *Plasmodium chabaudi adami* in nude mice with splenic T cells. *Infection and Immunity* **52**, 637–643.

Champsi, J. and McMahon-Pratt, D. (1988). Membrane glycoprotein M-2 protects against *Leishmania amazonensis* infection. *Infection and Immunity* **56**, 3272–3279.

Chan, J., Fujiwara, T., Brennan, P., McNeil, M., Turco, S. J., Sibille, J.-C., Snapper, M., Aisen, P. and Bloom, B. R. (1989). Microbial glycolipids: possible virulence factors that scavenge oxygen radicals. *Proceedings of the National Academy of Sciences of the USA* **86**, 2453–2457.

Channon, J. Y. and Blackwell, J. M. (1985a). A study of the sensitivity of *Leishmania donovani* promastigotes and amastigotes to hydrogen peroxide. I. Differences in sensitivity correlate with parasite-mediated removal of hydrogen peroxide. *Parasitology* **91**, 197–206.

Channon, J. Y. and Blackwell, J. M. (1985b). A study of the sensitivity of *Leishmania donovani* promastigotes and amastigotes to hydrogen peroxide. II. Possible mechanisms involved in protective H_2O_2 scavenging. *Parasitology* **91**, 207–217.

Charlab, R., Blaineau, C., Schechtman, D. and Barcinski, M. A. (1990). Granulocyte-macrophage colony-stimulating factor is a growth-factor for promastigotes of *Leishmania mexicana amazonensis*. *Journal of Protozoology* **37**, 352–357.

Charles, I. and Dougan, G. (1990). Gene expression and the development of live enteric vaccines. *TIBTECH* **8**, 117–121.

Chatelain, R., Varkila, K. and Coffman, R. L. (1992). IL-4 induces a Th2 response in *Leishmania major* infected mice. *Journal of Immunology* **148**, 1182–1187.

Chaudhuri, G., Chaudhuri, M., Pan, A. and Chang, K. P. (1989). Surface acid proteinase (gp63) of *Leishmania mexicana*: a metalloenzyme capable of protecting liposome-encapsulated proteins from phagolysosomal degradation by macrophages. *Journal of Biological Chemistry* **264**, 7483–7489.

Cher, D. J. and Mosmann, T. R. (1987). Two types of murine helper T cell clone. II. Delayed-type hypersensitivity is mediated by T_H1 clones. *Journal of Immunology* **138**, 3688–3694.

Cherwinski, H. M., Schumacher, J. H., Brown, K. D. and Mosmann, T. R. (1987). Two types of mouse helper T cell clone. III. Further differences in lymphokine synthesis between Th1 and Th2 clones revealed by RNA hybridization, functionally monospecific bioassays, and monoclonal antibodies. *Journal of Experimental Medicine* **166**, 1229–1244.

Chiplunkar, S., De Libero, G. and Kaufmann, S. H. E. (1986). *Mycobacterium leprae*-specific Lyt-2^+ T lymphocytes with cytolytic activity. *Infection and Immunity* **54**, 793–797.

Churchill, W. H., Piessens, W. F., Sulis, C. A. and David, J. R. (1975). Macrophages activated as suspension cultures with lymphocyte mediators devoid of antigen become cytotoxic for tumor cells. *Journal of Immunology* **115**, 781–786.

Cillari, E., Liew, F. Y. and Lelchuk, R. (1986). Suppression of interleukin-2 production by macrophages in susceptible BALB/c mice infected with *Leishmania major*. *Infection and Immunity* **54**, 386–394.

Cillari, E., Liew, F. Y., Lo Campo, P., Milano, S., Mansueto, S. and Salerno, A. (1988). Suppression of IL-2 production by cryopreserved peripheral blood mononuclear cells from patients with active visceral leishmaniasis in Sicily. *Journal of Immunology* **140**, 2721–2726.

Cillari, E., Dieli, M., Maltese, E., Milano, S., Salerno, A. and Liew, F. Y. (1989). Enhancement of macrophage interleukin-1 production by *Leishmania major* infection *in vitro* and its inhibition by IFN-gamma. *Journal of Immunology* **143**, 2001–2005.

Cillari, E., Milano, S., Dieli, M., Maltese, E., Di Rosa, S., Mansueto, S., Salerno, A. and Liew, F. Y. (1991a). Reduction in the number of UCHL-1^+ cells and IL-2 production in the peripheral blood of patients with visceral leishmaniasis. *Journal of Immunology* **146**, 1026–1030.

Cillari, E., Dieli, M., Lo Campo, P., Sireci, G., Caffareli, A., Maltese, E., Millott, S., Milano, S. and Liew, F. Y. (1991b). Protective effect of isoprinosine in genetically susceptible BALB/c mice infected with *Leishmania major*. *Immunology* **74**, 25–30.

Cillari, E., Milano, S., Dieli, M., Arcoleo, F., Perego, R., Leoni, F., Gromo, G., Severn, A. and Liew, F. Y. (1992). Thymopentin reduces the susceptibility of aged-mice to cutaneous leishmaniasis by modulating CD4 T cell subsets. *Immunology*, **76**, 362–366.

Clark, I. A. and Hunt, N. H. (1983). Evidence for reactive oxygen intermediates causing hemolysis and parasite death in malaria. *Infection and Immunity* **39**, 1–6.

Clark, I. A., Rockett, K. A. and Cowden, W. B. (1991). Proposed link between cytokines, nitric oxide and human cerebral malaria. *Parasitology Today* **7**, 205–207.

Coffman, R. L. and Carty, J. (1986). A T cell activity that enhances polyclonal IgE production and its inhibition by IFN-gamma. *Journal of Immunology* **136**, 949–954.

Coffman, R. L. and Mosmann, T. R. (eds) (1991). $CD4^+$ T-cell subsets: regulation of differentiation and function. *Research in Immunology* **142**, 1–79.

Coffman, R. L., Ohara, J., Bond, M. W., Carty, J., Zlotnik, A. and Paul, W. E. (1986). B cell stimulatory factor-1 enhances the IgE response of lipopolysaccharide-activated B cells. *Journal of Immunology* **136**, 4538–4541.

Coffman, R. L., Seymour, B. W. P., Lebman, D. A., Hiraki, D. D., Christiansen, J. A., Shrader, B., Cherwinski, H. M., Savelkoul, H. F. J., Finkelman, F. D., Bond, M. W. and Mosmann, T. R. (1988). The role of helper T cell products in mouse B cell differentiation and isotype regulation. *Immunological Reviews* **102**, 5–28.

Coffman, R. L., Seymour, B. W. P., Hudak, S., Jackson, J. and Rennick, D. (1989). Antibody to interleukin-5 inhibits helminth-induced eosinophilia in mice. *Science* **245**, 308–310.

Coffman, R. L., Varkila, K., Scott, P. and Chatelain, R. (1991a). Role of cytokines in the differentiation of $CD4^+$ T-cell subsets *in vivo*. *Immunological Reviews* **123**, 189–207.

Coffman, R. L., Chatelain, R., Leal, L. M. C. C. and Varkila, K. (1991b). *Leishmania major* infection in mice: a model system for the study of $CD4^+$ T-cell subset differentiation. *Research in Immunology* **142**, 36–40.

Colomer-Gould, V., Quintao, L. G., Keithly, J. and Nogueira, N. (1985). A common major surface antigen on amastigotes and promastigotes of *Leishmania* species. *Journal of Experimental Medicine* **162**, 902–916.

Convit, J., Castellanos, P. L., Rondón, A., Pinardi, M. E., Ulrich, M., Castés, M., Bloom, B. R. and Garcia, L. (1987). Immunotherapy versus chemotherapy in localised cutaneous leishmaniasis. *Lancet* **i**, 401–404.

Convit, J., Castellanos, P. L., Ulrich, M., Castés, M., Rondón, A., Pinardi, M. E., Rodriquez, N., Bloom, B. R., Formica, S., Valecillos, L. and Bretana, A. (1989). Immunotherapy of localized, intermediate, and diffuse forms of American cutaneous leishmaniasis. *Journal of Infectious Diseases* **160**, 104–115.

Corcoran, L. M., Metcalf, D., Edwards, S. J. and Handman, E. (1988). GM-CSF produced by recombinant vaccinia virus or in GM-CSF transgeneic mice has no effect *in vivo* on murine cutaneous leishmaniasis. *Journal of Parasitology* **74**, 763–767.

Cornall, R. J., Prins, J.-B., Todd, J. A., Pressey, A., DeLarato, N. H., Wicker, L. S. and Peterson, L. B. (1991). Type 1 diabetes in mice is linked to the interleukin-1 receptor and *Lsh/Ity/Bcg* genes on chromosome 1. *Nature* **353**, 262–264.

Coutinho, S. G., Louis, J. A., Mauël, J. and Enger, H. D. (1984). Induction by specific T lymphocytes of intracellular destruction of *Leishmania major* in infected murine macrophages. *Parasite Immunology* **6**, 157–170.

Crawford, G. D., Wyler, D. J. and Dinarello, C. A. (1985). Parasite-monocyte interactions in human leishmaniasis: production of interleukin-1 *in vitro*. *Journal of Infectious Diseases* **152**, 315–322.

Crocker, P. R., Blackwell, J. M. and Bradley, D. J. (1984). Expression of the natural resistance gene *Lsh* in resident liver macrophages. *Infection and Immunity* **43**, 1033–1040.

Crocker, P. R., Davies, E. V. and Blackwell, J. M. (1987). Variable expression of the murine natural resistance gene *Lsh* in different macrophage populations infected *in vitro* with *Leishmania donovani*. *Parasite Immunology* **9**, 705–719.

Crowle, A. J. and Hu, C. C. (1966). Split tolerance affecting delayed hypersensitivity and induced in mice by preimmunization with protein antigens in solution. *Clinical and Experimental Immunology* **1**, 323–335.

Crowle, A. J. and Hu, C. C. (1970). Studies on the induction and time course of repression of delayed hypersensitivity in the mouse by low and high doses of antigen. *Clinical and Experimental Immunology* **6**, 363–374.

Cunha, F. Q., Moncada, S. and Liew, F. Y. (1992). Interleukin-10 (IL-10) inhibits the induction of nitric oxide synthase by interferon-gamma in murine macrophages. *Biochemical and Biophysical Research Communications* **182**, 1155–1159.

Cunha, F. Q., Weiser, W. Y., David, J. R., Moss, D. W., Moncada, S. and Liew, F. Y. (1993). Recombinant migration inhibitory factor induces nitric oxide synthase in murine macrophages. *Journal of Immunology* **150**, 1908–1912.

Curran, R., Billiar, T., Stuehr, D. J., Hofmann, K. and Simmons, R. (1989). Hepatocytes produce nitrogen oxides from L-arginine in response to inflammatory products of Kupffer cells. *Journal of Experimental Medicine* **170**, 1769–1774.

Czaja, M. J., Weiner, F. R., Takahashi, S., Giambrone, M. A., van der Meide, P. H., Schellekens, H., Biempica, L. and Zern, M. A. (1989). Gamma-interferon treatment inhibits collagen deposition in murine schistosomiasis. *Hepatology* **10**, 795–800.

De Libero, G. and Kaufmann, S. H. E. (1986). Antigen-specific Lyt-2^+ cytolytic T lymphocytes from mice infected with the intracellular bacterium *Listeria monocytogenes*. *Journal of Immunology* **137**, 2688–2694.

Del Prete, G. F., De Carli, M., Mastromauro, C., Biagiotti, R., Macchia, D., Falagiani, P., Ricci, M. and Romagnani, S. (1991). Purified protein derivative of *Mycobacterium tuberculosis* and excretory-secretory antigen(s) of *Toxocara canis* expand *in vitro* human T cells with stable and opposite (type 1 T helper or type 2 T helper) profile of cytokine production. *Journal of Clinical Investigation* **88**, 346–350.

Denis, M. (1991). Tumor necrosis factor and granulocyte macrophage-colony stimulating factor stimulate human macrophages to restrict growth of virulent *Mycobacterium avium:* killing effector mechanism depends on the generation of reactive nitrogen intermediates. *Journal of Leukocyte Biology* **49**, 380–387.

De Rossell, R. A., Bray, R. S. and Alexander, J. (1987). The correlation between delayed-type hypersensitivity, lymphocyte activation and protective immunity in experimental murine leishmaniasis. *Parasite Immunology* **9**, 105–115.

Descoteaux, A., Turco, S. J., Sacks, D. L. and Matlashewski, G. (1991). *Leishmania donovani* lipophosphoglycan selectively inhibits signal transduction in macrophages. *Journal of Immunology* **146**, 2747–2753.

De Titto, E. H., Catterall, J. R. and Remington, J. S. (1986). Activity of recombinant tumor necrosis factor on *Toxoplasma gondii* and *Trypanosoma cruzi*. *Journal of Immunology* **137**, 1342–1345.

DeTolla, L. J., Semprevivo, L. H., Polczuk, N. C. and Passmore, H. C. (1980). Genetic control of acquired resistance to visceral leishmaniasis in mice. *Immunogenetics* **10**, 353–361.

DeTolla, L. J., Scott, P. A. and Farrell, J. P. (1981). Single gene control of resistance to cutaneous leishmaniasis in mice. *Immunogenetics* **14**, 29–39.

De Vries, J. E., De Waal Malefyt, R., Yssel, H., Roncarolo, M.-G. and Spits, H. (1991). Do human Thl and Th2 $CD4^+$ clones exist? *Research in Immunology* **142**, 59–63.

De Waal Malefyt, R., Haanen, J., Spits, H., Roncarolo, M.-G., De Velde, A., Figdor, C., Johnson, K., Kastelein, R., Yssel, H. and de Vries, J. E. (1991a). Interleukin 10 (IL–10) and viral IL-10 strongly reduce antigen-specific human T cell proliferation by diminishing the antigen-presenting capacity of monocytes via downregulation of class II major histocompatibility complex expression. *Journal of Experimental Medicine* **174**, 915–924.

De Waal Malefyt, R., Abrams, J., Bennett, B., Figdor, C.G. and de Vries, J. E. (1991b). Interleukin 10 (IL–10) inhibits cytokine synthesis by human monocytes: an autoregulatory role of IL–10 produced by monocytes. *Journal of Experimental Medicine* **174**, 1209–1220.

Dhaliwal, J. S., Liew, F. Y. and Cox, F. E. G. (1985). Specific suppressor T cells for delayed-type hypersensitivity in susceptible mice immunized against cutaneous leishmaniasis. *Infection and Immunity* **49**, 417–423.

Ding, A. H., Nathan, C. F. and Stuehr, D. J. (1988). Release of reactive nitrogen intermediates and reactive oxygen intermediates from mouse peritoneal macrophages. Comparison of activating cytokines and evidence for independent production. *Journal of Immunology* **141**, 2407–2412.

Ding, A. H., Nathan, C. F., Graycar, J., Derynck, R., Stuehr, D. J. and Srimal, S. (1990). Macrophage deactivating factor and transforming growth factor-β_1, β_2 and β_3 inhibit induction of macrophage nitrogen oxide synthesis by IFN-gamma. *Journal of Immunology* **145**, 940–944.

Dorf, M. E. and Benacerraf, B. (1984). Suppressor cells and immunoregulation. *Annual Review of Immunology* **2**, 127–158.

Dostrovsky, A. and Sagher, F. (1946). The intracutaneous test in cutaneous leishmaniasis. *Annals of Tropical Medicine and Parasitology* **40**, 265–269.

Doughty, B. L. and Phillips, S. M. (1982). Delayed hypersensitivity granuloma formation and modulation around *Schistosoma mansoni* eggs *in vitro*. II. Regulatory T cell subsets. *Journal of Immunology* **128**, 37–42.

Drapier, J.-C. (ed.) (1991). L-Arginine-derived nitric oxide and the cell-mediated immune response. *Research in Immunology* **7**, 553–602.

Drapier, J.-C. and Hibbs, J. B. jr. (1986). Murine cytotoxic activated macrophages inhibit aconitase in tumor cells. Inhibition involves the iron-sulfur prosthetic group and is reversible. *Journal of Clinical Investigation* **78**, 790–797.

Drapier, J. C., Wietzerbin, J. and Hibbs, J. B., jr. (1988). Interferon-gamma and tumor necrosis factor induce the L-arginine-dependent cytotoxic effector mechanism in murine macrophages. *European Journal of Immunology* **18**, 1587–1592.

Duxbury, R. E. and Sadun, E. H. (1964). Fluorescent antibody test for the serodiagnosis of visceral leishmaniasis. *American Journal of Tropical Medicine and Hygiene* **13**, 525–529.

Dvorak, H. F., Billote, J. B., McCarthy, J. S. and Flax, M. H. (1965). Immunologic unresponsiveness in the adult guinea-pig. I. Suppression of delayed hypersensitivity and antibody formation to protein antigens. *Journal of Immunology* **94**, 966–975.

Dwyer, D. M. (1976). Antibody-induced modulation of *Leishmania donovani* surface membrane antigens. *Journal of Immunology* **117**, 2081–2091.

Editorial (1991). Nitric oxide in the clinical arena. *Lancet* **338**, 1560–1562.

Edrissian, G. H. and Darabian, P. (1979). A comparison of enzyme-linked immunosorbent assay and indirect fluorescent antibody test in the sero-diagnosis of cutaneous and visceral leishmaniasis in Iran. *Transactions of the Royal Society of Tropical Medicine and Hygiene* **73**, 289–292.

El Safi, S. H. and Evans, D. A. (1989). A comparison of the direct agglutination test and enzyme-linked immunosorbent assay in the sero-diagnosis of leishmaniasis in the Sudan. *Transactions of the Royal Society of Tropical Medicine and Hygiene* **83**, 334–337.

Else, K. J. and Grencis, R. K. (1991a). Helper T cell subsets in murine trichuriasis. *Parasitology Today* **7**, 313–316.

Else, K. J. and Grencis, R. K. (1991b). Cellular immune responses to the murine nematode parasite *Trichuris muris*. I. Differential cytokine production during acute or chronic infection. *Immunology* **72**, 508–513.

Esparza, I., Mannel, D., Ruppel, A., Falk, W. and Krammer, P. (1987). Interferon-gamma and lymphotoxin or tumor necrosis factor act synergistically to induce macrophage killing of tumor cells and schistosomula of *Schistosoma mansoni*. *Journal of Experimental Medicine* **166**, 589–594.

Etges, R. J., Bouvier, J. and Bordier, C. (1986). The major surface protein of *Leishmania* promastigotes is a protease. *Journal of Biological Chemistry* **261**, 9099–9101.

Farrell, J. P., Müller, I. and Louis, J. A. (1989). A role for Lyt-2^+ T cells in resistance to cutaneous leishmaniasis in immunized mice. *Journal of Immunology* **142**, 2052–2056.

Feldmann, M., Beverley, P. C. L., Dunkley, M. and Kontiainen, S. (1975). Different Ly antigen phenotypes of *in vitro* induced helper and suppressor cells. *Nature* **258**, 614–616.

Feng, Z. Y., Louis, J., Kindler, V., Pedrazzini, T., Eliason, J. F., Behin, R. and Vasalli, P. (1988). Aggravation of experimental cutaneous leishmaniasis in mice by administration of IL-3. *European Journal of Immunology* **18**, 1245–1251.

Fernandez-Botran, R., Sanders, V. M., Mosmann, T. R. and Vitetta, E. S. (1988). Lymphokine-mediated regulation of the proliferative response of clones of T helper 1 and T helper 2 cells. *Journal of Experimental Medicine* **168**, 543–558.

Finkelman, F. D., Katona, I. M., Urban, J. F., jr, Snapper, C. M., Ohara, J. and Paul, W. E. (1986). Suppression of *in vivo* polyclonal IgE responses by monoclonal antibody to the lymphokine B-cell stimulatory factor 1. *Proceedings of the National Academy of Sciences of the USA* **83**, 9675–9678.

Finkelman, F. D., Katona, I. M., Urban, J. F., jr, Holmes, J., Ohara, J., Tung, A. S., Sample, J. G. and Paul, W. E. (1988). IL-4 is required to generate and sustain *in vivo* IgE responses. *Journal of Immunology* **141**, 2335–2341.

Finkelman, F. D., Holmes, J., Katona, I. M., Urban, J. F., jr, Beckmann, M. P., Park, L. S., Schooley, K. A., Coffman, R. L., Mosmann, T. R. and Paul, W. E. (1990). Lymphokine control of *in vivo* immunoglobulin isotype selection. *Annual Review of Immunology* **8**, 303–333.

Finkelman, F. D., Pearce, E. J., Urban, J. F., jr, and Sher, A. (1991). Regulation and biological function of helminth-induced cytokine responses. *In* "Immunoparasitology Today" (C. Ash and R. B. Gallagher, eds), pp. A62–A66. Elsevier Trends Journals, Cambridge.

Fiorentino, D. F., Bond, M. W. and Mosmann, T. R. (1989). Two types of mouse T helper cell. IV. Th2 clones secrete a factor that inhibits cytokine production by Th1 clones. *Journal of Experimental Medicine* **170**, 2081–2095.

Fiorentino, D. F., Zlotnik, A., Viera, P., Mosmann, T. R., Howard, M., Moore, K. W. and O'Garra, A. (1991). IL-10 acts on the antigen presenting cell to inhibit cytokine production by Th1 cells. *Journal of Immunology* **146**, 3444–3451.

Firestein, G. S., Roeder, W. D., Laxer, J. A., Townsend, K. S., Weaver, C. T., Hom, J. T., Linton, J., Torbett, B. E. and Glasebrook, A. L. (1989). A new murine $CD4^+$ T cell subset with an unrestricted cytokine profile. *Journal of Immunology* **143**, 518–525.

Fong, T. A. T. and Mosmann, T. R. (1990). Alloreactive murine $CD8^+$ T cell clones secrete the Th1 pattern of cytokines. *Journal of Immunology* **144**, 1744–1752.

Fortier, A. H., Mock, B. A., Meltzer, M. S. and Nacy, C. A. (1987). *Mycobacterium bovis* BCG-induced protection against cutaneous and systemic *Leishmania major* infections of mice. *Infection and Immunity* **55**, 1707–1714.

Frankenburg, S., Leibovici, V., Mansbach, N., Turco, S. J. and Rosen, G. (1990). Effect of glycolipids of *Leishmania* parasites on human monocyte activity. Inhibition by lipophosphoglycan. *Journal of Immunology* **145**, 4284–4289.

Frommel, T. O., Button, L. L., Fujikura, Y. and McMaster, W. R. (1990). The major surface glycoprotein (GP63) is present in both life stages of *Leishmania*. *Molecular and Biochemical Parasitology* **38**, 25–32.

Gajewski, T. F. and Fitch, F. W. (1988). Anti-proliferative effect of IFN-gamma in immune regulation. I. IFN-gamma inhibits the proliferation of Th2 but not Th1 murine HTL clones. *Journal of Immunology* **140**, 4245–4252.

Gajewski, T. F. and Fitch, F. W. (1991). Differential activation of murine Th1 and Th2 clones. *Research in Immunology* **142**, 19–23.

Gajewski, T. F., Pinnas, M., Wong, T. and Fitch, F. W. (1991). Murine Th1 and Th2 clones proliferate optimally in response to distinct antigen-presenting cell populations. *Journal of Immunology* **146**, 1750–1758.

Galvao-Castro, B., SaFerreira, J. A., Marzochi, K. F., Marzochi, M. C., Coutinho, S. G. and Lambert, P. H. (1984). Polyclonal B-cell activation, circulating immune complexes and autoimmunity in human visceral leishmaniasis. *Clinical and Experimental Immunology* **56**, 58–66.

Gardner, I. D. and Remington, J. S. (1977). Age-related decline in the resistance of mice to infection with intracellular pathogens. *Infection and Immunity* **16**, 593–598.

Garside, P., Hutton, A. K., Severn, A., Liew, F. Y. and Mowat, A. McI. (1992). Nitric oxide mediates intestinal pathology in graft-vs-host disease. *European Journal of Immunology*, **22**, 2141–2145.

Gazzinelli, R. T., Hakim, F. T., Hieny, S., Shearer, G. M. and Sher, A. (1991). Synergistic role of $CD4^+$ and $CD8^+$ T lymphocytes in IFN-gamma production and protective immunity induced by an attenuated *Toxoplasma gondii* vaccine. *Journal of Immunology* **146**, 286–292.

Gazzinelli, R. T., Oswald, I. P., James, S. L. and Sher, A. (1992). IL-10 inhibits parasite killing and nitrogen oxide production by IFN-gamma activated macrophages. *Journal of Immunology* **148**, 1792–1796.
Gershon, R. K. and Kondo, K. (1970). Cell interactions in the induction of tolerance: the role of thymic lymphocytes. *Immunology* **18**, 723–737.
Ghose, A. C., Haldar, J. P., Pal, S. C., Mishra, B. P. and Mishra, K. K. (1979). Phytohaemagglutinin-induced lymphocyte transformation test in Indian kala-azar. *Transactions of the Royal Society of Tropical Medicine and Hygiene* **73**, 725–726.
Ghose, A. C., Haldar, J. P., Pal, S. C., Mishra, B. P. and Mishra, K. K. (1980). Serological investigations on Indian kala-azar. *Clinical and Experimental Immunology* **40**, 318–326.
Giedlin, M. A., Longnecker, B. M. and Mosmann, T. R. (1986). Murine T-cell clones specific for chicken erythrocyte alloantigens. *Cellular Immunology* **97**, 357–370.
Glaser, T. A., Moody, S. F., Handman, E., Bacic, A. and Spithill, T. W. (1991). An antigenically distinct lipophosphoglycan on amastigotes of *Leishmania major*. *Molecular and Biochemical Parasitology* **45**, 337–344.
Glazunova, Z. I. (1965). Allergic reactions in guinea-pigs upon repeated inoculation with *Leishmania enriettii*. *Mediysinskaya Parazitologiaya I Parazitarnye Bolezni* **34**, 582–585.
Golub, E. S. (1981). Suppressor T cells and their possible role in the regulation of autoreactivity. *Cell* **24**, 595–596.
Gorczynski, R. M. (1983). *In vitro* analysis of immune responses of inbred guinea-pigs infected *in vivo* with *Leishmania enriettii* and an investigation of active immunization of susceptible animals. *Cellular Immunology* **75**, 255–270.
Gorczynski, R. M. and Macrae, S. (1982). Analysis of subpopulations of glass-adherent mouse skin cells controlling resistance/susceptibility to infection with *Leishmania tropica* and correlation with the development of independent proliferative signals to Lyt-1.2$^+$/Lyt-2.1$^+$ T lymphocytes. *Cellular Immunology* **67**, 74–89.
Granger, D. L. and Lehninger, A. L. (1982). Sites of inhibition of mitochondrial electron transport in macrophage-injured neoplastic cells. *Journal of Cell Biology* **95**, 527–535.
Granger, D. L., Perfect, J. R. and Durack, D. T. (1986). Macrophage-mediated fungistasis: requirement for a macromolecular component in serum. *Journal of Immunology* **137**, 693–701.
Granger, D. L., Hibbs, J. B., jr, Perfect, J. R. and Durack, D. T. (1990). Metabolic fate of L-arginine in relation to microbiostatic capacity of macrophages. *Journal of Clinical Investigation* **85**, 264–273.
Grau, G. E., Fajardo, L. F., Piguet, P.-F., Allet, B., Lambert, P.-H. and Vassalli, P. (1987). Tumor necrosis factor (cachectin) as an essential mediator in murine cerebral malaria. *Science* **237**, 1210–1212.
Grau, G. E., Kindler, V., Piguet, P.-F., Lambert, P. H. and Vassalli, P. (1988). Prevention of experimental cerebral malaria by anticytokine antibodies. *Journal of Experimental Medicine* **168**, 1499–1504.
Grau, G. E., Taylor, T. E., Molyneux, M. E., Wirima, J. J., Vassalli, P., Hommel, M. and Lambert, P. H. (1989a). Tumor necrosis factor and disease severity in children with falciparum malaria. *New England Journal of Medicine* **320**, 1586–1591.
Grau, G. E., Heremans, H., Piquet, P. F., Pointaire, P., Lambert, P. H., Billiau, A. and Vassalli, P. (1989b). Monoclonal antibody against interferon gamma can

prevent experimental cerebral malaria and its associated overproduction of tumor necrosis factor. *Proceedings of the National Academy of Sciences of the USA* **86**, 5572–5574.

Green, D. R., Flood, P. M. and Gershon, R. K. (1983). Immunoregulatory T-cell pathways. *Annual Review of Immunology* **1**, 439–463.

Green, L. C., Ruiz de Luzuriaga, K., Wagner, D. A., Rand, W., Istfan, N., Young, V. R. and Tannenbaum, S. R. (1981). Nitrate biosynthesis in man. *Proceedings of the National Academy of Sciences of the USA* **78**, 7764–7768.

Green, M. S., Kark, J. D., Greenblatt, C. L., Londner, M. V., Frankenburg, S. and Jacobson, R. L. (1983). The cellular and humoral immune response in subjects vaccinated against cutaneous leishmaniasis using *Leishmania tropica major* promastigotes. *Parasite Immunology* **5**, 337–344.

Green, S. J., Meltzer, M. S., Hibbs, J. B., jr and Nacy, C. A. (1990a). Activated macrophages destroy intracellular *Leishmania major* amastigotes by an L-arginine-dependent killing mechanism. *Journal of Immunology* **144**, 278–283.

Green, S. J., Crawford, R. M., Hockmeyer, J. T., Meltzer, M. S. and Nacy, C. A. (1990b). *Leishmania* amastigotes initiate the L-arginine-dependent killing mechanism in IFN-gamma-stimulated macrophages by induction of tumor necrosis factor-α. *Journal of Immunology* **145**, 4290–4297.

Green, S. J., Nacy, C. A. and Meltzer, M. S. (1991). Cytokine-induced synthesis of nitrogen oxides in macrophages: a protective host response to *Leishmania* and other intracellular pathogens. *Journal of Leukocyte Biology* **50**, 93–103.

Greenblatt, C. L. (1980). The present and future of vaccination for cutaneous leishmaniasis. *In* "New Developments with Human and Veterinary Vaccines" (A. Mizrahi, I. Hestman and M.A. Klinberg, eds), pp. 259–285. Alan R. Liss, New York.

Greenblatt, C. L. (1988). Cutaneous leishmaniasis: the prospect for a killed vaccine. *Parasitology Today* **4**, 53–54.

Greil, J., Bodendorfer, B., Röllinghoff, M. and Solbach, W. (1988). Application of recombinant granulocyte-macrophage colony-stimulating factor has a detrimental effect in experimental murine leishmaniasis. *European Journal of Immunology* **18**, 1527–1533.

Grencis, R. K., Hültner, L. and Else, K. J. (1991). Host protective immunity to *Trichinella spiralis* in mice: activation of Th cell subsets and lymphokine secretion in mice expressing different response phenotypes. *Immunology* **74**, 329–332.

Griffiths, W. A. D. (1987). Old World cutaneous leishmaniasis. *In* "The Leishmaniases in Biology and Medicine" (W. Peters and R. Killick-Kendrick, eds), Vol. 2, pp. 617–636. Academic Press, London.

Grzych, J. M., Pearce, E., Cheever, A., Caulada, Z. A., Caspar, P., Heiny, S., Lewis, F. and Sher, A. (1991). Egg deposition is the major stimulus for the production of Th2 cytokines in murine schistosomiasis. *Journal of Immunology* **146**, 1322–1327.

Guirges, S. Y. (1971). Natural and experimental re-infection of man with oriental sore. *Annals of Tropical Medicine and Parasitology* **65**, 197–205.

Gunders, A. E. (1987). Vaccination: past and future role in control. *In* "The Leishmaniases in Biology and Medicine" (W. Peters and R. Killick-Kendrick, eds), Vol. 2, pp. 929–941. Academic Press, London.

Haidaris, C. G. and Bonventre, P. F. (1982). A role for oxygen-dependent mechanisms in killing of *Leishmania donovani* tissue forms by activated macrophages. *Journal of Immunology* **129**, 850–855.

Haldar, J. P., Ghose, S., Saha, K. C. and Ghose, A. C. (1983). Cell-mediated immune response in Indian kala-azar and post kala-azar dermal leishmaniasis. *Infection and Immunity* **42**, 702–707.

Hale, C. and Howard, J. G. (1981). Immunological regulation of experimental cutaneous leishmaniasis. 2. Studies with Biozzi high and low responder lines of mice. *Parasite Immunology* **3**, 45–55.

Handman, E. and Burgess, A. W. (1979). Stimulation by granulocyte-macrophage colony-stimulating factor of *Leishmania tropica* killing by macrophages. *Journal of Immunology* **122,** 1134–1137.

Handman, E. and Mitchell, G. F. (1985). Immunization with *Leishmania* receptor for macrophages protects ice against cutaneous leishmaniasis. *Proceedings of the National Academy of Sciences of the USA* **82**, 5910–5914.

Handman, E., El-On, J., Spira, D. T., Zuckerman, A. and Greenblatt, C. L. (1977). Protection of C3H mice against *L. tropica* by a non-living antigenic preparation. *Journal of Protozoology* **24**, 20A–21A.

Handman, E., Ceredig, R. and Mitchell, G. F. (1979). Murine cutaneous leishmaniasis: disease patterns in intact and nude mice of various genotypes and examination of some difference between normal and impaired macrophages. *Australian Journal of Experimental Biology and Medical Science* **57**, 9–29.

Handman, E., McConville, M. J. and Goding, J. W. (1987). Carbohydrate antigens as possible parasite vaccines. A case for the *Leishmania* glycolipid. *Immunology Today* **8**, 181–185.

Handman, E., Button, L. L. and McMaster, W. R. (1990). *Leishmania major*: production of recombinant gp63, its antigenicity and immunogenicity in mice. *Experimental Parasitology* **70**, 427–435.

Harith, A. E., Kolk, A. H. J., Kager, P. A., Leeuwenburg, J., Muigai, R., Kiugi, S. and Laarman, J. J. (1986). A simple and economical direct agglutination test for serodiagnosis and sero-epidemiological studies of visceral leishmaniasis. *Transactions of the Royal Society of Tropical Medicine and Hygiene* **80**, 583–587.

Hedrick, S. M., Germain, R. N., Bevan, M. J., Dorf, M., Engel, I., Fink, P., Gascoigne, N., Heber-Katz, E., Kapp, J., Kaufmann, Y., Kaye, J., Melchers, F., Pierce, C., Schwartz, R. H., Sorensen, C., Taniguchi, M. and Davis, M. M. (1985). Rearrangement and transcription of a T cell receptor β-chain gene in different T cell subsets. *Proceedings of the National Academy of Sciences of the USA* **82**, 531–535.

Heinzel, F. P., Sadick, M. D., Holaday, B. J., Coffman, R. L. and Locksley, R. M. (1989). Reciprocal expression of interferon-gamma or interleukin-4 during the resolution or progression of murine leishmaniasis. Evidence for expansion of distinct helper T cell subsets. *Journal of Experimental Medicine* **169**, 59–72.

Heinzel, F. P., Sadick, M. D., Mutha, S. S. and Locksley, R. M. (1991). Production of interferon gamma, interleukin 2, interleukin 4, and interleukin 10 by $CD4^+$ lymphocytes *in vivo* during healing and progressive murine leishmaniasis. *Proceedings of the National Academy of Sciences of the USA* **88**, 7011–7015.

Herman, R. (1980). Cytophilic and opsonic antibodies in visceral leishmaniasis in mice. *Infection and Immunity* **28**, 585–593.

Hibbs, J. B., jr, Taintor, R. R., Chapman, H. A., jr and Weinberg, J. B. (1977). Macrophage tumor killing: influence of the local environment. *Science* **197**, 279–282.

Hibbs, J. B., jr, Taintor, R. R. and Vavrin, Z. (1984). Iron depletion: possible cause of tumor cell cytotoxicity induced by activated macrophages. *Biochemical and Biophysical Research Communications* **123**, 716–723.

Hibbs, J. B. jr, Taintor, R. R. and Vavrin, Z. (1987a). Macrophage cytotoxicity: role for L-arginine deiminase and imino nitrogen oxidation to nitrite. *Science* **235**, 473–476.

Hibbs, J. B. jr, Vavrin, Z. and Taintor, R. R. (1987b). L-Arginine is required for expression of the activated macrophage effector mechanism causing selective metabolic inhibition in target cells. *Journal of Immunology* **138**, 550–565.

Hibbs, J. B., jr, Taintor, R. R., Vavrin, Z. and Rachlin, E. M. (1988). Nitric oxide: a cytotoxic activated macrophage effector molecule. *Biochemical and Biophysical Research Communications* **157**, 87–94.

Hill, J. O. (1991). Reduced numbers of $CD4^+$ suppressor cells with subsequent expansion of $CD8^+$ protective T cells as an explanation for the paradoxical state of enhanced resistance to *Leishmania* in T-cell deficient BALB/c mice. *Immunology* **72**, 282–286.

Hill, J. O., Awwad, M. and North, R. J. (1989). Elimination of $CD4^+$ suppressor T cells from susceptible BALB/c mice releases $CD8^+$ T lymphocytes to mediate protective immunity against *Leishmania. Journal of Experimental Medicine* **169**, 1819–1827.

Hirokawa, K. (1988). Aging and the immune system. *In* "Cutaneous Aging" (A. M. Kligman and Y. Takase, eds), pp. 61–87. University of Tokyo Press, Tokyo.

Ho, J. L., He, S. H., Rios, M. J. C. and Wick, E. A. (1992). Interleukin-4 inhibits human macrophage activation by tumor necrosis factor, granulocyte-monocyte colony-stimulating factor, and interleukin-3 for antileishmanial activity and oxidative burst capacity. *Journal of Infectious Diseases* **165**, 344–351.

Ho, M., Koech, D. K., Iha, D. M. and Bryceson, A. D. M. (1983). Immunosuppression in Kenyan visceral leishmaniasis. *Clinical and Experimental Immunology* **51**, 207–214.

Hogg, N., Darley-Usmar, V. M., Wilson, M. T. and Moncada, S. (1992). Production of hydroxyl radicals from the simultaneous generation of superoxide and nitric oxide. *Biochemical Journal* **281**, 419–424.

Hoogstraal, H. and Heyneman, D. (1969). Leishmaniasis in the Sudan Republic. Final epidemiological report. *American Journal of Tropical Medicine and Hygiene* **18**, 1091–1210.

Holaday, B. J., Sadick, M. D., Wang, Z.-E., Reiner, S. L., Heinzel, F. P., Parslow, T. G. and Locksley, R. M. (1991). Reconstitution of *Leishmania* immunity in severe combined immunodeficient mice using Thl- and Th2-like cell lines. *Journal of Immunology* **147**, 1653–1658.

Hommel, M., Peters, W., Ranque, J., Quilici, M. and Lanotte, G. (1978). The micro-ELISA technique in the serodiagnosis of visceral leishmaniasis. *Annals of Tropical Medicine and Parasitology* **72**, 213–218.

Hosieth, S. K. and Stocker, B. A. D. (1981). Aromatic-dependent *Salmonella typhimurium* are non-virulent and effective as live vaccines. *Nature* **291**, 238–239.

Howard, J. G. (1985). Host immunity to leishmaniasis. *In* "Leishmaniasis" (K.-P. Chang and R.S. Bray, eds), pp. 140–162. Elsevier Science Publishers, Amsterdam.

Howard, J. G., Hale, C. and Chan-Liew, W. L. (1980a). Immunological regulation of experimental cutaneous leishmaniasis. I. Immunogenetic aspects of susceptibility to *Leishmania tropica* in mice. *Parasite Immunology* **2**, 303–314.

Howard, J. G., Hale, C. and Liew, F. Y. (1980b). Genetically determined susceptibility to *Leishmania tropica* infection is expressed by haematopoietic donor cells in mouse radiation chimeras. *Nature* **288**, 161–162.

Howard, J. G., Hale, C. and Liew, F. Y. (1980c). Immunological regulation of experimental cutaneous leishmaniasis. III. Nature and significance of specific suppression of cell-mediated immunity in mice highly susceptible to *Leishmania tropica*. *Journal of Experimental Medicine* **152**, 594–607.

Howard, J. G., Hale, C. and Liew, F. Y. (1981). Immunological regulation of experimental cutaneous leishmaniasis. IV. Prophylactic effect of sublethal irradiation as a result of abrogation of suppressor T cell generation in mice genetically susceptible to *Leishmania tropica*. *Journal of Experimental Medicine* **153**, 557–568.

Howard, J. G., Nicklin, S., Hale, C. and Liew, F. Y. (1982). Prophylactic immunization against experimental leishmaniasis. I. Protection induced in mice genetically vulnerable to fatal *Leishmania tropica* infection. *Journal of Immunology* **129**, 2206–2211.

Howard, J. G., Liew, F. Y., Hale, C. and Nicklin, S. (1984). Prophylactic immunization against experimental leishmaniasis. II. Further characterization of the protective immunity against fatal *L. tropica* infection induced by irradiated promastigotes. *Journal of Immunology* **132**, 450–455.

Howard, M. and O'Garra, A. (1992). Biological properties of interleukin 10. *Immunology Today* **13**, 198–200.

Hughes, H. P. A. (1988). Oxidative killing of intracellular parasites mediated by macrophages. *Parasitology Today* **4**, 340–347.

Hunt, N. C. A. and Goldin, R. D. (1992). Nitric oxide production by monocytes in alcoholic liver disease. *Journal of Hepatology* **14**, 146–150.

Iyengar, R., Stuehr, D. J. and Marletta, M. A. (1987). Macrophage synthesis of nitrite, nitrate and N-nitrosamines: precursors and role of the respiratory burst. *Proceedings of the National Academy of Sciences of the USA* **84**, 6369–6373.

Jaffe, C. L. and McMahon-Pratt, D. (1987). Serodiagnostic assay for visceral leishmaniasis employing monoclonal antibodies. *Transactions of the Royal Society of Tropical Medicine and Hygiene* **81**, 587–594.

Jaffe, C. L. and Zalis, M. (1988). Use of purified proteins from *Leishmania donovani* for the rapid serodiagnosis of visceral leishmaniasis. *Journal of Infectious Diseases* **157**, 1212–1220.

Jaffe, C. L., Shor, R., Trau, H. and Passwell, J. H. (1990). Parasite antigens recognized by patients with cutaneous leishmaniasis. *Clinical and Experimental Immunology* **80**, 77–82.

James, S. L. and Glaven, J. (1989). Macrophage cytotoxicity against schistosomula of *Schistosoma mansoni* involves arginine-dependent production of reactive nitrogen intermediates. *Journal of Immunology* **143**, 4208–4212.

James, S. L. and Hibbs, J. B., jr (1990). The role of nitrogen oxides as effector molecules of parasite killing. *Parasitology Today* **6**, 303–305.

Janeway, C. A. (1975). Cellular cooperation during *in vivo* anti-hapten antibody response. I. The effect of cell number on the response. *Journal of Immunology* **114**, 1394–1401.

Janicki, B. W., Schechter, G. P. and Schultz, K. E. (1970). Cellular reactivity to tuberculin in immune and serologically-unresponsive guinea pigs. *Journal of Immunology* **105**, 527–530.

Jardim, A., Alexander, J., Teh, N. S., Ou, D. and Olafson, R. W. (1990). Immunoprotective *Leishmania major* synthetic T cell epitopes. *Journal of Experimental Medicine* **172**, 645–648.

Jardim, A., Tolson, D. L., Turco, S. J., Pearson, T. W. and Olafson, R. W. (1991). The *Leishmania donovani* lipophosphoglycan T lymphocyte-reactive component is a tightly associated protein complex. *Journal of Immunology* **147**, 3538–3544.

Kahl, L. P. and McMahon-Pratt, D. (1987). Structural and antigenic characterization of a species- and promastigote-specific *Leishmania mexicana amazonensis* membrane protein. *Journal of Immunology* **138**, 1587–1595.
Kahl, L. P., Scott, C. A., Lelchuk, R., Gregoriadis, G. and Liew, F. Y. (1989). Vaccination against murine cutaneous leishmaniasis by using *Leishmania major* antigen/liposomes. Optimisation and assessment of the requirement for intravenous immunization. *Journal of Immunology* **142**, 4441–4449.
Kaplan, G., Britton, W. J., Hancock, G. E., Theuvenet, W. J., Smith, K. A., Job, C. K., Roche, P. W., Molloy, A., Burkhardt, R., Barker, J., Pradhan, H. M. and Cohn, Z. A. (1991). The systemic influence of recombinant interleukin 2 on the manifestations of lepromatous leprosy. *Journal of Experimental Medicine* **173**, 993–1006.
Kaufmann, S. H. E. (1988). $CD8^+$ T lymphocytes in intracellular microbial infections. *Immunology Today* **9**, 168–174.
Kaye, P. M. and Blackwell, J. M. (1989). *Lsh*, antigen presentation and the development of CMI. *Research in Immunology* **140**, 810–815.
Kaye, P. M., Patel, N. K. and Blackwell, J. M. (1988). Acquisition of cell-mediated immunity to *Leishmania*. II. *Lsh* gene regulation of accessory cell function. *Immunology* **65**, 17–22.
Kaye, P. M., Curry, A. J. and Blackwell, J. M. (1991). Differential production of Th1 and Th2-derived cytokines does not determine the genetically controlled or vaccine induced rate of cure in murine visceral leishmaniasis. *Journal of Immunology* **146**, 2763–2770.
Kelso, A. and Glasebrook, A. L. (1984). Secretion of interleukin 2, macrophage-activating factor, interferon, and colony-stimulating factor by alloreactive T lymphocyte clones. *Journal of Immunology* **132**, 2924–2931.
Killar, L., MacDonald, G., West, J., Woods, A. and Bottomly, K. (1987). Cloned, Ia-restricted T cells that do not produce interleukin 4 (IL 4)/B-cell stimulatory factor-1 (BSF-1) fail to help antigen-specific B cells. *Journal of Immunology* **138**, 1674–1679.
Kim, J., Woods, A., Becker-Dunn, E. and Bottomly, K. (1985). Distinct functional phenotypes of cloned Ia-restricted helper T cells. *Journal of Experimental Medicine* **162**, 188–201.
King, C. L. and Nutman, T. B. (1991). Regulation of the immune response in lymphatic filariasis and onchocerciasis. *In* "Immunoparasitology Today" (C. Ash and R. B. Gallagher, eds), pp. A54–A58. Elsevier Trends Journals, Cambridge.
Kirkpatrick, C. E. and Farrell, J. P. (1982). Leishmaniasis in beige mice. *Infection and Immunity* **38**, 1208–1216.
Kirkpatrick, C. E. and Farrell, J. P. (1984). Splenic natural killer-cell activity in mice infected with *Leishmania donovani*. *Cellular Immunology* **85**, 201–214.
Kolb, H., Kiesel, U., Kröncke, K.-D. and Kolb-Bachofen, V. (1991). Suppression of low dose streptozotocin induced diabetes in mice by administration of a nitric oxide synthase inhibitor. *Life Science* **49**, 213–217.
Koufman, Z., Egoz, N., Greenblatt, C. L., Handman, E., Montillo, B. and Even-Paz, Z. (1978). Observations on immunization against cutaneous leishmaniasis in Israel. *Israel Journal of Medical Science* **14**, 218–222.
Kröncke, K.-D., Kolb-Bachofen, V., Berschick, B., Burkart, V. and Kolb, H. (1991). Activated macrophages kill pancreatic syngeneic islet cells via arginine-dependent nitric oxide generation. *Biochemical and Biophysical Research Communications* **175**, 752–758.

Kronenberg, M., Steinmetz, M., Kobori, J., Kraig, E., Kapp, J. A., Pierce, C. W., Sorensen, C. M., Suzuki, G., Tada, T. and Hood, L. (1983). RNA transcripts for I-J polypeptides are apparently not encoded between the I-A and I-E subregions of the murine major histocompatibility complex. *Proceedings of the National Academy of Sciences of the USA* **80**, 5704–5708.

Kumar, S., Good, M. F., Dontfraid, F., Vinetz, J. M. and Miller, L. H. (1989). Interdependence of $CD4^+$ T cells and malarial spleen in immunity to *Plasmodium vinckei vinckei*. Relevance to vaccine development. *Journal of Immunology* **143**, 2017–2023.

Kurt-Jones, E. A., Hamberg, S., Ohara, J., Paul, W. E. and Abbas, A. K. (1987). Heterogeneity of helper/inducer T lymphocytes. I. Lymphokine production and lymphokine responsiveness. *Journal of Experimental Medicine* **166**, 1774–1787.

Kwiatkowski, D. (1990). Tumour necrosis factor, fever, and fatality in falciparum malaria. *Immunology Letters* **25**, 213–216.

Kwon, N. S., Nathan, C. F. and Stuehr, D. J. (1989). Reduced biopterin as a cofactor in the generation of nitrogen oxides by murine macrophages. *Journal of Biological Chemistry* **264**, 20496–20501.

Lancaster, J. R., jr and Hibbs, J. B., jr (1990). EPR demonstration of iron-nitrosyl complex formation by cytotoxic activated macrophages. *Proceedings of the National Academy of Sciences of the USA* **87**, 1223–1227.

Langhorne, J., Simon-Haarhaus, B. and Meding, S. J. (1990). The role of $CD4^+$ T cells in the protective immune response to *Plasmodium chabaudi in vivo*. *Immunology Letters* **25**, 101–108.

Leal, L.M.C.C., Moss, D.W., Kuhn, R., Muller, W. and Liew, F.Y. (1993). Interleukin-4 transgenic mice of resistant background are susceptible to *Leishmania major* infection. *European Journal of Immunology* **23**, 566–569.

Lebman, D. A. and Coffman, R. L. (1988). Interleukin 4 causes isotype switching to IgE in T cell-stimulated clonal B cell cultures. *Journal of Experimental Medicine* **168**, 853–862.

Le Gros, G., Ben-Sasson, S. Z., Seder, R., Finkelman, F. D. and Paul, W. E. (1990). Generation of interleukin 4 (IL-4)-producing cells *in vivo* and *in vitro*: IL-2 and IL-4 are required for *in vitro* generation of IL-4-producing cells. *Journal of Experimental Medicine* **172**, 921–929.

Lehn, M., Weiser, W. Y., Engelhorn, S., Gillis, S. and Remold, H. G. (1989). IL-4 inhibits H_2O_2 production and anti-leishmanial capacity of human cultured monocytes mediated by IFN-gamma. *Journal of Immunology* **143**, 3020–3024.

Lelchuk, R., Graveley, R. and Liew, F. Y. (1988). Susceptibility to murine cutaneous leishmaniasis correlates with the capacity to generate interleukin-3 in response to leishmania antigen *in vitro*. *Cellular Immunology* **111**, 66–76.

Lelchuk, R., Carrier, M. J., Kahl, L. P. and Liew, F. Y. (1989). Distinct IL-3 activation profile induced by intravenous versus subcutaneous routes of immunisation. *Cellular Immunology* **122**, 338–349.

Lemma, A. and Yau, P. (1973). Course of development of *Leishmania enriettii* infection in immunosuppressed guinea pigs. *American Journal of Tropical Medicine and Hygiene* **22**, 477–481.

Lepoivre, M., Fieschi, F., Coves, J., Thelander, L. and Fontecave, M. (1991). Inactivation of ribonucleotide reductase by nitric oxide. *Biochemical and Biophysical Research Communications* **179**, 442–448.

Lezama-Davila, C. M., Williams, D. M., Gallagher, G. and Alexander, J. (1992). Cytokine control of *Leishmania* infection in the BALB/c mouse: enhancement and inhibition of parasite growth by local administration of IL-2 or IL-4 is species and time dependent. *Parasite Immunology* **14**, 37–48.

Li, Y., Severn, A., Rogers, M. V., Palmer, R. M. J., Moncada, S. and Liew, F. Y. (1992). Catalase inhibits nitric oxide synthesis and the killing of intracellular *Leishmania major* in murine macrophages. *European Journal of Immunology* **22**, 441–446.

Lichtman, A. H., Chin, J., Schmidt, J. A. and Abbas, A. K. (1988). Role of interleukin 1 in the activation of T lymphocytes. *Proceedings of the National Academy of Sciences of the USA* **85**, 9699–9703.

Liew, F. Y. (1977). Regulation of delayed type hypersensitivity. I. T suppressor cells for delayed type hypersensitivity to sheep erythrocytes in mice. *European Journal of Immunology* **7**, 714–718.

Liew, F. Y. (1983). Specific suppression of responses to *Leishmania tropica* by a cloned T-cell line. *Nature* **305**, 630–632.

Liew, F. Y. (1987). Functional analysis of host-protective and disease-promoting T cells. *Annales de l'Institut Pasteur—Immunologie* **138**, 749–755.

Liew, F. Y. (1989a). New strategy for control of leishmaniasis: Vaccination. *In* "Leishmaniasis. The Current Status and New Strategies for Control" (D. T. Hart, ed.), pp. 835–838. NATO ASI Series, Plenum Press, New York.

Liew, F.Y. (1989b). Functional heterogeneity of $CD4^+$ T cells in leishmaniasis. *Immunology Today* **10**, 40–45.

Liew, F. Y. (1990). Regulation of cell-mediated immunity in leishmaniasis. *Current Topics in Microbiology and Immunology* **155**, 53–64.

Liew, F. Y. (1992). Induction, regulation and function of T-cell subsets in leishmaniasis. *Chemical Immunology* **54**, 117–135.

Liew, F. Y. and Cox, F. E. G. (1991). Nonspecific defence mechanism: the role of nitric oxide. *In* "Immunoparasitology Today" (C. Ash and R. B. Gallagher, eds), pp. A17–A21. Elsevier Trends Journals, Cambridge.

Liew, F. Y. and Dhaliwal, J. S. (1987). Distinctive cellular immunity in genetically susceptible BALB/c mice recovered from *Leishmania major* infection or after subcutaneous immunization with killed parasites. *Journal of Immunology* **138**, 4450–4456.

Liew, F. Y. and Parish, C. R. (1974). Lack of correlation between cell-mediated immunity to the carrier and the carrier-hapten helper effect. *Journal of Experimental Medicine* **139**, 779–784.

Liew, F. Y. and Russell, S. M. (1980). Delayed-type hypersensitivity to influenza virus: induction of antigen specific suppressor T cells for delayed-type hypersensitivity of haemagglutinin during influenza virus infection in mice. *Journal of Experimental Medicine* **151**, 799–814.

Liew, F. Y., Hale, C. and Howard, J. G. (1982). Immunologic regulation of experimental cutaneous leishmaniasis. V. Characterization of effector and specific suppressor T cells. *Journal of Immunology* **128**, 1917–1922.

Liew, F. Y., Howard, J. G. and Hale, C. (1984). Prophylactic immunization against experimental leishmaniasis. III. Protection against fatal *Leishmania tropica* infection induced by irradiated promastigotes involves $Lyt\text{-}1^+2^-$ T cells that do not mediate cutanous DTH. *Journal of Immunology* **132**, 456–461.

Liew, F. Y., Hale, C. and Howard, J. G. (1985a). Prophylactic immunization against experimental leishmaniasis. IV. Subcutaneous immunization prevents the induction of protective immunity against fatal *Leishmania major* infection. *Journal of Immunology* **135**, 2095–2101.

Liew, F. Y., Singleton, A., Cillari, E. and Howard, J. G. (1985b). Prophylactic immunization against experimental leishmaniasis. V. Mechanism of the anti-protective blocking effect induced by subcutaneous immunization against *Leishmania major* infection. *Journal of Immunology* **135**, 2102–2107.

Liew, F. Y., Hodson, K. and Lelchuk, R. (1987). Prophylactic immunization against experimental leishmaniasis. VI. Comparison of protective and disease-promoting T cells. *Journal of Immunology* **139**, 3112–3117.

Liew, F. Y., Millott, S., Li, Y., Lelchuk, R., Chan, W. L. and Ziltener, H. (1989). Macrophage activation by interferon-gamma from host-protective T-cells is inhibited by interleukin (IL)-3 and IL-4 produced by disease promoting T-cells in leishmaniasis. *European Journal of Immunology* **19**, 1227–1232.

Liew, F. Y., Parkinson, C., Millott, S., Severn, A. and Carrier, M. J. (1990a). Tumour necrosis factor (TNFα) in leishmaniasis. I. TNFα mediates host-protection against cutaneous leishmaniasis. *Immunology* **69**, 570–573.

Liew, F. Y., Millott, S., Parkinson, C., Palmer, R. M. J. and Moncada, S. (1990b). Macrophage killing of *Leishmania* parasites *in vivo* is mediated by nitric oxide from L-arginine. *Journal of Immunology* **144**, 4794–4797.

Liew, F. Y., Li, Y. and Millott, S. (1990c). Tumour necrosis factor (TNFα) in leishmaniasis. II. TNFα-induced macrophage leishmanicidal activity is mediated by nitric oxide from L-arginine. *Immunology* **71**, 556–559.

Liew, F. Y., Li, Y. and Millott, S. (1990d). TNFα synergizes with IFN-gamma in mediating killing of *Leishmania major* through the induction of nitric oxide. *Journal of Immunology* **145**, 4306–4310.

Liew, F. Y., Millott, S. and Schmidt, J. A. (1990e). A repetitive peptide of *Leishmania* is a disease-promoting epitope activating Th2 cells. *Journal of Experimental Medicine* **172**, 1359–1365.

Liew, F. Y., Li, Y., Severn, A., Millott, S., Schmidt, J., Salter, M. and Moncada, S. (1991a). A possible novel pathway of regulation by murine T helper type-2 (T_h2) cells of a T_h1 cell activity via the modulation of the induction of nitric oxide synthase on macrophages. *European Journal of Immunology* **21**, 2489–2494.

Liew, F. Y., Li, Y., Moss, D., Parkinson, C., Rogers, M. V. and Moncada, S. (1991b). Resistance to *Leishmania major* infection correlates with the induction of nitric oxide synthase in murine macrophages. *European Journal of Immunology* **21**, 3009–3014.

Liew, F. Y., Li, Y., Yang, D. M., Severn, A. and Cox, F. E. G. (1991c). TNF-α reverses the disease-exacerbating effect of subcutaneous immunization against murine cutaneous leishmaniasis. *Immunology* **74**, 304–309.

Locksley, R. M. and Scott, P. (1991). Helper T-cell subsets in mouse leishmaniasis: induction, expansion and effector function. *In* "Immunoparasitology Today" (C. Ash and R.B. Gallagher, eds), pp. A58–A61. Elsevier Trends Journals, Cambridge.

Locksley, R. M., Heinzel, F. P., Sadick, M. D., Holaday, B. J. and Gardner, K. D., jr (1987). Murine cutaneous leishmaniasis. Susceptibility correlates with differential expansion of helper T-cell subsets. *Annales de l'Institut Pasteur—Immunologie* **138**, 744–749.

Locksley, R. M., Heinzel, F. P., Holaday, B. J., Mutha, S. S., Reiner, S. L. and Sadick, M. D. (1991). Induction of Th1 and Th2 $CD4^+$ subsets during murine *Leishmania major* infection. *Research in Immunology* **142**, 28–32.

Loewi, G., Holborow, E. J. and Temple, A. (1966). Inhibition of delayed hypersensitivity by preimmunization without complete adjuvant. *Immunology* **10**, 339–347.

Lohman, K. L., Langer, P. J. and McMahon-Pratt, D. (1990). Molecular cloning and characterization of the immunologically protective surface glycoprotein GP46/M-2 of *Leishmania amazonensis*. *Proceedings of the National Academy of Sciences of the USA* **87**, 8393–8397.

Lohoff, M., Dingfelder, J. and Röllinghoff, M. (1991). A search for cells carrying the gamma/γ T cell receptor in mice infected with *Leishmania major*. *Current Topics in Microbiology and Immunology* **173**, 285–289.

Lopes, U. G. and Wirth, D. F. (1986). Identification of visceral *Leishmania* species with cloned sequences of kinetoplast DNA. *Molecular and Biochemical Parasitology* **20**, 77–84.

Lukic, M. L., Stosic-Grujicic, S., Ostojic, N., Chan, W. L. and Liew, F. Y. (1991). Inhibition of nitric oxide generation affects the induction of diabetes by streptozocin in mice. *Biochemical and Biophysical Research Communications* **178**, 913–920.

Lynch, N. R., Yarzabal, L., Verdel, O., Avila, J. L., Monzon, H. and Convit, J. (1982). Delayed-type hypersensitivity and immunoglobulin E in American cutaneous leishmaniasis. *Infection and Immunity* **38**, 877–881.

Lyons, C. R., Orloff, G. J. and Cunningham, J. M. (1992). Molecular cloning and functional expression of an inducible nitric oxide synthase from a murine macrophage cell line. *Journal of Biological Chemistry* **267**, 6370–6374.

Madden, K. B., Urban, J. F., jr, Ziltener, H. J., Schrader, J. W., Finkelman, F. D. and Katona, I. M. (1991). Antibodies to IL-3 and IL-4 suppress helminth induced intestinal mastocytosis. *Journal of Immunology* **147**, 1387–1391.

Maggi, E., Del Prete, G. F., Macchia, D., Parronchi, P., Tiri, A., Chretien, I., Ricci, M. and Romagnani, S. (1988). Profiles of lymphokine activities and helper function for IgE in human T cell clones. *European Journal of Immunology* **18**, 1045–1050.

Maggi, E., Biswas, P., Del Prete, G., Parronchi, P., Macchia, D., Simonelli, C., Emmi, L., De Carli, M., Tiri, A., Ricci, M. and Romagnani, S. (1991). Accumulation of Th-2-like helper T cells in the conjunctiva of patients with vernal conjunctivitis. *Journal of Immunology* **146**, 1169–1174.

Magilavy, D. B., Fitch, F. W. and Gajewski, T. F. (1989). Murine hepatic accessory cells support the proliferation of Th1 but not Th2 helper T lymphocyte clones. *Journal of Experimental Medicine* **170**, 985–990.

Manson-Bahr, P. E. C. (1959). East African kala-azar with special reference to the pathology, prophylaxis and treatment. *Transactions of the Royal Society of Tropical Medicine and Hygiene* **53**, 123–136.

Manson-Bahr, P. E. C. (1961). Immunity in kala-azar. *Transactions of the Royal Society of Tropical Medicine and Hygiene* **55**, 550–555.

Manson-Bahr, P. E. C. (1963). Active immunization in leishmaniasis. *In* "Immunity to Protozoa" (P. C. C. Garnham, A. E. Pierce and J. Roitt, eds), pp. 246–252. Blackwell Scientific Publications, Oxford.

Marletta, M. A. (1989). Nitric oxide: biosynthesis and biological significance. *Trends in Biochemical Sciences* **14**, 488–492.

Marletta, M. A., Yoon, P. S., Iyengar, R., Leaf, C. D. and Wishnok, J. S. (1988). Macrophage oxidation of L-arginine to nitrite and nitrate: nitric oxide is an intermediate. *Biochemistry* **27**, 8706–8711.

Marrack, P. C. and Kappler, J. W. (1975). Antigen-specific and non-specific mediators of T cell/B cell cooperation. I. Evidence for their production by different T cells. *Journal of Immunology* **114**, 1116–1125.

Marzinowsky, E. I. and Schurenkowa, A. (1924). Oriental sore and immunity against it. *Transactions of the Royal Society of Tropical Medicine and Hygiene* **18**, 67–69.

Mauël, J. and Behin, R. (1987). Immunity: clinical and experimental. *In* "The

Leishmaniases in Biology and Medicine" (W. Peters and R. Killick-Kendrick, eds), Vol. 2, pp. 731–791. Academic Press, London.

Mauël, J., Buchmüller, Y. and Behin, R. (1978). Studies on the mechanisms of macrophage activation. I. Destruction of intracellular *Leishmania enriettii* in macrophages activated by cocultivation with stimulated lymphocytes. *Journal of Experimental Medicine* **148**, 393–407.

Mauël, J., Corradin, S. B. and Rouiller, Y. B. (1991a). Nitrogen and oxygen metabolites and the killing of *Leishmania* by activated murine macrophages. *Research in Immunology* **7**, 577–580.

Mauël, J., Ransign, A. and Buchmüller-Rouiller, Y. (1991b). Killing of *Leishmania* parasites in activated murine macrophages is based on an L-arginine-dependent process that produces nitrogen derivatives. *Journal of Leukocyte Biology* **49**, 73–82.

Mayrink, W., Melo, M. N., da Costa, C. A., Magalhães, P. A., Dias, M., Coelho, M. V., Araújo, F. G., Williams, P., Pigueiredo, Y. P. and Batista, S. M. (1976). Intradermorreaçao de Montenegro na leishmaniose tegumentar americana após terapêutica antimonial. *Revista do Instituto de Medicina Tropical de São Paulo* **19**, 182–185.

Mayrink, W., Da Costa, C. A., Magalhães, P. A., Melo, M. N., Dias, M., Oliveira Lima, A., Michalick, M. S. and Williams, P. (1979). A field trial of a vaccine against American dermal leishmaniasis. *Transactions of the Royal Society of Tropical Medicine and Hygiene* **73**, 385–387.

Mayrink, W., Williams, P., Da Costa, C. A., Magalhães, P. A., Melo, M. N., Dias, M., Oliveira-Lima, A., Michalick, M. S. M., Carvalho, E. F., Barros, G. C., Sessa, P. A. and De Alencar, J. T. A. (1985). An experimental vaccine against American dermal leishmaniasis: experience in the state of Espírito Santo, Brazil. *Annals of Tropical Medicine and Parasitology* **79**, 259–269.

Mazingue, C., Cottrez-Detoeuf, F., Louis, J., Kweider, M., Auriault, C. and Capron, A. (1989). *In vitro* and *in vivo* effects of interleukin 2 on the protozoan parasite leishmania. *European Journal of Immunology* **19**, 487–491.

McCall, T. B., Boughton-Smith, N. K., Palmer, R. M. J., Whittle, B. J. R. and Moncada, S. (1989). Synthesis of nitric oxide from neutrophils: release and interaction with superoxide anion. *Biochemical Journal* **261**, 293–296.

McCall, T. B., Palmer, R. M. J. and Moncada, S. (1991). Induction of nitric oxide synthase in rat peritoneal neutrophils and its inhibition by dexamethasone. *European Journal of Immunology* **21**, 2523–2527.

McConville, M. J. and Bacic, A. (1989). A family of glycoinositol phospholipids from *Leishmania major*. Isolation, characterisation and antigenicity. *Journal of Biological Chemistry* **264**, 757–766.

McGurn, M., Boon, T., Louis, J. A. and Titus, R. G. (1990). *Leishmania major*: nature of immunity induced by immunisation with a mutagenized avirulent clone of the parasite in mice. *Experimental Parasitology* **71**, 81–89.

McMahon-Pratt, D., Rodriguez, D., Rodriguez, J.-R., Zhang, Y., Manson, K., Bergman, C., Rivas, L., Rodriguez, J. F., Lohman, K. L., Ruddle, N. H. and Esteban, M. (1993). Recombinant GP46/M-2 vaccinia viruses protect against *Leishmania* infection. *Journal of Immunology*, in press.

McNeely, T. B. and Turco, S. J. (1987). Inhibition of protein kinase C activity by the *Leishmania donovani* lipophosphoglycan. *Biochemical and Biophysical Research Communications* **148**, 653–657.

McNeely, T. B., Rosen, G., Londner, M. V. and Turco, S. J. (1989). Inhibitory effects on protein kinase C activity by lipophosphoglycan fragments and glycosylphosphatidylinositol antigens of the protozoan parasite *Leishmania*. *Biochemical Journal* **259**, 601–604.

Medina-Acosta, E., Karess, R. E., Schwartz, H. and Russell, D. G. (1989). The promastigote surface protease (gp63) of *Leishmania* is expressed but differentially processed and localized in the amastigote stage. *Molecular and Biochemical Parasitology* **37**, 263–274.

Mellouk, S., Green, S. J., Nacy, C. A. and Hoffman, S. L. (1991). IFN-gamma inhibits development of *Plasmodium berghei* exoerythrocytic stages in hepatocytes by an L-arginine-dependent effector mechanism. *Journal of Immunology* **146**, 3971–3976.

Mendonça, S. C. F., Russell, D. G. and Coutinho, S. G. (1991). Analysis of the human T cell responsiveness to purified antigens of *Leishmania*: lipophosphoglycan (LPG) and glycoprotein 63 (gp63). *Clinical and Experimental Immunology* **83**, 472–478.

Menzel, S. and Bienzle, U. (1978). Antibody responses in patients with cutaneous leishmaniasis of the Old World. *Tropenmedizin und Parasitologie* **29**, 194–197.

Midgley, S., Grant, I. S., Haynes, W. G. and Webb, D. J. (1991). Nitric oxide in liver failure. *Lancet* **338**, 1590.

Miller, R. H., Reed, S. G. and Parsons, M. (1990). *Leishmania* gp63 molecule implicated in cellular adhesion lacks an Arg-Gly-Asp sequence. *Molecular and Biochemical Parasitology* **39**, 267–278.

Milon, G., Titus, R. G., Cerottini, J.-C., Marchal, G. and Louis, J. A. (1986). Higher frequency of *Leishmania major*-specific L3T4$^+$ T cells in susceptible BALB/c as compared with resistant CBA mice. *Journal of Immunology* **136**, 1467–1471.

Miltenburg, A. M. M., Van Laar, J. M., De Kuiper, R., Daha, M. R. and Breedveld, F. C. (1992). T cells cloned from human rheumatoid synovial membrane functionally represent the Thl subset. *Scandinavian Journal of Immunology* **35**, 603–610.

Mirkovich, A. M., Galelli, A., Allison, A. C. and Modabber, F. (1986). Increased myelopoiesis during *Leishmania major* infection in mice: generation of "safe target", a possible way to evade the effector immune mechanism. *Clinical and Experimental Immunology* **64**, 1–7.

Mitchell, G. F. and Handman, E. (1986). The glycoconjugate derived from a *Leishmania major* receptor for macrophages is a suppressogenic, disease-promoting antigen in murine cutaneous leishmaniasis. *Parasite Immunology* **8**, 255–263.

Mitchell, G. F., Curtis, J. M., Handman, E. and McKenzie, I. F. C. (1980). Cutaneous leishmaniasis in mice: disease patterns in reconstituted nude mice of several genotypes infected with *Leishmania tropica*. *Australian Journal of Experimental Biology and Medical Science* **58**, 521–532.

Mitchell, G. F., Curtis, J. M. and Handman, E. (1981a). Resistance to cutaneous leishmaniasis in genetically susceptible BALB/c mice. *Australian Journal of Experimental Biology and Medical Science* **59**, 555–565.

Mitchell, G. F., Curtis, J. M., Scollay, R. G. and Handman, E. (1981b). Resistance and abrogation of resistance to cutaneous leishmaniasis in reconstituted BALB/c nude mice. *Australian Journal of Experimental Biology and Medical Science* **59**, 539–554.

Mitchell, G. F., Anders, R. F., Brown, G. V., Handman, E., Roberts-Thompson, I. C., Chapman, C. B., Forsyth, K. P., Kahl, L. P. and Cruise, K. M. (1982). Analysis of infection characteristics and antiparasite immune responses in resistant compared with susceptible hosts. *Immunological Reviews* **61**, 137–188.

Mitchell, G. F., Handman, E. and Spithill, T. W. (1984). Vaccination against cutaneous leishmaniasis in mice using nonpathogenic cloned promastigotes of *L.*

major and importance of route of injection. *Australian Journal of Experimental Biology and Medical Science* **62**, 145–153.

Mock, B. A., Fortier, A. H., Potter, M., Blackwell, J. and Nacy, C. A. (1985a). Genetic control of systemic *Leishmania major* infection: identification of subline differences for susceptibility to disease. *Current Topics in Microbiology and Immunology* **122**, 115–121.

Mock, B. A., Russek-Cohen, E., Hilgers, J. and Nacy, C. A. (1985b). Discriminant function analysis of genetic traits associated with *Leishmania major* infection in the CXS recombinant inbred strains. *Progress in Leukocyte Biology* **3**, 83–95.

Modabber, F. (1987). The leishmaniases. *In* "Tropical Disease Research, a Global Partnership, Eighth Programme Report, TDR" (J. Maurice and A. M. Pearce, eds), pp. 99–112. World Health Organization, Geneva.

Modabber, F. (1990). Development of vaccines against leishmaniasis. *Scandinavian Journal of Infectious Diseases*, supplement no. 76, 72–78.

Modlin, R. L., Pirmez, C., Hofman, F. M., Torigian, V., Uyemura, K., Rea, T. H., Bloom, B. R. and Brenner, M. B. (1989). Lymphocytes bearing antigen-specific gamma-γ T-cell receptors accumulate in human infectious disease lesions. *Nature* **339**, 544–548.

Moll, H., Scollay, R. G. and Mitchell, G. F. (1988). Resistance to cutaneous leishmaniasis in nude mice injected with $L3T4^+$ T cells but not with $Lyt2^+$ T cells. *Immunology and Cellular Biology* **66**, 57–63.

Moll, H., Binöder, K., Bogdan, C., Solbach, W. and Röllinghoff, M. (1990). Production of tumour necrosis factor during murine cutaneous leishmaniasis. *Parasite Immunology* **12**, 483–494.

Moncada, S., Palmer, R. M. J. and Higgs, E. A. (1991). Nitric oxide: physiology, pathophysiology and pharmacology. *Pharmacological Reviews* **43**, 109–142.

Montenegro, J. (1926). Cutaneous reaction in leishmaniasis. *Archives in Dermatology and Syphilis* **13**, 187.

Moore, K. W., Vieira, P., Fiorentino, D. F., Trounstine, M. L., Khan, T. A. and Mosmann, T. R. (1990). Homology of cytokine synthesis inhibitory factor (IL-10) to the Epstein-Barr virus gene BCRF1. *Science* **248**, 1230–1234.

Mosmann, T. R. and Coffman, R. L. (1987). Two types of mouse helper T-cell clone—implications for immune regulation. *Immunology Today* **8**, 223–227.

Mosmann, T. R. and Coffman, R. L. (1989). TH1 and TH2 cells: different patterns of lymphokine secretion lead to different functional properties. *Annual Review of Immunology* **7**, 145–173.

Mosmann, T. R. and Moore, K. W. (1991). The role of IL-10 in crossregulation of Th1 and Th2 responses. *In* "Immunoparasitology Today" (C. Ash and R. B. Gallagher, eds), pp. A49–A53. Elsevier Trends Journals, Cambridge.

Mosmann, T. R., Cherwinski, H., Bond, M. W., Giedlin, M. A. and Coffman, R. L. (1986). Two types of murine helper T cell clone. I. Definition according to profiles of lymphokine activities and secreted proteins. *Journal of Immunology* **136**, 2348–2357.

Mosser, D. M. and Edelson, P. J. (1984). Activation of the alternative complement pathway by *Leishmania* promastigotes: parasite lysis and attachment to macrophages. *Journal of Immunology* **132**, 1501–1505.

Müller, I., Pedrazzini, T., Farrell, J. P. and Louis, J. A. (1989). T-cell responses and immunity to experimental infection with *Leishmania major*. *Annual Review of Immunology* **7**, 561–578.

Müller, I., Pedrazzini, T., Kropf, P., Louis, J. A. and Milon, G. (1991). Establishment of resistance to *Leishmania major* infection in susceptible BALB/c mice

requires parasite-specific CD8$^+$ T cells. *International Immunology* **3**, 587–597.
Munoz-Fernandez, M. A., Fernandez, M. A. and Fresno, M. (1992). Synergism between tumor necrosis factor-α and interferon-gamma on macrophage activation for the killing of intracellular *Trypanosoma cruzi* through a nitric oxide-dependent mechanism. *European Journal of Immunology* **22**, 301–307.
Murphy, D. B. (1987). The I-J puzzle. *Annual Review of Immunology* **5**, 405–427.
Murphy, D. B., Herzenberg, L. A., Okumura, K. and McDevitt, H. O. (1976). A new I subregion (I-J) marked by a locus (Ia-4) controlling surface determinants on suppressor T lymphocytes. *Journal of Experimental Medicine* **144**, 699–712.
Murray, H. W. (1981). Susceptibility of *Leishmania* to oxygen intermediates and killing by normal macrophages. *Journal of Experimental Medicine* **153**, 1302–1315.
Murray, H. W. (1982). Cell-mediated immune response in experimental visceral leishmaniasis. II. Oxygen-dependent killing of intracellular *Leishmania donovani* amastigotes. *Journal of Immunology* **129**, 351–357.
Murray, H. W. and Cartelli, D. M. (1983). Killing of intracellular *Leishmania donovani* by human mononuclear phagocytes. Evidence for oxygen-dependent and -independent leishmanicidal activity. *Journal of Clinical Investigation* **72**, 32–44.
Murray, H. W., Rubin, B. Y. and Rothermel, C. D. (1983). Killing of intracellular *Leishmania donovani* by lymphokine-stimulated human mononuclear phagocytes: evidence that interferon-gamma is the activating lymphokine. *Journal of Clinical Investigation* **72**, 1506–1510.
Murray, H. W., Stern, J. J., Welte, K., Rubin, B. Y., Carriero, S. M. and Nathan, C. F. (1987). Experimental visceral leishmaniasis: production of interleukin 2 and interferon-gamma, tissue immune reaction, and response to treatment with interleukin 2 and interferon-gamma. *Journal of Immunology* **138**, 2290–2297.
Murray, H. W., Squires, K. E., Miralles, C. D., Stoeckle, M. Y., Granger, A. M., Granelli-Piperno, A. and Bogdan, C. (1992). Acquired resistance and granuloma formation in experimental visceral leishmaniasis. Differential T cell and lymphokine roles in initial versus established immunity. *Journal of Immunology* **148**, 1858–1863.
Murray, P. J., Spithill, T. W. and Handman, E. (1989). The PSA-2 glycoprotein complex of *Leishmania major* is a glycosylphosphatidylinositol-linked promastigote surface antigen. *Journal of Immunology* **143**, 4221–4226.
Nacy, C. A., Meltzer, M. S., Leonard, E. J. and Wyler, D. J. (1981). Intracellular replication and lymphokine-induced destruction of *Leishmania tropica* in C3H/HeN mouse macrophages. *Journal of Immunology* **127**, 2381–2386.
Nacy, C. A., Fortier, A. H., Meltzer, M. S., Buchmeier, N. A. and Schreiber, R. D. (1985). Macrophage activation to kill *Leishmania major*: activation of macrophages for intracellular destruction of amastigotes can be induced by both recombinant interferon-gamma and non-interferon lymphokines. *Journal of Immunology* **135**, 3505–3511.
Nacy, C. A., Nelson, B. J., Meltzer, M. S. and Green, S. J. (1991). Cytokines that regulate macrophage production of nitrogen oxides and expression of anti-leishmanial activities. *Research in Immunology* **7**, 573–576.
Nathan, C. F. (1983). Mechanisms of macrophage antimicrobial activity. *Transactions of the Royal Society of Tropical Medicine and Hygiene* **77**, 620–630.
Nathan, C. F. and Hibbs, J. B., jr (1991). Role of nitric oxide synthesis in macrophage antimicrobial activity. *Current Opinion in Immunology* **3**, 65–70.

Nathan, C. F., Karnovsky, M. L. and David, J. R. (1971). Alterations of macrophage functions by mediators from lymphocytes. *Journal of Experimental Medicine* **133**, 1356–1376.

Nathan, C. F., Remold, H. G. and David, J. R. (1973). Characterization of a lymphocyte factor which alters macrophage functions. *Journal of Experimental Medicine* **137**, 275–290.

Nelson, B. J., Ralph, P., Green, S. J. and Nacy, C. A. (1991). Differential susceptibility of activated macrophage cytotoxic effector reactions to the suppressive effects of transforming growth factor-β1. *Journal of Immunology* **146**, 1849–1857.

Neva, F. A., Wyler, D. and Nash, T. (1979). Cutaneous leishmaniasis—a case with persistent organisms after treatment in presence of normal immune response. *American Journal of Tropical Medicine and Hygiene* **28**, 467–471.

Nüssler, A., Drapier, J.-C., Renia, L., Pied, S., Miltgen, F., Gentilini, M. and Mazier, D. (1991). L-Arginine-dependent destruction of intrahepatic malaria parasites in response to tumor necrosis factor and/or interleukin 6 stimulation. *European Journal of Immunology* **21**, 227–230.

Ockenhous, C. F., Schulman, S. and Shear, H. L. (1984). Induction of crisis forms in the human malaria parasite *Plasmodium falciparum* by gamma-interferon-activated, monocyte-derived macrophages. *Journal of Immunology* **133**, 1601–1608.

Olobo, J. O., Handman, E., Curtis, J. M. and Mitchell, G. F. (1980). Antibodies to *Leishmania tropica* promastigotes during infection in mice of various genotypes. *Australian Journal of Experimental Biology and Medical Science* **58**, 595–601.

Paliard, X., De Waal Malefyt, R., Yssel, H., Blanchard, D., Chretien, I., Abrams, J., De Vries, J. and Spits, H. (1988). Simultaneous production of IL-2, IL-4, and IFN-gamma by activated human $CD4^+$ and $CD8^+$ T cell clones. *Journal of Immunology* **141**, 849–855.

Pampiglione, S., La Placa, M. and Schlick, G. (1974). Studies on Mediterranean leishmaniasis. I. An outbreak of visceral leishmaniasis in northern Italy. *Transactions of the Royal Society of Tropical Medicine and Hygiene* **68**, 349–359.

Paraense, W. L. (1953). The spread of *Leishmania enriettii* through the body of the guinea pig. *Transactions of the Royal Society of Tropical Medicine and Hygiene* **47**, 556–560.

Parish, C. R. (1971). Immune response to chemically modified flagellin. II. Evidence for a fundamental relationship between humoral and cell-mediated immunity. *Journal of Experimental Medicine* **134**, 21–47.

Parish, C. R. and Liew, F. Y. (1972). Immune response to chemically modified flagellin. III. Enhanced cell-mediated immunity during high and low zone antibody tolerance to flagellin. *Journal of Experimental Medicine* **135**, 298–311.

Parronchi, P., Macchia, D., Piccinni, M.-P., Biswas, P., Simonelli, C., Maggi, E., Ricci, M., Ansari, A. A. and Romagnani, S. (1991). Allergen- and bacterial antigen-specific T-cell clones established from atopic donors show a different profile of cytokine production. *Proceedings of the National Academy of Sciences of the USA* **88**, 4538–4542.

Patel, P. J. (1981). Aging and antimicrobial immunity. Impaired production of mediator T cells as a basis for the decreased resistance of senescent mice to listeriosis. *Journal of Experimental Medicine* **154**, 821–831.

Pearce, E. J., Caspar, P., Grzych, J.-M., Lewis, F. A. and Sher, A. (1991). Downregulation of Thl cytokine production accompanies induction of Th2 responses by a parasitic helminth, *Schistosoma mansoni*. *Journal of Experimental Medicine* **173**, 159–166.

Pearson, R. D. and Steigbigel, R. T. (1980). Mechanism of lethal effect of human serum upon *Leishmania donovani*. *Journal of Immunology* **125**, 2195–2201.

Pearson, R. D., Harcus, J. L., Symes, P. H., Romito, R. and Donowitz, G. R. (1982). Failure of the phagocytic oxidative response to protect human monocyte-derived macrophages from infection by *Leishmania donovani*. *Journal of Immunology* **129**, 1282–1286.

Pearson, R. D., Harais, J. L., Roberts, D. and Donowitz, G. R. (1983). Differential survival of *Leishmania donovani* amastigotes in human monocytes. *Journal of Immunology* **131**, 1994–1999.

Peleman, R., Wu, J., Fargeas, C. and Delespesse, G. (1989). Recombinant interleukin 4 suppresses the production of interferon gamma by human mononuclear cells. *Journal of Experimental Medicine* **170**, 1751–1756.

Pellat, C., Henry, Y. and Drapier, J.-C. (1990). IFN-gamma activated macrophages: detection by electron paramagnetic resonance of complexes between L-arginine-derived nitric oxide and non-heme iron proteins. *Biochemical and Biophysical Research Communications* **166**, 119–125.

Peters, B. S., Fish, D., Golden, R., Evans, D. A., Bryceson, A. D. M. and Pinching, A. J. (1990). Visceral leishmaniasis in HIV infection and AIDS: clinical features and response to therapy. *Quarterly Journal of Medicine* **77**, 1101–1111.

Petersen, E. A., Neva, F. A., Barral, A., Correa-Coronas, R., Bogaert-Diaz, H., Martinez, D. and Ward, F. E. (1984). Monocyte suppression of antigen-specific lymphocyte responses in diffuse cutaneous leishmaniasis patients from the Dominican Republic. *Journal of Immunology* **132,** 2603–2606.

Petros, A., Bennett, D. and Vallance, P. (1991). Effect of nitric oxide synthase inhibitors on hypotension in patients with septic shock. *Lancet* **338,** 1557–1558.

Pfeilschifter, J. and Vosbeck, K. (1991). Transforming growth factor β_2 inhibits interkeukin 1β- and tumour necrosis factor α-induction of nitric oxide synthase in rat renal mesangial cells. *Biochemical and Biophysical Research Communications* **175,** 372–379.

Pham, T. and Mauël, J. (1987). Studies on intracellular killing of *Leishmania major* and lysis of host macrophages by immune lymphoid cells *in vitro*. *Parasite Immunology* **9,** 721–736.

Pinching, A. J. (1988). Factors affecting the natural history of human immunodeficiency virus infection. *Immunodeficiency Review* **1,** 23–38.

Plant, J. E. and Glynn, A. A. (1979). Locating *Salmonella* resistance gene on mouse chromosome 1. *Clinical and Experimental Immunology* **37,** 1–6.

Plant, J. E., Blackwell, J. M., O'Brien, A. D., Bradley, D. J. and Glynn, A. A. (1982). Are the *Lsh* and *Ity* disease resistant genes at one locus on mouse chromosome 1? *Nature* **297,** 570–571.

Plata, F., Wietzerbin, F., Pons, F. G., Falcoff, E. and Eisen, H. (1984). Synergistic protection by specific antibodies and interferon against infection by *Trypanosoma cruzi in vitro*. *European Journal of Immunology* **14,** 930.

Playfair, J. H. (1990). Non-specific killing mechanisms effective against blood stage malaria parasites. *Immunology Letters* **25,** 173.

Poirier, T. P., Kehoe, M. A. and Beachey, E. H. (1988). Protective immunity evoked by oral administration of attenuated *aroA Salmonella typhimurium* expressing cloned streptococcal M protein. *Journal of Experimental Medicine* **168,** 25–32.

Pond, L., Wassom, D. L. and Hayes, C. E. (1989). Evidence for differential induction of helper T cell subsets during *Trichinella spiralis* infection. *Journal of Immunology* **143,** 4232–4237.

Powers, G. D., Abbas, A. K. and Miller, R. A. (1988). Frequencies of IL-2 and IL-4-

secreting T cells in naive and antigen-stimulated lymphocyte populations. *Journal of Immunology* **140,** 3352–3357.
Pozzi, L. M. and Weiser, W. Y. (1991). Recombinant human macrophage migration inhibitory factor activates human monocyte derived macrophages to kill tumor cells. *FASEB Journal* **5,** A1092.
Preston, P. M. and Dumonde, D. C. (1976a). Experimental cutaneous leishmaniasis. V. Protective immunity in subclinical and self-healing infection in the mouse. *Clinical and Experimental Immunology* **23,** 126–138.
Preston, P. M. and Dumonde, D. C. (1976b). Immunology of clinical and experimental leishmaniasis. *In* "Immunology of Parasitic Infections" (S. Cohen and E. H. Sadun, eds), pp. 167–202. Blackwell Scientific Publications, Oxford.
Preston, P. M., Carter, R. L., Leuchars, E., Davies, A. J. S. and Dumonde, D. C. (1972). Experimental cutaneous leishmaniasis. III. Effects of thymectomy on the course of infection of CBA mice with *Leishmania tropica. Clinical and Experimental Immunology* **10,** 337–357.
Preston, P. M., Behbehani, K. and Dumonde, D. C. (1978). Experimental cutaneous leishmaniasis. VI. Anergy and allergy in the cellular immune response during non-healing infection in different strains of mice. *Journal of Clinical and Laboratory Immunology* **1,** 207–219.
Prystowsky, M. B., Ely, J. M., Beller, D. I., Eisenberg, L., Goldman, J., Goldman, M., Goldwasser, E., Ihle, J., Quintans, J., Remold, H., Vogel, S. N. and Fitch, F. W. (1982). Alloreactive cloned T cell lines. VI. Multiple lymphokine activities secreted by helper and cytolytic cloned T lymphocytes. *Journal of Immunology* **129,** 2337–2344.
Radomski, M. W., Palmer, R. M. J. and Moncada, S. (1990). An L-arginine/nitric oxide pathway present in human platelets regulates aggregation. *Proceedings of the National Academy of Sciences of the USA* **87,** 5193–5197.
Ramshaw, I. A., Bretscher, P. A. and Parish, C. R. (1976). Regulation of the immune response. I. Suppression of delayed-type hypersensitivity by T cells from mice expressing humoral immunity. *European Journal of Immunology* **6,** 674–679.
Raziuddin, S., Telmasani, A. W., El-Awad, M. E.-H., Al-Amari, O. and Al-Janadi, M. (1992). Gamma-γ T cells and the immune response in visceral leishmaniasis. *European Journal of Immunology* **22,** 1143–1148.
Reed, S. G. (1988). *In vivo* administration of recombinant IFN-gamma induces macrophage activation, and prevents acute disease, immune suppression, and death in experimental *Trypanosoma cruzi* infections. *Journal of Immunology* **140,** 4342–4347.
Reed, S. G., Badaro, R. and Lloyd, R. M. C. (1987a). Identification of specific and cross-reactive antigens of *Leishmania donovani chagasi* by human infection sera. *Journal of Immunology* **138,** 1596–1601.
Reed, S. G., Nathan, C. F., Pihl, D. L., Rodricks, P., Shanebeck, K., Conlon, P. J. and Grabstein, K. H. (1987b). Recombinant granulocyte/macrophage colony-stimulating factor activates macrophages to inhibit *Trypanosoma cruzi* and release hydrogen peroxide. Comparison to interferon-gamma. *Journal of Experimental Medicine* **166,** 1734.
Reed, S. G., Carvalho, E. M., Sherbert, C. H., Sampaio, D. P., Russo, D. M., Bacelar, O., Pihl, D. L., Scott, J. M., Barral, A., Grabstein, K. H. and Johnson, W. D., jr (1990a). *In vitro* responses to *Leishmania* antigens by lymphocytes from patients with leishmaniasis or Chagas' disease. *Journal of Clinical Investigation* **85,** 690–696.

Reed, S. G., Shreffler, W. G., Burns, J. M., jr, Scott, J. M., Da Gloria Orge, M., Ghalib, H. W., Siddig, M. and Badaro, R. (1990b). An improved serodiagnostic procedure for visceral leishmaniasis. *American Journal of Tropical Medicine and Hygiene* **43**, 632–639.
Rees, P. H. and Kager, P. A. (1987). Visceral leishmaniasis and post-kala-azar dermal leishmaniasis. *In* "The Leishmaniases in Biology and Medicine" (W. Peters and R. Killick-Kendrick, eds), Vol. 2, pp. 583–615. Academic Press, London.
Reiner, N. E. (1987). Parasite accessory cell interactions in murine leishmaniasis. I. Evasion and stimulus-dependent suppression of the macrophage interleukin-1 response by *Leishmania donovani*. *Journal of Immunology* **138**, 1919–1925.
Reiner, N. E. and Finke, J. H. (1983). Interleukin 2 deficiency in murine *Leishmania donovani* and its relationship to depressed spleen cell responses to phytohaemagglutinin. *Journal of Immunology* **131**, 1487–1491.
Reiner, N. E., Ng, W., Wilson, C. B., McMaster, W. R. and Burchett, S. K. (1990). Modulation of *in vitro* monocyte cytokine responses to *Leishmania donovani*. Interferon-gamma prevents parasite-induced inhibition of interleukin 1 production and primes monocytes to respond to *Leishmania* by producing both tumor necrosis factor-α and interleukin 1. *Journal of Clinical Investigation* **85**, 1914–1924.
Rezai, H. R., Ardekali, S. M., Amirhakimi, G. and Kharazmi, A. (1978). Immunological features of kala-azar. *American Journal of Tropical Medicine and Hygiene* **27**, 1079–1083.
Rezai, H. R., Farrell, J. and Soulsby, E. J. L. (1980). Immunological responses of *L. donovani* infection in mice and significance of T cell in resistance to experimental leishmaniasis. *Clinical and Experimental Immunology* **40**, 508–514.
Ridel, P.-R., Esterre, P., Dedet, J.-P., Pradinaud, R., Santoro, F. and Capron, A. (1988). Killer cells in human cutaneous leishmaniasis. *Transactions of the Royal Society of Tropical Medicine and Hygiene* **82**, 223–226.
Ridley, D. S., Marsden, P. D., Cuba, C. C. and Barreto, A. C. (1980). A histological classification of mucocutaneous leishmaniasis in Brazil and its clinical evaluation. *Transactions of the Royal Society of Tropical Medicine and Hygiene* **74**, 508–514.
Roach, T. I. A., Wakelin, D., Else, K. J. and Bundy, D. A. P. (1988). Antigenic cross-reactivity between the human whipworm, *Trichuris trichiura*, and the mouse trichuroids *Trichuris muris* and *Trichinella spiralis*. *Parasite Immunology* **10**, 279–291.
Roach, T. I. A., Kiderlen, A. F. and Blackwell, J. M. (1991). Role of inorganic nitrogen oxides and tumor necrosis factor alpha in killing *Leishmania donovani* amastigotes in gamma interferon-lipopolysaccharide-activated macrophages from Lsh^s and Lsh^r congenic mouse strains. *Infection and Immunity* **59**, 3935–3944.
Roberts, M., Alexander, J. and Blackwell, J. M. (1990). Genetic analysis of *Leishmania mexicana* infection in mice: single gene (*Scl-2*) controlled predisposition to cutaneous lesion development. *Journal of Immunogenetics* **17**, 89–100.
Rodgers, M. R., Popper, S. J. and Wirth, D. F. (1990). Amplification of kinetoplast DNA as a tool in the detection and diagnosis of *Leishmania*. *Experimental Parasitology* **71**, 267–275.
Rodrigues, M. M., Xavier, M. T., Previato, L. M. and Barcinski, M. A. (1986). Characterization of cellular immune response to chemically defined glycoconjugates from *Leishmania mexicana* subsp. *amazonensis*. *Infection and Immunity* **51**, 80–86.

Rodrigues, M. M., Mendonca-Previato, L., Charlab, R. and Barcinski, M. A. (1987). The cellular immune response to a purified antigen from *Leishmania mexicana* subsp. *amazonensis* enhances the size of the leishmanial lesion on susceptible mice. *Infection and Immunity* **55**, 3142–3148.

Rodrigues, M. M., Xavier, M. T., Mendonca-Previato, L. and Barcinski, M. A. (1988). Novel 17-kilodalton *Leishmania* antigen revealed by immunochemical studies of a purified glycoprotein fraction recognized by murine T lymphocytes. *Infection and Immunity* **56**, 1766–1770.

Roffi, J., Dedet, J.-P., Desjeux, P. and Garré, M.-T. (1980). Detection of circulating antibodies in cutaneous leishmaniasis by enzyme-linked immunosorbent assay (ELISA). *American Journal of Tropical Medicine and Hygiene* **29**, 183–189.

Rogers, W. O., Burnheim, P. F. and Wirth, D. F. (1988). Detection of *Leishmania* within sandflies by kinetoplast DNA hybridization. *American Journal of Tropical Medicine and Hygiene* **39**, 434–439.

Romagnani, S. (1990). Regulation and deregulation of human IgE synthesis. *Immunology Today* **11**, 316–321.

Romagnani, S. (1991). Human Th1 and Th2 subsets: doubt no more. *Immunology Today* **12**, 256–257.

Rose, N. R., Kong, Y.-C. M., Okayasu, I., Giraldo, A. A., Beisel, K. and Sundick, R. S. (1981). T-cell regulation in autoimmune thyroiditis. *Immunological Reviews* **55**, 299–314.

Rosen, G., Londner, M. V., Greenblatt, C. L., Morsey, T. A. and El-On, J. (1986). *Leishmania major*: solid phase radioimmunoassay for antibody detection in human cutaneous leishmaniasis. *Experimental Parasitology* **62**, 79–84.

Rosen, G., Londner, M. V., Sevlever, D. and Greenblatt, C. L. (1988). *Leishmania major*: glycolipid antigens recognized by immune human sera. *Molecular and Biochemical Parasitology* **27**, 93–100.

Russell, D. G. and Alexander, J. (1988). Effective immunization against cutaneous leishmaniasis with defined membrane-antigens reconstituted into liposomes. *Journal of Immunology* **140**, 1274–1279.

Russell, D. G. and Wilhelm, H. (1986). The involvement of the major surface glycoprotein (gp63) of *Leishmania* promastigotes in attachment to macrophages. *Journal of Immunology* **136**, 2613–2620.

Russo, D. M., Burns, J. M., jr, Carvalho, E. M., Armitage, R. J., Grabstein, K. H., Button, L. L., McMaster, W. R. and Reed, S. G. (1991). Human T cell responses to gp63, a surface antigen of *Leishmania*. *Journal of Immunology* **147**, 3575–3580.

Russo, D. M., Turco, S. J., Burns, J. M., jr and Reed, S. G. (1992). Stimulation of human T lymphocytes by *Leishmania* lipophosphoglycan-associated proteins. *Journal of Immunology* **148**, 202–207.

Sacks, D. L., Scott, P. A., Asofsky, R. and Sher, A. (1984). Cutaneous leishmaniasis in anti-IgM-treated mice: enhanced resistance due to functional depletion of a B cell-dependent T cell involved in the suppressor pathway. *Journal of Immunology* **132**, 2072–2077.

Sacks, D. L., Lat, S. L., Shrivastava, S. N., Blackwell, J. and Neva, F. A. (1987). An analysis of T cell responsiveness in Indian kala-azar. *Journal of Immunology* **138**, 908–913.

Sacks, D. L., Louis, J. A. and Wirth, D. F. (1993). Immunology and molecular biology of leishmaniasis. *In* "Immunology of Parasitic Infections" (K. Warren, ed.), Blackwell Scientific Publications, Oxford, in press.

Sadick, M. D., Locksley, R. M., Tubbs, C. and Raff, H. V. (1986). Murine cutaneous leishmaniasis: resistance correlates with the capacity to generate interferon-gamma in response to *Leishmania* antigens *in vitro*. *Journal of Immunology* **136**, 655–661.

Sadick, M. D., Heinzel, F. P., Shigekane, V. M., Fisher, W. L. and Locksley, R. M. (1987). Cellular and humoral immunity to *Leishmania major* in genetically susceptible mice after *in vitro* depletion of L3T4$^+$ T cells. *Journal of Immunology* **139**, 1303–1309.

Sadick, M. D., Heinzel, F. P., Holaday, B. J., Pu, R. T., Dawkins, R. S. and Locksley, R. M. (1990). Cure of murine leishmaniasis with anti-interleukin 4 monoclonal antibody. Evidence for a T cell-dependent, interferon gamma-independent mechanism. *Journal of Experimental Medicine* **171**, 115–127.

Sadick, M. D., Street, N., Mosmann, T. R. and Locksley, R. M. (1991). Cytokine regulation of murine leishmaniasis: interleukin 4 is not sufficient to mediate progressive disease in resistant C57BL/6 mice. *Infection and Immunity* **59**, 4710–4714.

Sadoff, J. C., Ballou, W. R., Baron, L. S., Majarian, W. R., Brey, R. N., Hockmeyer, W. T., Young, J. F., Cryz, J. J., Ou, J., Lowell, G. H. and Chulay, J. D. (1988). Oral *Salmonella typhimurium* vaccine expressing circumsporozoite protein protects against malaria. *Science* **240**, 336–338.

Salerno, J. C., Ohnishi, T., Lim, J. and King, T. E. (1976). Tetranuclear and binuclear iron-sulfur clusters in succinate dehydrogenase: a method of iron quantitation by formation of paramagnetic complexes. *Biochemical and Biophysical Research Communications* **73**, 833–840.

Salgame, P., Abrams, J. S., Clayberger, C., Goldstein, H., Convit, J., Modlin, R. L. and Bloom, B. R. (1991). Differing lymphokine profiles of functional subsets of human CD4 and CD8 T cell clones. *Science* **254**, 279–282.

Salvemini, D., Masini, E., Anggard, E., Mannaioni, P. F. and Vane, J. (1990). Synthesis of a nitric oxide-like factor from L-arginine by rat serosal mast cells: stimulation of guanylate cyclase and inhibition of platelet aggregation. *Biochemical and Biophysical Research Communications* **169**, 596–601.

Savelkoul, G. F. J., Lebman, D. A., Brenner, R. and Coffman, R. L. (1988). Increase of precursor frequency and clonal size of murine IgE-secreting cells by IL-4. *Journal of Immunology* **141**, 749–755.

Schmidt, H. H. H. W., Seifert, R. and Bohme, E. (1989). Formation and release of nitric oxide from human neutrophils and HL-60 cells induced by a chemotactic peptide, platelet activating factor and leukotriene B_4. *FEBS Letters* **244**, 357–360.

Schnur, L. F., Morsey, T. A., Feinsod, F. M. and El Missiry, A. G. (1985). Is *Leishmania major* the cause of infantile kala-azar in Alexandria, Egypt? *Transactions of the Royal Society of Tropical Medicine and Hygiene* **79**, 134–135.

Scott, P. (1990). T-cell subsets and T-cell antigens in protective immunity against experimental leishmaniasis. *Current Topics in Microbiology and Immunology* **155**, 35–52.

Scott, P. (1991a). Host and parasite factors regulating the development of CD4$^+$ subsets in experimental cutaneous leishmaniasis. *Research in Immunology* **142**, 32–36.

Scott, P. (1991b). IFN-gamma modulates the early development of Th1 and Th2 responses in a murine model of cutaneous leishmaniasis. *Journal of Immunology* **147**, 3149–3155.

Scott, P. A. and Farrell, J. P. (1981). Experimental cutaneous leishmaniasis. I. Nonspecific immunosuppression in BALB/c mice infected with *Leishmania tropica*. *Journal of Immunology* **127**, 2395–2400.
Scott, P., James, S. and Sher, A. (1985). The respiratory burst is not required for killing of intracellular and extracellular targets by a lymphokine-activated macrophage cell line. *European Journal of Immunology* **15**, 553–558.
Scott, P., Natovitz, P. and Sher, A. (1986). B lymphocytes are required for the generation of T cells that mediate healing of cutaneous leishmaniasis. *Journal of Immunology* **137**, 1017–1021.
Scott, P., Pearce, E., Natovitz, P. and Sher, A. (1987). Vaccination against cutaneous leishmaniasis in a murine model. I. Induction of protective immunity with a soluble extract of promastigotes. *Journal of Immunology* **139**, 221–227.
Scott, P., Natovitz, P., Coffman, R. L., Pearce, E. and Sher, A. (1988). Immunoregulation of cutaneous leishmaniasis. T cell lines that transfer protective immunity or exacerbation belong to different T helper subsets and respond to distinct parasite antigens. *Journal of Experimental Medicine* **168**, 1675–1684.
Severn, A., Wakelam, M.J.O. and Liew, F.Y. (1992). The role of protein kinase C in the induction of nitric oxide synthesis by murine macrophages. *Biochemical and Biophysical Research Communications* **188**, 997–1002.
Sher, A. and Coffman, R. L. (1992). Regulation of immunity to parasites by T cells and T cell-derived cytokines. *Annual Review of Immunology* **10**, 385–409.
Sher, A., Coffmann, R. L., Hieny, S. and Cheever, A. W. (1990). Ablation of eosinophil and IgE responses with anti-IL-5 or anti-IL-4 antibodies fails to affect immunity against *Schistosoma mansoni* in the mouse. *Journal of Immunology* **145**, 3911–3916.
Sher, A., Fiorentino, D. F., Caspar, P., Pearce, E. and Mosmann, T. (1991). Production of IL-10 by $CD4^+$ T lymphocytes correlates with down-regulation of Th1 cytokine synthesis in helminth infection. *Journal of Immunology* **147**, 2713–2716.
Silva, J. S., Twardzik, D. R. and Reed, S. G. (1991). Regulation of *Trypanosoma cruzi* infections *in vitro* and *in vivo* by transforming growth factor β (TGF-β). *Journal of Experimental Medicine* **174**, 539–545.
Silva, J. S., Morrissey, P. J., Grabstein, K. H., Mohler, K. M., Anderson, D. and Reed, S. G. (1992). Interleukin 10 and interferon gamma regulation of experimental *Trypanosoma cruzi* infection. *Journal of Experimental Medicine* **175**, 169–174.
Silver, J. and Benacerraf, B. (1974). Dissociation of T cell helper function and delayed hypersensitivity. *Journal of Immunology* **113**, 1872–1875.
Skamene, E., Gros, P., Forget, A., Kongshavn, P. A. L., St Charles, C. and Taylor, B. A. (1982). Genetic regulation of resistance to intracellular pathogens. *Nature* **297**, 506–509.
Skov, C. B. and Twohy, D. W. (1974a). Cellular immunity to *L. donovani* I. The effect of T cell depletion on resistance to *L. donovani* in mice. *Journal of Immunology* **113**, 2004–2011.
Skov, C. B. and Twohy, D. W. (1974b). Cellular immunity to *Leishmania donovani*. II. Evidence for synergy between thymocytes and lymph node cells in reconstitution of acquired resistance to *L. donovani* in mice. *Journal of Immunology* **113**, 2012–2019.
Smith, D. F., Searle, S., Ready, P. D., Gramiccia, M. and Ben, I. R. (1989). A kinetoplast DNA probe diagnostic for *Leishmania major*: sequence homologies between regions of *Leishmania* minicircles. *Molecular and Biochemical Parasitology* **37**, 213–223.

Smith, L. E., Rodrigues, M. and Russell, D. G. (1991). Cytotoxic T cells and *Leishmania* infected macrophages. *Journal of Experimental Medicine* **174**, 499–506.

Snapper, C. M. and Paul, W. E. (1987). Interferon-gamma and B cell stimulatory factor-1 reciprocally regulate Ig isotype production. *Science* **236**, 944–947.

Soares, L. R. B. and Barcinski, M. A. (1992). Differential production of granulocyte-macrophage colony-stimulating factor by macrophages from mice susceptible and resistant to *Leishmania mexicana amazonensis*. *Journal of Leukocyte Biology* **51**, 220–224.

Solbach, W., Forberg, K., Kammerer, E., Bogdan, C. and Röllinghoff, M. (1986). Suppressive effect of cyclosporine A on the development of *Leishmania tropica*-induced lesions in genetically susceptible BALB/c mice. *Journal of Immunology* **137**, 702–707.

Solbach, W., Greil, J. and Röllinghoff, M. (1987). Anti-infectious responses in *Leishmania major*-infected BALB/c mice injected with recombinant granulocyte-macrophage colony-stimulating factor. *Annales de l'Institut Pasteur–Immunologie* **138**, 759–762.

Stavnezer, J., Radcliffe, G., Lin, Y.-C., Nietupski, J., Berggren, L., Sitia, R. and Severinson, E. (1988). Immunoglobulin heavy-chain switching may be directed by prior induction of transcripts from constant-region genes. *Proceedings of the National Academy of Sciences of the USA* **85**, 7704–7708.

Stenger, S., Solbach, W., Röllinghoff, M. and Bogdan, C. (1991). Cytokine interactions in experimental cutaneous leishmaniasis. II. Endogenous tumor necrosis factor-α production by macrophages is induced by the synergistic action of interferon (IFN)-gamma and interleukin (IL) 4 and accounts for the antiparasitic effect mediated by IFN-gamma and IL 4. *European Journal of Immunology* **21**, 1669–1675.

Stern, J. J., Oca, M. J., Rubin, B. Y., Anderson, S. L. and Murray, H. W. (1988). Role of L3T4$^+$ and Lyt-2$^+$ cells in experimental visceral leishmaniasis. *Journal of Immunology* **140**, 3971–3977.

Stevens, T. L., Bossie, A., Sanders, V. M., Fernandez-Botran, R., Coffman, R. L., Mosmann, T. R. and Vitetta, E. S. (1988). Regulation of antibody isotype secretion by subsets of antigen-specific helper T cells. *Nature* **334**, 255–258.

Stover, C. K., de la Cruz, V. F., Fuerst, T. R., Burlein, J. E., Benson, L. A., Bennett, L. T., Bansal, G. P., Young, J. F., Lee, M. H., Hatfull, G. F., Snapper, S. B., Barletta, R. G., Jacobs, jr, W. R. and Bloom, B. R. (1991). New use of BGG for recombinant vaccines. *Nature* **351**, 456–460.

Street, N. E., Schumacher, J. H., Fong, T. A. T., Bass, H., Fiorentino, D. F., Leverah, J. A. and Mosmann, T. R. (1990). Heterogeneity of mouse helper T cells. Evidence from bulk cultures and limiting dilution cloning for precursors of Th1 and Th2 cells. *Journal of Immunology* **144**, 1629–1639.

Stuehr, D. J. and Marletta, M. A. (1985). Mammalian nitrate biosynthesis: mouse macrophages produce nitrite and nitrate in response to *Escherichia coli* lipopolysaccharide. *Proceedings of the National Academy of Sciences of the USA* **82**, 7738–7742.

Stuehr, D. J. and Marletta, M. A. (1987a). Induction of nitrite/nitrate synthesis in murine macrophages by BCG infection, lymphokines, or interferon-gamma. *Journal of Immunology* **139**, 518–525.

Stuehr, D. J. and Marletta, M. A. (1987b). Synthesis of nitrite and nitrate in murine macrophage cell lines. *Cancer Research* **47**, 5590–5594.

Stuehr, D. J. and Nathan, C. F. (1989). Nitric oxide: a macrophage product responsible for cytostasis and respiratory inhibition in tumor target cells. *Journal of Experimental Medicine* **169**, 1543–1545.

Stuehr, D. J., Gross, S., Sakuma, I., Levi, R. and Nathan, C. F. (1989). Activated murine macrophages secrete a metabolite of arginine with the bioactivity of endothelium-derived relaxing factor and the chemical reactivity of nitric oxide. *Journal of Experimental Medicine* **169**, 1011–1020.

Stuehr, D. J., Kwon, N. S. and Nathan, C. F. (1990). FAD and GSH participate in macrophage synthesis of nitric oxide. *Biochemical and Biophysical Research Communications* **168**, 558–565.

Suzuki, Y. and Remington, J. S. (1988). Dual regulation of resistance against *Toxoplasma gondii* infection by Lyt-2$^+$ and L3T4$^+$ T cells in mice. *Science* **240**, 516–519.

Swain, S. L., Weinberg, A. D., English, M. and Huston, G. (1990). IL-4 directs the development of Th2-like helper effectors. *Journal of Immunology* **145**, 3796–3806.

Sypek, J. P. and Wyler, D. J. (1991). Antileishmanial defense in macrophages triggered by tumor necrosis factor expressed on CD4$^+$ T lymphocytes plasma membrane. *Journal of Experimental Medicine* **174**, 755–759.

Szuro-Sudol, A., Murray, H. W. and Nathan, C. F. (1983). Suppression of macrophage antimicrobial activity by a tumor cell product. *Journal of Immunology* **131**, 384–397.

Tada, T. and Okumura, K. (1979). The role of antigen-specific T cell factors in the immune response. *Advances in Immunology* **28**, 1–87.

Tada, T., Taniguchi, M. and David, C. S. (1976). Properties of the antigen-specific suppressive T-cell factor in the regulation of antibody response of the mouse. IV. Special subregion assignment of the gene(s) that codes for the suppressive T-cell factor in the H-2 histocompatibility complex. *Journal of Experimental Medicine* **144**, 713–725.

Thompson-Snipes, L., Dhar, V., Bond, M. W., Mosmann, T. R., Moore, K. W. and Rennick, D. (1991). Interleukin-10: a novel stimulatory factor for mast cells and their progenitors. *Journal of Experimental Medicine* **173**, 507–512.

Titus, R. G., Lima, G. C., Engers, H. D. and Louis, J. A. (1984a). Exacerbation of murine cutaneous leishmaniasis by adoptive transfer of parasite-specific helper T cell population capable of mediating *Leishmania major* specific delayed type hypersensitivity. *Journal of Immunology* **133**, 1594–1600.

Titus, R. G., Kelso, A. and Louis, J. A. (1984b). Intracellular destruction of *Leishmania tropica* by macrophages activated with macrophage activating factor/interferon. *Clinical and Experimental Immunology* **55**, 157–165.

Titus, R. G., Ceredig, R., Cerottini, J. C. and Louis, J. A. (1985a). Therapeutic effect of anti-L3T4 monoclonal antibody GK1.5 on cutaneous leishmaniasis in genetically susceptible BALB/c mice. *Journal of Immunology* **135**, 2108–2114.

Titus, R. G., Marchand, M., Boon, T. and Louis, J. A. (1985b). A limiting dilution assay for quantifying *Leishmania major* in tissues of infected mice. *Parasite Immunology* **7**, 545–555.

Titus, R. G., Milon, G., Marchal, G., Vassalli, P., Cerottini, J.-C. and Louis, J. A. (1987). Involvement of specific Lyt-2$^+$ T cells in the immunological control of experimentally induced murine cutaneous leishmaniasis. *European Journal of Immunology* **17**, 1429–1433.

Titus, R. G., Kimsey, P., Theodos, C. and Louis, J. A. (1988). Induction of resistance to experimental cutaneous leishmaniasis in genetically-susceptible BALB/c mice by immunization with chemically-mutagenized non-infective clones of *Leishmania major*. *FASEB Journal* **2**, A887.

Titus, R. G., Sherry, B. and Cerami, A. (1989). Tumor necrosis factor plays a protective role in experimental murine cutaneous leishmaniasis. *Journal of Experimental Medicine* **170**, 2097–2104.

Titus, R. G., Müller, I., Kinsey, P., Cerny, A., Behin, R., Zinkernagel, R. M. and Louis, J. A. (1991). Exacerbation of experimental murine cutaneous leishmaniasis with $CD4^+$ *Leishmania major*-specific T cell lines or clones which secrete interferon-gamma and mediate parasite specific delayed-type hypersensitivity. *European Journal of Immunology* **21**, 559–567.

Titus, R. G., Theodos, C. M., Kimsey, P. B., Shankar, A., Hall, L., McGurn, M. and Povinelli, L. (1993). Role of T cells in immunity to the intracellular pathogen, *Leishmania major*. *In* "Subcellular Biochemistry", vol. 18 "*Intracellular Parasites*" (Avila and Harris (eds), p. 99–129. Plenum Press, New York.

Tremonti, L. and Walton, B. C. (1970). Blast transformation and migration-inhibition in toxoplasmosis and leishmaniasis. *American Journal of Tropical Medicine and Hygiene*, **19**, 49–56.

Turco, S. J. (1988). The lipophosphoglycan of *Leishmania*. *Parasitology Today* **4**, 255–257.

Turk, J. L. and Bryceson, A. D. M. (1971). Immunological phenomena in leprosy and related diseases. *Advances in Immunology* **13**, 209–266.

Ulczak, O. M. and Blackwell, J. M. (1983). Immunoregulation of genetically controlled acquired responses to *Leishmania donovani* infection in mice: the effects of parasite dose, cyclophosphamide and sublethal irradiation. *Parasite Immunology* **5**, 449–463.

Ulczak, O. M., Ghadirian, E., Skamene, E., Blackwell, J. M. and Kongshavn, P. A. L. (1989). Characterization of protective T cells in the acquired response to *Leishmania donovani* in genetically determined cure ($H\text{-}2^b$) and noncure ($H\text{-}2^d$) mouse strains. *Infection and Immunity* **57**, 2892–2899.

Umetsu, D. T., Jabara, H. H., DeKruyff, R. H., Abbas, A. K., Abrams, J. S. and Geha, R. S. (1988). Functional heterogeneity among human inducer T cell clones. *Journal of Immunology* **140**, 4211–4216.

Urban, J. F., jr, Katona, I. M., Paul, W. E. and Finkelman, F. D. (1991). Interleukin 4 is important in protective immunity to a gastrointestinal nematode infection in mice. *Proceedings of the National Academy of Sciences of the USA* **88**, 5513–5517.

Uyemura, K., Klotz, J., Pirmez, C., Ohmen, J., Wang, X.-H., Ho, C., Hoffman, W. L. and Modlin, R. L. (1992). Microanatomic clonality of gamma-γ T cells in human leishmaniasis lesions. *Journal of Immunology* **148**, 1205–1211.

Vallance, P. and Moncada, S. (1991). Hyperdynamic circulation in cirrhosis: a role for nitric oxide? *Lancet* **337**, 776–778.

Van, E. G. J., Schoone, G. J., Ligthart, G. S., Laarman, J. J. and Terpstra, W. J. (1987). Detection of *Leishmania* parasites by DNA *in situ* hybridization with non-radioactive probes. *Parasitology Research* **73**, 199–202.

Varkila, K., Chatelain, R., Leal, L.M.C.C. and Coffman, R.L. (1993). Reconstitution of C.B-17 *scid* mice with BALB/c T cells initiates a T helper type-1 response and renders them capable of healing *Leishmania major* infection. *European Journal of Immunology* **23**, 262–268.

Vincendeau, P. and Daulouede, S. (1991). Macrophage cytostatic effect on *Trypanosoma musculi* involves an L-arginine-dependent mechanism. *Journal of Immunology* **146**, 4338–4343.

Wallis, A. E. and McMaster, W. R. (1987). Identification of *Leishmania* genes encoding proteins containing tandemly repeating peptides. *Journal of Experimental Medicine* **166**, 724–729.

Walton, B. C. (1987). American cutaneous and mucocutaneous leishmaniasis. *In* "The Leishmaniases in Biology and Medicine" (W. Peters and R. Killick-Kendrick, eds), Vol. 2, pp. 637–664. Academic Press, London.

Walton, B. C. and Valverde, L. (1979). Racial differences in espundia. *Annals of Tropical Medicine and Parasitology* **73**, 23–29.

Walton, B. C., Brooks, W. H. and Arjona, I. (1972). Serodiagnosis of American leishmaniasis by indirect fluorescent antibody test. *American Journal of Tropical Medicine and Hygiene* **21**, 296–299.

Weinberg, A. D., English, M. and Swain, S. L. (1990). Distinct regulation of lymphokine production is found in fresh versus *in vitro* primed murine helper T cells. *Journal of Immunology* **144**, 1800–1807.

Weinberg, J. B., Chapman, H. A., jr and Hibbs, J. B., jr (1978). Characterization of the effects of endotoxin on macrophage tumor cell killing. *Journal of Immunology* **121**, 72–80.

Weiser, W. Y., van Niel, A., Clark, S. C., David, J. R. and Remold, H. G. (1987). Recombinant human granulocyte/macrophage colony-stimulating factor activates intracellular killing of *Leishmania donovani* by human monocyte-derived macrophages. *Journal of Experimental Medicine* **166**, 1436–1446.

Weiser, W. Y., Pozzi, L.-A. M. and David, J. R. (1991). Human recombinant migration inhibitory factor activates human macrophages to kill *Leishmania donovani*. *Journal of Immunology* **147**, 2006–2011.

Wenyon, C. M. (1911). Oriental sore in Bagdad [*sic*], together with observations on a gregarine in *Stegomyia fasciata*, the haemogregarine of dogs and the flagellates of house flies. *Parasitology* **4**, 273–344.

Werner-Felmayer, G., Werner, E. R., Fuchs, D., Hausen, A., Reibnegger, G. and Wachter, H. (1991). On multiple forms of NO synthase and their occurrence in human cells. *Research in Immunology* **7**, 555–561.

Wierenga, E. A., Snoek, M., De Groot, C., Chretien, I., Bos, J. D., Jansen, H. M. and Kapsenberg, M. L. (1990). Evidence for compartmentalization of functional subsets of $CD4^+$ T lymphocytes in atopic patients. *Journal of Immunology* **144**, 4651–4656.

Winther, M. D. and Dougan, G. (1984). The impact of new technologies on vaccine development. *Biotechnology and Genetic Engineering Reviews* **2**, 1–39.

Wirth, D. F. and McMahon-Pratt, D. (1982). Rapid identification of *Leishmania* species by specific hybridization of kinetoplast DNA in cutaneous lesions. *Proceedings of the National Academy of Sciences of the USA* **79**, 6999–7003.

Wirth, D. F., Rogers, W. O., Barker, R. J., Dourado, L., Suesebang, L. and Albuquerque, B. (1986). Leishmaniasis and malaria: new tools for epidemiologic analysis. *Science* **234**, 975–979.

Wirth, J. J., Kierszenbaum, F., Sonnenfeld, G. and Zlotnik, A. (1985). Enhancing effects of gamma interferon on phagocytic cell association with and killing of *Trypanosoma cruzi*. *Infection and Immunity* **49**, 61–66.

Wright, C. D., Mulsch, A., Busse, R. and Osswald, H. (1989). Generation of nitric oxide by human neutrophils. *Biochemical and Biophysical Research Communications* **160**, 813–819.

Wyler, D. J. (1982). Circulating factor from a kala-azar patient suppresses *in vitro* antileishmanial T cell proliferation. *Transactions of the Royal Society of Tropical Medicine and Hygiene* **76**, 304–306.

Wyler, D. J., Weinbaum, F. I. and Herrod, H. R. (1979). Characterization of *in vitro* proliferative responses of human lymphocytes to leishmanial antigens. *Journal of Infectious Diseases* **140**, 215–221.

Xie, Q.-W., Cho, H. J., Calaycay, J., Mumford, R. A., Swiderek, K. M., Lee, T. D., Ding, A., Troso, T. and Nathan, C. (1992). Cloning and characterization of inducible nitric oxide synthase from mouse macrophages. *Science* **256**, 225–228.

Yamamura, M., Uyemura, K., Deans, R. J., Weinberg, K., Rea, T. H., Bloom, B. R. and Modlin, R. L. (1991). Defining protective responses to pathogens: cytokine profiles in leprosy lesions. *Science* **254**, 277–279.

Yang, D. M., Fairweather, N., Button, L., McMaster, W. R., Kahl, L. P. and Liew, F. Y. (1990). Oral *Salmonella typhimurium* ($AroA^-$) vaccine expressing a major leishmanial surface protein (gp63) preferentially induced Th1 cells and protective immunity against leishmaniasis. *Journal of Immunology* **145**, 2281–2285.

Yang, D. M., Rogers, M. V. and Liew, F. Y. (1991). Identification and characterisation of host-protective T-cell epitopes of a major surface glycoprotein (gp63) from *Leishmania major*. *Immunology* **72**, 3–9.

Yang, D.M., Rogers, M.V., Brett, S.J. and Liew, F.Y. (1993). Immunological analysis of the zinc-binding peptides of surface metalloproteinase (gp63) of *Leishmania major*. *Immunology* **78**, 582–585.

Zeledón, R., De Ponce, E. and Ponce, C. (1977). The Montenegro and MIF tests in three cured and one chronic case of human leishmaniasis. *Transactions of the Royal Society of Tropical Medicine and Hygiene* **70**, 536–537.

Zwilling, B. S., Vespa, L. and Massie, M. (1987). Regulation of I-A expression by murine peritoneal macrophages: differences linked to the Bcg gene. *Journal of Immunology* **138**, 1372–1376.

Zwingenberger, K., Harms, G., Pedrosa, C., Omena, S., Sandkamp, B. and Neifer, S. (1990). Determinants of the immune response in visceral leishmaniasis: evidence for predominance of endogenous interleukin 4 over interferon-gamma production. *Clinical Immunology and Immunopathology* **57**, 242–249.

Transport of Nutrients and Ions across Membranes of Trypanosomatid Parasites

DAN ZILBERSTEIN

Department of Biology, Technion–Israel Institute of Technology, Haifa 32000, Israel

I. Introduction

Leishmania and *Trypanosoma* are protozoan parasites that live inside mammalian hosts, either intracellularly or extracellularly. Successful adaptation

ADVANCES IN PARASITOLOGY VOL. 32
ISBN 0-12-031732-X

of these organisms to hostile environments within their vectors and hosts depends on their ability to maintain intracellular homeostasis of ions and nutrients. The interface between parasites and their hosts occurs ultimately at the level of the parasite surface membrane. Thus, the occurrence of controlled transport of nutrients and ions into and from parasite cells is important for the maintenance of intracellular homeostasis. Transporters are the means by which movement of molecules across membranes is facilitated. The level of expression and function of such membrane transporters is therefore critical for parasite survival inside their hosts.

The aim of this review is to shed light on a relatively neglected area that only recently started to receive increasing attention in the field of parasitology, namely, membrane transport mechanisms in *Leishmania* and *Trypanosoma*. This article focuses on transport of four groups of substrates: amino acids, sugars, protons and calcium, and will summarize the information accumulated in the last two decades on the structure and function of membrane transporters of these substances in *Leishmania* and *Trypanosoma*.

II. Proton Transport

A. PROTON-TRANSLOCATING ATPases IN *LEISHMANIA*

1. *Characterization of H^+-ATPase activity*

During its life cycle *Leishmania* undergoes extreme changes in environmental pH, from the relatively alkaline (pH >7.5) of the sandfly vector's gut to the neutral pH of the bloodstream and to the acidic pH of the lysosomes of macrophages (pH 4.5–5.5). Furthermore, the differentiation of promastigotes to amastigotes occurs during phagocytosis by host macrophages with exposure of the parasites to an acidic environment (Chang and Dwyer, 1976; Chang *et al.*, 1985; Zilberstein, 1991). It is therefore critical for these organisms to have a mechanism by which they can regulate their intracellular pH (pH_i). In various microorganisms, pH_i is maintained by means of proton pumps. Such pumps are able to move protons in and out of cells efficiently, thus controlling the cellular concentration of protons (Padan *et al.*, 1981; Padan and Schuldiner, 1987; Madshus, 1988). The presence of proton carriers in protozoan parasites was first demonstrated in *L. donovani* promastigotes by Zilberstein and Dwyer (1985). They showed that active transport of L-proline and D-glucose was driven by the proton motive force. A proton electrochemical gradient is created across the promastigote plasma membrane and this is coupled to transport by maintaining symport translocation of the specific substrates with protons. Electrochemical gradients across plasma membranes are created by primary proton pumps such as the proton-translocating adenosine triphosphatase (ATPase) (H^+-ATPase) (Perlin *et al.*, 1984; Pedersen and Carafoli, 1987). In a variety of cells and organelles, membrane bound H^+-ATPases have been demonstrated to be energy transducers which utilize the energy of adenosine triphosphate (ATP) hydrolysis to generate a proton electrochemical gradient (Goffeau and Slayman, 1981; Pedersen and Carafoli, 1987). It was therefore expected that an H^+-ATPase would be present in the plasma membrane of *Leishmania* cells.

Zilberstein and co-workers (Zilberstein *et al.*, 1987; Zilberstein and Dwyer, 1988) have identified an H^+-ATPase activity in plasma membranes of *L. donovani* promastigotes. This enzyme has been characterized as Mg^{2+}-dependent, has optimal activity at pH 6.5, and is very sensitive to orthovanadate (IC_{50} = 7.5 μM). The H^+-ATPase possesses high affinity and specificity for ATP (Michaelis-Menten constant $[K_m] = 1$ mM, $V_{max} = 225$ nmol min^{-1} (mg protein)$^{-1}$). More importantly, this enzyme is a proton pump, as was demonstrated using membrane vesicles derived from promastigote plasma membranes (Zilberstein and Dwyer, 1988). This H^+-ATPase is the primary proton pump in *L. donovani* and is responsible for the creation of the proton electrochemical gradient across the parasites' plasma membranes (Zilberstein and Dwyer, 1985; Zilberstein *et al.*, 1987, 1989; Zilberstein 1991). As outlined above, this gradient drives energy transduction processes across parasites' plasma membranes. The H^+-ATPase also has a role in regulating intracellular pH while keeping the chemiosmotic energy constant in both promastigotes and amastigotes (Zilberstein and Dwyer, 1988; Zilberstein *et al.*, 1989; Zilberstein, 1991; Glaser *et al.*, 1992).

The enzymatic activity of the *L. donovani* H^+-ATPase resembles that of the H^+-ATPase of yeast and fungi (Goffeau and Slayman, 1981). Furthermore, antibodies raised against the H^+-ATPase of *Saccharomyces cerevisiae* reacted with a 66 kDa membrane protein of *L. donovani* promastigotes (Liveanu *et al.*, 1991) and, when immunoprecipitated, this protein possessed an ATPase activity similar to that of the parasite H^+-ATPase. Indirect immunofluorescence assays and cryosection analysis localized the H^+-ATPase to the plasma membrane of promastigotes, including that of the cell and flagellar pocket (Zilberstein and Dwyer, 1988; Liveanu *et al.*, 1991).

2. *Molecular cloning of the leishmanial H^+-ATPase gene*

Meade *et al.* (1987) used an oligonucleotide probe containing the sequence of the phosphate binding site that is highly conserved between all plasma membrane cation-translocating ATPases (P-type ATPases) (Serrano *et al.*, 1986) to clone a pair of tandemly linked genes that encode a 107 kDa protein. The putative protein had all the characteristics of a P-type ATPase most closely related to yeast and fungi H^+-ATPases. The upstream gene displayed about 45% homology to *S. cerevisiae*, as well as plant H^+-ATPases, but much less to mammalian P-type ATPases (<27%). The nucleotide sequences of the pair of *Leishmania* ATPase genes were only slightly different, predicting differences in 20 amino acid residues (Meade *et al.*, 1989). Furthermore, these putative H^+-ATPases are developmentally regulated: transcripts from the upstream gene of the pair were present in both promastigotes and amastigotes. However, the second gene of this pair was expressed predominantly in amastigotes (Meade *et al.*, 1989). It was concluded by Meade *et al.* (1989) that the differential expression of these genes in the two developmental stages suggested that at least some of the amino acid differences between the two putative proteins have functional significance. Genes homologous to these ATPase genes also appeared in "new world" *Leishmania* species, *L. mexicana* and *L. braziliensis* (Meade *et al.*, 1990). Of interest was the observation that the downstream transcript increased in abundance in promastigotes that were allowed to reach the stationary phase of growth but it was most abundant in amastigotes (Meade *et al.*, 1987, 1989, 1990). On the other hand, the upstream gene was transcribed more abundantly by promastigotes in the exponential phase of growth than by stationary cells or amastigotes. The *L. donovani* H^+-ATPase gene was also used as a probe to identify homologous gene(s) in other related trypanosomatid genera. Meade *et al.* (1989) found that the leishmanial gene strongly hybridized on Southern blots to deoxyribonucleic acid (DNA) from *T. cruzi* and bloodstream forms of *T. brucei brucei*. Although the results are interesting one must be aware that, up to date, no direct evidence has been presented to link the cloned gene and the *L. donovani* H^+-ATPase activity described above.

B. P-TYPE ATPases IN *TRYPANOSOMA BRUCEI*

As outlined above, the *L. donovani* H^+-ATPase gene cross hybridized with genomic DNA from *T. brucei*, indicating that a similar gene exists in the latter organism. This was confirmed by Revelard and Pays (1991), who used a *kpnI* fragment from the H^+-ATPase gene of *Saccharomyces pombe* to clone a 3033 base pair (bp) gene that encodes a 110 kDa protein. However, the putative protein showed only 29% homology to the putative H^+-ATPase gene of *L. donovani* but had much higher homology (49%) to Ca^{2+}-ATPase of the sarcoplasmic reticulum of rats. The cloned gene appears in a single copy in the parasite genome and is expressed equally in procyclic and bloodstream forms of this organism. It is most likely that, although the cloned gene represents a P-type ATPase, it is not an H^+-ATPase but rather an ATPase that translocates another cation.

C. REGULATION OF INTRACELLULAR pH AND PROTON MOTIVE FORCE

1. *In* Leishmania

As outlined above, proton pumps and proton carriers can work together to maintain pH_i homeostasis. Here, I would like to present evidence for such a mechanism in *Leishmania* species. Intracellular pH was determined by various methods, mostly with *L. donovani* (Glaser *et al.*, 1988, 1992; Zilberstein *et al.*, 1989). Zilberstein *et al.* (1989) used the fluorescence pH indicator 2′,7′-bis(carboxyethyl)-5,6-carboxyfluorescein (BCECF) to measure the pH_i of *L. donovani* promastigotes. This dye was introduced into promastigotes as a tetraacetoxymethyl ester, which permeates the cells and is subsequently hydrolysed by cytoplasmic esterases, thus entrapping the dye in the cytosol (Rink *et al.*, 1982). Using this method it was found that *L. donovani* promastigotes regulate their pH_i at 6.4–6.7 throughout a wide extracellular pH (pH_o) range of 5–7.5 (Zilberstein *et al.*, 1989). Similar results were also obtained using two other independent methods: pH null-point assays, and the determination of the distribution across the parasite plasma membrane of the fluorescent amine acridine orange and the weak acid 5,5-dimethyl-2-4-oxazolidinedione (DMO). pH_i was also estimated by Glaser *et al.* (1988) using phosphate nuclear magnetic resonance and the DMO method. They measured a pH_i value of 6.8–7.4 throughout a pH_o range of 5–7.4. These values are higher than those determined using BCECF. This is most probably due to the accumulation of DMO in the mitochondrion of promastigotes, which might lead to an overestimation of the pH_i values (Zilberstein *et al.*, 1989). Nevertheless, Glaser *et al.* (1988) showed that amastigotes of *L. donovani* maintained a pH_i value similar to that of promastigotes throughout a similar pH_o range.

The foregoing observations imply that, when in the vector's midgut, promastigotes have a cytosolic pH which is more acidic than in their environment, whereas in the acidic environment of lysosomes the pH_i of amastigotes is more alkaline than their environment. This means that, as pH_o increases, the chemical gradient of protons across the parasite plasma membrane (ΔpH) decreases. This raises the question of how active transport processes that are driven by the proton motive force are able to function at the various environmental pHs. Both L-proline and D-glucose are actively accumulated by promastigotes throughout a pH_o range of 4.5–8.0 (Zilberstein *et al.*, 1989; Zilberstein, 1993; Zilberstein and Gepstein, 1993).

L-Proline is also actively transported in amastigotes of this organism throughout the same pH_o range and, moreover, there is evidence that suggests that the driving force of transport in amastigotes is also the proton motive force (Glaser and Mukkada, 1992). This indicates that a proton electrochemical gradient (directed inward) exists across the parasite's plasma membrane throughout the pH range indicated above. How then, is it possible for the parasite to maintain both pH_i and the proton motive force concomitantly?

The proton electrochemical gradient (Δp) is composed of two components, the pH gradient (ΔpH) and membrane potential ($\Delta\psi$), according to:

$$\Delta p = \Delta\psi + (2.3RT/F)\Delta pH$$

where R is the gas constant, T is the temperature in Kelvin and F is the Faraday constant. In order to keep both Δp and pH_i constant, the membrane potential has to compensate for the change in ΔpH at the various external pHs (for a review see Padan *et al.*, 1981), as has been shown in *L. donovani* promastigotes (Zilberstein *et al.*, 1989). Furthermore, Glaser *et al.* (1992) have recently determined the proton membrane potential in hamster-derived amastigotes of *L. donovani*. The membrane potential values (90–113 mV) were similar to those determined in promastigotes of this organism (Zilberstein *et al.*, 1989). But, more importantly, these results indicated that membrane potential at pH 7 was higher than at pH 5.5. Hence, $\Delta\psi$ can compensate for the change in ΔpH and as a result Δp remains constant throughout a wide range of environmental pHs. In conclusion, the foregoing observations indicate that, in *Leishmania*, proton pumps are capable of maintaining both pH_i and the chemiosmotic energy required to drive active transport of nutrients in environments of widely different pHs.

2. *In* T. brucei

Since *T. brucei* is an extracellular parasite that lives in the bloodstream of its mammalian host, it encounters an environment which is totally different from that of *Leishmania*. It was therefore interesting to compare energy transduction precesses as well as the assessment of how the two parasites regulate their pH_i.

The presence of a membrane potential in bloodstream form *T. brucei* was first demonstrated by Midgley (1978) using the lipophilic fluorescent cationic dye 2-(q-N,N-dimethylaminostyryl)-1-ethylpyridinium and Cs^+ in the presence of valinomycin. He estimated the membrane potential at 100–160 mV across the parasites' plasma membrane. No internal compartment that possessed membrane potential was observed. These results were in agreement with the notion that mitochondria of bloodstream forms of *T. brucei* are not functional. A different conclusion was presented by Nolan and Voorheis (1991). Using the lipophilic cation tetraphenylphosphonium (TPP^+), they demonstrated that, in addition to plasma membranes, membrane potential created by electrogenic movement of H^+ also exists across the membrane of intracellular organelles. Unlike plasma membranes, these organelles possessed no permeability to K^+ ions. Nevertheless, using TPP^+, a membrane potential of 155 mV was determined at pH 7.5 in bloodstream forms of *T. brucei* (Thissen and Wang, 1991).

Bloodstream forms of *T. brucei* are able to regulate their pH_i at 7.0–7.2 throughout a pH_o range of 6–8 (Thissen and Wang, 1991). The pH gradient was sensitive to the proton ionophore carbonyl cyanide *m*-chlorophenylhydrazone (CCCP) in the acidic range but not at pH 8.0. Membrane potential across parasites' plasma membranes was also pH-dependent, ranging between 76 mV at pH_o 6.0 and 160 mV at pH7 8.0. Thus, *T. brucei*, like *Leishmania*, is able to regulate its Δp constant throughout a wide range of pH_o.The regulation of pH_i in this organism has recently been confirmed by Ter Kuile *et al.* (1992), who also showed that membrane potential increased with an increase of pH from 80mV at pH_o 6 to 150 mV at pH_o 8. However, they noted that all previous experiments were done at 30°C, a subphysiological temperature, and that at 37°C only a small membrane potential and pH gradient were detected, i.e. the pH_i was totally dependent on the external pH. Ter Kuile *et al.* (1992) elegantly showed that, at around 26°C, the lipids of parasite membranes undergo phase transition which results in increased-leakiness to ions. These observations put into question the significance of ΔpH and membrane potential to energy transduction processes for bloodstream trypanosomes *in situ* and they are in agreement with previous observations on the mechanisms of transport of sugar which are active in procyclic forms but passive in bloodstream forms (Munoz-Antonia *et al.*, 1991; Ter Kuile and Opperdoes, 1991; Zilberstein, 1993). Ter Kuile *et al.* (1992) speculated that the phenomenon of change in membrane permeability to ions with temperature might play a role in triggering differentiation in this organism. Further studies should be done to assess the role of membrane permeability in the trypanosomes' development.

III. D-GLUCOSE TRANSPORT

Catabolism of D-glucose in *Trypanosoma* and *Leishmania* occurs in closed organelles called glycosomes (Opperdoes and Borst, 1977; Opperdoes, 1987; Fairlamb, 1989). These organelles contain most of the glycolytic enzymes including hexokinase, the enzyme responsible for the first step in glycolysis. Therefore, each of the glucose molecules that enters parasite cells must transfer into this organelle. The glycosomal membrane serves as a permeability barrier to most glycolytic intermediates (Opperdoes, 1987) and therefore may not allow simple diffusion. Thus, each D-glucose molecule must cross two membranes before it can be metabolized. However, all studies on D-glucose transport in trypanosomatids have focused on uptake over the plasma membrane, with none on the entry of glucose into glycosomes.

A. SPECIFICITY OF D-GLUCOSE TRANSPORTERS IN TRYPANOSOMATIDAE

Table 1 summarizes substrate recognition by D-glucose transport systems of *T. brucei* trypanosomes and *Leishmania.* The substrates are ranked in the order of their affinities for each transport system. All transport systems show a wide range of substrate specificities, which are similar in both *Leishmania* and *Trypanosoma.* The D-glucose transporters of both groups of organisms are not specific for the configuration at C-2; substrates like 2-deoxy-D-glucose (2DG), D-mannose, D-glucosamine and N-acetyl D-glucosamine are taken up by these transport systems (Table 1). *T. brucei* can also accept D-glucose analogues in which the alcohol group at C-6 is exchanged for a proton (6-deoxy-D-glucose, 6DG). On the other hand, none of the transporters of trypanosomes can accommodate changes in the configuration of the C-3 hydroxyl residue, nor is the addition of methyl groups in either the α or β configuration at C-1 tolerated by these transporters. Therefore, D-glucose analogues such as 3-0-methyl-D-glucose or α or β methyl-D-glucose are not recognized by the D-glucose transporters of organisms of either genus. It is interesting that the foregoing characteristics of trypanosomatids' glucose transporters resemble both the D-galactose transport system of *Escherichia coli* (Table 1) (Henderson *et al.*, 1977; Henderson, 1990) and the D-glucose transporter of erythrocytes and adipocytes of mammals (Table 1) (Barnett *et al.*, 1973; Rees and Holman, 1981; Baldwin and Henderson, 1989). These observations further support the notion of the evolutionary link between sugar transporters of unicellular organisms (prokaryotes as well as eukaryotes) and multicellular organisms.

Characterization of a transport system requires that the transport process be separated from its subsequent metabolic pathway. This is most crucial when uptake is the rate-limiting step in metabolism of the transporting substance, as in the case of glucose in *Leishmania* and *Trypanosoma* (Zilberstein, 1993; Ter Kuile and Opperdoes, 1991). Separation of transport from metabolism can be achieved in either of the following two ways. The first is the use of membrane-derived vesicles (Cohen *et al.*, 1986; Urbina *et al.*, 1988; Zilberstein and Dwyer, 1988; Benaim and Romero, 1990). These

TABLE 1 *Molecular recognition by sugar transporters in* Leishmania, Trypanosoma, *erythrocytes and* Escherichia coli

Leishmania *Promastigotes*[a]	T. brucei *Procyclic forms*[b]	*Bloodstream forms*[c]	*Erythrocytes* D-*glucose transporter*[d]	E. coli *Galactose transporter*[e]
D-glucose	2-deoxy-D-glucose	2-deoxy-D-glucose	2-deoxy-D-glucose	D-glucose
2-deoxy-D-glucose	D-glucose	D-glucose	D-glucose	2-deoxy-D-glucose
D-fructose	D-mannose	D-mannose		D-galactose
			α-methyl-D-glucose	
D-mannose	D-galactose	6-deoxy-D-glucose	3-0-methyl-D-glucose	
	D-fructose		6-deoxy-D-glucose	6-deoxy-D-glucose
		1-deoxy-D-glucose	D-mannose	D-mannose
N-acetyl-D-glucosamine	N-acetyl-D-glucosamine		D-galactose	
D-glucosamine	D-glucosamine	D-glucosamine		
D-galactose				

[a]Schaefer and Mukkada, 1976; D. Zilberstein and A. Gepstein, unpublished observations.
[b]Parsons and Nielsen, 1990.
[c]Game *et al.*, 1986; Eisenthal *et al.*, 1989.
[d]Barnett *et al.*, 1973; Baly and Horuk, 1988.
[e]Henderson *et al.*, 1977; Henderson, 1990.

are closed vesicles that are free of cytosolic enzymes. The second way is to use analogues that are carried by the transport system but are not metabolized. In mammalian cells, α-methyl-D-glucose and 3-0-methyl-D-glucose have been used to study glucose transport (Baly and Horuk, 1988) but these analogues are not taken up by either *Leishmania* or *Trypanosoma* (Table 1). Consequently, 2DG was used as the second choice (Schaefer *et al.*, 1974; Zilberstein and Dwyer, 1984; Zilberstein, 1993). Unlike the former two compounds, 2DG is phosphorylated by hexokinase and subsequently accumulated as 2 deoxy-D-glucose-6-P (Schaefer *et al.*, 1974; Eisenthal *et al.*, 1989; Zilberstein, 1991). Although the accumulated 2-deoxy-D-glucose-6-P does not inhibit hexokinase, the proportion of non-phosphorylated 2DG in the cytosol increases with time (Munoz-Antonia *et al.*, 1991). Thus, although 2DG undergoes phosphorylation, it is currently the best analogue available for the analysis of glucose transport mechanisms in *Leishmania* and *Trypanosoma*.

B. D-GLUCOSE TRANSPORT IN *LEISHMANIA*

1. *Characterization of transport activity*

Schaefer *et al.* (1974) were the first to describe D-glucose transport in *L. major*. Using 2DG as a substrate, they showed that the uptake was carrier-mediated and suggested that the transport was active in nature. The carrier had high affinity for D-glucose with an apparent K_m of 0.16 mM and a V_{max} of 3.2 nmol min^{-1} $(mg\ cells)^{-1}$. The D-glucose transport systems of both *L. major* and *L. donovani* have broad substrate specificities (Table 1), with the following order of apparent affinity: D-glucose ≈ 2-deoxy-D-glucose ≈ 2 D-fructose > D-mannose > N-acetyl-D-glucosamine > D-galactose (Schaefer and Mukkada, 1976; D. Zilberstein and A. Gepstein, unpublished results). Apparently, substitution at C-2, which affects the length of the side chain at least to a limited extent, can be tolerated without loss of recognition by the carrier. While there is little specificity for the group at C-2, changes in carbon 1 and carbon 3 result in loss of carrier affinity; α-methyl-D-glucoside and 3-0-methyl-D-glucoside do not compete for 2DG uptake (Schaefer and Mukkada, 1976; D. Zilberstein and A. Gepstein, unpublished results). Furthermore, none of the *Leishmania* species transports either 3-0-methyl-D-glucoside or α-methyl-D-glucoside.

Although D-glucose transporters of *L. donovani* and *L. major* show similar molecular recognition patterns, the *L. donovani* affinity (as indicated by the Michaelis-Menten constant, K_m) for 2DG (K_m = 24 μM) and D-glucose (K_m = 19 μM) is much higher than that for *L. major* (K_m = 0.16 mM)

(Zilberstein and Dwyer, 1984; Schaefer *et al.*, 1974, respectively). Nevertheless, a much lower affinity for D-glucose (K_m = 0.1 mM) was observed with promastigotes of *L. infantum* that were grown in continuous culture in a chemostat using glucose as substrate at concentrations at which transport was the rate-limiting step in metabolism (Ter Kuile and Opperdoes, 1992a). These studies also indicated that, as long as the promastigotes were not depleted of energy, they maintained intracellular concentrations of glucose constant at 50 mM. The mechanism of D-glucose transport was characterized in more detail in *L. donovani* promastigotes (Mukkada, 1985; Zilberstein and Dwyer, 1985). This transport system exhibited sugar/H^+ symport translocation and was driven by the proton motive force. The proton electrochemical gradient that drives the transport is created across the parasite plasma membranes by H^+-ATPase that acts as a primary proton pump (see Section II). Zilberstein *et al.* (1989) showed that *Leishmania* parasites can maintain the proton electrochemical gradient constant throughout a pH range of 5–8. This is significant because it means that the parasites are able to maintain the chemiosmotic energy required to drive active transport of D-glucose at both alkaline pH, which is similar to the pH in the vector's midgut, and acidic pH, similar to that inside lysosomes of the host macrophage (Geisow *et al.*, 1981; Lukacs *et al.*, 1991). As a result, the transport of D-glucose and of L-proline, both of which are proton motive force-driven, remains constant throughout this range of environmental pHs (Zilberstein *et al.*, 1989; D. Zilberstein and A. Gepstein, unpublished results).

2. *Affinity labelling of the D-glucose transporter*

Cytochalasin B is a fungal product that inhibits mammalian D-glucose transport proteins (Jung and Rampal, 1977) and binds to them when irradiated with ultraviolet (u.v.) light (Shanahan *et al.*, 1982). It inhibited the uptake of 2DG by *L. donovani* promastigotes, indicating that the D-glucose transporter contains a cytochalasin B binding site (Zilberstein *et al.*, 1986). This observation was consistent with those on similar carrier proteins from mammals, although the concentration required to inhibit 2DG transport in *L. donovani* (50% inhibitory concentration = 50 μM) was about 1000 times higher than those that inhibited glucose uptake in mammalian cells (Jung and Rampal, 1977; Baly and Horuk, 1988). However, these findings are consistent with data showing that the arabinose and galactose transporters in *Escherichia coli* also contain a cytochalasin B binding site and that their binding capacity is similar to that of the glucose transporter of *L. donovani* promastigotes (Zilberstein *et al.*, 1986; Cairns *et al.*, 1989a; Charalambous *et al.*, 1989; Henderson, 1990). Hence, these observations, together with the transporters' substrate specificities, suggest that the leishmanial glucose transporter is more closely related to the bacterial one than to the erythrocyte sugar transporters.

Using the u.v. cross-linking method, a single band with an apparent molecular mass between 20 and 30 kDa was demonstrated by sodium dodecyl sulphate polyacrylamide gel electrophoresis. The specificity of the labelling with cytochalasin B was indicated by the inhibition of the labelling (58%) in the presence of 500 mM D-glucose (Zilberstein *et al.*, 1986). However, the apparent molecular mass of the *L. donovani* cytochalasin B-labelled protein was less than that of both the mammalian D-glucose transporter (Baly and Horuk, 1988) and the *E. coli* galactose transporter (Cairns *et al.*, 1989a; Henderson, 1990). It is possible that the 20–30 kDa carrier might represent a subunit of a larger protein or one that has been enzymatically cleaved by endogenous protease activity.

3. *Molecular cloning of putative D-glucose transporters in* Leishmania

Cairns *et al.* (1989b) have cloned a gene (*pro*1) from *L. enriettii* promastigotes that encodes a protein of 61.4 kDa which has all the characteristics of a D-glucose transporter. This gene is developmentally regulated, being expressed more abundantly in promastigotes than in amastigotes (Cairns *et al.*, 1989b). The *pro*1 leishmanial gene displays about 44.4% homology with the human erythrocyte D-glucose transporter and to a lesser extent with the *AraE*, *GalP* and *XylE* proteins of *E. coli* (see Cairns *et al.*, 1989a, Henderson, 1990). The transporter is encoded by a single family of tandemly clustered genes containing eight copies of a 3.6 kb repeat unit (Stein *et al.*, 1990). The first gene in this tandem repeat encodes a unique isoform of the transporter that contains a distinct hydrophilic amino-terminal domain, which was suggested to be localized in the cytoplasm. The remainder of the gene is identical to the second gene (Stack *et al.*, 1990).

Sequences homologous to the *L. enriettii* D-glucose transporter gene (*pro*1) are also present in other species such as *L. donovani* and *L. major* promastigotes (Stein *et al.*, 1990). Recently, several genes homologous to *pro*1 have been identified and cloned from *L. donovani* promastigotes, and are currently being characterized (Langford *et al.*, 1992).

C. D-GLUCOSE TRANSPORT IN *TRYPANOSOMA*

1. *In procyclic trypomastigotes*

The D-glucose transport in procyclic trypomastigotes of *T. brucei* resembles that of *L. donovani* promastigotes. The trypanosome's transporter has a broad substrate specificity with the following order of apparent affinity: D-glucose = 2-deoxy-D-glucose = D-mannose > D-galactose > D-fructose > D-glucosamine (Table 1) (Parsons and Nielsen, 1990). Two independent studies determined similar kinetic characteristics of *T. brucei* procyclic forms with 2DG, i.e., $K_m = 23$ μM and 38 μM and $V_{max} = 11.4$ and 7 nmol min^{-1} (mg protein)$^{-1}$, respectively (Parsons and Nielsen, 1990; Munoz-Antonia, 1991). A K_m of 60 μM for D-glucose was determined using cells grown in a chemostat (Ter Kuile and Opperdoes, 1992a). As in *Leishmania*, neither 3-*O*-methyl-D-glucose nor α-methyl-D-glucose are substrates of the *T. brucei* procyclic form's glucose transport system. The active transport of glucose is inhibited by ionophores [carbonyl *p*-(trifluoromethoxy) phenylhydrazone, CCCP] and by metabolic inhibitors (KCN, dicyclohexylcarbodiimide).

Using cultures of procyclic forms grown in a chemostat, Ter Kuile and Opperdoes (1992a) demonstrated that, unlike *Leishmania* promastigotes, procyclic trypomastigotes did not maintain a constant intracellular concentration of glucose; rather, it was dependent on the growth rate. They also suggested that under the chemostat conditions glucose was not actively accumulated against its gradient, but that the mechanism of transport was facilitated diffusion. Since the experiments in the chemostat were done using glucose as a substrate and since transport was measured indirectly, a more direct approach should be taken before the idea that glucose is taken up by facilitated diffusion can be accepted.

2. *In bloodstream trypomastigotes*

Trypanosomes, both procyclic and bloodstream forms, do not store polysaccharides (Opperdoes, 1987; Fairlamb, 1989). However, the bloodstream forms of *T. brucei* are totally dependent upon D-glucose for their energy supply (Opperdoes, 1987). Indeed, bloodstream trypanosomes possess a high rate of glycolysis, about 10 times the rate in mammalian cells (Flynn and Brown, 1973; Visser and Opperdoes, 1980; Kiaira and Njogu, 1988; Eisenthal *et al.*, 1989). The high efficiency of glycolysis is facilitated by compartmentalizing the glycolytic enzymes in glycosomes, which are microbody-like organelles (Opperdoes, 1987). Glucose is phosphorylated by hexokinase within the glycosomes and further metabolized to 3-phosphoglycerate which is released to the cytosol and further metabolized to pyruvate. This process requires each glucose molecule taken up from the medium to cross two membranes before it is metabolized.

Glucose transport was studied in bloodstream forms of *T. lewisi*, *T. b. gambiense* and *T. b. brucei*. In all of these, glucose transport is carrier-mediated by a facilitated diffusion mechanism. Table 2 summarizes the kinetics of the glucose transport system in these different species. The apparent K_m values for glucose in all species tested falls in the mM range (Sanchez and Read, 1969; Southworth and Read, 1969, 1970; Sanchez, 1974; Gruenberg *et al.*, 1978; Wayne and Waughan, 1979; Game *et al.*, 1986; Eisenthal *et al.*, 1989; Seyfang and Duszenko, 1991; Ter Kuile and Opperdoes, 1991). In *T. brucei*, for example, the apparent K_m is between 0.5 and 4 mM, depending on the sugar analogue used. The rate of transport is relatively high and is similar to the rate of glucose metabolism in these organisms (Sanchez and Read, 1969; Eisenthal *et al.*, 1989; Ter Kuile and Opperdoes, 1991). Based on these figures it was suggested that the transport into bloodstream trypanosomes is the rate-limiting step in glucose metabolism. The glucose transport system of bloodstream trypanosomes recognizes, in addition to D-glucose, 2DG, D-mannose, N-acetyl-D-glucosamine and D-glucosamine (Table 1). Also, 1DG and 6DG are transported and competitively inhibit glucose transport (Game *et al.*, 1986; Eisenthal *et al.*, 1989). Using these analogues as substrates for transport, Eisenthal *et al.* (1989) have shown that neither D-galactose nor D-xylose is recognized or transported by bloodstream forms of *T. brucei*. This is significant because (i) both D-galactose and D-xylose are substrates of the D-galactose transport system in *E. coli* (see Baldwin and Henderson, 1989; Henderson, 1990), and (ii) 1DG is not a substrate of the *E. coli* galactose transporter. The data therefore suggest that bloodstream trypanosomes express a glucose transporter which is closely related to the mammalian erythrocyte glucose transporter but different from that expressed in the procyclic forms. Furthermore, recent studies have indicated that phloritin as well as cytochalasin B, both of which inhibit glucose transport in mammalian erythrocytes, inhibit glucose transport in bloodstream forms of *T. brucei* (see Munoz-Antonia *et al.*, 1991; Seyfang and Duszenko, 1991). It is conceivable, however, that since this form of *T. brucei* lives in the bloodstream of its mammalian host, it has adapted to life in a glucose-rich environment and it has therefore evolved a glucose transporter that is more similar to the erythrocyte sugar transporter than to the bacterial sugar transporter.

The apparent K_m for D-glucose transport in bloodstream forms of *T. brucei* is about 10–100 times higher than the K_m in procyclic forms and the V_{max} is 3–5 times higher. These characteristics, together with the differences observed in substrate specificities outlined above, suggest that procyclic and bloodstream trypomastigotes of *T. brucei* have two different glucose transporters and, furthermore, that their expression is developmentally regulated (Parsons and Nielsen, 1990; Munoz-Antonia *et al.*, 1991; Bringaud and Baltz, 1992).

TABLE 2 *Kinetic characteristics of glucose transport in bloodstream forms of various* Trypanosoma *species*

Species	*Substrate*	K_m [a]	V_{max}	*Reference*
T. lewisi	D-Glucose	1.25	3.7 nmol min^{-1} (mg dry weight)$^{-1}$	Eisenthal *et al.*, 1989
T. lewisi	D-Glucose	0.3	4.0 nmol min^{-1} 1×10^{-8} cells	Gruenberg *et al.*, 1978
	D-Glucose	—	169.0 nmol min^{-1} (mg cells)$^{-1}$	Sanchez and Read, 1969
T. b. gambiense	D-Glucose	1.05	24.0 nmol min^{-1} (mg dry weight)$^{-1}$	Wayne and Vaughan, 1979; Sanchez, 1974
T. b. brucei	1-Deoxy-D-glucose	4.03	18.3 nmol min^{-1} 1×10^{-8} cells	Southworth and Read, 1969
	1-Deoxy-D-glucose	3.41[b]	23.0 nmol min^{-1} 1×10^{-8} cells	Southworth and Read, 1970
	2-Deoxy-D-glucose	1.0	18.0 nmol min^{-1} (mg dry weight)$^{-1}$	Munoz-Antonia *et al.*, 1991
	2-Deoxy-D-glucose	0.237	10.0 nmol min^{-1} (mg protein)$^{-1}$	Game *et al.*, 1986
	6-Deoxy-D-glucose	1.54[b]	28.8 nmol min^{-1} 1×10^{8} cells	Southworth and Read, 1970
	D-Glucose	1.92	—	Ter Kuile and Opperdoes, 1991
	D-Glucose	0.49[c]	112.0 nmol min^{-1} 1×10^{8} cells	Seybang and Duszenko, 1991

[a] mM
[b] Determined at 20°C; all other measurements were at 37°C.
[c] Determined by following external D-glucose.

3. *Molecular cloning of a putative glucose transporter in* T. brucei

Attempts were made by a number of groups to clone the gene that encodes the glucose transporter in trypanosomes. But, as with the leishmanial glucose transporter gene, identification of the corresponding gene in *T. brucei* was done by accident. Bringaud and Baltz (1992) have recently used a 14-mer oligonucleotide probe that contained a sequence common to all variable surface glycoproteins to clone genes that were differentially expressed in bloodstream forms of *T. brucei.* One of these genes coded for a putative membrane protein of 527 amino acids which was found to be very similar to the *pro*1 gene, the putative glucose transporter gene of *L. enriettii*, i.e., 68% homology and 43% identity (Cairns *et al.*, 1989b). The *T. brucei* gene is also similar to the human erythrocyte glucose transporter; 19% of its amino acids are identical and 42% are similar to the erythrocyte glucose transporter. The *T. brucei* gene is also similar to other sugar transporting genes such as the *araE* and *XylE* in *E. coli* (see Maiden *et al.*, 1987; Henderson, 1990), hexose-carrier from *Chlorella* (see Saur and Tanner, 1989), and yeast glucose and galactose transporters (Celenza *et al.*, 1988; Nehlin *et al.*, 1989). Interestingly, among these homologous genes there were 15 amino acids that were strictly conserved in their sequence and specific location in the gene. On the other hand, a few highly conserved sequences such as ATP-binding sites that appear in most sugar transport genes did not appear in the *T. brucei* putative glucose transporter gene. This shares some characteristics with the corresponding *Leishmania* gene. Similar to the *pro*1 protein (Cairns *et al.*, 1989b), the *T. brucei* gene contains in its largest extracellular loop a large number of cysteine residues (six in 50 amino acids), which were suggested to be involved in disulphide bonds (Bringaud and Baltz, 1992). The putative *T. brucei* glucose transporter belongs to a cluster of multigene families that contain at least six copies (Bringaud and Baltz, 1992). The foregoing observation thus strongly suggests that the gene cloned by Bringaud and Baltz (1992) encodes a glucose transporter.

IV. Amino Acid Transport

A. L-PROLINE METABOLISM

L-Proline is often the main amino acid constituent of haemolymph in various insects (Raghupath; *et al.*, 1969; Konji *et al.*, 1988). In tsetse flies, for example, it appears at a concentration of 60 mM (Brusell, 1970; Konji *et al.*, 1988). Many flies, including tsetse, utilize oxidation of L-proline as an energy source for flight. Insect stages of both *Leishmania* and *Trypanosoma* have adapted to the L-proline-rich environment in their vectors by using this amino acid as the main metabolic source of energy.

Krassner (1969) reported that promastigotes of *L. tarentolae* can respire using L-proline and L-glutamate instead of, and as efficiently as, D-glucose. Furthermore, these organisms can grow in a glucose-free medium if L-proline is present. Using continuous cultures in a chemostat, Ter Kuile and Opperdoes (1992b) have recently shown that *L. donovani* promastigotes can grow in medium with L-proline as the only carbon source. Utilization of D-glucose by promastigotes was delayed to a late exponential phase of growth (Mukkada *et al.*, 1974).

Utilization of L-proline by procyclic culture forms of *T. brucei* resembled that of *Leishmania* promastigotes (Evans and Brown, 1972). Exponentially dividing cultures of these organisms used L-proline as an energy source even in the presence of 2DG, an inhibitor of glycolysis. Furthermore, Evans and Brown (1972) observed that differentiation of bloodstream forms of *T. brucei* to procyclic forms was followed by a dramatic increase in the activity of L-proline oxidase to levels that are similar to those found in insect muscles.

More than 50% of the accumulated L-proline was catabolized to CO_2 and this process was suppressed by D-glucose (Krassner and Flory, 1977). A significant amount of the accumulated L-proline was converted into L-alanine (Krassner, 1969; Evans and Brown, 1972; Krassner and Flory, 1972, 1977). This amino acid was synthesized in *Leishmania* via pyruvate from either L-proline or D-glucose. However, the synthesis of L-alanine was more efficient when the precursor was D-glucose. Moreover, the presence of D-glucose suppressed the conversion of L-proline into L-alanine (Krassner and Flory, 1977). A similar observation was made with culture forms of *T. brucei* and *T. scelopori* (Krassner and Flory, 1977). The significance of L-alanine was recently demonstrated by Burrows and Blum (1991), who showed that it was used as an osmotic buffer by *Leishmania*.

B. L-PROLINE TRANSPORT

1. *L-Proline transport in* Leishmania

Law and Mukkada (1979) were the first to report the active accumulation of L-proline in *L. major* via a carrier-mediated system. This transporter showed a wide range of specificity for other amino acids such as L-alanine, methionine, valine, α-aminoisobutyric acid and various L-proline analogues. Bonay and Cohen (1983) suggested that L-proline was taken up by promastigotes via a neutral amino acid transporter. They also suggested that promastigotes possessed two transport systems for neutral amino acids, one of which carried L-proline and the other L-alanine. Both systems were similar in their affinity and rate of uptake. L-Proline is also actively accumulated against its concentration gradient by *L. donovani*, by both promastigotes and amastigotes (Zilberstein and Dwyer, 1985; Glaser and Mukkada, 1992).

Zilberstein and Dwyer (1985) showed that the active transport of L-proline was driven by the proton motive force by maintaining symport translocation of L-proline with protons. The active transport of L-proline was further demonstrated in additional studies by this and other groups (Zilberstein *et al.*, 1989; Glaser and Mukkada, 1992; Zilberstein and Gepstein, 1993). A different approach to studying the relation between L-proline transport and its metabolism was taken recently by Ter Kuile and Opperdoes (1992b), who reported that they succeeded in growing *L. donovani* promastigotes in continuous cultures in a chemostat in a D-glucose-free medium with varying concentrations of L-proline. Under these conditions they found that intracellular concentrations of L-proline at steady state transport were proportional to its concentration in the growth medium. They found no indication of active transport of this amino acid.

Kinetic analysis of L-proline transport indicated some differences between *L. major* and *L. donovani* promastigotes. The K_m for L-proline in *L. major* promastigotes is 60 μM, and in *L. donovani* it was reported to be 0.65 mM (Law and Mukkada, 1979; Bonay and Cohen, 1983; Zilberstein and Gepstein, 1993). For *L. donovani* promastigotes grown in a chemostat, the K_m was 48 μM (Ter Kuile and Opperdoes, 1992b). The rate of uptake also varied from as much as 64 nmol min^{-1} (mg protein)$^{-1}$ in *L. donovani* to 12 nmol min^{-1} (mg protein)$^{-1}$ in *L. major* (Law and Mukkada, 1979; Zilberstein and Gepstein, 1993). The rate of uptake by *L. donovani* resembled that of D-glucose uptake by bloodstream forms of *T. brucei* (for reviews, see those by Fairlamb, 1989 and Ter Kuile, 1991).

Transport of L-proline has also been determined recently in amastigotes of *L. donovani* (Glaser and Mukkada, 1992). Amastigotes derived from hamsters accumulate L-proline at a rate and extent that are much lower than those in promastigotes. Furthermore, transport activity in the amastigotes was optimal at about pH 5.5, whereas in promastigotes it was optimal at pH 7–7.5 (Zilberstein *et al.*, 1989; Glaser and Mukkada, 1992). This is in agreement with the metabolism of L-proline in amastigotes, where incorporation and metabolism of this amino acid were optimal at a pH value of about 5 (Mukkada *et al.*, 1985). These differences in transport activity were further demonstrated by the observation that rapid acidification of the assay medium from pH 7 to pH 5 enhanced transport activity in amastigotes but suppressed it in promastigotes. On the other hand, rapid alkalinification of the assay medium from pH 5 to pH 7 increased L-proline transport activity in promastigotes and suppressed it in amastigotes (Glaser and Mukkada,

1992). The foregoing observation suggests, although indirectly, that promastigotes and amastigotes possess different transporters for L-proline. This conclusion raised the question of the possible role of environmental pH on L-proline transport activity in *Leishmania*. This question is intriguing in the light of recent observations that growth of promastigotes at acidic pH can trigger their differentiation to amastigotes (Zilberstein *et al.*, 1991).

In order to address this question, promastigotes of *L. donovani* were adapted to grow in media adjusted to various pHs ranging from 4.5 to 7.5 (Zilberstein *et al.*, 1991). The characteristics of L-proline transport in *L. donovani* promastigotes were strongly influenced by culture pH (Zilberstein and Gepstein, 1993). Reducing the pH of the growth medium from 7 to 4.5 caused a dramatic decrease in both the rate and extent of L-proline transport. In promastigotes grown at pH 4.5, the V_{max} of transport was one-tenth, and the affinity for L-proline was twice, that in cells grown at pH 7 (Zilberstein and Gepstein, 1993). Furthermore, while the optimum pH for transport in promastigotes grown at pH 7 was 7–7.5, it decreased as the culture pH was lowered, reaching an optimal pH of 5.5 in cells grown at pH 4.5. Steady state levels of transport also decreased with the pH of the growth medium. Hence, in promastigotes grown at pH 4.5 steady state transport was at least three times lower than in the cells grown at pH 7. Moreover, these characteristics of L-proline transport in promastigotes responded reversibly to changes in culture pH. It took 48–72 h after transferring promastigotes to acidic pH for L-proline transport to switch to the transport level in the acid-grown cells. This parallels the time it takes promastigotes to differentiate to amastigotes *in vivo* (Chang and Dwyer, 1976; Jaffe and Rachamim, 1989). Upon switching the culture pH from 4.5 to 7, transport resumed the activity of pH 7 cells in less than 24 h by a process dependent on protein synthesis (Zilberstein and Gepstein, 1993). It was clear from these experiments that the transport at low pH resembled the L-proline transport described in amastigotes of the same species (Glaser and Mukkada, 1992). The foregoing observations can be explained in either of the following ways: (i) promastigotes and amastigotes of *L. donovani* possess two distinct transporters for L-proline, the expression of which is regulated by culture pH, or (ii) the pH-dependent expression of L-proline transporters reflects modification of a single transporter which results in a shift to its optimal pH. Recent studies indicated that *L. donovani* promastigotes grown at pH 5.0 exhibited optimal activities of DNA and protein synthesis at pH 5.5 (R. Asamoa and D. Zilberstein, paper in preparation).

2. *L-Proline transport in* Trypanosoma

There is a paucity of studies concerning L-proline transport in *T. brucei* and *T. cruzi*. In preliminary work, D. Zilberstein and P. Gardiner (unpublished results) measured uptake in procyclic and bloodstream forms of *T. brucei*. They found that L-proline transport in procyclic forms resembled transport activity in *L. donovani* promastigotes, i.e. high K_m and V_{max} [1 mM and 52 nmol min^{-1} (mg protein)$^{-1}$, respectively]. Furthermore, uptake in bloodstream forms of this organism was the same as in procyclic forms. Thus, unlike *Leishmania*, L-proline transport in *T. brucei* does not appear to be developmentally regulated.

C. UPTAKE OF OTHER AMINO ACIDS

1. *In* Leishmania

α-Aminoisobutyric acid is a non-metabolizable analogue of L-proline which is sometimes used to analyse transport of neutral amino acids. α-Aminoisobutyric acid was actively accumulated against its concentration gradient by "*L. tropica*" (= *L. major*), demonstrating two different K_m one at 10 μM and the other at 250 μM (Lepely and Mukkada, 1983). Competition analysis indicated a wide range of specificity of the transport of α-aminoisobutyric acid, including neutral and polar amino acids. α-Aminoisobutyric acid transport with broad amino acid specificity has also been demonstrated in *Crithidia fasciculata* (Midgley, 1978). The transport in this organism was also active and an H^+-symport translocation with α-aminoisobutyric acid was suggested, leading to the conclusion that the transport was driven by the proton motive force.

Uptake of neutral amino acids by promastigotes was also investigated by Bonay and Cohen (1983). They found that promastigotes of *Leishmania* possess at least two systems for the transport of neutral amino acids. The transport of amino acids such as L-proline, L-alanine, and L-phenylalanine are active in nature and have high affinity to their substrates. The amino acid methionine is also taken up by *L. tropica* via a carrier-mediated system and the transport is active (Mukkada and Simon, 1977; Simon and Mukkada, 1977).

2. *In* Trypanosoma (Trypanozoon) *species*

Bloodstream forms of *T. b. gambiense* take up lysine, arginine, glutamate, phenylalanine, threonine, glycine and alanine via carrier-mediated systems (Hansen, 1977). Five distinct transport proteins mediate the uptake of these amino acids. These transporting proteins were defined as follows: (i) locus A, specific for the uptake of glutamate, arginine and lysine. The binding of either glutamate or arginine stimulates the uptake of lysine; (ii) locus B transports threonine, glycine and alanine and appears to be partially sensitive to sodium and ouabain; (iii) locus C transports glutamate; (iv) locus D transports phenylalanine and methionine and (v) locus E transports lysine and arginine. The affinities of these transport systems for their substrates ranges from a minimal K_m of 0.23 mM for methionine through to a maximum K_m of 1.94 mM for alanine. *T. equiperdum* also possesses several transport systems for amino acids, which probably resemble those described for *T. b. gambiense* (see Jackson and Fisher, 1977). Threonine transport was further analysed in *T. brucei* by Voorheis (1977), by assessing the effect of H^+ concentration on transport activity. He showed that pH affected transport activity due to changes in ionic groups in both the carrier and its substrate. However, no H^+- or Na^+-coupled uptake of threonine was observed.

V. Calcium Transport

It is well established that calcium has an important role in the control of cellular metabolism (Hesketh *et al.*, 1983, 1985; Thomas *et al.*, 1984; Berridge, 1986) and in signal transduction mechanisms in eukaryotic cells (Berridge, 1986; Carafoli, 1987). Most of the calcium-dependent activities derived from periodic increases in cellular free calcium, which subsequently activated cellular proteins such as calmodulin, phospholipase C, protein kinases and others. Carafoli (1987) once postulated that, in order for calcium to act as a secondary messenger of signals, its concentration in the cytosol must be kept at a very low level. All living cells tested, prokaryotes as well as eukaryotes, maintain their cytosolic free calcium at concentrations as low as 100 nM (Carafoli, 1987). Three compartments are responsible for the low level of calcium homeostasis in eukaryotes. These are endoplasmic reticulum, mitochondria and plasma membranes (Burgess *et al.*, 1983; Somlyo, 1984; Carafoli, 1987). In plasma membranes and endoplasmic reticulum, calcium is transported via Ca^{2+} transporting ATPases (Gill and Cheuh, 1985; Berridge, 1986), whereas mitochondria possess electrogenic calcium transport systems (McCormack and Denton, 1986).

Up to date very little is known about the role of calcium in signal transduction and differentiation in trypanosomatids or any other parasitic protozoa. Several studies have been made to elucidate calcium homeostasis mechanisms in both *Leishmania* and *Trypanosoma*, in the hope that such observations will lead to a better understanding of the role of calcium in parasite development and metabolism.

A. Calcium Homeostasis and Transport in *Leishmania*

Cytosolic concentrations of free calcium in promastigotes were determined by using the fluorescence Ca^{2+} indicators fura 2 and quin 2 (Philosoph and Zilberstein, 1989; Benaim *et al.*, 1990). These observations indicated that free calcium in the cytoplasm of *L. donovani* promastigotes was maintained at 73–95 nM throughout the extracellular [Ca^{2+}] range of 0–1 mM (Philosoph and Zilberstein, 1989). In *L. braziliensis* promastigotes intracellular [Ca^{2+}] was maintained at 50 nM in a calcium-free medium (Benaim *et al.*, 1990). The maintenance of low intracellular [Ca^{2+}] was energy-dependent and the regulation was executed by mobilizing calcium into intracellular pools.

Intracellular traffic of calcium was examined by measuring its transport in digitonin-permeabilized promastigotes (Philosoph and Zilberstein, 1989; Benaim *et al.*, 1990; Vercesi and Docampo, 1992). As has been shown in a number of organisms, including protozoa (Philosoph and Zilberstein, 1989; Vercesi *et al.*, 1990), digitonin renders the cholesterol-rich plasma membrane permeable to ions and small molecules, while leaving intracellular organelles relatively intact. Three transport systems for calcium were identified in these cells (Philosoph and Zilberstein, 1989; Vercesi and Docampo, 1992). One was respiration-dependent and was located in the mitochondria. The uptake of calcium into this organelle was sensitive to KCN and required addition of inorganic phosphate to the reaction mixture. The uptake of calcium into mitochondria of promastigotes was electrogenic as it was driven by the proton membrane potential. Calcium transport into mitochondria was further analysed using mitochondrial membrane vesicles from *L. braziliensis* promastigotes. These experiments indicated that mitochondria mediate electrogenic transport of Ca^{2+} at high levels of calcium (>1 μM). These observations also indicated that mitochondria of promastigotes store large quantities of calcium, as do the mitochondria of higher eukaryotic cells.

A second transport system for calcium was identified in the digitonin-permeabilized promastigotes. It was respiration-independent but required either endogenous or externally added ATP (Philosoph and Zilberstein, 1989; Vercesi and Docampo, 1992). This ATP-dependent Ca^{2+} transport was optimal at pH 7.1 and had high affinity for calcium (K_m = 92 nM, V_{max} = 1 nmol min^{-1} [mg protein]$^{-1}$). These properties of calcium transport suggested the presence of Ca^{2+}-ATPase similar to that found in endoplasmic reticulum of mammals.

A third calcium transporter was identified in plasma membranes of *L. donovani* and *L. braziliensis* promastigotes (Benaim and Romero, 1990; Ghosh *et al.*, 1990). In *L. donovani*, plasma membranes contained a high affinity (K_m = 35 nM) Ca^{2+}-ATPase, which had an apparent molecular mass of 215 000 Da, consisting of two subunits of 55 000 and 57 000 Da. The activity of this enzyme was Ca^{2+}-dependent and was inhibited by orthovanadate. Moreover, the affinity of the Ca^{2+}-ATPase to calcium increased in the presence of heterologous calmodulin (K_m = 12 nM). Benaim and

Romero (1990) have identified (Ca^{2+},Mg^{2+})-ATPase activity in plasma membranes of *L. braziliensis* which had high affinity for calcium (K_m = 0.7 μM) and was sensitive to orthovanadate. The foregoing observations, with both *L. donovani* and *L. braziliensis*, further indicated that intracellular [Ca^{2+}] was regulated at low concentrations by a mechanism similar to those found in higher eukaryotic cells.

B. CALCIUM TRANSPORT AND HOMEOSTASIS IN *TRYPANOSOMA*

Bloodstream forms of *T. brucei* regulate their intracellular free Ca^{2+} at a concentration of about 98 nM (Ruben *et al.*, 1991). This low level of Ca^{2+} is maintained by an ATP-independent, pH sensitive intracellular pool of calcium. Increasing the pH_i with proton ionophores resulted in a concomitant decrease in the level of Ca^{2+}, whereas conditions that caused a decrease in pH_i were followed by an increase in cellular free calcium.

Fura 2 was used to determine [Ca^{2+}]$_i$ in epimastigote and amastigote forms of *T. cruzi* (see Moreno *et al.*, 1991; Vercesi *et al.*, 1991). In epimastigotes, intracellular free [Ca^{2+}] was determined to be 100–200 nM and the maintenance of [Ca^{2+}]$_i$ homeostasis was due to activity of intracellular pools, resembling the mechanism described for *L. donovani* and *T. brucei* (see Philosoph and Zilberstein, 1989; Ruben *et al.*, 1991). *T. cruzi* amastigotes contained a lower level of cytosolic free Ca^{2+} (40–45 nM) than epimastigotes and its regulation was sensitive to orthovanadate. The mitochondria of amastigotes, like those of epimastigotes, possessed an electrogenic transport system for calcium which was sensitive to the proton ionophore CCCP.

Intracellular transport of calcium in *T. cruzi* was analysed using digitonin-permeabilized cells (Docampo and Vercesi, 1989a; Vercesi *et al.*, 1990). Two intracellular Ca^{2+} transport systems were identified; one was respiration-dependent and the other was ATP-dependent. The ATP-dependent system had high affinity for calcium, and was probably localized in the endoplasmic reticulum. This system takes up calcium until intracellular [Ca^{2+}] reaches 150 nM. Vercesi *et al.* (1990) suggested that this system was responsible for the maintenance of cellular calcium at 0.1–1.0 μM. The mitochondrial calcium transport system was effective at concentrations of 1 μM and above (Docampo and Vercesi, 1989a,b; Vercesi *et al.*, 1990). This transport system resembles that of *Leishmania* and mammalian mitochondria: it is active in nature and driven by mitochondrial membrane potential (Docampo and Vercesi, 1989a,b; Vercesi *et al.* 1990, 1991).

A high affinity (Ca^{2+}-Mg^{2+})-ATPase activity has been identified in plasma membranes of vesicles within *T. cruzi* epimastigotes. These studies indicated that calcium transport resembled Ca^{2+}-ATPase activity: the uptake was Mg^{2+}-dependent and inhibited by orthovanadate. Moreover, both Ca^{2+} uptake and ATPase activity were stimulated by calmodulin, suggesting that calmodulin plays a role in regulating intracellular calcium concentration in *T. cruzi* (Benaim *et al.*, 1991). These observations of (Ca^{2+}-Mg^{2+})ATPase activities were in contrast with the Ca^{2+}-ATPase activity found in *Leishmania* promastigotes, which was Mg^{2+}-independent (Ghosh *et al.*, 1990).

C. IS CALCIUM A SECONDARY MESSENGER OF EXTERNAL SIGNALS IN TRYPANOSOMATIDS?

Docampo and Pignataro (1991) have recently presented evidence for the existence of Ca^{2+}-induced phospholipase C that is responsible for activating the inositol phosphate cycle in epimastigotes of *T. cruzi*. The components of the inositol phosphate cycle have also been identified in amastigotes of this species (Moreno *et al.*, 1991). Observations have also been made by A. Dahan and D. Zilberstein (paper in preparation) that showed that *L. donovani* promastigotes contained all the components of the inositol phosphate cycle and that Ca^{2+} activated phospholipase C in these organisms. Furthermore, these studies indicated that addition of inositol triphosphate to digitonin-permeabilized *L. donovani* promastigotes caused a significant release of calcium from the endoplasmic reticulum. These observations suggested that Ca^{2+} can play a role in signal transduction pathways in trypanosomatids.

The limiting step in the study of signal transduction mechanisms in trypanosomatids is the lack of specific ligands. It is easy, but premature, to postulate that signal transduction mechanisms in this group of organisms resemble those of higher eukaryotic organisms. For example, the possible function of the inositol phosphate cycle in *T. cruzi* and *L. donovani* (see Docampo and Pignataro, 1991; A. Dahan and D. Zilberstein, paper in preparation). Another example is the observation that *L. donovani* and *T. cruzi* have proteins in their plasma membranes that resemble in structure and function the mammalian G-proteins (Eisenschols *et al.*, 1986; Cassel *et al.*, 1991). However, the question of whether they are G-proteins of the signal transduction pathway still has to be answered. Revealing such mechanisms would be important for a better understanding of the interaction of parasites with their environments. However, the main focus should be on the identification of specific ligands that can activate signal transduction pathways.

Acknowledgements

I thank Drs Amira Gepstein and John D. Lonsdale-Eccles for critical discussions, and Joshua Saar, Dina Katz and Maria Mulindi for help during the preparation of the manuscript.

References

Baldwin, S.A. and Henderson, P.J.F. (1989). Homology between sugar transporters from eukaryotes and prokaryotes. *Annual Reviews of Physiology* **51**, 459–471.

Baly, D.L. and Horuk, R. (1988). The biology and biochemistry of the glucose transporter. *Biochimica et Biophysica Acta* **947**, 571–590.

Barnett, J.E.G., Holman, G.D. and Muday, K.A. (1973). Structural requirements for binding of the sugar transport system of the human erythrocyte. *Biochemical Journal* **131**, 211–231.

Benaim, G. and Romero, P.J. (1990). A calcium pump in plasma membrane vesicles from *Leishmania braziliensis*. *Biochimica et Biophysica Acta* **1027**, 79–84.

Benaim, G., Bermudez, R. and Urbina, J.A. (1990). Ca^{2+} transport in isolated mitochondrial vesicles from *Leishmania braziliensis* promastigotes. *Molecular and Biochemical Parasitology* **38**, 61–68.

Benaim, G., Losada, S., Gadelha, F.R. and Docampo, R. (1991). A calmodulin-activated (Ca^{2+}-Mg^{2+})-ATPase is involved in Ca^{2+} transport by plasma membrane vesicles from *Trypanosoma cruzi*. *Biochemical Journal* **280**, 715–720.

Berridge, M.J. (1986). Inositol triphosphate and calcium mobilization. *Journal of Cardiovascular Pharmacology* **8**, S85–S90.

Bonay, P. and Cohen, E.B. (1983). Neutral amino acid transport in *Leishmania* promastigotes. *Biochimica et Biophysica Acta* **731**, 222–228.

Bringaud, F. and Baltz, T. (1992). A potential hexose transporter gene expressed predominantly in the bloodstream form of *Trypanosoma brucei*. *Molecular and Biochemical Parasitology* **52**, 111–122.

Brusell, E. (1970). "An Introduction to Insect Physiology", pp. 31–43. Academic Press, London.

Burgess, G.M., McKinney, J.S., Fabiato, A., Laslie, B.A. and Putney, J.W. (1983). Calcium pools in saponin-permeabilized guinea pig hepatocytes. *Journal of Biological Chemistry* **258**, 15336–15345.

Burrows, C.M. and Blum, J.J. (1991). Effects of hyperosmotic stress on alanine content of *Leishmania major* promastigotes. *Journal of Protozoology* **38**, 47–52.

Cairns, M.T., Smith, G., Henderson, P.J.F. and Baldwin, S.A. (1989a). Photoaffinity labelling of the GalP D-galactose transport protein of *Escherichia coli* with cytochalasin B. *Biochemical Society Transactions* **17**, 552–553.

Cairns, B.R., Collard, M.W. and Landfear, S.M. (1989b). Developmentally regulated gene from *Leishmania* encodes a putative membrane transport protein. *Proceedings of the National Academy of Sciences of the USA* **86**, 7682–7686.

Carafoli, E. (1987). Intracellular calcium homeostasis. *Annual Review of Biochemistry* **56**, 395–433.

Cassel, D., Shoubi, S., Glusman, G., Cukierman, E., Rotman, M. and Zilberstein, D. (1991). *Leishmania donovani*: characterization of a 38 kDa membrane protein that cross-reacts with the mammalian G-protein transducin. *Experimental Parasitology* **72**, 411–417.

Celenza, J.L., Marshall-Carlson, L. and Carlson, M. (1988). The yeast SNF3 gene encodes a glucose transporter homologous to the mammalian protein. *Proceedings of the National Academy of Sciences of the USA* **85**, 2130–2134.

Chang, K.P. and Dwyer, D.M. (1976). Multiplication of a human parasite (*Leishmania donovani*) in phagolysosomes of hamster macrophages *in vitro*. *Science* **193**, 678–680.

Chang, K.P., Fong, D. and Bray, R.S. (1985). Biology of *Leishmania* and leishmaniasis. In: "Leishmaniasis" (K.P. Chang and R.S. Bray, eds), pp. 1–30. Elsevier, Amsterdam.

Charalambous, B.M., Maiden, M.C.J., McDonald, T.P., Cunningham, I.J. and Henderson, P.J.F. (1989). Detection of proton-linked sugar proteins in Enterobacteriaceae. *Biochemical Society Transactions*, 17, 441–444.

Cohen, B.E., Ramos, H., Camargo, M. and Urbina, J. (1986). The water and ionic permeability induced by polyene antibiotics across plasma membrane vesicles from *Leishmania* sp. *Biochimica et Biophysica Acta* **860**, 57–65.

Docampo, R. and Pignataro, O. P. (1991). The inositol phosphate/diacylglycerol signalling pathway in *Trypanosoma cruzi*. *Biochemical Journal* **275**, 407–411.

Docampo, R. and Vercesi, A.E. (1989a). Ca^{2+} transport in coupled *Trypanosoma cruzi* mitochondria *in situ*. *Journal of Biological Chemistry* **264**, 108–111.

Docampo, R. and Vercesi, A.E. (1989b). Characteristics of Ca^{2+} transport by *Trypanosoma cruzi in situ*. *Archives in Biochemistry and Biophysics* **272**, 122–129.

Eisenschols, C., Paladini, A., Vedia, M.A., Torres, L. and Flawia, M. (1986). Evidence for the existence of a Gs-type regulatory protein in *Trypanosoma cruzi* membranes. *Biochemical Journal* **237**, 913–917.

Eisenthal, R., Game, S. and Holman, G.D. (1989). Specificity and kinetics of hexose transport in *Trypanosoma brucei*. *Biochimica et Biophysica Acta* **985**, 81–89.

Evans, D.A. and Brown, R.C. (1972). The utilization of glucose and proline by culture forms of *Trypanosoma brucei*. *Journal of Protozoology* **19**, 686–690.

Fairlamb, A.H. (1989). Novel biochemical pathways in parasitic protozoa. *Parasitology* **99**, S93–S112.

Flynn, I.W. and Brown, I.B.R. (1973). The metabolism of carbohydrates by pleotrophic African trypanosomes. *Comparative Biochemistry and Physiology* **45B**, 25–33.

Game, S., Holman, G.D. and Eisenthal, R. (1986). Sugar transport in *Trypanosoma brucei*: a suitable kinetic probe. *FEBS Letters* **194**, 126–130.

Geisow, M.J., Hart, P.D. and Young, M.R. (1981). Temporal changes of lysosome pH during phagolysosome formation in macrophages: study by fluorescence spectroscopy. *Journal of Cell Biology* **89**, 645–651.

Ghosh, J., Ray, M., Sarkar, S. and Amar, B. (1990). A high affinity Ca^{2+}-ATPase on the surface of *Leishmania donovani* promastigotes. *Journal of Biological Chemistry* **265**, 11345–11351.

Gill, D.L. and Cheuh, S.H. (1985). An intracellular (ATP-Mg^{2+})-dependent calcium pump. *Journal of Biological Chemistry* **260**, 9289–9297.

Glaser, T.A. and Mukkada, A.J. (1992). Proline transport in *Leishmania donovani* amastigotes: dependence on pH gradients and membrane potential. *Molecular and Biochemical Parasitology* **51**, 1–8.

Glaser, T.A., Baatz, J.E., Kreishman, G.P. and Mukkada, A.J. (1988). pH homeostasis in *Leishmania donovani* amastigotes and promastigotes. *Proceedings of the National Academy of Sciences of the USA* **85**, 7602–7606.

Glaser, T.A., Utz, G.L. and Mukkada, A.J. (1992). The plasma membrane electrical gradient (membrane potential) in *Leishmania donovani* promastigotes and amastigotes. *Molecular and Biochemical Parasitology* **51**, 9–16.

Goffeau, A. and Slayman, C.W. (1981). The proton translocation ATPase of the fungal plasma membrane. *Biochimica et Biophysica Acta* **639**, 197–223.

Gruenberg, J., Sharma, P.R. and Deshusses, J. (1978). D-Glucose transport is the rate limiting step in its metabolism. *European Journal of Biochemistry* **89**, 461–469.

Hansen, B.D. (1977). *Trypanosoma gambiense*: membrane transport of amino acids. *Experimental Parasitology* **48**, 296–304.

Henderson, P.J.F. (1990). Proton-linked sugar transport systems in bacteria. *Journal of Bioenergetics and Biomembranes* **22**, 525–569.

Henderson, P.J.F., Giddens, R.A. and Jones-Mortimer, M.C. (1977). Transport of galactose, glucose and their molecular analogues by *Escherichia coli* K12. *Biochemical Journal* **162**, 309–320.

Hesketh, T.R., Smith, G.A., Moore, J.P., Taylor, M.V. and Metcalf, J.C. (1983). Free cytosolic calcium concentration and the nitrogenic stimulation of lymphocytes. *Journal of Biological Chemistry* **258**, 4876–4882.

Hesketh, T.R., Moore, J.P., Morris, J.D.H., Taylor, M.V., Rogers, J. and Metcalf, J.C. (1985). A common sequence of calcium and pH signals in the mitogenic stimulation of eukaryotic cells. *Nature* **313**, 481–484.

Jackson, P.R. and Fisher, F.M., jr (1977). Carbohydrate effects on amino acid transport by *Trypanosoma equiperdum. Journal of Protozoology* **24**, 345–353.

Jaffe, C.L. and Rachamim, N. (1989). Amastigote stage-specific monoclonal antibodies against *Leishmania major. Infection and Immunity* **57**, 3770–3777.

Jung, C.Y. and Rampal, A.L. (1977). Cytochalasin B binding sites and transport carriers in human erythrocyte ghosts. *Journal of Biological Chemistry*, **252**, 5456–5463.

Kiaira, J.K. and Njogu, R.M. (1988). *Trypanosoma brucei brucei*: the catabolism of glycolytic intermediates by digitonin-permeabilized bloodstream trypomastigotes and some aspects of regulation of anaerobic glycolysis. *International Journal of Biochemistry* **20**, 1165–1170.

Konji, V.N., Olembo, N.K. and Pearson D.J. (1988). Proline synthesis in the fat body of the tsetse fly *Glossina morsitans*, and its stimulation by isocitrate. *Insect Biochemistry*, **18**, 449–452.

Krassner, S.M. (1969). Proline metabolism in *Leishmania tarentolae. Experimental Parasitology* **24**, 348–363.

Krassner S.M. and Flory, B. (1972). Proline metabolism in *Leishmania donovani* promastigotes. *Journal of Protozoology* **19**, 682–685.

Krassner, S.M. and Flory, B. (1977). Physiologic interactions between L-proline and D-glucose in *Leishmania tarentolae, L. donovani* and *Trypanosoma scelopori* culture forms. *Acta Tropica* **34**, 157–166.

Langford, C. K., Ewbank, S. A., Hanson, S. S., Ullman, B. and Landfear, S.M. (1992). Molecular characterization of two genes encoding members of the glucose transporter superfamily in the parasitic protozoan *Leishmania donovani. Molecular and Biochemical Parasitology* **55**, 51–64.

Law, S.S. and Mukkada, A.J. (1979). L-proline transport and regulation in *Leishmania tropica* promastigotes. *Journal of Protozoology* **26**, 295–301.

Lepely, P.R. and Mukkada, A.J. (1983). Characteristics of the uptake system for α-aminoisobutyric acid in *Leishmania tropica* promastigotes. *Journal of Protozoology* **30**, 41–46.

Liveanu, V., Webster, P. and Zilberstein, D. (1991). Localization of the plasma membrane and mitochondrial H^+-ATPase in *Leishmania donovani* promastigotes. *European Journal of Cell Biology* **54**, 95–101.

Lukacs, G.L., Rotstein, O.D. and Grinstein, S. (1991). Determination of the phagolysosomal pH in macrophages: *in situ* assessment of vacuolar H^+-ATPase activity, counterion conductance, and H^+ "leak". *Journal of Biological Chemistry* **266**, 24540–24548.

Madshus, I.H. (1988). Regulation of intracellular pH in eukaryotic cells. *Biochemical Journal* **250**, 1–8.

Maiden, M.C.J., Davis, E.O., Baldwin, S.A., Moore, D.C.M. and Henderson, P.J.F. (1987). Mammalian and bacterial sugar transport proteins are homologous. *Nature* **325**, 641–643.

McCormack, J.G. and Denton, R.M. (1986). Ca^{2+} as a second messenger within mitochondria. *Trends in Biochemical Sciences* **11**, 259–262.

Meade, J.C., Shae, J., Lemaster, S., Gallagher, G. and Stringer, J.R. (1987). Structure and expression of a tandem gene pair in *Leishmania donovani* that encodes a protein structurally homologous to eukaryotic ATPases. *Molecular and Cellular Biology* **7**, 3937–3946.

Meade, J.C., Hudson, K.M., Stringer, S.L. and Stringer, J.R. (1989). A tandem pair of *Leishmania donovani* cation transporting ATPase genes encode isoforms that are differentially expressed. *Molecular and Biochemical Parasitology* **33**, 81–92.

Meade, J.C., Coombs, G.H., Mottram, J.C., Steele, P.E. and Stringer, J.R. (1990). Conservation of cation-transporting ATPase genes in *Leishmania. Molecular and Biochemical Parasitology* **45**, 29–38.

Midgley, M. (1978). The transport of α-aminoisobutyrate into *Crithidia fasciculata. Biochemical Journal* **174**, 191–202.

Moreno, S.N.J., Vercesi, A.E., Pignataro, O.P. and Docampo, R. (1991). Calcium homeostasis in *Trypanosoma cruzi* amastigotes: presence of inositol phosphates and lack of an inositol 1,4,5-trisphosphate-sensitive calcium pool. *Molecular and Biochemical Parasitology* **52**, 251–262.

Mukkada, A.J. (1985). Energy coupling in active transport of substrates in *Leishmania. In* "Transport Process, Endo- and Osmoregulation" (G. Gilles and M. Gilles-Baillien, eds), pp. 326–333. Springer, Berlin.

Mukkada, A.J., Schaefer, F.W., III, Simon, M.W. and Neu, C. (1974). Delayed *in vitro* utilization of glucose by *Leishmania tropica* promastigotes. *Journal of Protozoology* **21**, 393–397.

Mukkada, A.J. and Simon, M.W. (1977). *Leishmania tropica*: uptake of methionine by promastigotes. *Experimental Parasitology* **42**, 87–96.

Mukkada, A.J., Meade, J.C., Glaser, T.A. and Bonventre, P.F. (1985). Enhanced metabolism of *Leishmania donovani* amastigotes at acidic pH: an adaptation for intracellular growth. *Science* **229**, 1099–1101.

Munoz-Antonia, T., Richards, F.F. and Ullu, E. (1991). Differences in glucose transport between bloodstream and procyclic forms of *Trypanosoma brucei rhodesiense. Molecular and Biochemical Parasitology* **47**, 73–82.

Nehlin, J.O., Carlberg, M. and Ronne, H. (1989). Yeast galactose permease is related to yeast and mammalian glucose transporters. *Gene* **85**, 313–319.

Nolan, D.P. and Voorheis, H.P. (1991). The distribution of permeant ions demonstrates the presence of at least two distinct electrical gradients in bloodstream forms of *Trypanosoma brucei. European Journal of Biochemistry* **202**, 411–420.

Opperdoes, F.R. (1987). Compartmentation of carbohydrate metabolism. *Annual Reviews of Microbiology* **41**, 127–151.

Opperdoes, F.R. and Borst, P. (1977). Localization of nine glycolytic enzymes in a microbody-like organelle in *Trypanosoma brucei*: the glycosome. *FEBS Letters* **80**, 360–364.

Padan, E. and Schuldiner, E. (1987). Intracellular pH and membrane potential as regulators in the prokaryotic cell. *Journal of Membrane Biology* **95**, 189–198.

Padan, E., Zilberstein, D. and Schuldiner, S. (1981). pH homeostasis in bacteria. *Biochimica et Biophysica Acta* **650**, 151–166.

Parsons, M. and Nielsen, B. (1990). Active transport of 2-deoxy-D-glucose in *Trypanosoma brucei* procyclic forms. *Molecular and Biochemical Parasitology* **42**, 197–204.

Pedersen, P.L. and Carafoli, E. (1987). Ion motive ATPases. I. Ubiquity, properties and significance to cell function. *Trends in Biochemical Sciences* **12**, 146–150.

Perlin, D.A., Kasamo, K., Brooker, R.J. and Slayman, C.W. (1984). Electrogenic H^+-translocation by the plasma membrane ATPase of *Neurospora*: studies on plasma membrane vesicles and reconstituted enzyme. *Journal of Biological Chemistry* **259**, 7884–7892.

Philosoph, H. and Zilberstein, D. (1989). Regulation of intracellular calcium in promastigotes of the human protozoan parasite *Leishmania donovani*. *Journal of Biological Chemistry* **264**, 10420–10424.

Raghupathi, S., Reddy, R. and Campbell, J.W. (1969). Arginine metabolism in insects. *Biochemical Journal* **115**, 495–503.

Rees, W.D. and Holman, G.D. (1981). Hydrogen binding requirements for insulin-sensitive sugar transport system of rat adipocytes. *Biochimica et Biophysica Acta* **646**, 251–260.

Revelard, P. and Pays, E. (1991). Structure and transcription of a P-ATPase gene from *Trypanosoma brucei*. *Molecular and Biochemical Parasitology* **46**, 241–252.

Rink, T.J., Tsein, R.Y. and Pozzan, T. (1982). Cytoplasmic pH and free Mg^{2+} in lymphocytes. *Journal of Cell Biology* **95**, 189–196.

Ruben, L., Hutchinson, A. and Moehlman, J. (1991). Calcium homeostasis in *Trypanosoma brucei*. *Journal of Biological Chemistry* **266**, 24351–24358.

Sanchez, G. (1974). Effect of some amino acids on carbohydrate uptake by *Trypanosoma lewisi*. *Comparative Biochemistry and Physiology* **47A**, 553–559.

Sanchez, G. and Read, C.P. (1969). Carbohydrate transport in *Trypanosoma lewisi*. *Comparative Biochemistry and Physiology* **28**, 931–945.

Saur, N. and Tanner, W. (1989). The hexose carrier from *Chlorella*. cDNA cloning of a eukaryotic H^+-cotransporter. *FEBS Letters* **259**, 43–46.

Schaefer, F.W., III and Mukkada, A.J. (1976). Specificity of the glucose transport system in *Leishmania tropica* promastigotes. *Journal of Protozoology* **23**, 446–449.

Schaefer, F.W., III, Martin, E. and Mukkada, A.J. (1974). The glucose transport system in *Leishmania tropica*. *Journal of Protozoology* **21**, 592–596.

Serrano, R., Kielland-Brandt, M.C. and Fink, G.R. (1986). Yeast plasma membrane ATPase is essential for growth and has homology with (Na^+-K^+)- and Ca^{2+}-ATPases. *Nature* **319**, 689–693.

Seyfang, A. and Duszenko, M. (1991). Specificity of glucose transport in *Trypanosoma brucei*. Effective inhibition by phloretin and cytochalasin B. *European Journal of Biochemistry* **202**, 191–196.

Shanahan, M.F., Olson, S.A., Weber, M.J., Linhard, G.E. and Gorga, J.C. (1982). Photoaffinity of glucose-sensitive cytochalasin B binding proteins in erythrocyte, fibroblast and adipocyte membrane. *Biochemical and Biophysical Research Communications* **107**, 38–43.

Simon, M.W. and Mukkada, A.J. (1977). *Leishmania tropica*: regulation and specificity of the methionine transport system in promastigotes. *Experimental Parasitology* **42**, 97–105.
Somlyo, A.P. (1984). Cell physiology: cellular site of calcium regulation. *Nature* **309**, 516–517.
Southworth, G.C. and Read, C.P. (1969). Carbohydrate transport in *Trypanosoma gambiense*. *Journal of Protozoology* **16**, 720–723.
Southworth, G.C. and Read, C.P. (1970). Specificity of sugar transport in *Trypanosoma gambiense*. *Journal of Protozoology* **17**, 396–399.
Stack, S.P., Stein, D.A. and Landfear, S.M. (1990). Structural isoform of a membrane transport protein from *Leishmania enriettii*. *Molecular and Cellular Biology* **10**, 6785–6790.
Stein, D.A., Cairns, B.R. and Landfear, S.M. (1990). Developmentally regulated transporter in *Leishmania* is encoded by a family of clustered genes. *Nucleic Acid Research* **18**, 1549–1557.
Ter Kuile, B.H. (1991). Glucose uptake mechanism as potential target for drugs against trypanosomatids. *In* "Biochemical Protozoology" (G.H. Coombs and M.J. North, eds), pp. 359–366. Taylor and Francis, London.
Ter Kuile, B.H. and Opperdoes, F.R. (1991). Glucose uptake by *Trypanosoma brucei*. *Journal of Biological Chemistry* **266**, 857–862.
Ter Kuile, B.H. and Opperdoes, F.R. (1992a). Comparative physiology of two protozoan parasites, *Leishmania donovani* and *Trypanosoma brucei*, grown in chemostat. *Journal of Bacteriology* **174**, 2929–2934.
Ter Kuile, B.H. and Opperdoes, F.R. (1992b). Proline metabolism in *Leishmania donovani*. *Journal of Protozoology*, **39**, 555–558.
Ter Kuile, B.H., Wiemer, E.A.C., Michels P.A.M. and Opperdoes, F.R. (1992). The electrochemical proton gradient in the bloodstream form of *Trypanosoma brucei* is dependent on the temperature. *Molecular and Biochemical Parasitology* **55**, 21–27.
Thissen, J.A. and Wang, C.C. (1991). Maintenance of internal pH and an electrochemical gradient in *Trypanosoma brucei*. *Experimental Parasitology* **72**, 243–251.
Thomas, A.P., Alexander, J. and Williamson, J.R. (1984). Relationship between inositol phosphate production and the increase of cytosolic free Ca^{2+} induced by vasopression in isolated hepatocytes. *Journal of Biological Chemistry* **259**, 5574–5584.
Urbina, J.A., Vivas, J., Ramos, H., Larrale, G., Aguilar, Z. and Avilan, L. (1988). Alteration of lipid order profile and permeability of plasma membranes from *Trypanosoma cruzi* epimastigotes grown in the presence of ketoconazole. *Molecular and Biochemical Parasitology* **30**, 185–196.
Vercesi, A.E. and Docampo, R. (1992). Ca^{2+} transport by digitonin-permeabilized *Leishmania donovani*. Effects of Ca^{2+}, pentamidine and WR-6026 on mitochondrial membrane potential *in situ*. *Biochemical Journal* **285**, 463–467.
Vercesi, A.E., Macedo, D.V., Lima, S.A., Gadelha, F.R. and Docampo, R. (1990). Ca^{2+} transport in digitonin-permeabilized trypanosomatids. *Molecular and Biochemical Parasitology* **42**, 119–124.
Vercesi, A.E., Hoffman, M.E., Bernardes, C.F. and Docampo, R. (1991). Regulation of intracellular calcium in *Trypanosoma cruzi*. Effects of calmidazolium and trifluoroperazine. *Cell Calcium* **12**, 316–369.
Visser, N. and Opperdoes, F.R. (1980). Glycolysis in trypanosomes. *European Journal of Biochemistry* **103**, 623–629.

Voorheis, H.P. (1977). Changes in the kinetic behavior of threonine transport into *Trypanosoma brucei* elicited by variation in hydrogen ion concentration. *Biochemical Journal* **164**, 15–25.

Wayne, P.S. and Vaughan, G.L. (1979). *Trypanosoma lewisi*: alterations in membrane function in the rat. *Experimental Parasitology* **48**, 15–21.

Zilberstein, D. (1991). Adaptation of *Leishmania* species to an acidic environment. *In* "Biochemical Protozoology" (G.H. Coombs and M.J. North, eds), pp. 349–358. Taylor and Francis, London.

Zilberstein, D. (1993). Glucose transport in protozoa. *In* "Ion Coupled Sugar Transport in Microorganisms" (M.G.P. Page and P.J.F. Henderson, eds). CRC Press, Boca Raton, Florida, in press.

Zilberstein, D. and Dwyer, D.M. (1984). Glucose transport in *Leishmania donovani* promastigotes. *Molecular and Biochemical Parasitology* **12**, 327–336.

Zilberstein, D. and Dwyer, D.M. (1985). Proton motive force-driven active transport of D-glucose and L-proline in the protozoan parasite *Leishmania donovani*. *Proceedings of the National Academy of Sciences of the USA* **82**, 1716–1720.

Zilberstein, D. and Dwyer, D.M. (1988). Identification of a surface membrane proton-translocating ATPase in promastigotes of the parasitic protozoan *Leishmania donovani*. *Biochemical Journal* **256**, 13–21.

Zilberstein, D. and Gepstein, A. (1993). Regulation by extracellular pH of L-proline transport in *Leishmania donovani*. *Molecular and Biochemical Parasitology*, in press.

Zilberstein, D., Dwyer, D.M., Matthaei, S. and Horuk, R. (1986). Identification and biochemical characterization of the plasma membrane glucose transporter of *Leishmania donovani*. *Journal of Biological Chemistry* **261**, 15053–15057.

Zilberstein, D., Sheppard, H.W. and Dwyer, D.M. (1987). The plasma membrane H^+-ATPase of *Leishmania donovani* promastigotes. *In* "Host–Parasite Cellular and Molecular Interactions in Protozoal Infections" (K.P. Chang and D. Snary, eds), Vol. H11, pp. 183–188 Springer, Berlin. (NATO ASI Series).

Zilberstein, D., Philosoph, H. and Gepstein, A. (1989). Maintenance of cytoplasmic pH and proton motive force in promastigotes of *Leishmania donovani*. *Molecular and Biochemical Parasitology* **36**, 109–118.

Zilberstein, D., Blumenfeld, N., Liveanu, V., Gepstein, A. and Jaffe, C.L. (1991). Growth at acidic pH induces an amastigote stage-specific protein in *Leishmania* promastigotes. *Molecular and Biochemical Parasitology* **45**, 175–178.

The Biology of Fish Coccidia

A. J. DAVIES

School of Life Sciences, Kingston University, Penrhyn Road, Kingston upon Thames, Surrey, KT1 2EE, UK

AND

S. J. BALL

Department of Life Sciences, University of East London, Romford Road, London, E15 4LZ, UK

ADVANCES IN PARASITOLOGY VOL. 32
ISBN 0-12-031732-X

I. Introduction

A bewildering array of coccidia exists in fishes from freshwater streams, rivers, lakes, brackish water, coastal seawater pools, shallow seas and from deeper salt water. Large numbers of teleosts and several elasmobranchs each have one or more kinds of coccidia and, not infrequently, host specificity is low, so that fish share these parasites. Except for Antarctica, infected fishes have been found in and around all continents of the world, although records from some areas (e.g. Africa, Australia, Indian subcontinent, New Zealand and South America) are few compared with those from others (e.g. Canada, China, Europe, North America and Russia).

According to Levine (1988), fish coccidia as a group include not only members of the family Eimeriidae Minchin, 1903, but others such as haemogregarines and dactylosomes. Historically however, the term "coccidia" has referred mainly to obligate intracellular protozoa of the genera *Eimeria* (Schneider, 1875) and *Isospora* (Schneider, 1881) in the family Eimeriidae. There are many *Eimeria* species in fish but few *Isospora*, and there are parasites from three additional families (Table 1). All these families are placed in the suborder Eimeriorina Léger, 1911 by Levine (1988), and Levine (1973) prefers to regard only members of this suborder as true coccidia. The fish parasites recorded here can therefore be regarded as true coccidia, except for two enigmatic examples, *Sarcocystis* and *Rhabdospora*, which are mentioned briefly.

Table 1 *Classification scheme for fish coccidia based on Levine (1982, 1983, 1988)*

Phylum Apicomplexa Levine, 1970
Apical complex present at some stage, usually comprising polar ring(s), rhoptries, micronemes, conoid and subpellicular microtubules; micropore(s) generally present at some stage; sexuality by syngamy; all species parasitic

Class Sporozoasida Leuckart, 1879
Conoid usually present and forms a complete cone; reproduction usually asexual and sexual; oocysts contain infective sporozoites which result from sporogony; locomotion by body flexion, gliding, undulation or flagella; flagella, when present, on microgametes only; homoxenous or heteroxenous

Subclass Coccidiasina Leuckart, 1879
Gamonts usually present, small and intracellular; conoid not modified into mucron or epimerite; syzygy generally absent, if present involves gametes; anisogamy marked; life cycle characteristically involves merogony, gamogony and sporogony; most species in vertebrates

Order Eucoccidiorida Léger and Duboscq, 1910
Merogony, gamogony, and sporogony present; in invertebrates and vertebrates

Suborder Eimeriorina Léger, 1911

Macrogamete and microgamont develop independently; syzygy absent; microgamont typically produces many microgametes; zygote not motile; sporozoites typically enclosed in a sporocyst

Family Cryptosporidiidae Léger, 1911

Development just under surface membrane of host cell or within its brush border and not in cell proper; developmental stages form attachment organelle that anchors parasite to base of the parasitophorous vacuole; microgametes without flagella; homoxenous; oocysts with or without sporocysts, with 4 naked sporozoites

Genus *Crytosporidium* Tyzzer, 1907

With characteristics of the family; 4 or 5 valid species, 1 in fish

Family Eimeriidae Minchin, 1903

Oocysts with 0, 1, 2, 4 or more sporocysts, each with 1 or more sporozoites; sporocysts univalved, without dehiscence line (Fig. 1a); homoxenous; merogony and gamogony within host, sporogony typically outside; microgametes with 2 or 3 flagella

Genus *Eimeria* Schneider, 1875

Oocysts with 4 sporocysts, each with 2 sporozoites; merogony intracellular; see Table 2

Genus *Epieimeria* Dyková and Lom, 1981

Oocysts with 4 sporocysts, each with 2 sporozoites; merogony and gamogony extracellular; sporogony intracellular; in fish only; 4 named species

Genus *Isospora* Schneider, 1881

Oocysts with two sporocysts, each with 4 sporozoites; 3 examples in fish

Genus *Octosporella* Ray and Raghavachari, 1942

Oocysts with 8 sporocysts, each with 2 sporozoites; 6 named species, 3 in fish

Family Barrouxiidae Léger, 1911

Homoxenous; oocysts with different numbers of sporocysts depending upon genus; sporocysts bivalved, with dehiscence suture between the 2 halves

Genus *Goussia* Labbé, 1896

Oocysts with 4 sporocysts, each with 2 sporozoites; dehiscence suture longitudinal (Fig. 1b); see Table 3

Genus *Crystallospora* Labbé, 1896

Oocysts with 4 sporocysts, each with 2 sporozoites; sporocysts resemble crystals, comprising 2 pyramids with hexagonal bases, joined base to base; with equatorial suture line (Fig. 1c); only 1 named species, and that in fish

Family Calyptosporidae Overstreet, Hawkins and Fournie, 1984

Heteroxenous; oocysts with 4 sporocysts, each with 2 sporozoites; sporocysts with sporopodia (Fig. 1d).

Genus *Calyptospora* Overstreet, Hawkins and Fournie, 1984

with characteristics of the family; 4 named species, these in fish

The history of the coccidia has been reviewed by Stunkard (1969) and Levine (1973), and several excellent annotated lists of fish coccidia have been drawn up, including those of Shul'man and Shtein (1962), Pellérdy (1974), Dyková and Lom (1983) and Levine (1982, 1983, 1988). The first few of these parasites to be named from fish were reported a century ago. Thélohan (1890) described *Coccidium gasterostei* from the liver of *Gasterosteus aculeatus*, and *Coccidium sardinae* in the testes of *Sardina pilchardus*, while in 1892 (Thélohan 1892a) he named *Coccidium cruciatum* from the liver of *Caranx trachurus*, and *Coccidium minutum* in the liver, kidneys and spleen of *Tinca tinca*. Labbé (1896) transferred the latter two parasites to the genus *Goussia*, which he created, and he also devised the genus *Crystallospora* for *Coccidium crystalloides* (Thélohan, 1893) from the intestine and pyloric caeca of *Motella tricirrata*. As Dyková and Lom (1981) noted, Labbé's (1896) genera *Goussia* and *Crystallospora* were later reduced to synonyms of *Eimeria* by Doflein (1909), while others continued to use *Goussia* (Léger and Hesse, 1919; Stankovitch, 1920), but by 1953 (Grassé, 1953; Reichenow, 1953) all fish coccidia were included in the genus *Eimeria*.

Since this beginning, the study of fish coccidia has lagged behind studies on mammalian and avian coccidiosis, probably for commercial reasons. In farm animals, serious disease can result from intensive husbandry which can lead to a build-up of the infective stages (oocysts) on litter or pasture. The result has been an enormous amount of work on the biology and control of the coccidia of veterinary importance. Until relatively recently, by contrast, most of the work on fish coccidia has been descriptive, based on oocyst appearance, and for most species only this stage and the site of infection are known. However, a growing realization of the role of coccidia in fish diseases, together with expansion in aquaculture, has helped to improve knowledge of this group.

When Pellérdy compiled his classic text, "Coccidia and Coccidiosis" (Pellérdy, 1974), he included in the genus *Eimeria* 68 named species of fish coccidia and 11 unnamed species. While many fish coccidia are still retained within this genus, other genera have since been resurrected or created (Table 1). The genera *Goussia* Labbé 1896 and *Crystallospora* Labbé 1896 were revived by Dyková and Lom (1981), who also created the new genus *Epieimeria* Dyková and and Lom 1981. Levine (1984a) united *Epieimeria* with *Cryptosporidium* but this decision was later revised (Levine, 1988). A second new genus, *Calyptospora* Overstreet, Hawkins and Fournie was established in 1984, while Daoudi (1987) (see Molnár, 1989) and Daoudi *et al.* (1989) proposed the genera *Nucleoeimeria* and *Nucleogoussia*. In addition, two genera thought not to occur in fish until recently, *Cryptosporidium* and *Octosporella*, were noted by Levine (1988), and there is one account of *Isospora* in fish (Davronov, 1987). Two records of a

Sarcocystis-like fish parasite exist (Fantham and Porter, 1943; Kent *et al.*, 1989), and there are several accounts of *Rhabdospora*, but there is no firm evidence that either is a true fish coccidian.

Long and Joyner (1984) discussed the problem of identification of species of *Eimeria*, with particular reference to those of chickens, and they indicated the limitations of using morphological data derived from oocysts and the necessity for using other characteristics. However, the oocyst and, in particular the structure of its contained sporocysts, are considered important features in differentiating genera and species of coccidia in fishes. For example, Upton *et al.* (1984) produced a taxonomic key for the identification of the Eimeriidae in North American fishes using oocyst and sporocyst characteristics. Except for three genera, *Cryptosporidium, Octosporella* and *Isospora*, fish coccidia have oocysts containing four sporocysts, each with two sporozoites (tetrasporocystic, dizoic), and therefore they conform to the genus *Eimeria*. Dyková and Lom's (1981) decision to revive the genera *Goussia* and *Crystallospora* was based mainly on the excystation structure of the sporocyst and patterns of development, while the genus *Calyptospora* was created for similar reasons (Overstreet *et al.*, 1984) (see Fig.1). The genera *Epieimeria, Nucleoeimeria* and *Nucleogoussia* were established or suggested for *Eimeria* and *Goussia* developing outside traditional (intra-cytoplasmic) sites within the host cell (Dyková and Lom, 1981; Daoudi *et al.*, 1987, 1989).

As with coccidia from other groups of vertebrates, the gut is often the favoured site for development, but fish coccidia are remarkable for their extraintestinal development (Ball *et al.*, 1989). Levine (1988) listed 134 species of fish coccidia in seven genera. Of the 98 species of *Eimeria*, only seven occur in the Chondrichthyes; six of these are in the gut and one is extraintestinal. Twenty-two of 91 *Eimeria* species found in the Osteichthyes develop in extraintestinal sites, namely liver, kidney, spleen, testes, ovary, gall bladder, swim bladder, peritoneum, adipose tissue and gills (see Table 2). One species of *Goussia* is recorded in the Chondrichthyes. Thirteen of 26 *Goussia* species from Osteichthyes occur in sites other than the intestines (see Table 3). This high proportion (nearly 30%) of coccidia developing extrain-testinally is one principal difference between those in fish and those in homeotherms. In addition, fish coccidia show tissue specificity which is sometimes so low that identical stages may develop in several different organs and tissue types within a single host.

Development follows a eucoccidian pattern of merogony, gamogony, oogony and sporogony. Merogony, gamogony and oogony are endogenous. Sporogony, however, may contrast with that of eimerians of birds and mammals where, except in isosporans that sporulate endogenously (e.g. *Isospora papionis* McConnell, Vos, Basson and de Vos, 1971, from chacma

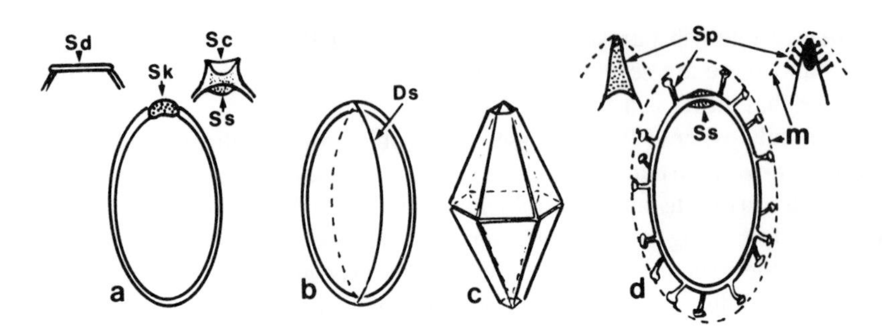

FIG. 1. Diagrammatic representations of examples of sporocysts of various genera of fish coccidia (sporozoites not shown). (a) *Eimeria*; *Epieimeria*. Sd, Stieda body disc-like, e.g. *Eimeria esoci*, *Eimeria truttae*; or shallow collar-like, e.g. *Eimeria ivanae*, *Eimeria lairdi*; Sk, knob- or cap-like Stieda body, e.g. *Eimeria jiroveci*, *Eimeria moronei*, *Eimeria nucleocola*. *Eimeria variabilis*; Sc, collar-like Stieda body, e.g., *Eimeria hexagona*; with sub-Stieda body (Ss), e.g. *Epieimeria isabellae* (after Lom and Dykova, 1982). (b) *Goussia*. Ds, dehiscence suture between two valves. (c) *Crystallospora* (after Thélohan, 1893, 1894). (d) *Calyptospora*. m, membrane. Sporopodia (Sp) over entire surface, e.g. *Calyptospora funduli* (after Duszynski *et al.*, 1979); single sporopodium, e.g. *Calyptospora tucunarensis* (after Bekesi and Molnar, 1991); single sporopodium with numerous minute lateral projections (not shown), e.g. *Calyptospora empristica*; 8–10 sporopodia, e.g. *Calyptospora serrasalmi* (after Cheung *et al.*, 1986).

TABLE 2 *Host and site of development of* Eimeria *of fishes*

Species of Eimeria	*Host*	*Location in host*[a]	*Original reference*[b]
sp.'*a*'	*Notemigonus crysoleucas*	Sb	Li and Desser (1985a)
acerinae	*Acerina cernuo*	A	Pellérdy and Molnár (1971)
ambassi	*Barbus ambassis*	Ab	Patnaik and Archarya (1972)
amudarinica	*Scardinius erythrophthalmus*	I	Davronov (1987)
amurensis	*Pseudorasbora parva*	K,L	Achmerov (1959)
	Sarcochilichthys sinensis		
aristichthysi	*Aristichthys nobilis*	A	Lee and Chen (1964)
	Hypophthalmichthys molitrix		

TABLE 2 *Continued*

Species of Eimeria	*Host*	*Location in host*[a]	*Original reference*[b]
ashburneri	*Macquaria ambigua*	I,Pc	Molnár and Rohde (1988b)
atherinae	*Atherina boyeri*	I	Daoudi *et al.* (1987)
aurati	*Carassius auratus*	F	Hoffman (1965)
banyulensis	*Crenilabrus mediterraneus*	A	Lom and Dyková (1982)
barbi	*Barbus capito conocephalus*	I	Davronov (1987)
baueri	*Carcassius carassius*	Gb, H, K, L, S, U	Alvarez-Pellitero and Gonzalez-Lanza (1986)
bouixi	*Dicentrarchus labrax*	Pc	Daoudi and Marquès (1986–1987)
branchiphila	*Rutilus rutilus*	G, Gf, K, S	Dyková *et al.* (1983)
brevoortiana	*Brevoortia tyrannus*	Pc, T	Hardcastle (1944)
capoetobramae	*Capoetobrama kuschakowitschi*	K, U	Allamaratov and Iskov (1970)
carassii	*Carassius carassius*	F	Yakimoff and Gousseff (1935)
carassius-aurati[c]	*Carassius auratus*	A	Romero-Rodriguez (1978)
catalana	*Crenilabrus mediterraneus*	A	Lom and Dyková (1981)
catostomi	*Catostomus commersoni* *Hypentelium nigricans*	A	Molnár and Hanek (1974)
cheilodactyli	*Cheilodactylus fuscus*	A, Pc	Molnár and Rohde (1988a)
cheisini	*Gobio gobio* *Hemiculter leucisculus* *Hemibarbus labeo*	I, Gb, Me, Sb	Shul'man and Zaika (1962)
cheni	*Mylopharyngodon piceus*	I	Shul'man and Zaika (1962)
chollaensis	*Urolophus halleri*	I Sv	Upton *et al.* (1988)
ciliatae	*Sillago ciliata*	I, Pc	Molnár and Rohde (1988a)
citriformis	*Tilesina gibbosa*	A	Dogiel (1948)
clini	*Clinus superciliosus*	I	Fantham (1932)

TABLE 2 *Continued*

Species of Eimeria	*Host*	*Location in host*[a]	*Original reference*[b]
cobitis	*Cobitis taenia*	L	Stankovitch (1924)
cotti	*Cottus gobio*	I, Pc	Gauthier (1921)
culteri	*Culter erythropterus*	I	Lee and Chen (1964)
cylindrospora	*Alburnus alburnus*	I	Stankovitch (1921)
cyprini	*Carassius carrassius* *Cyprinus carpio*	I	Plehn (1924)
cyprinorum	*Abramis brama* *Barbus barbus* and others	I	Stankovitch (1921)
dicentrarchi	*Dicentrarchus labrax*	Pc	Daoudi and Marquès (1968–1987)
dingleyi	*Blennius pholis*	I	Davies (1978)
duszynskii	*Cottus bairdi*	I	Conder *et al.* (1980)
dykovae	*Cheilodactylus fuscus*	I, Pc	Molnár and Rohde (1988a)
escoi	*Esox lucius*	I	Shul'man and Zaika (1962)
	Lepomis gibbosus	G, Gb, I, K, L, M, S, Sb	Li and Desser (1985a)
etheostomae	*Etheostoma exile* *E. nigrum*	I	Molnár and Hanek (1974)
etrumei	*Etrumeus micropus*	T	Dogiel (1984)
euzeti	*Myliobatus aquila*	L	Daoudi *et al.* (1987)
evaginata	*Sebastodes taczanowskii* *Myoxocephalus stelleri*	Pc	Dogiel (1948)
fernandoae	*Catastomus commersoni* *Hypentelium nigricans*	I	Molnár and Hanek (1974)
freemani	*Notropis conutus*	K	Molnár and Fernando (1974)
gasterostei	*Gasterosteus aculeatus* *G. clupeatus*	L	Thélohan (1890)

TABLE 2 *Continued*

Species of Eimeria	*Host*	*Location in host*[a]	*Original reference*[b]
E.(?) gigantea	*Lamna cornubica*	Sv	Labbé (1896)
glenorensis	*Morone americana*	I	Molnár and Fernando (1974)
glossogobii	*Glossogobius giuris*	A	Mukherjee and Haldar (1980)
gobii	*Gobius nudiceps*	I	Fantham (1932)
haichengensis	*Cyprinus carpio*	I	Chen (1962)
halleri	*Urolophus halleri*	F	Upton *et al.* (1986)
haneki	*Culaea inconstans*	I	Molnár and Fernando (1974)
harpodoni	*Harpodon nehereus*	I	Setna and Bana (1935)
hemibarba	*Hemibarbus maculatus*	I	Su and Chen (1987)
hemiculterii	*Hemiculter leucisculus*	I	Chen and Hsieh (1964)
hexagona	*Onos tricirratus*	I	Lom and Dyková (1981)
hoffmani	*Umbra limi*	I	Molnár and Hanek (1974)
huanggangensis	*Misgurnus anguillicaudatus*	K	Su and Chen (1987)
hupehensis	*Carassius auratus*	I	Chen and Hsieh (1964)
hybognathi	*Hybognathus hankinsoni*	I	Molnár and Fernando (1974)
hypoph-thalmichthys	*Hypothalmichthys molitrix*	K	Achmerov (1959)
ictaluri	*Ictalurus nebulosus*	I	Molnár and Fernando (1974)
insignis	*Scorpaena notata*	Pc	Lom and Dyková (1982)
invanae	*Serranus cabrilla*	Pc	Lom and Dyková (1981)
jiroveci	*Raja clavata*	I	Lom and Dyková (1981)
kassaii	*Umbra krameri*	A, F	Molnár (1978)
kotorensis	*Spicara maena*	I	Daoudi *et al.* (1987)
kwangtun-gensis	*Channa argus* *C. maculata*	I	Chen (1960)
lairdi	*Myoxocephalus scorpius*	Pc	Lom and Dyková (1981)

TABLE 2 *Continued*

Species of Eimeria	*Host*	*Location in host*[a]	*Original reference*[b]
leucisci	*Leuciscus leuciscus baicalensis* and others	K	Shul'man and Zaika (1964)
maggieae	*Pagellus erythrinus*	I	Lom and Dyková (1981)
matskasii	*Umbra krameri*	A, F	Molnár (1978)
merlangi	*Odontogadus merlangus euxinus*	I, Gb	Zaika (1966)
meszarosi	*Umbra krameri*	A, F	Molnár (1978)
micropteri	*Micropterus dolomieni* *M. salmoides*	I	Molnár and Hanek (1974)
misgurni	*Misgurnus fossilis*	I	Stankovitch (1924)
molnari[c]	*Gobio gobio*	A	Jastrzebski (1982)
moronei	*Morone americana*	I	Molnár and Fernando (1974)
muraiae	*Misgurnus fossilis*	K	Molnár (1978)
mylopharyngodoni	*Mylopharyngodon piceus*	I, K, L	Chen (1956a)
myoxocephali	*Myoxocephalus polycanthocephalus*	I	Fitzgerald (1975)
nemethi	*Alburnus alburnus*	K, L, S	Molnár (1978)
nicollei	*Carassius carassius*	F	Yakimoff and Gousseff (1935)
nishin	*Clupea harengus*	T	Fujita (1934)
notopteri	*Notopterus notopterus*	I	Chakravarty and Kar (1944)
nucleocola	*Myoxocephalus scorpius*	Pc	Lom and Dyková (1981)
ochetobii	*Ochetobius elongatus*	I	Lee and Chen (1964)
odontobutis	*Odontobutis obscura*	I	Su and Chen (1987)
ojibwana	*Cottus bairdi*	I	Molnár and Fernando (1974)
ophiocephali	*Ophiocephalus argus* *O. maculatus*	I	Chen and Hsieh (1960)
orientalis	*Misgurnus anguillicaudatus*	K	Chen (1984)

TABLE 2 *Continued*

Species of Eimeria	*Host*	*Location in host*[a]	*Original reference*[b]
osmeri	*Osmerus mordax*	I	Molnár and Fernando (1974)
ottojiroveci	*Raja clavata*	I, Sv	Dyková and Lom (1983)
parasiluri	*Parasilurus asotus*	Gb	Lee and Chen (1964)
pastuszkoi	*Nemachilus barbatulus*	A	Jastrzebski (1982)
patersoni	*Lepomis gibbosus*	L, R, S	Lom *et al.* (1989)
percae	*Perca fluviatilis*	I (L?)	Dujarric de la Rivière (1914)
petrovici	*Symphodus ocellatus*	I	Daoudi *et al.* (1987)
philypnodoni	*Philypnodon grandiceps*	I	Molnár and Rohde (1988b)
piraudi	*Cottus gobio*	I	Gauthier (1921)
pleurostici	*Sphaeroides pleurosticus*	I	Molnár and Rohde (1988a)
pneumatophori	*Scomber japonicus*	L	Dogiel (1948)
pungitii	*Pungitius pungitius*	I	Molnár and Hanek (1974)
quentini[c]	*Aetobatis narinari*	Pe	Boulard (1977)
raiarum	*Raja batis*	I	van den Berghe (1937)
raibauti	*Trisopterus minutus*	Pc	Daoudi *et al.* (1989b)
roussillona	*Labrus turdus*	I	Lom and Dyková (1981)
rouxi	*Tinca tinca*	I	Elmassian (1909)
rutili	*Rutilus rutilus caspicus*	K	Dogiel and Bykhovskii (1939)
salvelini	*Salvenlinus fontinalis*	I	Molnár and Hanek (1974)
sardinae	*Clupea pilcharda* (and others since)	T	Thélohan (1890)
saurogobii	*Ctenopharyngodon idella*	I	Chen (1964)
scardinii	*Scardinius erythrophthalmus*	K	Pellérdy and Molnár (1968)
schizothoraci	*Schizothorax intermedius*	I	Davronov (1987)
schulmani[c]	*Leuciscus idus*	I	Kulemina (1969)
scorpaenae	*Scorpaena porcus*	I	Zaika (1966)
sericei	*Rhodeus sericeus*	I	Mikhailov (1975)
sillaginis	*Sillago ciliata*	I, Pc	Molnár and Rohde (1988a)

TABLE 2 *Continued*

Species of Eimeria	*Host*	*Location in host*[a]	*Original reference*[b]
siluri	*Silurus glanis*	I	Davronov (1987)
sinensis[c]	*Hypophthalmichthys moltrix* *Aristichthys nobilis*	I	Chen (1956b)
smaris	*Spicara smaris*	I	Daoudi *et al.* (1989b)
soufiae	*Leuciscus soufia agassizi*	I	Stantovitch (1921)
southwelli	*Aetobatis narinari*	Sv, L	Halawani (1930)
sphaerica	*Opisthocentrus ocellatus*	K	Dogiel (1948)
squali	*Squalus acanthias*	Sv	Fitzgerald (1975)
strelkovi	*Pseudorasbora parva*	K	Shul'man and Zaika (1962)
symphodi	*Symphodus rostratus*	I	Daoudi *et al.* (1989b)
syngnathi	*Syngnathus nigrolineatus*	I contents	Yakimoff and Gousseff (1936)
syrdarinica	*Ctenophargyngodon idella*	I	Davronov (1987)
tedlai	*Perca flavescens*	I	Molnár and Fernando (1974)
triglae	*Trigla lucerna*	Pc	Daoudi *et al.* (1989b)
truttae	*Salmo fario*	I, Pc	Léger and Hesse (1919)
vanasi	*Oreochromis aureus* and others	I	Landsberg and Paperna (1987)
variabilis	*Cottus bubalis* and others	I, Pc	Thélohan (1893, 1894)
varicorhini	*Varicorhinus capoeta heratensis*	I	Davronov (1987)
zarnowskii[c]	*Gasterosteus aculeatus*	A	Jastrzebski (1982)
zygaenae	*Sphyrna blochii*	I	Mandal and Chakravarty (1965)

[a]Abbreviations: A, alimentary tract; Ab, abscess in shoulder; F, faeces; G, gonad; Gb, gall bladder; Gf, gill filaments; H, heart; I, intestine; K, kidney; L, liver; M, muscle; Me, mesentery; P, pancreas; Pc, pyloric caeca; Pe, peritoneum; R, renal tubule; S, spleen; Sb, swim bladder; Sv, spiral valve; T, testis; U, urinary duct; Ub, urinary bladder.
[b]These references give the original authority for the coccidian species named and the original site in the host; the species may have been described since from other host organs. Generic and specific names for fish hosts are those quoted in the original descriptions of the coccidia. Some host names may have been revised subsequently.
[c]These species are referred to as *Goussia* in some recent texts.

TABLE 3 *Host and site of development of* Goussia *of fishes*

Species of Goussia	*Host*	*Location in host*[a]	*Original reference*[b]
sp. '*a*'	*Notropus cornutus*	Sb	Li and Desser (1985a)
acipenseris	*Acipenser ruthenus*	I, Pc	Molnár (1986)
aculeati	*Gasterosteus aculeatus*	I	Jastrzebski (1984), Jastrzebski *et al.* (1988)
alburni	*Alburnus lucidus* and others	I, Pf	Stankovitch (1920)
arrawarra	*Sillago ciliata*	I	Molnár and Rohde (1988a)
auxidis	*Auxis maru*	L	Dogiel (1948)
sp. '*b*'	*Nototropus cornutus*	Sb	Li and Desser (1985a)
balatonica	*Blicca bjoerkna*	I	Molnár (1989)
bigemina	*Ammodytes tobianus*	I	Labbé (1896)
callinani	*Hypseleotris compressa*	I	Molnár and Rohde (1988b)
carassici	*Carassius auratus*	L	Su (1987)
carpelli	*Cyprinus carpio*	I	Léger and Stankovitch (1921)
caseosa	*Macrourus berglax*	Gb, Gg, I, Mb Sb	Lom and Dykóva (1982)
cichlidarum	*Sarotherodon galilaeus* and others	Sb	Landsberg and Paperna (1985)
clupearum	*Clupea harengus* and others	L	Thélohan (1894)
cruciata	*Caranx trachurus*	L	Thélohan (1892a)
degiustii	*Notropus cornutus* and others	S	Molnár and Fernando (1974)
ethmalotis	*Ethmalosa fimbriata*	L	Obiekezie (1986)
gadi	*Melanogrammus aeglefinus*	Sb	Fiebiger (1913)
girellae	*Girella nigricans*	E, I, L, S	Kent *et al.* (1988)
iroquoina	*Notropis cornutus* and others since	I	Molnár and Fernando (1974)
janae	*Leuciscus leuciscus L. cephalus*	I	Lukeš and Dyková (1990)
langdoni	*Macquaria ambigua*	I, Pc	Molnár and Rohde (1988b)

TABLE 3 *Continued*

Species of Goussia	*Host*	*Location in host*[a]	*Original reference*[b]
laureleus	*Perca flavescens*	I	Molnár and Fernando (1974)
legeri	*Alburnis lucidus* *Scardinius erythrophthalmus* and others	I	Stankovitch (1920)
lomi	*Maccullochella peeli*	I	Molnár and Rohde (1988b)
luciae	*Mullus barbatus*	I	Lom and Dyková (1982)
lucida	*Mustelus canis* and others	Sv	Labbé (1893)
metchnikovi	*Gobio gobio*	I, K, L, S	Laveran (1897)
microcanthi	*Microcanthus stigatus*	I	Molnár and Rohde (1988a)
minuta	*Tinca tinca*	K, L, S	Thélohan (1892a)
motellae	*Motella tricirrata*	I, Pc	Labbé (1893)
notemigonica	*Notemigonus crysoleucas*	K, S, Sb, U	Li and Desser (1985a)
notropicum	*Notropus cornutus*	I	Li and Desser (1985a)
pannonica	*Blicca bjoerkna*	I	Molnár (1989)
polylepidis	*Chondrostoma polylepis*	K, Pe, Sb, U	Alvarez-Pellitero and Gonzalez-Lanza (1985)
siliculiformis	*Gobio albipinnatus tenuicorpus*	I, K, Sb	Shul'man and Zaika (1962)
sp.	*Tinca tinca*	I	Molnár (1982)
sp.	*Rhabdosargus sarba*	L	Molnár and Rohde (1988a)
sp.	*Anguilla reinhardti* *A. australis*	F, I	Molnár and Rohde (1988b)
sp.	*Gnathopogon chankaensis*	I	Dogiel and Achmerov (1959)
sp.I	*Abramis brama*	I	Molnár (1989)
sp.II	*Rutilus rutilus*	I	Molnár (1989)
sp.III	*Leuciscus cephalus*	I	Molnár (1989)
sp.IV	*Barbus barbus*	I	Molnár (1989)
sp.V	*Alburnus alburnus*	I	Molnár (1989)
sp.VI	*Gobio albipinnatus*	I	Molnár (1989)
sp.VII	*Abramis brama*	I	Molnár (1989)
sp.VIII	*Rutilus rutilus*	I	Molnár (1989)
sp.IX	*Leuciscus cephalus*	I	Molnár (1989)

TABLE 3 *Continued*

Species of Goussia	*Host*	*Location in host*[a]	*Original reference*[b]
sp.X	*Scardinius erythrophthalmus*	I	Molnár (1989)
spraguei	*Gadus morhua* *Melanogrammus aeglefinus*	K	Morrison and Poynton (1989a,b)
stankovitchi	*Alburnus alburnus* and others	I	Pinto (1928)
subepithelialis	*Cyprinus carpio*	I	Moroff and Fiebiger (1905)
thelohani	*Labrus* sp.	L	Labbé (1896)
vargai	*Acipenser ruthenus*	I, Pc	Molnár (1986)

[a]Abbreviations: as in Table 2 plus E, endothelium of gill blood vessels; Gg, gas gland; Mb, mesenteric blood vessels; Pf, peritoneal fat.
[b]See footnote b to Table 2.

baboons), exogenous sporulation is normally the rule. In fish coccidia sporogony is often endogenous, though it may be exogenous, or both endogenous and exogenous within one species. Endogenous development may be classified as intracellular, extracellular, intercellular, intranuclear, epiplasmal or epicellular. Polyxenous species are not uncommon. Some coccidia infect up to seven or eight separate fish species (Dyková and Lom, 1983), and one has been reported from 17 different fishes (Lukeš *et al.*, 1991). With one exception (Solangi and Overstreet, 1980), evidence for a truly heteroxenous life cycle for fish coccidia is lacking, although paratenic hosts are involved in transmission of some species (Molnár, 1979; Kent and Hedrick, 1985; Steinhagen and Körting, 1990; Lukeš *et al.*, 1991).

During the past two decades knowledge of fish coccidia has progressed significantly. Recent information, together with stimulating articles on fish coccidia, such as those by Dyková and Lom (1981), Overstreet (1981) and Desser (1981), prompted this review and undoubtedly influenced our approach.

II. LIFE CYCLES

A. GENERAL

When the life cycles of fish coccidia have been studied, their sequential development apparently is similar to that seen in coccidia from other vertebrates. Sporozoites initiate infection probably when released in the lumen of the gut from an ingested sporulated oocyst or free sporocyst (homoxenous), or from an ingested intermediate host (heteroxenous). Sporozoites may then enter host gut epithelial cells, or travel to extraintestinal sites, and there transform into trophozoites which become meronts. These divide by asexual proliferation, merogony, to form merozoites which may enter new host cells to produce more merogonous cycles. Usually, in coccidia, there are fixed numbers of asexual generations. Merozoites of the final merogony enter host cells to initiate the sexual cycle, and become either macrogamonts or microgamonts. The former develop into macrogametes without further division whilst microgamonts divide by multiple fission to produce, usually, large numbers of flagellated microgametes. Microgametes fertilize macrogametes to produce zygotes. In fish, zygotes may eventually be surrounded by very thin (e.g. 13–100 nm, see Lom, 1971; Upton *et al.*, 1984) or thick resistant walls (e.g. 1–2 μm, see Duszynski *et al.*, 1979; Upton *et al.*, 1984) and become oocysts, which eventually sporulate.

B. SPECIFIC

The diagnostic features of each genus noted below are recorded in Table 1, except for *Sarcocystis* and *Rhabdospora.*

1. *Homoxenous genera*

(a) *Cryptosporidiidae*

(i) *Cryptosporidium. Cryptosporidium* species develop within the microvillous region of the intestinal epithelial cells of many vertebrates and occasionally in the respiratory tract of birds. Development is intracellular but extracytoplasmic, and is referred to as epicellular. One enteric species has been named from fish, *Cryptosporidium nasorum* Hoover, Hoerr, Carlton, Hinsman and Ferguson, 1981 which occurred in the naso tang, *Naso lituratus* (see Hoover *et al.*, 1981). Unnamed species of *Cryptosporidium* have been reported from the stomach of the cichlid hybrid *Oreochromis aureus* × *Oreochromis niloticus* by Landsberg and Paperna (1986), and from *Oreochromis aureus* by Paperna (1987).

The taxonomy of this genus is uncertain because cross transmission experiments have shown little host specificity between isolates from mammals. On this basis, Levine considered only four (Levine, 1984b) or five (Levine, 1988) of the 19 named species valid, one in reptiles, one in birds, two in mammals, and one, *Cryptosporidium nasorum*, in a fish. Current and Blagburn (1990) added one further valid species infecting birds. The life cycle of a *Cryptosporidium* in fish has not been fully described. It may follow a pattern similar to that established for this genus in mammals (Fig. 2d.) (see Goodwin, 1989; Current and Blagburn, 1990). If so, thin-walled autoinfective oocysts as well as thick-walled resistant oocysts will be present, together with merozoites that have the ability to recycle.

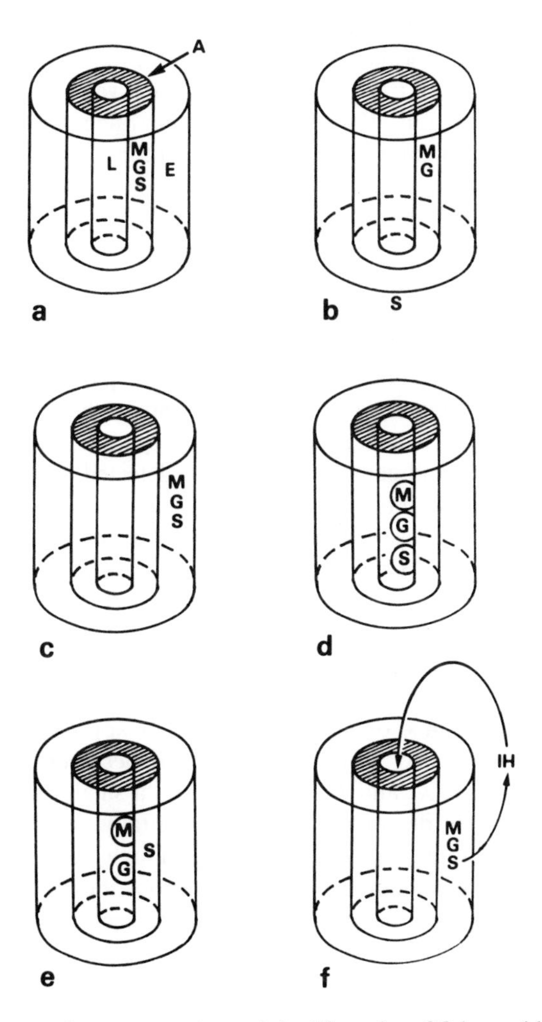

FIG. 2. Diagrammatic representations of the life cycles of fish coccidia. A, alimentary tract; L, lumen; E, extraintestinal sites; IH, intermediate host; M, merogony; G, gamogony; S, sporogony (S outside host denotes exogenous sporogony). Encircled letters indicate epicellular development.

(b) *Eimeriidae*

(i) *Eimeria* (Fig. 1a, Table 2). The life cycles of the majority of fish eimerians follow a pattern of merogony and gamogony within an intracytoplasmic parasitophorous vacuole. Those species developing within epithelial cells of the alimentary tract may sporulate endogenously (Fig. 2a), e.g. *Eimeria variabilis* (Thélohan, 1893) Reichenow, 1921, *Eimeria catostomi* Molnár and Hanek, 1974, *Eimeria fernandoae* Molnár and Hanek, 1974, *Eimeria myoxocephali* Fitzgerald, 1975 and *Eimeria banyulensis* Lom and Dyková, 1982. Some sporulate exogenously (Fig. 2b), e.g. *Eimeria syngnathi* Yakimoff and Gousseff, 1936, *Eimeria raiarum* van den Berghe, 1937, *Eimeria aurati* Hoffmann, 1965, *Eimeria squali* Fitzgerald, 1975, *Eimeria dingleyi* Davies, 1978, and *Eimeria halleri* Upton, Bristol, Gardner and Duszynski, 1986.

Different stages may develop in separate organs. In *Eimeria brevoortiana* Hardcastle, 1944, merogony and gamogony were found to occur in the epithelium of the pyloric caeca of menhaden, *Brevoortia tyrannus*, and sporogony was noted in the testes (Hardcastle, 1944). In another example, Dyková *et al.* (1983) described stages of *Eimeria branchiphila* Dyková, Lom and Grupcheva, 1983, from the gills, kidney and spleen of a single roach, *Rutilus rutilus*. Only mature oocysts were found in the spleen, whereas in both the gill secondary lamellae and the kidney, zygotes and sporogony were observed. Merogony and gamogony were not detected.

Some species have been recorded with life cycles that apparently converge (Figs 2a and b), because they have both endogenous and exogenous sporogony. Mandal and Chakravarty (1965) reported rectal contents from the hammer-headed shark, *Zygaena blochii*, with both immature and mature oocysts of *Eimeria zygaenae*. Similarly, Molnár and Hanek (1974) described new species of *Eimeria* from Canadian freshwater fishes in which only a proportion of oocysts sporulated in the host, and the others passed with the faeces either unsporulated or semisporulated. The approximate percentages sporulating endogenously were as follows: *Eimeria etheostomae*, 50%; *Eimeria micropteri*, 10%; and *Eimeria pungitii*, "a small proportion". Such development requires demonstrating experimentally, because, as Dyková and Lom (1981) noted, infections severe enough to destroy intestinal epithelium could release oocysts before endogenous sporogony can occur. In addition, oocysts passed in an unsporulated or semisporulated condition may not always complete sporogony outside the host.

Almost all *Eimeria* species develop within the cytoplasm of their host cells, but some invade the nucleus. Merogony, gamogony, and sporogony of *Eimeria quentini* Boulard, 1977 (subsequently transferred to *Goussia*) occurred in the nuclei of peritoneal cells of the spotted eagle ray, *Aetobatis narinari* (Fig. 2c). Similarly, intranuclear development was described for

Eimeria bouixi Daoudi and Marques, 1986–1987 which parasitized the epithelial cells of the pyloric caeca of *Dicentrarchus labrax*, and gamogony and sporogony of *Eimeria nucleocola* were observed within enlarged epithelial cell nuclei of the pyloric caeca of the short-horned sculpin, *Myoxocephalus scorpius* (see Lom and Dyková, 1981).

(ii) *Epieimeria*. This genus was proposed by Dyková and Lom (1981) for *Eimeria* of fish in which merogony and gamogony are epicellular and sporogony is intracellular (Fig. 2e). Levine (1988) noted extracellular merogony and gamogony, and intracellular sporogony, as characteristics of *Epieimeria*. Four species have been described: *Epieimeria anguillae* (Léger and Hollande, 1922) Dyková and Lom, 1981 from the intestine of eels such as *Anguilla anguilla* and *Anguilla rostata; Epieimeria isabellae* Lom and Dyková, 1982 from the intestine of the conger eel, *Conger conger; Epieimeria lomae* Daoudi, Radujković, Marquès and Bouix, 1987, from the pyloric caeca of *Scorpaena porcus*; and *Epieimeria puytoraci* Daoudi, Radujković, Marquès and Bouix, 1989b from the anterior intestine of *Symphodus tinca*.

Epieimeria isabellae, *Epieimeria lomae* and *Epieimeria puytoraci* conform to a pattern of epieimerian development. Gamonts were the only developmental stages of *Epieimeria isabellae* detected by Lom and Dyková (1982), and of *Epieimeria lomae* found by Daoudi *et al.* (1987). In both species these stages were epicellular, while sporulating oocysts were intracellular. This was also true for *Epieimeria puytoraci*, in which epicellular trophozoites and meronts were also described (Daoudi *et al.*, 1989b).

Epieimeria anguillae is rather different. Molnár and Baska (1986) demonstrated, by light and electron microscopy, intracellular (epiplasmal) merogony and gamogony of *Epieimeria anguillae*, while sporogony was occasionally intercellular but mostly exogenous. In this instance *Epieimeria anguillae* did not behave strictly as an epieimerian, according to the definition of Dyková and Lom (1981), and its merogony and gamogony were not extracellular, as required by Levine (1988). However, on other occasions *Epieimeria anguillae* sporulates both deep in the intestinal epithelium and intercellularly (Léger and Hollande, 1922; Molnár and Hanek, 1974; Hine, 1975; Lacey and Williams, 1983), thereby fulfilling Dyková's and Lom's (1981) requirement for an epieimerian.

In *Eimeria pigra* Léger and Bory, 1932 and *Eimeria catalana* Lom and Dyková, 1981 both gamogony and oocyst formation are normally epicellular, although occasionally *Eimeria pigra* undergoes intracellular development. Strictly, then, these two species should not be placed in the genus *Epieimeria*. *Eimeria* (*s.l.*) *vanasi* Landsberg and Paperna, 1987 undergoes both intra- and epiepithelial merogony and gamogony. Like *Eimeria pigra* it sporulates exogenously, and sporocysts apparently lack a Stieda body, although *Eimeria* (*s.l.*) *vanasi* sporocysts have not been shown to have sutures either. Clearly, more information is needed before final decisions on the status of all three parasites can be reached (Dyková and Lom, 1981; Lom and Dyková, 1981; Landsberg and Paperna, 1987).

(iii) *Octosporella*. Three species of *Octosporella* were named from fish by Li and Desser (1985b). They are: *Octosporella notropis* in the intestinal epithelium, spleen and swim bladder of the common shiner, *Notropis conutus*; and *Octosporella opeongoensis* and *Octosporella sasajewunensis*, both from the swim bladder of the golden shiner, *Notemigonus crysoleucas*. Oocysts were the only stages recorded, and the life cycle of *Octosporella* in fish has not been described.

(c) *Barrouxiidae*

(i) *Goussia* (Fig. 1b). Species of this genus invade the digestive tract, but a proportion develop in other organs (Table 3). Some of the gut inhabiting species have endogenous sporulation (Fig. 2a), e.g. *Goussia schulmani* Kulemina, 1969, *Goussia carassiusaurati* Romero-Rodriguez, 1978, *Goussia molnari* Jastrzebski, 1984, and *Goussia carpelli* Léger and Stankovitch, 1921. Others have been reported to excrete oocysts in an unsporulated state (Fig. 2b), e.g. *Goussia vargai* Molnár, 1986 and *Goussia acipenseris* Molnár, 1986. The exact sequence of development in some *Goussia* species seems difficult to discover. For example, in experimental infections of *Goussia carpelli* in the common carp, *Cyprinus carpio*, different merogonic generations could not be distinguished by light microscopy (Steinhagen *et al.*, 1989). In another example, Lom and Dyková (1982) were unclear about the significance of intranuclear macrogamonts of *Goussia lucida* (Labbé, 1893) Labbé, 1896, in the intestine of the lesser spotted dogfish, *Scyliorhinus canicula*.

Several *Goussia* develop in extraintestinal sites, and of these six species occur in the swim bladder (Fig. 2c). Four of these latter are known only by their oocysts. These are *Goussia siliculiformis* Shul'man and Zaika, 1962, *Goussia caseosa* Lom and Dyková, 1982, *Goussia notemigonica* Li and Desser, 1985a, and *Goussia polylepidis* Alvarez-Pellitero and Gonzalez-Lanza, 1985. For the remaining two species, more information is available. Fiebiger (1913) illustrated a membranous coat around the gamonts of *Goussia gadi* Fiebiger, 1913, from the swim bladder of gadoid fish, while Landsberg and Paperna (1985) noted swim bladder cells, infected with *Goussia cichlidarum* Landsberg and Paperna, 1985, reduced to membranous sacs at an early stage of parasite development. In *Goussia cichlidarum* merogony and gamogony were intracellular, and infected cells were hypertrophied and gradually displaced above the epithelium of the swim bladder, although still attached. Landsberg and Paperna (1985) recorded the similarities between their description, that of Fiebiger (1913), and Lom and Dyková's (1982) account of oocysts of *Goussia caseosa* within a "honeycomb" host substrate. They noted a relationship of the parasite with dead host cells among *Goussia* infecting fish swim bladders, which was quite distinct from coccidian development in active host cells.

Goussia caseosa is important for another reason. Overstreet *et al.* (1984) selected this as the type-species for a new subgenus (*Plagula*) that they created. *Goussia (Plagula) caseosa* has a diagnostic veil around the sporocysts. Other members of the subgenus apparently include *Goussia (Plagula) gadi, Goussia (Plagula) degiustii* Molnár and Fernando, 1974, and *Goussia (Plagula) subepithelialis* (Moroff and Fiebiger, 1905) Dyková and Lom, 1981, which have similar veils.

(ii) *Crystallospora* (Figs 1c, 2a). The type species, *Crystallospora cristalloides* (Thélohan, 1893) Labbé, 1896, was recorded from the intestine and pyloric caeca of *Motella tricirrata* by Thélohan (1893), who illustrated the sporocyst. A year later, he amplified the description and depicted sporulated oocysts in a villus of the pyloric caecum of the same host (Thélohan, 1894). Labbé (1893) confirmed Thélohan's original finding. In 1896 Labbé changed both generic and specific names from *Coccidium cristalloides* Thélohan, 1893 to *Crystallospora thelohani* (Thélohan, 1893) Labbé 1896, and he also illustrated sporulated oocysts in tissue and free sporocysts. According to Dyková and Lom (1981), Doflein (1909) was responsible for placing the parasite within the genus *Eimeria*. Both these authors and Levine (1988) note that Doflein also restored its original specific name. The parasite thus became *Eimeria cristalloides*. Since Doflein's time the parasite has been recorded by Dogiel (1948). Dyková and Lom (1981) proposed the reinstatement of Labbé's genus, and this has been accepted generally (see Levine 1984c, 1988). The sporocysts of other coccidia, namely *Epieimeria anguillae* Léger and Hollande, 1922, *Eimeria hexagona* Lom and Dyková, 1981, and *Eimeria raibuuti* Daoudi, Radujković, Marquès and Bouix, 1989b are also hexagonal in cross section, and Upton *et al.* (1984) suggested that a more detailed examination of sporocysts was needed to confirm the true status of *Crystallospora* as a genus.

2. *Heteroxenous genera*

(a) *Calyptosporidae*

(i) *Calyptospora* (Fig. 1d). Overstreet *et al.* (1984) erected the genus *Calyptospora* for the fish coccidian originally named *Eimeria funduli* Duszynski, Solangi and Overstreet, 1979. They also established the family Calyptosporidae on the basis of a heteroxenous life cycle (Fig. 2f).

Since then, three more species of *Calyptospora* have been named: *Calyptospora empristica* Fournie, Hawkins and Overstreet 1985 from the liver of starhead topminnows, *Fundulus notti; Calyptospora serrasalmi* Cheung,

Nigrelli and Ruggieri, 1986 from the liver of the black piranha, *Serrasalmus niger*; and *Calyptospora tucunarensis* Békési and Molnár, 1991 from the liver of tucunare, *Cichla ocellaris* (see Fig. 1d). Preliminary studies on *Calyptospora empristica* suggested that a crustacean intermediate host is needed for transmission (Fournie *et al.*, 1985), as with *Calyptospora funduli*, but the life cycles of the other two species are unknown. A problem exists therefore with the classification of *Calyptospora serrasalmi* and *Calyptospora tucunarensis* within the Calyptosporidae. Their taxonomic position awaits further study on the life histories and evidence of an intermediate host.

Sporulated oocysts of *Calyptospora funduli* were discovered in hepatocytes of Gulf killifish, *Fundulus grandis*, throughout the estuaries of the Mississippi and in Louisiana (Duszynski *et al.*, 1979). Solangi and Overstreet (1980) described the development of this parasite in the liver and pancreas of both *Fundulus grandis* and *Fundulus similis*, and noted the need for an intermediate host, the grass shrimp, *Palaemonetes pugio*. Later, the life cycle was reported in another killifish, *Fundulus heteroclitus* (see Upton and Duszynski, 1982).

Solangi and Overstreet (1980) found young meronts in parasitophorous vacuoles within the cytoplasm of hepatocytes and pancreatic cells 5 days after infection (d.a.i.) in fish maintained at approximately 24°C (counting from the time when fish were fed grass shrimps from epizootic areas). First generation merozoites developed apparently by 10 d.a.i., whereas sexual stages were observed between 15 and 20 d.a.i. Nuclear division of the zygote occurred between 19 and 26 d.a.i. and sporozoites developed by about day 60. Upton and Duszynski (1982) observed two types of meronts which could not be separated temporally by the methods used, and found the life cycle to be completed in 44 days. It is not known whether differences in results between the two groups of workers resulted from the use of different fish species, from different diets, or from the slight difference in the temperature at which the fish were maintained (22°C and 24°C).

3. *Miscellaneous*

(a) *Isospora*. Davronov (1987) described the sporulated oocysts of three *Isospora* from fish, but did not give them specific names. Drawings of the oocysts were provided, and these appear to be the first records of *Isospora* from fish. *Isospora* species have been associated mainly with mammals, birds, reptiles and amphibians, with the type species, *Isospora rara* Schneider, 1881, having been recorded from an unnamed species of *Limax*. The fish parasites and hosts are: *Isospora* species 1, from *Gobio gobio lepidolaemus*; *Isospora* species 2, from *Barbus capito conocephalus*; and *Isospora* species 3, from *Hypophthalmichthys molitrix*.

(b) *Sarcocystis*. *Sarcocystis salvelini* was described and illustrated by Fantham and Porter (1943) from the Canadian speckled trout, *Salvelinus fontinalis*. The parasite was observed in only one of several hundred trout, and it occurred as minute, whitish threads in the abdominal muscles. Levine and Tadros (1980) mentioned that the validity of this species had been questioned, but they considered it acceptable at that time. Fantham and Porter (1943) also recorded a *Sarcocystis* from the muscle of an eel pout, *Zoarces angularis*, and Kent *et al.* (1989) noted a possible member of the family Sarcocystidae from hardy head fish, *Atherinomorus capricornensis* collected in Queensland. All these parasites need further examination.

(c) *Rhabdospora thelohani* Laguesse, 1895. Thélohan (1892b, c) reported an enigmatic cell from several fish, which was subsequently named *Rhabdospora thelohani* (Figs 3, 4). Since then there has been controversy whether *Rhabdospora thelohani* is an apicomplexan parasite or a host "rodlet cell". Publications concerning the nature of these cells were reviewed by Morrison and Odense (1978) and Mayberry *et al.* (1979). The true nature of "rhabdosporans" is still unresolved because they possess characteristics of both parasites and fish cells. Some authors (e.g. Mayberry *et al.*, 1979, 1986) favour the interpretation that *Rhabdospora thelohani* might be parasitic, while others (e.g. Desser and Lester, 1975; Paterson and Desser, 1981d) consider that the presence of characteristic apicomplexan structures cannot be concluded from the findings to date. All agree that more information is required to determine the true character of *Rhabdospora thelohani*.

III. Transmission

In the life cycles of fish coccidia, transmission is thought to be of two main types (Desser, 1981). The simpler method is by direct transmission involving faecal contamination. The second mode of transmission is indirect, and involves an invertebrate host (Desser, 1981; Overstreet, 1981). Some fish coccidia apparently employ both methods.

A. Direct

Direct transmission has been demonstrated on several occasions. For example, Marinček (1973a) transmitted *Goussia subepithelialis* to carp by feeding sporulated oocysts. Landsberg and Paperna (1985) transferred *Goussia cichlidarum* to cichlid fish by feeding oocysts in paste, and seven of nine hybrid fish (*Oreochromis aureus* × *Oreochromis niloticus*) and all of

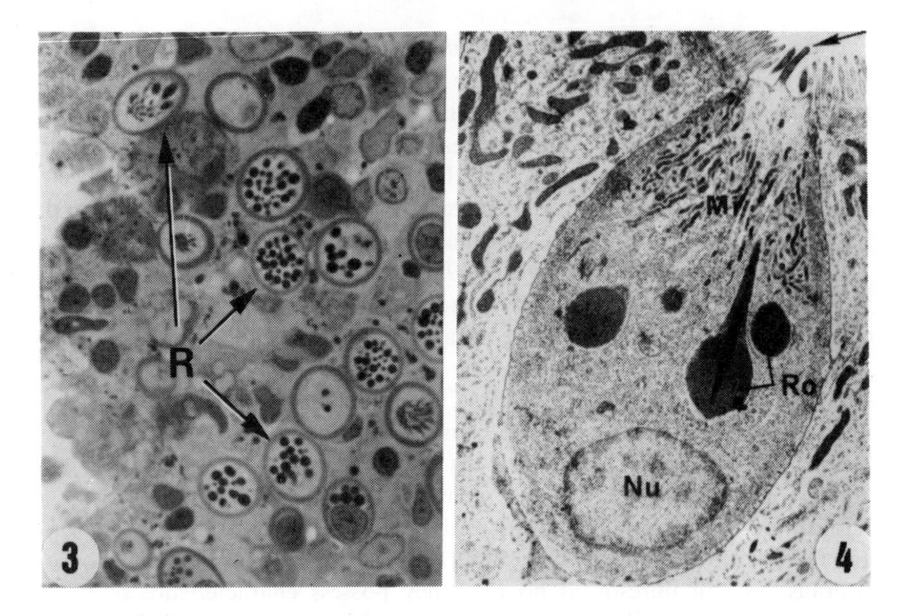

FIGS 3 and 4. *Rhabdospora thelohani* in the tissues of fishes. FIG. 3. Photomicrograph of kidney of *Rutilus rubilio* showing *Rhabdospora* (R) in tissue below the kidney capsule (× 840). (Kindly supplied by L. Mayberry from Mayberry *et al.*, 1979, and reproduced by permission of the editor.) FIG. 4. Electron micrograph of *Rhabdospora* in intestinal epithelium of the common shiner, *Notropis cornutus*. Longitudinal section illustrating characteristic features including basal nucleus (Nu), flask-shaped rodlets (Ro), small dense mitochondria (Mi). Dense microvilli (arrow) of *Rhabdospora* contrast with less dense microvilli of adjacent epithelial cells (× 14 200). (Kindly supplied by S. S. Desser from Paterson and Desser, 1981d, and reproduced by permission of the editor, B. B. Nickol.)

three "gold tilapia" (undefined hybrid of *Oreochromis mossambicus*) became heavily infected 95 days later. Transmission was also demonstrated for *Goussia iroquoina* when the sources of oocysts were infected faeces or mud from a pond used by fathead minnows (Paterson and Desser, 1981c, 1982). Molnár (1979) and Steinhagen and Körting (1988) showed that *Goussia carpelli* could be transmitted directly, as well as indirectly (see below). Steinhagen and Körting (1988) demonstrated direct transfer to uninfected *Cyprinus carpio* by feeding either infected minced intestinal tissue or faeces, or by placing uninfected and infected carp in the same tank.

B. INDIRECT

The first heteroxenous life cycle to be demonstrated experimentally for fish coccidia was that of *Calyptospora funduli* from killifish (Fournie and

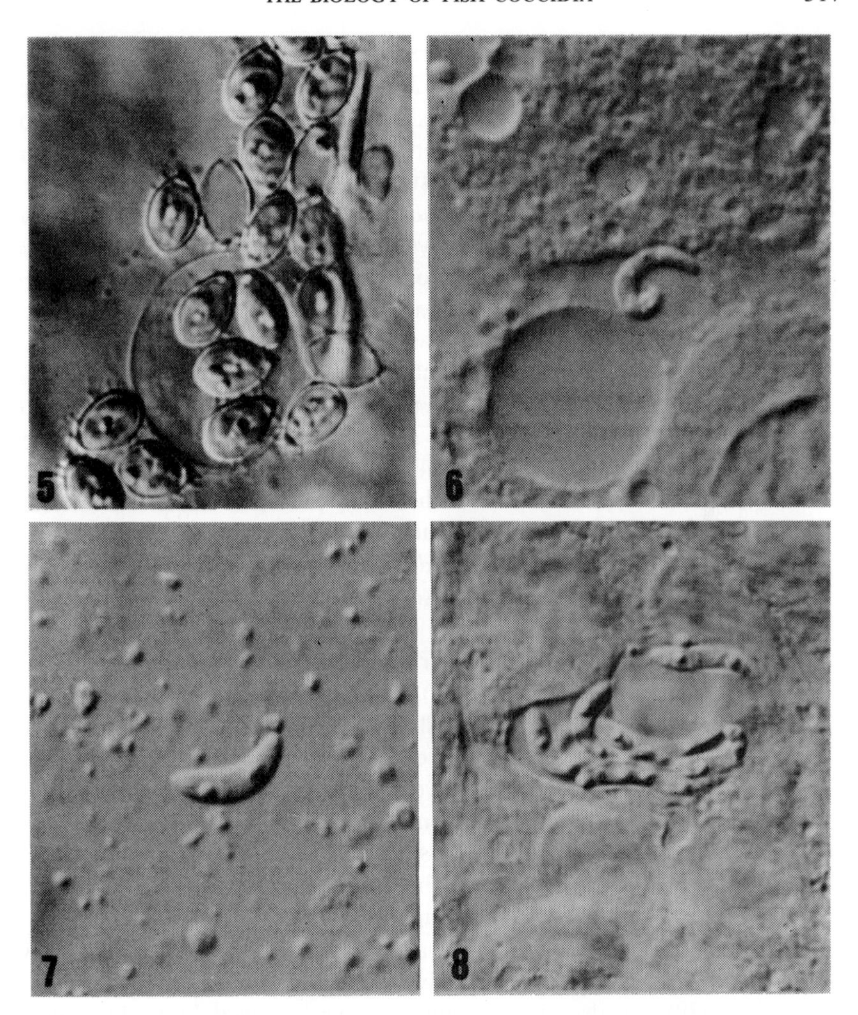

FIGS 5–8. *Calyptospora funduli* in experimentally infected *Palaemonetes pugio* (photomicrographs, Nomarski optics). FIG. 5. Intact oocyst, free sporocysts with sporozoites, and empty sporocysts in gastric mill (10 h after infection (a.i.)) (× 1050). FIG. 6. Uninfective sporozoite in intestine (17 h a.i.) (× 1400). FIG. 7. Infective sporozoite from intestinal contents (8 days after infection (d.a.i)). Note prominent refractile bodies (× 1050). FIG. 8. Nest of infective sporozoites among intestinal cells (8 d.a.i.) (× 900). (Kindly supplied by J. W. Fournie from Fournie and Overstreet, 1983, and reproduced by permission of the editor.)

Overstreet, 1983), after Solangi and Overstreet (1980) failed to infect *Fundulus grandis* by direct feeding of sporulated oocysts. The natural intermediate hosts of *Calyptospora funduli* from killifish were the palaemonid shrimps *Palaemonetes pugio, Palaemonetes vulgaris, Palaemonetes paludosus, Palaemonetes kadiakensis* and *Macrobrachium ohione*. In these, sporocysts excysted to release sporozoites that changed morphologically before they became infective to fish (Solangi and Overstreet, 1980; Upton and Duszynski, 1982; Fournie and Overstreet, 1983).

When infected liver from *Fundulus grandis* was fed to grass shrimp, *Palaemonetes pugio*, sporulated oocysts were mechanically ruptured by the shrimps' gastric mill (Fig. 5). Sporozoites excysted through an opening in the sporocyst, and from 5 to 83 d.a.i. sporozoites were seen free in the alimentary tract of the shrimp or between intestinal cells (Figs 6, 7, 8). Some sporozoites also migrated to the tubules of the hepatopancreas. Infected grass shrimps failed to infect killifish at 2 or 4 d.a.i. but did so from 5 to 201 d.a.i. A patent period of 5 days in shrimps, when sporozoites underwent change corresponding to their ability to infect fish, indicated that the shrimps were true intermediate rather than paratenic hosts (Fournie and Overstreet, 1983). In contrast to palaemonids, the following crustaceans did not infect *Fundulus grandis*: *Mysidopsis bahia* (mysidacean), *Gammarus mucronatus* (amphipod), *Callinectes sapidus* (blue crab) and *Penaeus setiferus* (white shrimp).

Goussia carpelli has also been transmitted in the laboratory by grass shrimps (Kent and Hedrick, 1985), although these are unlikely natural vectors of *Goussia carpelli* since fish and shrimps do not occur together. Molnár (1979) transmitted the same coccidian to carp using tubificid worms, although Kent and Hedrick (1985) were able to achieve this to only one in 10 goldfish, *Carassius auratus*. In their experiments, minced intestines with oocysts were introduced with mud substrate to aquaria containing several hundred tubificid worms, some of which were then fed daily to goldfish. The single infection produced was very light. However, three goldfish developed moderate infections after eating grass shrimp, *Palaemonetes kadiakensis*, which had previously fed on heavily infected intestines and viscera. Since Molnár (1979) was also able to transmit *Goussia carpelli* by feeding oocysts homogenized in chicken livers (direct transmission, confirmed by Steinhagen and Körting, 1988), Kent and Hedrick (1985) suggested that the natural vector may be paratenic rather than a true intermediate host.

In recent experiments by Steinhagen and Körting (1990) on the transmission of *Goussia carpelli*, tubificid oligochaetes (*Tubifex tubifex* and *Limnodrilis hoffmeisteri*) were fed intestinal tissue from carp containing sporulated oocysts. Later, infection was produced in laboratory-reared carp fed on the infected tubificid worms. Motile sporozoites were found in the intestinal contents of the oligochaetes. Sporozoites also invaded intestinal epithelial cells and survived for at least 9 weeks (Steinhagen, 1991a). They did not develop or degenerate (Figs 9, 10), although they were sometimes surrounded by phagocyte-like cells. Sporozoites of *Goussia subepithelialis* were also examined for up to 3 weeks in tubificid worms. They invaded worm epithelial cells but, unlike *Goussia carpelli* sporozoites, they were found often in phagocytic cells and showed signs of degeneration.

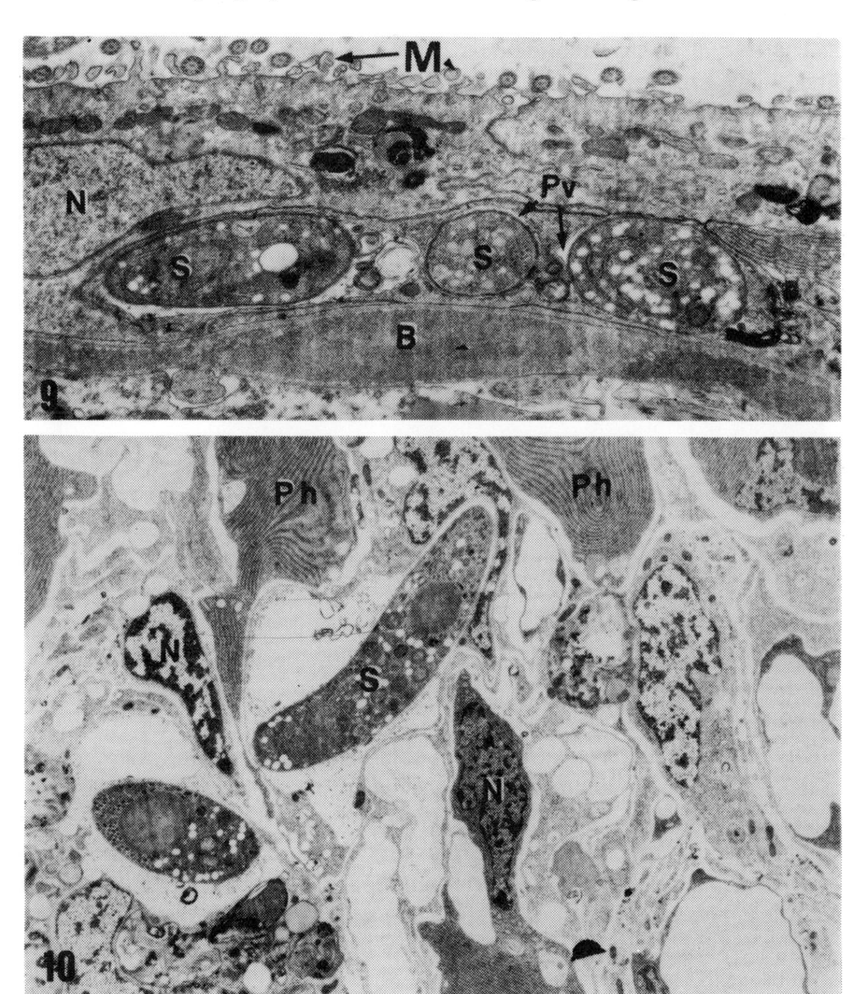

FIGS 9 and 10. Transmission electron micrographs of *Goussia carpelli* sporozoites in *Tubifex tubifex*. FIG. 9. Sporozoites (S) within membrane-bound parasitophorous vacuoles (Pv) in intestinal cells, located near a blood vessel (B). N, host cell nucleus; M, microvilli of intestinal epithelial cells (× 11 200). FIG. 10. Sporozoites (S) in intestinal epithelium surrounded by phagocyte-like cells (Ph); N, host cell nucleus (× 5200). (Kindly supplied by D. Steinhagen from Steinhagen, 1991a, and reproduced by permission of the editor.)

Indirect cross transmission of *Goussia carpelli* between different species of carp was achieved by Lukes *et al.* (1991). They were able to transfer *Goussia carpelli* from common carp (*Cyprinus carpio*) to goldfish (*Carassius auratus*) using tubificids as paratenic hosts. These experiments resulted in weak infections in four of 15 goldfish used.

The role of an invertebrate host in the transmission of fish coccidia was first suggested by Landau *et al.* (1975). They observed sporozoites in the intestinal cells of mysids that had been in water containing *Eimeria* oocysts derived from moray eels, *Gymnothorax moringa*. These authors suggested that the shrimps acted as paratenic hosts, although experiments were not performed to prove this. Heteroxenous life cycles have been postulated also for *Goussia gadi* (see Overstreet, 1981; Fournie and Overstreet, 1983) and for *Goussia degiustii* (see Desser, 1981; Paterson and Desser, 1982). In the case of *Goussia degiustii*, this was suggested because Paterson and Desser (1982) found no direct transfer of the parasite to laboratory reared fathead minnows, *Pimephales promelas*, that were fed infected spleens from the common shiner (*Notropis cornutus*).

IV. Structure and Development of Asexual Stages and Gamonts

A. TROPHOZOITES, MEROGONY AND MEROZOITES

Considering the size of the aquatic environment into which oocysts are often discharged, host fish presumably acquire relatively few infective stages at one time (unless these are concentrated in an intermediate host). If this is compared with the extent of some coccidian infections within fish, it suggests highly efficient multiplication, probably involving several merogonous cycles. However, in common with coccidia from other vertebrates, it is the oocyst and sporocysts that attract most attention in fish coccidia, while trophozoites, merogony, and the numbers of merozoites and generations produced, usually receive only brief mention, or all too frequently are not recorded. This is understandable since trophozoites and merozoites are often difficult to find, and natural infections give little indication which generation of merogony is present. In addition, the location and morphology of trophozoites and merozoites are unlikely to govern the taxonomic status of the coccidian, although merogonic stages of *Goussia degiustii* may be essential for establishing whether morphological variations in this coccidian are elicited by development in different organs (Lom *et al.*, 1989).

Descriptions of trophozoites and meronts have usually been limited to recording their size, and the number of merozoites produced. The location of these stages varies considerably. Intracytoplasmic merogony is common, but extracytoplasmic merogony occurs in *Eimeria* (*s.l.*) *vanasi* (see Paperna and Landsberg, 1987), and meronts of *Epieimeria anguillae* are epiplasmal (Molnár and Baska, 1986). The epithelial lining of the pyloric caeca, intestine, or rectum are favoured sites for many species, e.g. *Eimeria brevootiana, Goussia carpelli, Eimeria dingleyi*, and *Eimeria variabilis* (see Hardcastle, 1944; Steinhagen, 1991b and Davies, 1978, 1990, respectively). Trophozoites and successful merogony of *Epieimeria anguillae* and *Goussia sinensis* also occur occasionally in goblet cells (Molnár and Baska, 1986; Baska and Molnár, 1989). Meronts are found in hepatocytes in *Calyptospora funduli* (see Upton and Duszynski, 1982), and connective tissue surrounding the swim bladder is the location of merozoites of *Goussia gadi* (Fiebiger, 1913).

The early trophozoite (derived from the sporozoite) is usually rounded. That of *Goussia iroquoina* contains a refractile body, remnants of micronemes, mitochondria, and a nucleus with a prominent nucleolus, and lies within a parasitophorous vacuole (Paterson and Desser, 1981c). The youngest trophozoites of *Epieimeria anguillae* (Fig. 11) are surrounded by two unit membranes (Molnár and Baska, 1986). Later trophozoites of *Goussia iroquoina*, derived from merozoites of the previous generation, usually contain the remnants of the pellicular complex, micronemes, a nucleus with a prominent nucleolus, mitochondria, endoplasmic reticulum and vesicles (Paterson and Desser, 1981c). Before merogony, the trophozoite may round up, lose its pellicle and undergo nuclear division. The meront of *Goussia iroquoina* is limited by a single membrane, and it contains several nuclei, Golgi complexes, mitochondria, endoplasmic reticulum and lipid (Paterson and Desser, 1981c).

The site of merogony may be the same as that in which gamonts, oocysts, and early sporogony occur. This is true of *Eimeria variabilis, Calyptospora funduli* and *Eimeria* (*s.l.*) *vanasi* (see Davies, 1978; Hawkins *et al*., 1984; Landsberg and Paperna, 1987). Sometimes, however, its location is very different from the site of gamonts or subsequent stages. In *Eimeria brevoortia*, for example, merozoites and gamonts apparently occur in the intestine, while sporogony occurs in the swim bladder (Hardcastle, 1944).

In several instances the study of experimentally infected hosts has done much to improve understanding of the sequential asexual development of fish coccidia. Trophozoites of *Goussia iroquoina* in the intestinal epithelial cells of experimentally infected fathead minnows, *Pimephales promelas* (see Paterson and Desser, 1981c), and first generation merogonic stages of *Goussia carpelli* (see Steinhagen, 1991b) were identified by the presence of refractile bodies. Conversely, the absence of such bodies from meronts of *Goussia carpelli* in experimentally infected common carp, *Cyprinus carpio*, probably indicated that these parasites belonged to later generations (Stein-

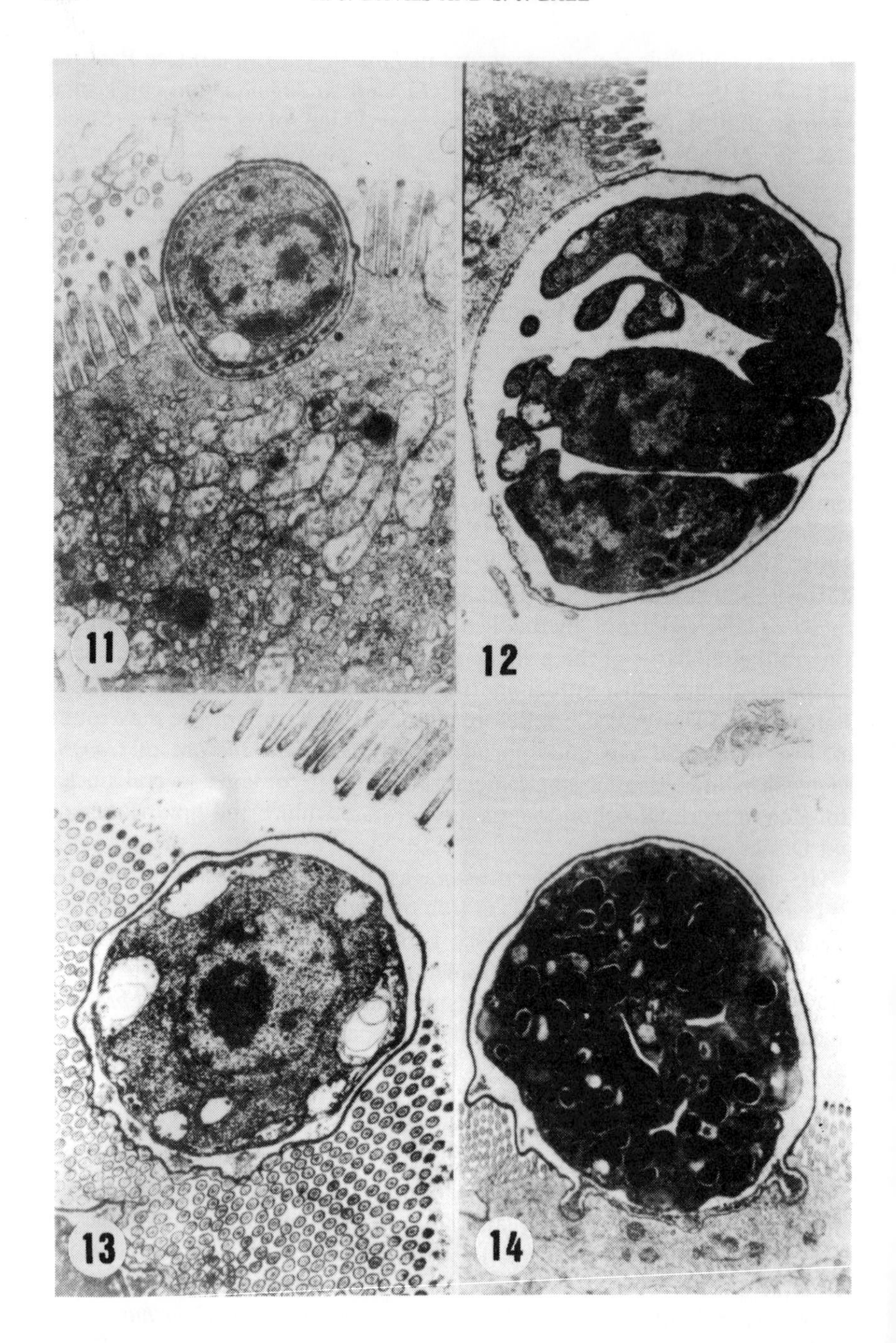
11
12
13
14

hagen, 1991b). Refractile bodies have usually been identified by transmission electron microscopy in fish coccidia (e.g. Steinhagen, 1991b), whereas in *Eimeria* species from mammals specific histological staining techniques have been applied (Peirce, 1980).

Merogony in several fish coccidia is similar to that seen in homeotherm hosts. Meronts of *Goussia iroquoina* (Figs 15,16), *Eimeria* (*s.l.*) *vanasi*, *Goussia sinensis*, and *Goussia carpelli* develop within parasitophorous vacuoles (Paterson and Desser, 1981c; Landsberg and Paperna, 1987; Baska and Molnár, 1989; Steinhagen, 1991b), but no parasitophorous vacuole was evident surrounding two meronts of *Eimeria variabilis* examined ultrastructurally by Davies (1990). In *Epieimeria anguillae* meronts develop within parasitophorous vacuoles in the apical part of intestinal epithelial cells and these are large enough to protrude into the intestinal lumen beyond the microvilli (Fig. 12). In this species the parasitophorous vacuole adjoining the cell ectoplasm was bordered by a single unit membrane, whereas the part directed towards the intestinal lumen had an additional covering layer that appeared to be intestinal cell membrane, that is, a continuation of the membrane of the microvilli (Molnár and Baska, 1986).

Calyptospora funduli meronts in the hepatocytes of experimentally infected longnose killifish, *Fundulus similis*, were described by Hawkins *et al.* (1984). Second generation merozoites at 9 d.a.i. lay free in the cell cytoplasm. A membrane-lined parasitophorous vacuole then formed around the the developing meront. In most coccidians the parasitophorous vacuole membrane is produced by the host cell, but its origin varies. The host cell plasmalemma often forms the vacuole membrane during the process of phagocytic entry by the sporozoite or the merozoite of the previous generation although, after more active penetration, membranes appear to form *de novo* (Chobotar and Scholtyseck, 1982; Ball and Pittilo, 1990). Unlike the single parasitophorous vacuole membrane seen in the majority of coccidians, that around *Calyptospora funduli* meronts (and macro- and microgamonts) was double (Hawkins *et al.*, 1984). The outer membrane (in contact with the host cytoplasm) was studded with ribosomes and possibly originated from

---

FIGS 11–14. Transmission electron micrographs of *Epieimeria anguillae* in the anterior intestine of the eel, *Anguilla anguilla*. FIG. 11. Young trophozoite in the apical part of an intestinal epithelial cell (× 12 200). FIG. 12. Mature merozoites in a parasitophorous vacuole (× 8900). FIG. 13. Young macrogamont transversely sectioned through its extracellular part. The gamont contains a large nucleus and numerous peripheral mitochondria (× 12 500). FIG. 14. Macrogamete with numerous polysaccharide and lipid granules developing on the surface of a flattened host cell. Protrusions of the parasitophorous vacuole establish a close connection with the host cell (× 7300). (Kindly supplied by K. Molnár from Molnár and Baska, 1986, and reproduced by permission of the editor.)

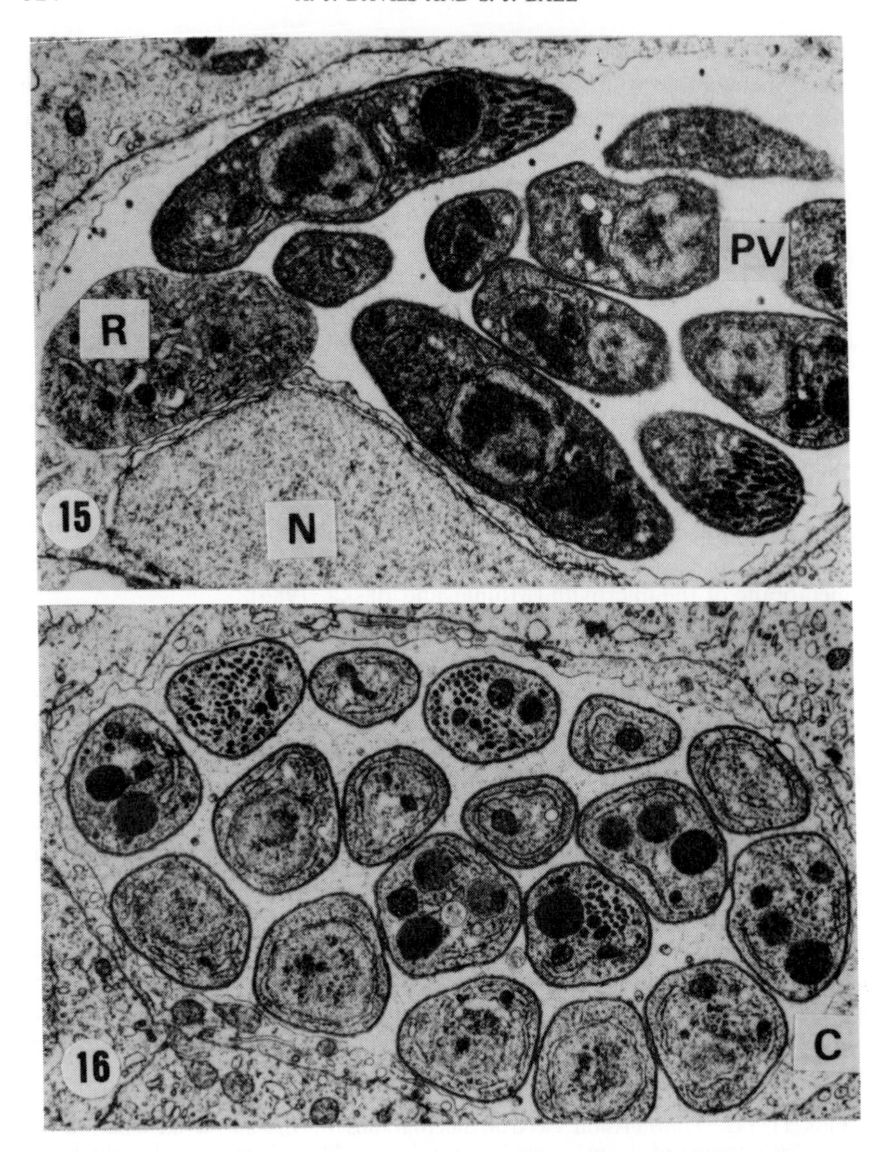

FIGS 15 and 16. Transmission electron micrographs of merozoites of *Goussia iroquioina* in intestinal epithelium cells of fathead minnows, *Pimephales promelas*, 6–10 days after infection. FIG. 15. Longitudinal and oblique sections of merozoites free within the parasitophorous vacuole (PV). R, residual meront cytoplasm; N, host cell nucleus (× 13 200). FIG. 16. Eighteen transversely sectioned merozoites. C, host cell cytoplasm (× 11 300). (Kindly supplied by S. S. Desser from Paterson and Desser, 1981c, and reproduced by permission of the editor.)

rough endoplasmic reticulum, whereas the inner membrane (limiting the parasitophorous vacuole) had no ribosomes. Where the outer membrane was in close association with that of the hepatocyte nucleus, the two membranes were continuous. In this region the typical ribosome-studded outer membrane was absent.

Numbers of merozoites recorded vary. In *Eimeria* (*s.l.*) *vanasi* 2–32 merozoites, or rosettes of 4–16 develop within parasitophorous vacuoles (Landsberg and Paperna, 1987). In *Goussia iroquoina* there are 13–18 merozoites (Paterson and Desser, 1981c), in *Goussia carpelli* 12–20 (Steinhagen, 1991b), and in *Goussia sinensis* 8–16 (Baska and Molnár, 1989).

Merozoites appear to be produced by three methods: exogenesis (ectomerogony), in which merozoites are budded from the surface of the meront; endomerogony, in which several merozoites are produced within the meront; and endodyogeny, in which paired merozoites are produced internally. Some merozoites of *Goussia iroquoina*, and those of *Goussia aculeatus* and *Goussia sinensis*, develop by exogenesis (ectomerogony) (Paterson and Desser, 1981c; Jastrzebski, 1989; Baska and Molnár, 1989). Final generation merozoites of *Goussia iroquoina* and *Goussia carpelli* develop by endomerogony (Paterson and Desser 1981a; Steinhagen, 1991b). Endodyogeny has been reported in *Eimeria* (*s.l.*) *vanasi* by Landsberg and Paperna (1987). Endomerogony was described in some detail by Steinhagen (1991b) for *Goussia carpelli*. Merozoite pellicle formation began with the appearance of curved vacuoles within the meront cytoplasm which were partially underlain by an inner membrane of the pellicle and associated subpellicular tubules. Formation of merozoites began apically and proceeded posteriorly as a budding process. Merozoite formation utilized almost all of the meront cytoplasm so that only a small residuum remained.

The ultrastructure of merozoites of piscine coccidia resembles that of other closely related coccidia parasitic in birds and mammals (Scholtyseck, 1973; Chobotar and Scholtyseck, 1982; Ball and Pittilo, 1990). Merozoites of *Eimeria* (*s.l.*) *vanasi* are bound by two unit membranes (Paperna, 1990); those of *Goussia sinensis* are bound by a three-layered pellicle with 22 subpellicular microtubules (Baska and Molnár, 1989). An identical number of subpellicular microtubules occurs in the merozoites of *Goussia carpelli* (see Steinhagen, 1991b), and 22–24 exist in the merozoites of *Goussia iroquoina* (see Paterson and Desser, 1981c). Micropores, micronemes, Golgi complex between the nucleus and apical region, mitochondria, endoplasmic reticulum, ribosomes, rhoptries and polar rings are apparently similar to those seen in other Apicomplexa (Paterson and Desser, 1981c; Davies, 1990; Steinhagen, 1991b).

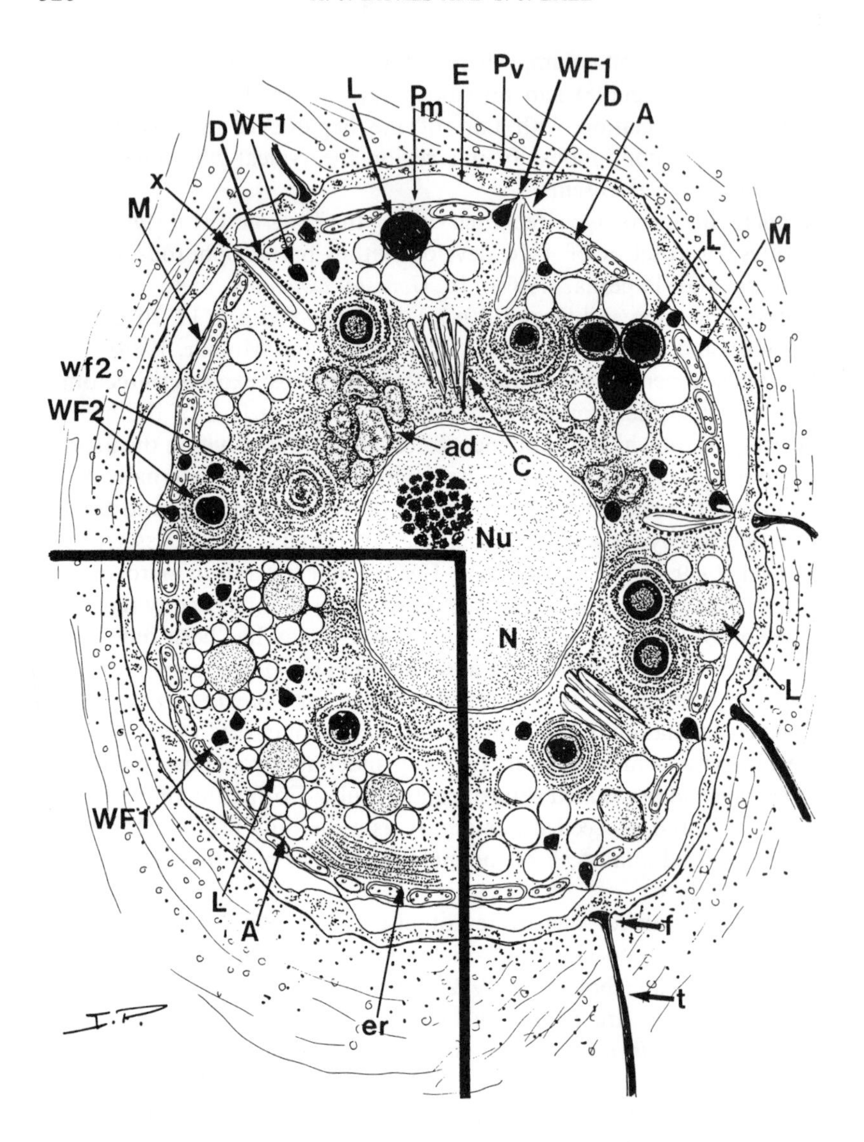

FIG. 17. *Eimeria* (*s.l.*) *vanasi*. Diagram of mature macrogamont showing A, amylopectin (polysaccharide) granules; ad, adnuclear bodies; C, canaliculi; D, ribosome-lined ducts; E, envelope; er, endoplasmic reticulum; f, funnels; L, lipid vacuoles; M, mitochondria; N, nucleus; Nu, nucleolus; P_m, macrogamont cell boundary; P_v, wall of parasitophorous vacuole; (t) tubular system; WF1, bodies resembling type 1 wall-forming bodies; WF2, bodies resembling type 2 wall-forming bodies; wf2, aggregating WF2; x, attachment points of envelope. (Kindly supplied by I. Paperna from Paperna, 1990, and reproduced by permission of Inter-Research.)

B. MACROGAMONTS AND MACROGAMETES

Young macrogamonts are usually 5.0–26.0 μm in diameter. They are mostly spherical or ellipsoidal structures surrounded by one or more delicate membranes and they commonly lie within a parasitophorous vacuole. In *Eimeria* (*s.l.*) *vanasi* the parasitophorous vacuole of macrogamonts is bounded by a unit membrane interspersed with funnels (Fig. 17) connected to an elaborate tubular system extending into the host cytoplasm (Paperna, 1990) (see Section VI.A.1). In *Goussia spraguei* a double parasitophorous membrane surrounds the developing parasite as it separates from the surface of the kidney epithelial cells of cod, and the same membrane serves as an oocyst membrane until sporocysts mature (Morrison and Poynton, 1989a).

The nucleus of macrogamonts is normally large and roughly central, and a nucleolus is prominent (Figs 13, 17). The cytoplasm, particularly of older macrogamonts, may contain granules of polysaccharide, protein, and lipid, e.g. in *Goussia subepithelialis* (see Marinček, 1973a), *Eimeria variabilis* (see Davies, 1978, 1990), *Goussia cichlidarum* (see Paperna *et al.*, 1986) and *Epieimeria anguillae* (see Molnár and Baska, 1986) (Fig. 14). Adnuclear bodies and canaliculi also occur (Figs 17–19). Young macrogamonts of *Goussia iroquoina* have elaborately shaped mitochondria, granular endoplasmic reticulum, and Golgi complexes (Paterson and Desser, 1981a).

Electron microscopical studies have shown that the surface of some macrogamonts bears micropores, e.g., that of *Goussia cichlidarum* (see Paperna *et al.*, 1986). Early macrogamonts of *Goussia cichlidarum* are bound by two unit membranes, whereas those of mature macrogamonts (macrogametes?) have a single unit membrane surrounded by two additional envelopes (Paperna *et al.*, 1986). In contrast, macrogamonts of *Goussia iroquoina* are bound by a single membrane throughout their development (Paterson and Desser, 1981a). Macrogametes of *Goussia spraguei* are also limited by a unit membrane but with a variable number of membranes external and internal to this (Morrison and Poynton, 1989a).

Organelles that resemble the wall-forming bodies of terrestrial coccidia have been recorded in the cytoplasm of macrogamonts and macrogametes of a number of fish coccidia, e.g. in *Goussia iroquoina* (see Paterson and Desser, 1981a), *Calyptospora funduli* (see Hawkins *et al.*, 1983a), *Goussia laureleus* (see Desser and Li, 1984), *Goussia clupearum* and *Eimeria sardinae* (see Morrison and Hawkins, 1984), *Goussia cichlidarum* (see Paperna *et al.*, 1986), *Goussia spraguei* (see Morrison and Poynton, 1989a), *Eimeria variabilis* (see Davies, 1990), *Goussia zarnowskii* (see Jastrzebski and Komorowski, 1990), and *Eimeria* (*s.l.*) *vanasi* (see Paperna, 1990).

In *Goussia iroquoina*, membrane-bound vesicles with electron-dense matrices (the so-called Mv inclusions) coalesced into large spheroidal packets of electron-dense material (Mx) in maturing macrogamonts and macrogametes (Paterson and Desser, 1981a). The Mv were not considered homologous to the wall-forming bodies of coccidia of terrestrial vertebrates because they coalesced into larger masses (Mx), and they did not participate in the formation of the oocyst wall. Mv were probably proteinaceous, and perhaps were precursors of the refractile bodies of sporozoites (Paterson and Desser, 1981a). A similar function was proposed for the dense membrane-bound inclusions in the macrogamonts of *Goussia laureleus*, which were thought to form the residual bodies of sporoblasts and sporocysts (Desser and Li, 1984). In *Calyptospora funduli* three kinds of inclusions were present in the cytoplasm of mature macrogamonts. Two types were identified as wall-forming bodies but they were not seen to release their contents at the oocyst wall (Hawkins *et al.*, 1983a). Similarly, Morrison and Hawkins (1984) and Morrison and Poynton (1989a) could not firmly assign wall-forming functions to bodies they found in the sporonts of *Goussia clupearum* and *Eimeria sardinae* and in the macrogametes of *Goussia spraguei*. Granular bodies occurring in *Goussia cichlidarum* appeared to form within endoplasmic reticulum membranes, and they reached full differentiation in mature (apparently fertilized) macrogametes (Paperna *et al.*, 1986). Mature macrogametes of *Goussia cichlidarum*, which were probably already zygotes, were distinguished by a predominance of granular bodies, a reduction in the number and size of mitochondria, reduced volume of the Golgi apparatus, and an extensive subpellicular endoplasmic reticulum plexus (Paperna *et al.*, 1986). In *Eimeria variabilis* wall-forming bodies could not be demonstrated cytochemically in macrogametes by light microscopy (Davies, 1978), and both large and smaller electron-dense bodies (possibly proteinaceous) that were demonstrated later by electron microscopy were thought to participate in sporocyst wall formation (Davies, 1990) (see Section V.B). Jastrzebski and Komorowski (1990) identified wall-forming bodies of two densities in the macrogamete of *Goussia zarnowskii*, and they provided evidence that some were discharged outside the pellicle towards the newly created oocyst wall membranes.

FIGS 18 and 19. Transmission electron micrographs of mature macrogamonts of *Eimeria* (*s.l.*) *vanasi*. FIG. 18. General view showing: N, nucleus; C, canaliculi; ad, adnuclear bodies; er, endoplasmic reticulum; L, lipid; A, amylopectin (× 9500). FIG. 19. Macrogamont with: A, L, scattered storage material granules; D, ribosome-lined ducts; WF1 and WF2, two kinds of wall-forming bodies; Pm, macrogamont boundary; E, envelope with attachment points (x); Pv, wall of parasitophorous vacuole; ad, adnuclear bodies; er, endoplasmic reticulum (× 21 200). (Kindly supplied by I. Paperna from Paperna, 1990, and reproduced by permission of Inter-Research.)

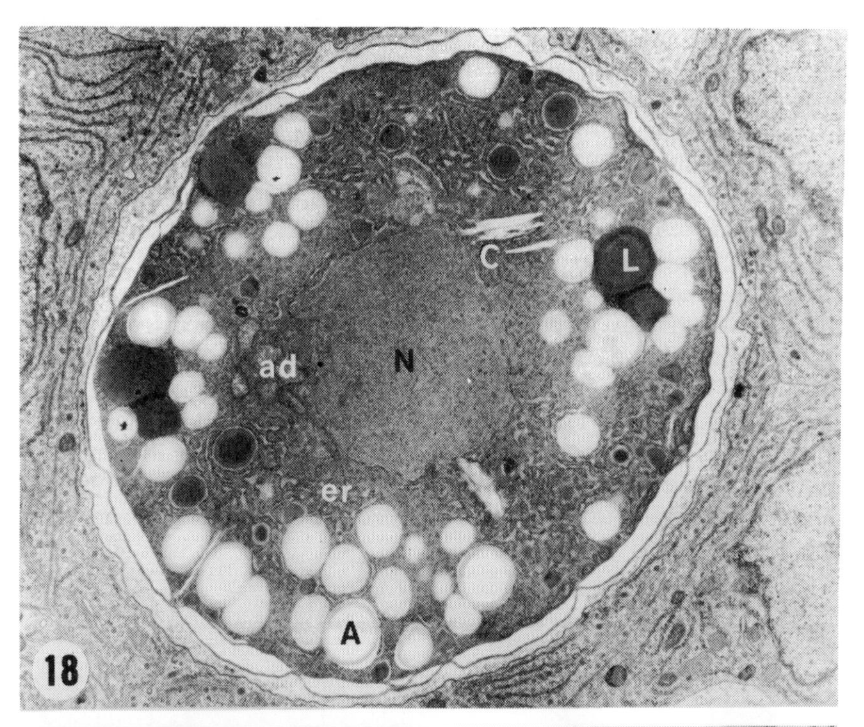
C
L
ad
N
er
A
18

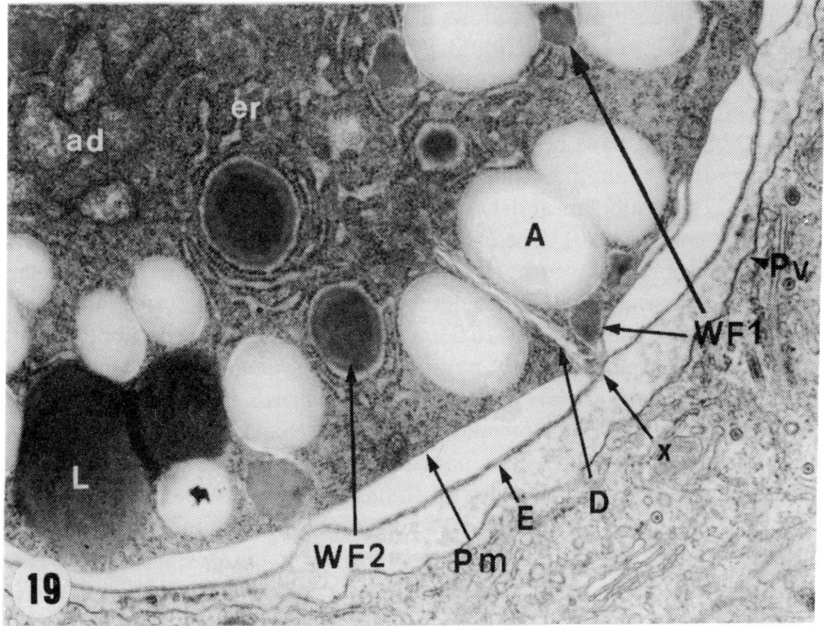
er
ad
A
Pv
WF1
X
D
L
E
WF2
Pm
19

Perhaps the most convincing evidence for the presence of structures resembling wall-forming bodies within macrogametes of fish coccidia comes from a fine study by Paperna (1990) (Figs 17, 18, 19). In this case, two kinds of bodies were identified within *Eimeria* (*s.l.*) *vanasi*. Small electron-dense vesicles resembling type 1 wall-forming bodies (WF1) of other coccidia, and larger bodies with an electron-dense core, located within rough endoplasmic cisternae, and resembling type 2 wall-forming bodies (WF2), were described within the cytoplasm of macrogametes. Ribosome-lined ducts and organelles resembling type 1 wall-forming bodies appeared to be involved in the formation of the envelope around the macrogametes. However, Paperna (1990) noted that, since there is such variation in the occurrence and fine structure of organelles resembling wall-forming bodies among piscine coccidia, they are unlikely to assume the same function as the wall-forming bodies of coccidia of higher vertebrates, which have hard oocyst walls. Paperna's type 2 bodies were considered earlier by Paperna and Landsberg (1985) (in *Goussia cichlidarum*) to be involved in sporocyst wall formation, and this is discussed further in Section V.B.

C. MICROGAMETOGENESIS AND MICROGAMETES

Microgametogenesis usually follows a basic pattern similar to that described for coccidia from other vertebrates. A phase of nuclear division associated with growth of the microgamont precedes differentiation of microgametes (Scholtyseck, 1973: Chobotar and Scholtyseck, 1982; Ball and Pittilo, 1990). Fish coccidia that have been shown to conform to this pattern of development include, for example, *Goussia iroquoina, Goussia aculeati*, and *Goussia zarnowskii* (see Paterson and Desser, 1981b; Jastrzebski, 1989; Jastrzebski and Komorowski, 1990). One exception is apparently *Eimeria* (*s.l.*) *vanasi*, in which the microgamont nucleus does not subdivide before microgamete formation (Paperna, 1990), although the process was not described in detail.

Microgamonts measure 6.0–26.0 μm in diameter and are round or oval in section. They develop within a parasitophorous vacuole, which in *Goussia iroquoina, Calyptospora funduli* and *Goussia sinensis* is bound by a single limiting membrane (Paterson and Desser, 1981b; Hawkins *et al.*, 1983b; Baska and Molnár, 1989). The parasitophorous vacuole of *Eimeria* (*s.l.*) *vanasi* microgamonts is fringed with funnels like that of the macrogamont (Paperna, 1990) (see Section IV.B). The cytoplasm of young microgamonts is usually bound by a single limiting membrane, e.g. in *Goussia iroquoina* (see Paterson and Desser, 1981b) and in *Calyptospora funduli* (see Hawkins *et al.*, 1983b). However, the pellicle of *Goussia zarnowskii* microgamonts during nuclear division is characteristically formed from two membranes (Jastrzebski and Komorowski, 1990). Micropores may be present in the limiting membranes of some species, e.g. in *Goussia laureleus* (see Desser and Li, 1984) and *Calyptospora funduli* (see Hawkins *et al.*, 1983b).

In *Goussia iroquoina* and *Goussia subepithelialis* the microgamont cytoplasm contains free ribosomes, cisternae of rough endoplasmic reticulum, mitochondria, Golgi complexes, lipid and amylopectin (Paterson and Desser, 1981b; Steinhagen *et al.*, 1990). Multimembranous vacuoles, lipoid bodies and floccular material are found within the cytoplasm of other species (Hawkins *et al.*, 1983b; Davies, 1990).

Growth and nuclear division usually occur simultaneously. Young microgamonts have often numerous small nuclei that become aligned peripherally within the cytoplasm, e.g. in *Calyptospora funduli, Goussia sinensis* and *Goussia carpelli* (see Hawkins *et al.*, 1983b; Baska and Molnár, 1989; Steinhagen, 1991b) (Fig. 20). As these nuclei mature they frequently show marginated chromatin. *Eimeria variabilis* microgamonts are unusual in producing relatively small numbers of peripheral microgamete nuclei (Davies, 1978, 1990). In other species, such as *Goussia iroquoina* and *Goussia subepithelialis*, nuclei are also located beneath the surface membranes of deep invaginations within the microgamonts (Paterson and Desser, 1981b; Steinhagen *et al.*, 1990).

Microgamete formation has been described in a few species. In *Goussia iroquoina* and *Goussia carpelli* protrusion of the nuclei into the parasitophorous vacuole was accompanied by partition of the nucleoplasm into an electron-dense portion, which became the nucleus of the microgamete, and an inner less electron-dense region that remained in the microgamont to form residual nuclei (Paterson and Desser, 1981b; Steinhagen, 1991b) (Fig. 21). In *Goussia carpelli* microgamete development started with the formation of the perforatorium anlage and flagella in close association with the centriole–centrocone complex of the final nuclear division (Steinhagen, 1991b). In *Eimeria* (*s.l.*) *vanasi* a small mitochondrion accompanied each peripheral nucleus. These lay adjacent to microgamete primordia and to flagella, which emerged into the parasitophorous vacuole before the body of the microgamete (Paperna, 1990).

Microgametes of *Goussia iroquoina, Calyptospora funduli, Goussia subepithelialis* and *Goussia carpelli* have two flagella each (Paterson and Desser, 1981b; Hawkins *et al.*, 1983b), Steinhagen *et al.*, 1990; Steinhagen, 1991b; whereas those of *Goussia spraguei* in the kidney tubules of cod, *Gadus morhua*, appear to have a single flagellum (Morrison and Poynton, 1989a). In addition, microgametes have a dense elongate nucleus, one mitochondrion, basal bodies, an apical dense perforatorium and several microtubules (4–12 have been reported) in close association with the mitochondrion and

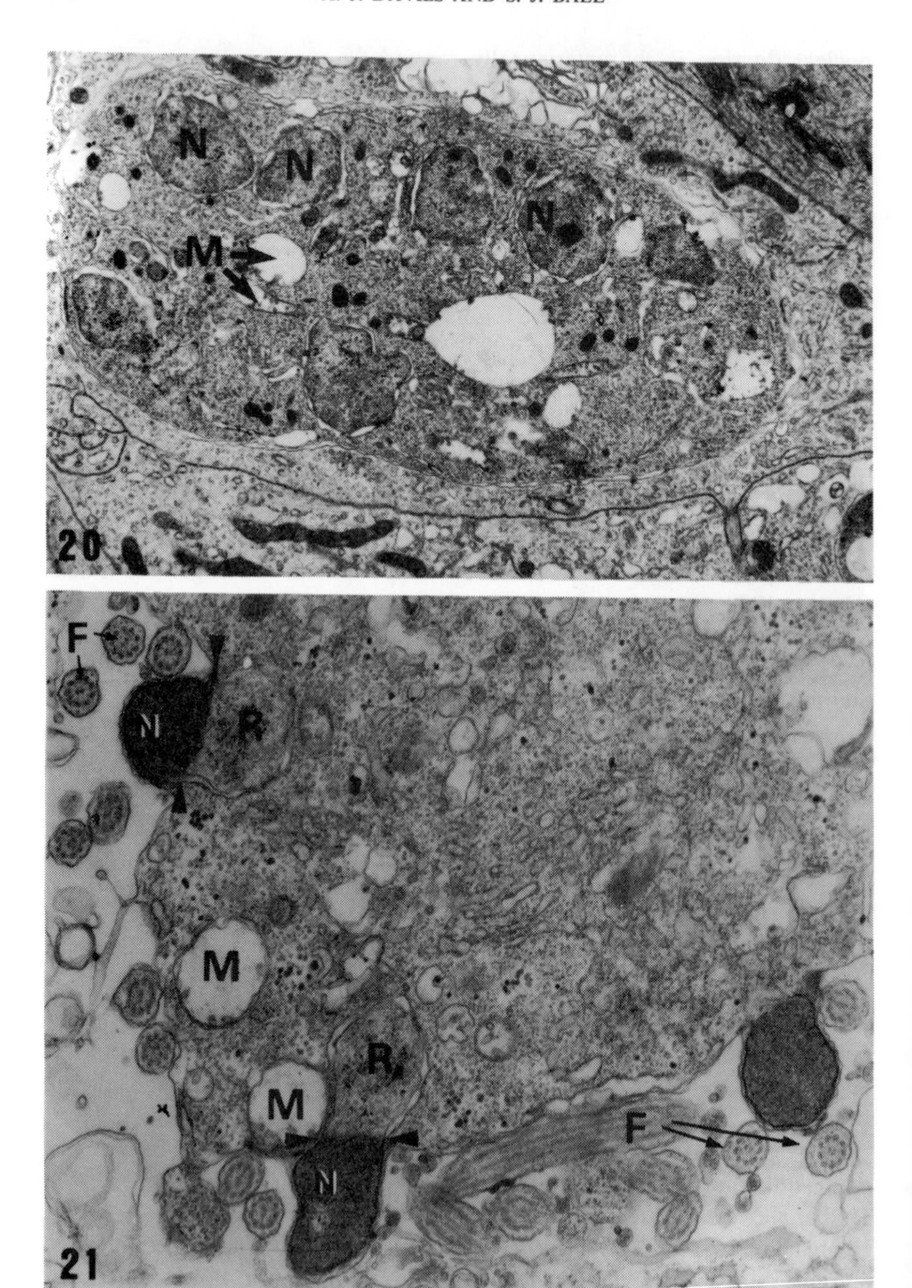
N
N
N
M
20
F
N
R
M
M
R
N
F
21

nucleus. Microgametes of *Calyptospora funduli* also have dense rod-like structures (Hawkins *et al.*, 1983b).

When microgamete production is complete in *Calyptospora funduli*, the limiting membrane of the microgamont may disintegrate. At this stage the microgamont cytoplasm contains multimembranous vacuoles, residual nuclei, polysaccharide deposits, vacuoles and swollen rough endoplasmic reticulum (Hawkins *et al.*, 1983b).

D. FERTILIZATION

Observations on fertilization in coccidian parasites generally are rare (Ball and Pittilo, 1990).

Paterson and Desser (1984) observed microgametes within the cytoplasm of macrogametes of *Goussia iroquoina* from the intestinal epithelium of experimentally infected fathead minnows. In this case the entire microgamete was within the macrogamete. The number of membranes surrounding microgametes appeared to correlate with maturity of the macrogamete. In the immature macrogamete the cytoplasm was separated from that of the microgamete by the plasmalemma of each gamete plus two, or rarely four, additional membranes. In mature macrogametes the cytoplasm of the gametes was separated by only two plasmalemmas. Fusion of the limiting membranes of both gametes occurred eventually.

In another account (Desser and Li, 1984), microgametes of *Goussia laureleus* were seen in the parasitophorous vacuole enclosing mature macrogamonts. In one case, two microgametes appeared to depress the surface of the macrogamete. Projections from the macrogamete contacted and indented the boundary membrane of the parasitophorous vacuole on either side of the microgametes.

Hawkins *et al.* (1983b) also recorded microgametes of *Calyptospora funduli* in the parasitophorous vacuoles of mature macrogamonts. Only once was a microgamont found in an invagination of the limiting membrane of a developing macrogamont. External to the plasma membrane of the microgamete were ribosome-studded and smooth membranes that were probably

---

FIGS 20 and 21. Transmission electron micrographs of *Goussia carpelli*. FIG. 20. Microgamont with peripherally located nuclei. N, nucleus; M, mitochondrion (× 10 500). FIG. 21. Nuclei protruding along the limiting membrane of the microgamont. Nucleoplasm divided into two parts (arrowheads). F, flagellum; M, mitochondrion; N, portion of nucleus to become nucleus of microgamete; R, residual portion of nucleus (× 21 500). (Kindly supplied by D. Steinhagen from Steinhagen, 1991b, and reproduced by kind permission of the author and Gustav Fischer Verlag.)

host membranes acquired during passage of the microgamete into the parasitophorous vacuole of the macrogamont.

One other observation concerns *Goussia spraguei* from the kidney tubules of cod. On a single occasion the nucleus of a microgamete was found inside the parasitophorous membrane of the macrogamete (Morrison and Poynton, 1989a). The microgamete appeared to have lost its surrounding membrane, and no flagellum was seen.

V. INFECTIVE STAGES

A. OOCYST MORPHOLOGY AND SPORULATION

1. *Morphology*

Oocysts are the most frequently described features of fish coccidia and, together with their contained sporocysts, are the structures that largely determine the taxonomic status of the parasite (see Table 1).

Oocysts of fish coccidia tend to be spherical or ellipsoid structures (generally 4.5–70 μm in diameter, depending on species), although some are cylindrical, e.g. those of *Eimeria southwelli* Halawani, 1930 and *Eimeria quentini* Boulard, 1977. Their finely granular cytoplasm contains polysaccharide reserves, protein and some lipid (Fig. 22). Before sporulation, a single nucleus occupies a roughly central position within this cytoplasm. The oocyst walls are commonly thin, and one species, *Goussia sinensis*, is reported to have a micropyle (Chen, 1956a). Some thin oocyst walls tend to collapse on to the sporocysts following sporulation. This thin-walled feature is thought to reflect an aquatic existence, where mechanical insult and desiccation are lesser problems than on land (Lom, 1971; Molnár, 1977). However fish coccidia with thick-walled oocysts are known, e.g. *Eimeria southwelli* (see Pellérdy, 1974) and *Goussia cichlidarum* (see Paperna and Landsberg, 1985; Paperna and Cross, 1985), and Davies (1978) commented on the difficulties in relating oocyst wall thickness to either environmental conditions or to the permeability of the structure. Occasionally, oocyst walls in fish coccidia appear coloured (yellow, brown, lavender), and pitted (Pellérdy, 1974; Fitzgerald, 1975).

The nature of the oocyst wall of fish coccidia and its mode of construction have been the subject of much speculation. In coccidia of homeotherms the oocyst wall tends to be thick, and is formed from two types of wall-forming bodies (WF1 and WF2) that are discharged around the fertilized macrogamete, but in fish coccidia these bodies are not always recognized. Hawkins *et al.* (1984) have commented that the paucity or absence of wall-forming

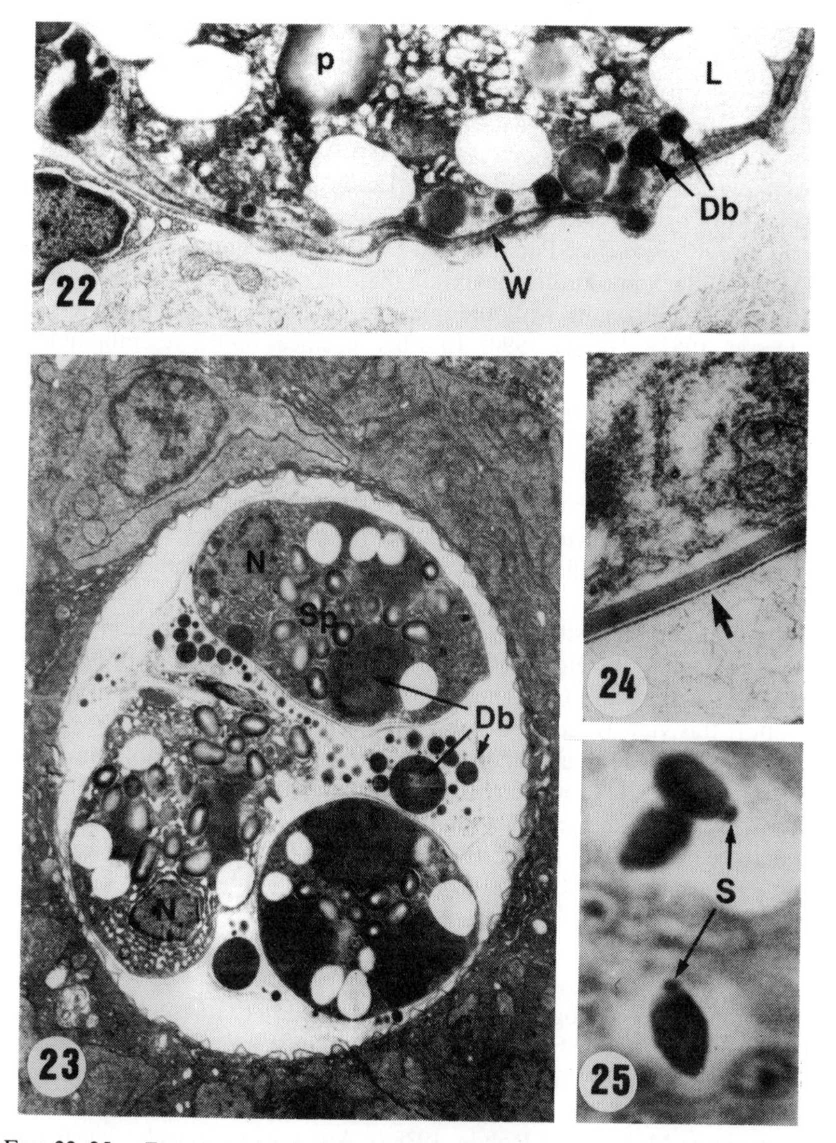

FIGS 22–25. *Eimeria variabilis* from the pyloric caeca of *Cottus (Taurulus) bubalis*. FIGS 22–24. Transmission electron micrographs. FIG. 22. Portion of oocyst showing small peripheral dense bodies (Db), polysaccharide granule (p) and lipid (L). Oocyst wall (W) comprises two closely apposed membranes (× 12 000). FIG. 23. Division of oocyst contents into sporoblasts. Note that development is asynchronous. Sporoblast (Sp), nuclei (N) and dense bodies (Db) are visible. Dense bodies also occur in the cavity of the oocyst (× 6500). FIG. 24. Sporocyst wall consisting of a fine dense layer (arrow) outside the broad dense layer and separated from it by an electron-lucent layer (× 60 000). FIG. 25. Photomicrograph. Sporocysts stained with Giemsa-colophonium showing Stieda bodies (S) (× 1400). (Figs 22–24 from Davies, 1990, and reproduced by permission of the editor.)

bodies is not surprising in view of the characteristic thinness of the oocyst wall of piscine coccidia. In such coccidia, the oocyst wall may consist of a single membrane as in *Calyptospora funduli* (see Hawkins *et al.*, 1983a), *Goussia laureleus* (see Desser and Li, 1984) and *Goussia gadi* (see Odense and Logan, 1976), or two layers as in *Eimeria variabilis* (see Davies, 1990) (Fig. 22), *Goussia iroquoina* (see Paterson and Desser, 1984), and *Goussia degiustii* (see Lom, 1971). Some authors maintain that the oocyst wall is formed from membranes of the zygote with endoplasmic reticulum additions (Paterson and Desser, 1981a; Davies, 1990). In other instances bodies resembling the wall-forming bodies of coccidia of homeotherms have been described, but some of these may be involved in sporocyst wall formation (see Sections IV.B and V.B).

Oocysts vary in their site of development. As for other stages of fish coccidia, the gut is a favoured location but extraintestinal sites where oocysts may be found include liver, kidney, spleen, pancreas, testes, ovary, peritoneum, swim bladder, gall bladder, adipose tissue and gill filaments. Within the same species, oocysts may occur in the same location as merozoites and gamonts, or at different sites, suggesting that in some cases migration occurs. Some oocysts may therefore behave like ookinetes (e.g. in *Eimeria branchiphila* Dykova, Lom and Grupcheva, 1983), although experimental evidence to support this view is required.

Oocysts are usually surrounded by a parasitophorous vacuole and develop intracellularly. However, in *Cryptosporidium* they occur either just under the host cell surface, or within its brush border, while in *Epieimeria* they may be intercellular (Hine, 1975; Molnár and Baska, 1986), epicellular or intracellular (Levine, 1984a).

2. *Sporulation*

Sporulation is often endogenous, but this is not the rule, and several coccidia that sporulate exogenously are known. These include, e.g., *Eimeria squali* (see Fitzgerald, 1975), *Eimeria dingleyi* (see Davies, 1978), *Goussia vargai* and *Goussia acipenseris* (see Molnár, 1986), *Eimeria pleurostici* and *Goussia arrawarra* (see Molnár and Rohde, 1988a). Attempts to sporulate such fish coccidian oocysts in potassium dichromate or formaldehyde solutions frequently end in greater failure (Fitzgerald, 1975; Davies, 1978; Molnár, 1977) than attempts using water alone (Molnár, 1986, 1989; Molnár and Baska, 1986). Antibiotic solutions in saline or water may improve the process (Davies, 1978; Molnár and Rohde, 1988a,b). Perhaps considerable physiological or biochemical differences exist between the oocysts of fish coccidia and those of homeotherms, the thick-walled oocysts of which can be stored satisfactorily in potassium dichromate.

Remarkably, some oocysts sporulate both endogenously and exogenously. This is true of some coccidia of carp and tench (Marinček, 1973a; Molnár, 1982, 1989). Molnár (1989) believed this reflects the characteristics of an immunological host reaction to the parasite rather than the properties of the coccidian itself (see Section VI.B).

The process of sporulation subdivides the contents of the oocyst into four sporocysts in *Calyptospora*, *Crystallospora*, *Eimeria*, *Epieimeria*, *Goussia*, and presumably in the proposed genera *Nucleoeimeria* and *Nucleogoussia*. In *Cryptosporidium* one or no sporocyst is formed, while there are eight in *Octosporella*.

In *Goussia cichlidarum* sporoblasts are formed endogenously, within an outer rim of oocyst cytoplasm (Paperna and Landsberg, 1985). Exogenous sporulation is seen in other species. In *Eimeria variabilis* sporulation has been studied in some detail (Davies, 1978, 1990). In this species the oocyst nucleus divides into four, and this is followed by complete cytoplasmic cleavage to give four uninucleate sporoblasts. Subsequently, these sporoblasts become binucleate, although sporoblast development may be asynchronous (Fig. 23), and finally they mature into sporocysts (Fig. 25). Similar development occurs in *Goussia iroquoina* (see Paterson and Desser, 1984) but in *Goussia laureleus* the initial stage of sporulation cleaves the zygote into equal halves (Desser and Li, 1984). Sporulation in *Goussia spraguei* apparently follows zygote cleavage into either halves or quarters, so that two patterns of development are present (Morrison and Poynton, 1989a).

B. SPOROCYST FORMATION AND MORPHOLOGY

Sporocyst walls of fish coccidia are often substantial structures. Those of *Goussia auxidis* were found to consist of three layers (Fig. 26). The outer layer was a laminar envelope with 2–12 laminations parallel to the wall, and with 9–10 nm between laminations. Beneath this layer was an electron dense layer 14 nm thick, and beneath that was a transversely laminated wall 160–180 nm thick. Laminations in this section of the wall were repeated every 14 nm (Jones, 1990). Sporocyst walls of *Goussia clupearum* comprised an outer layer of closely apposed membranes (about 40 nm total thickness), separated from a transversely striated inner layer (about 200 nm wide) by a narrow electron-lucent space (Morrison and Hawkins, 1984). Substantial sporocyst walls (0.1–0.5 μm thick) were also present in *Goussia gadi* (see Odense and Logan, 1976). Most of the wall comprised uniform material containing many dark longitudinal lines of irregular length perpendicular to a regular transverse striated pattern with a periodicity of 12.5 nm. The wall was not uniformly rounded but consisted of segments or plates of various lengths abutting at angles of 10–20°.

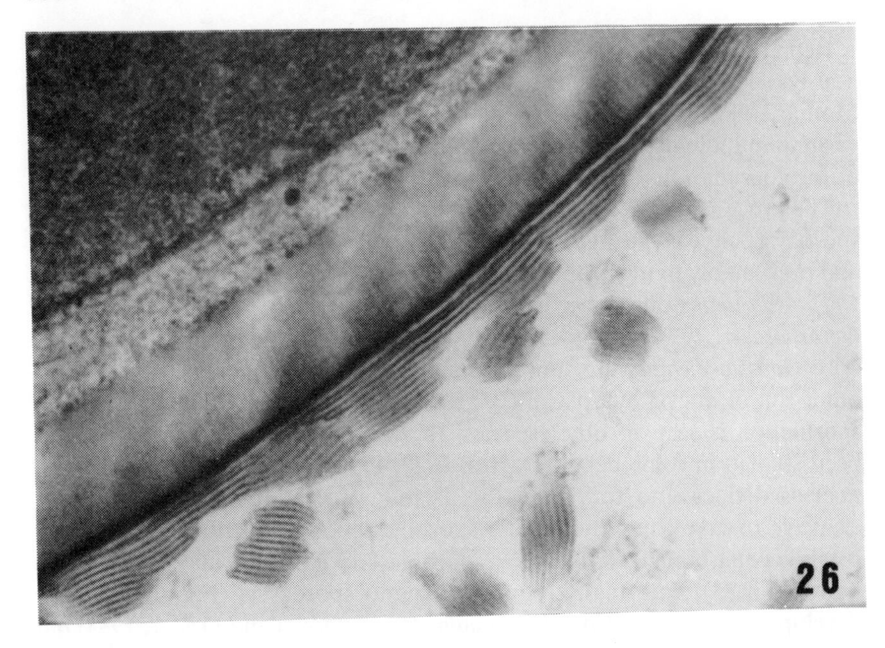

FIG. 26. Transmission electron micrograph of sporocyst wall of *Goussia auxidis*. Outer layer (which may be sloughing off) with laminations parallel to the surface, intermediate dense layer, and transversely laminated inner layer (× 100 000). (Kindly supplied by J. B. Jones from Jones, 1990, and reproduced by permission of the editor.)

Sporocyst walls of *Calyptospora funduli* were 130-150 nm thick, with a thin outer layer (about 10 nm wide) and thicker inner layer (120–130 nm in width) with striations in some planes of section. Sporopodia (Figs 27–30) (outgrowths of the sporocyst wall) had similar layers to the sporocyst wall but did not appear transversely striated, and they terminated in knob-like swellings that often apposed a sporocyst veil (Hawkins *et al.*, 1983c).

Thinner walls were recorded for *Goussia laureleus*. These were approximately 50 nm thick, with an outer electron-lucent layer (about 30 nm), and an inner dense zone (about 20 nm) with striations oriented at right angles to the sporocyst surface (Desser and Li, 1984). Sporocyst walls of similar thickness were seen in *Eimeria variabilis* (see Davies, 1978, 1990). These were 51 nm thick, with an outer layer 3 nm wide, an intermediate electron-lucent layer (7 nm wide), and an inner broad dense layer 41 nm wide (Fig. 24). Corrugations were noted on the inner surface of the sporocyst wall in some planes of section.

The origin of the sporocyst wall of fish coccidia is unresolved. In some species the source may be endoplasmic reticulum membranes (Paterson

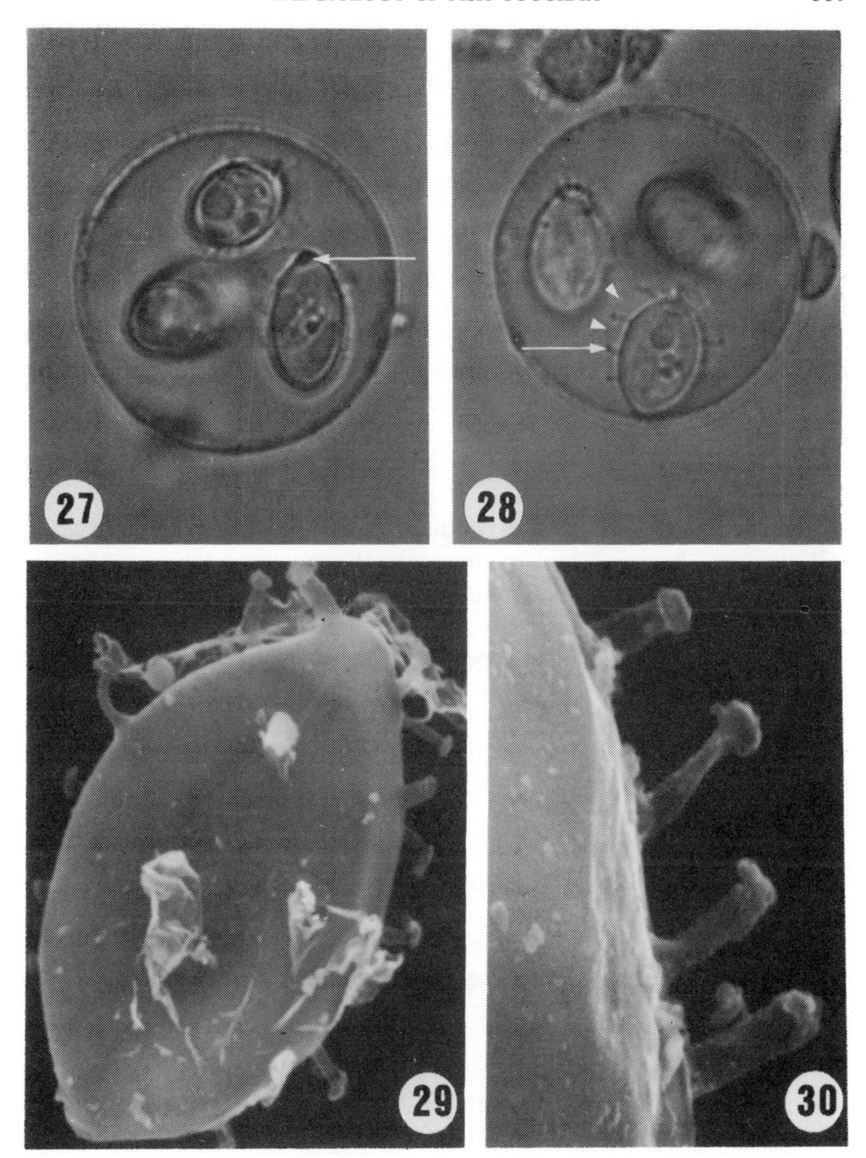

FIGS 27–30. Sporulated oocysts and sporocysts of *Calyptospora funduli*. FIG 27. Substieda body (arrow) (× 2220). FIG. 28. Sporopodia projecting from sporocyst wall (arrow). Note transparent membrane (arrow heads) (× 2220) (Photomicrographs kindly supplied by D. W. Duszynski from Duszynski *et al.*, 1979, and reproduced by kind permission of the editor). FIG. 29. Sporocyst showing sporopodia (× 10 000). FIG 30. Close-up of sporopodia showing resemblance to long-stalked mushrooms (× 20 000). (Figs 29 and 30. Unpublished scanning electron micrographs by courtesy of D. W. Duszynski.)

and Desser, 1984), but in addition, if thick-walled oocysts of the coccidia of terrestrial vertebrates result from the activity of wall-forming bodies, it may be that equivalent structures in fish coccidia give rise to sporocyst walls, especially since these tend to be thicker than the walls of oocysts.

Structures resembling wall-forming bodies have been found in the macrogametes and oocysts of several fish coccidia and, when they occur, they have usually been assigned a role in oocyst wall formation. However, some authors have questioned this role (Morrison and Hawkins, 1984; Paperna and Landsberg, 1985; Paperna *et al.*, 1986; Paperna, 1990; Davies, 1990). In particular, Paperna *et al.* (1986) and Paperna (1990) have been most cautious about attributing an oocyst wall-forming function to all these bodies in fish parasites, noting that the wall-forming bodies of some coccidia of invertebrates and of *Sarcocystis* may be concerned with formation of sporocyst walls.

Paperna and Landsberg (1985) recorded bodies resembling type 2 wall-forming bodies of coccidia of homeotherms within the mature macrogamonts of *Goussia cichlidarum*. These granules were thought to load the cytoplasm of the sporoblasts, and their intimate association with the limiting membranes at sporogony suggested their participation in wall formation of the sporocyst. The thick sporocyst wall appeared to be formed by incorporation of at least three thin, hard layers or plates formed in sequence in the subpellicular cytoplasm of the sporocyst.

In *Eimeria variabilis* dense bodies were found in the macrogamete and the oocyst (Fig. 22), but they persisted when oocyst wall formation was complete, and they also occurred in sporoblasts and in the cavity of the oocyst at sporulation (Davies, 1990) (Fig. 23). Davies (1990) tentatively suggested their involvement in sporocyst wall formation in this species, and noted their association with fine filamentous material that lay parallel, but external, to the walls of developing sporocysts. Similar, numerous lamellated filaments were illustrated extending from the walls of developing sporoblasts of *Goussia clupearum*, but their presence was not explained in detail (Morrison and Hawkins, 1984). The outer layers of the sporocysts of *Goussia auxidis* were also associated with detached pieces of sporocyst wall within the cavity of the sporulating oocyst (Jones, 1990) and these were interpreted as sloughing off (Fig. 26). Veils (membranaceous collars) extended from the sporocyst sutures of *Goussia (Plagula) subepithelialis* (see Lom, 1971), but, their mode of formation was not recorded.

Major problems in determining the development of sporocysts result mainly from difficulties encountered in inducing electron microscopy fixatives and embedding media to penetrate their thick walls properly. Until satisfactory fixation and impregnation is achieved the true development of sporocyst walls will remain an enigma.

While the mode of construction of sporocyst walls is largely unknown, the diversity which they achieve at their formation is remarkable (Table 1; Fig.1). In *Eimeria* species sporocyst walls are smooth and plugged by a Stieda-like body at one pole (Fig. 25), whereas in *Goussia* species they are sutured longitudinally and open like a pea pod. In *Plagula*, a subgenus of *Goussia*, a veil is present. In *Crystallospora* sutures run horizontally around the equator of the sporocyst, which has the appearance of two hexagonal pyramids fitted base-to-base. *Calyptospora* sporocysts have a thin veil supported by one or more sporopodia and a membrane covering an apical opening often associated with a suture, but usually no Stieda or sub-Stieda body (Figs 27–30). Since classification of fish coccidia is based primarily on morphological differences between oocysts and sporocysts, this allows the tetrasporocystic-dizoic genera *Eimeria, Goussia, Crystallospora*, and *Calyptospora* to be distinguished. However, intracellular location has also been considered an important criterion for classification, and on this basis Dyková and Lom (1981) proposed the genus *Epieimeria*, and Daoudi *et al.* (1987) suggested creating the genera *Nucleoeimeria* and *Nucleogoussia*. As Molnár (1989) noted, it would be possible to create the genus *Epigoussia* as a counterpart to *Epieimeria*, and the species described by him as *Goussia pannonica* would be an example. Presumably, *Goussia zarnowskii*, an epicellular parasite of sticklebacks (Jastrzebski and Komorowski, 1990), and *Goussia janae* from dace (Lukeš and Dyková, 1990), might be others. In addition, some eimerian species from other animals, e.g. *Eimeria ranarum* (Labbé, 1894) Dolflein, 1909, which develops in the epithelial cell nuclei of the digestive tract of *Rana esculenta* and *Rana temporaria*, should perhaps be transferred to *Nucleoeimeria*. Molnár (1989) suggested caution in accepting location within the cell as a generic character, for other factors such as development with or without an intermediate host, or development on a yearly cycle instead of continuously, might serve equally as a basis for classification.

C. SPOROZOITE FORMATION AND STRUCTURE

Sporocysts of *Calyptospora, Crystallospora, Eimeria, Epieimeria, Goussia, Isospora* and *Octosporella* each contain two sporozoites, while oocysts (or sporocysts) of *Cryptosporidium* contain four. Sporozoite formation in fish coccidia has been difficult to follow because of the problems associated with fixing and embedding sporocysts noted in Section V.B. However, freezing sporocysts in liquid carbon dioxide may help to open sutures and improve preservation, and addition of dimethyl sulphoxide to fixatives may also be important (Morrison and Hawkins, 1984).

Essentially, sporozoites resemble those of other coccidia except, as Dyková and Lom (1981) noted, they are somewhat more elongate. Mature sporozoites of *Goussia subepithelialis* within the sporocyst are striated anteriorly, and this may result from folding of the pellicle of quiescent sporozoites, since it disappears if sporozoites are activated with sodium bicarbonate saturated with carbon dioxide (Lom, 1971).

Sporozoite development has probably been best described for *Goussia spraguei* from the kidney tubules of cod (Morrison and Poynton, 1989a), and sporozoite structure for *Goussia subepithelialis* and *Goussia carpelli* (Figs 9, 10) from the intestine of tubificid oligochaetes (which may be paratenic hosts for these species) (Steinhagen, 1991a). Accounts of the sporozoites of *Goussia gadi, Goussia iroquoina, Goussia laureleus* and *Goussia cichlidarum* are less detailed (Odense and Logan, 1976; Paterson and Desser, 1984; Desser and Li, 1984; Paperna and Landsberg, 1985), but conform generally to the descriptions given below.

In *Goussia spraguei* sporozoite formation began with division of the sporoblast nucleus, forming two sporozoites. These sporozoites then elongated and were attached initially to a residuum containing amylopectin granules. Sporozoites then became banana-shaped and were surrounded by a pellicle, and each contained crystalloid material that was also seen in the residuum. As sporozoites matured and a conoid apparatus appeared, the residuum was reduced in size. Mature sporozoites contained one, or sometimes two, refractile bodies, and the conoid apparatus associated with micronemes and rhoptries (Morrison and Poynton, 1989a).

Goussia carpelli sporozoites from the intestinal epithelium of tubificids had a three-layered pellicle with a micropore. Both *Goussia carpelli* and *Goussia subepithelialis* sporozoites from this source contained refractile bodies, a central nucleus, mitochondria, lipid, amylopectin inclusions, dense bodies, rhoptries, and micronemes. *Goussia subepithelialis* sporozoites had two apical polar rings and 28 subpellicular tubules, whereas 24 microtubules were counted in *Goussia carpelli*. The pellicle showed signs of destruction in *Goussia subepithelialis* sporozoites found in cells with extended endoplasmic reticulum. In *Goussia carpelli*, sporozoites were incorporated in membrane-bound parasitophorous vacuoles in the basal regions of the epithelium near blood vessels (Steinhagen, 1991a) (Fig. 9).

D. DISPERSAL OF OOCYSTS

Dispersal of oocysts from the gut of fish occurs presumably in the faeces, where oocysts may be passed unsporulated, semisporulated, or fully sporulated. Tissue-inhabiting coccidia such as *Eimeria sardinae* from the testes of herring (*Clupea harengus harengus*) are thought to be released at spawning, and the development of the parasite is linked to the reproductive dynamics of the male fish (McGladdery, 1987). Coccidia inhabiting the urinary tract may be released in a similar fashion, although *Goussia spraguei* may occlude kidney nephrons in cod through granuloma formation (Morrison and Poynton, 1989a), which might prevent discharge of oocysts.

One unusual coccidian is *Eimeria branchiphilia* from the gill filaments of roach. This may be dispersed by disruption of gill lamellae with the resultant release of oocysts (Dykova *et al.*, 1983).

Other tissue coccidia present a problem. It is known that *Calyptospora funduli*, for example, has a requirement for an intermediate host (see Section III.B), but how is this parasite and others like *Goussia auxidis* and *Goussia clupearum* able to escape from the liver and infect new hosts? Are such parasites released down the bile ducts to the gut or must these parasites wait until the death of their hosts? Similar questions can be asked about swim bladder coccidia like *Goussia cichlidarum, Goussia gadi* and *Goussia caseosa*. The possibility that oocysts of fish coccidia can survive passage through the alimentary tract of carnivores was noted by Molnár (1981). This author cited Shul'man (1962), who recorded the passage of morphologically well-preserved *Goussia clupearum* and *Eimeria sardinae* oocysts through the gut of humans. Are these species also able to survive transit through the gut of carnivorous fish or marine scavengers? Odense and Logan (1976) noted that young haddock infected with *Goussia gadi* are eaten by other gadoids that can act as hosts for the coccidian, although in their study swim bladders of gadoids other than haddock were uninfected. Alternatively, perhaps the oocysts of fish coccidia excyst in the gastro-intestinal tract of non-specific fish hosts, with sporozoites surviving in tissues as a potential source of infection to predator fish.

E. "MIGRATION" AND BLOOD STAGES

A number of fish coccidia apparently undergo "migration" during development, but there is little to indicate how this occurs.

Goussia subepithelialis, for example, begins development in the epithelial lining of the intestine of carp and then oocysts move to the subepithelium (Marinček, 1973a,b). Meronts and gamonts of *Eimeria variabilis* from *Cottus bubalis* lie in the epithelial lining of the pyloric caeca or rectum, whereas oocysts may occur epithelially or subepithelially (Davies, 1990). In a similar manner, merogony and gamogony of *Goussia sinensis* and *Goussia carpelli* occur in epithelial cells of the gut but developing gamonts and early oocysts receded into the lamina propria (Molnár, 1979). Fertilized macrogametes of *Epieimeria anguillae* were assumed by Léger and Hollande (1922) to migrate intracellularly as ookinetes. Molnár (1981) and Baska and Molnár (1989) suggested that cells with macrogamonts and oocysts cannot perform a covering function at the gut surface and in consequence, as a defensive host reaction, they are overgrown by intact epithelial cells, pushing infected cells into deeper layers.

While this explanation may indicate how stages that are normally immobile, like macrogamonts, macrogametes and oocysts, reach the deeper layers of the gut, it does not account for the extraordinary migration seen in some fish coccidia, and the low tissue specificity shown by others. For example, Hardcastle (1944) recorded sporogony of *Eimeria brevoortiana* in the swim bladder of menhaden (*Brevoortia tyrannus*), while merozoites and gamonts apparently were found in the intestine. *Eimeria patersoni* develops within kidney, spleen and liver in its host fish *Lepomis gibbosus* (see Lom *et al.*, 1989). How are both parasites able to reach, and develop in, such diverse organs? Even more remarkable is the highly pleomorphic *Goussia degiustii*, which Lom *et al.* (1989) noted is able to infect gall bladder, gut, kidney, liver, pancreatic islets, spleen and swim bladder of common shiners in Canada.

One means of transport might be within tissue macrophages. On one occasion Molnár and Baska (1986) found a merozoite of *Epieimeria anguillae* within a macrophage in the gut lumen of an eel.

Alternatively, perhaps blood stream stages of fish coccidia exist. In one heavily infected specimen of haddock, for example, Odense and Logan (1976) noted sporozoites of *Goussia gadi* within and around a major hepatic vein, though not in peripheral blood films. Dyková *et al.* (1983) suggested that *Eimeria branchiphilia* may reach the gills of roach by migrating from elsewhere as a free ookinete, possibly in the bloodstream. The occurrence of a high prevalence of *Goussia clupearum* and an intraleucocytic haemogregarine has been noted in Atlantic mackerel (MacLean and Davies, 1990), and perhaps they are the same parasite. Solangi and Overstreet (1980) suggested that *Calyptospora funduli* might reach the liver via the blood stream, since infections were absent from the bile duct and intestine of killifish, and occasional extrahepatic development occurred.

In other vertebrates there is strong evidence that dissemination of coccidia occurs through the blood stream (Ball *et al.*, 1989). In mammals there are examples of *Eimeria* species developing in the bile ducts of liver, in gall bladder, lymph nodes, mammary glands, placenta, and reproductive tracts. These are relatively rare in comparison with intestinal species and in some cases may be aberrant. On the other hand, renal coccidiosis is common in water fowl (Gajadhar *et al.*, 1983 a,b). In some species of *Eimeria* infecting chickens sporozoites are carried within intraepithelial lymphocytes from the intestinal surface to enterocytes of the intestinal crypts. It is thought that these lymphocytes can also transport sporozoites throughout the body to various organs, even though their sites of development lie in the intestine (see Ball *et al.*, 1989).

VI. Host–Parasite Interactions

A. PATHOLOGY AND SEASONALITY

Limited information exists on the importance of coccidiosis in fish. Pellérdy (1974) considered intestinal coccidiosis to be largely a disease of cultured fish brought about by factors that decrease host resistance, such as overcrowding. As in other vertebrates, it is generally assumed that under natural conditions piscine coccidia cause little disease unless the host–parasite balance is upset. However, when wild-caught fish are taken, perhaps only those best able to survive coccidiosis are seen. In addition, as Dyková and Lom (1981) noted, the impact of pathological changes in the host may be underestimated because attention is paid often to only severe host tissue reactions.

Gregory (1990) has recorded many factors that may contribute to the pathology of coccidiosis in mammals and birds, including species of parasite, infecting oocyst dose, frequency of infection, concurrent infections (of coccidia and other parasites), age, immune status, stress and nutrition. For fish coccidia, information on these contributing factors is mostly lacking, as was emphasized by Dyková and Lom (1981).

1. *Gut-inhabiting stages*

The pathology of some gut-inhabiting species of *Goussia* has been studied fairly extensively, including that of *Goussia sinensis, Goussia carpelli* and *Goussia subepithelialis*. In these the extent of damage, and character of the tissue reaction, depend on the intensity of infection and depth of parasite penetration into the intestinal wall (see Dyková and Lom, 1981).

In *Goussia sinensis* from silver carp and bigheads (*Aristichthys nobilis*) held in ponds, the earliest appearance of oocysts was in 18-day-old fish (Molnár, 1976). Massive infections were found in fry and "one-summer" fish (presumably those less than 1 year old). Mortality occasionally reached 60–70% among fry 4–6 weeks old, which had many coccidial stages in the intestine. Molnár noted that the coccidian was always associated with other parasites, so that only indirect conclusions could be drawn concerning the pathogenesis of *Goussia sinensis*. Infected fish of both species were darker and thinner than healthy fish. At autopsy, oedema of the serous membranes of the intestinal wall, absence of fat, and swollen intestinal mucosa were recorded. Injury resulted chiefly from meront development. When large numbers of macrogametes and oocysts were present, the intestinal wall of fry was dark, and both intestinal epithelium and contents were coated with yellowish, viscous, mucous exudate. Histopathological changes were usually milder than might be expected for severe infections. For example, when approximately half of the epithelial cells and most cells of the lamina propria were invaded, no inflammation or haemorrhagic lesions were recorded.

Kent and Hedrick (1985) recorded goldfish, 30–45 days old and naturally infected with *Goussia carpelli*, that exhibited chronic enteritis with loss of intestinal villar structure; the lamina propria was filled with mononuclear inflammatory cells, yellow bodies, and intact and degenerating oocysts. In 8-day-old fish, small meronts were identified in the intestine, and gamonts and oocysts were first observed at 10 days. Significant lesions were not associated with these stages. Sporulated oocysts, often surrounded by yellow bodies, were first seen in fish 15 days old. Yellow bodies were more prominent in older fish, in some forming areas of focal necrosis in spleen and kidney interstitium, although *Goussia carpelli* itself was confined to the intestine. The nature of yellow bodies, which have also been recorded in infections due to *Eimeria acerinae, Eimeria cheni, Goussia sinensis* and *Calyptospora funduli*, is unresolved. Kent and Hedrick (1985) proposed that these bodies formed from degenerating membranes within secondary lysosomes and contained lipofuscin. Baska and Molnár (1989) suggested two origins for the yellow bodies of *Goussia sinensis*, the granular part being derived from parasite (zygote) residue, and the homogeneous portion from the necrotic host cell (or several aggregated host cells).

In *Goussia subepithelialis* infections, the main lesions were caused by gamogony in the epithelium and extending into the muscular mucosa (Pellérdy and Molnár, 1968). Oocysts were frequently encapsulated by proliferative inflammatory cells. Impaired host condition, clinical signs and histopathological changes provided good evidence that parts of the intestine were non-functional (Marinček, 1973a: Dyková and Lom, 1981).

In contrast to the situation with *Goussia sinensis, Goussia carpelli* and *Goussia subepithelialis*, Philbey and Ingram (1991) considered that the lack of clinical signs and the presence of only a minimal inflammatory response in the intestine, even with high levels of infection detected histologically, indicated that *Goussia lomi* in Murray cod fry, *Maccullochella peeli*, was not highly pathogenic.

Seasonality of infection has been reported in natural intestinal infections with *Goussia iroquoina* in common shiner (*Notropis cornutus*) and fathead minnow (*Pimephales promelas*). Infection is acquired soon after fish begin feeding at about 4 weeks, and in heavy infections large areas of epithelium are destroyed (Paterson and Desser, 1982). In Ontario, all developmental stages were found in spring, but by early summer the majority of fish harboured only oocysts. *Goussia iroquoina* has a seasonal periodicity and cold temperature shock is required before some oocysts become infective. For example, oocysts from fish kept at 24–26°C were non-infective for 89 days, and presumably require cold shock to become infective, but oocysts from fathead minnows kept at 21°C were directly infective. Therefore, oocysts formed at the cooler temperatures of spring and summer presumably would be infective immediately, whereas oocysts formed at higher temperatures would become infective only in spring after the cold shock of winter.

Seasonality also occurs in *Goussia* species in cyprinid fish in Hungary (Molnár, 1989). Merogony occurred until March and gamogony, which lasted no longer than 2 months, took place from April. Molnár also demonstrated that cyprinid fish could be infected with different *Goussia* concurrently, some causing diffuse coccidiosis, others producing nodular (see below) or epicellular coccidiosis. However, despite high intensities of coccidial infection, there was no evidence of pathogenicity, and fish showed no clinical signs of disease. Occasionally, 70% of the intestine was infected, but there was no evidence of general epithelial necrosis. Pathological effects were limited to destruction of infected epithelial cells, and mucus released from them, which contained many oocysts, lined the gut in a tube-like fashion. Rapid regeneration of the intestine was typical and even intensive infections disappeared without trace (Molnár, 1989).

Most of the intestinal coccidia of fish develop evenly within a given area of epithelium and cause diffuse coccidiosis. However, Molnár (1982, 1989) has drawn attention to the lesser known nodular coccidiosis, with special reference to *Goussia* species. In a nodular type of infection, such as that caused by *Goussia balatonica* in the white bream, *Blicca bjoerkna*, and an unnamed species in the tench, *Tinca tinca* (see Molnár, 1982), merogonic and gamogonic stages infected epithelial cells in areas extending over three to five intestinal folds (Fig. 31). Almost every cell was infected in these areas which appeared as white, pinhead-sized nodules. Towards the edge of such areas, parasites were usually found only at the tips of folds, and this lighter infection could be adjacent to the uninfected epithelium which surrounded the nodules (Molnár, 1982). Molnár (1981, 1982) recorded similar nodules caused by *Goussia subepithelialis* in the gut of the common carp. In this case the epithelial cells contained micro- and macrogamonts and, later, sporulated oocysts occurred in the submucosa. Marinček (1973 a,b) and Molnár (1981) noted that this type of coccidiosis of carp began in early spring (March–April) with oocysts either being shed unsporulated from April, or remaining in the intestinal epithelium and then sinking into the submucosa (see Section V.E).

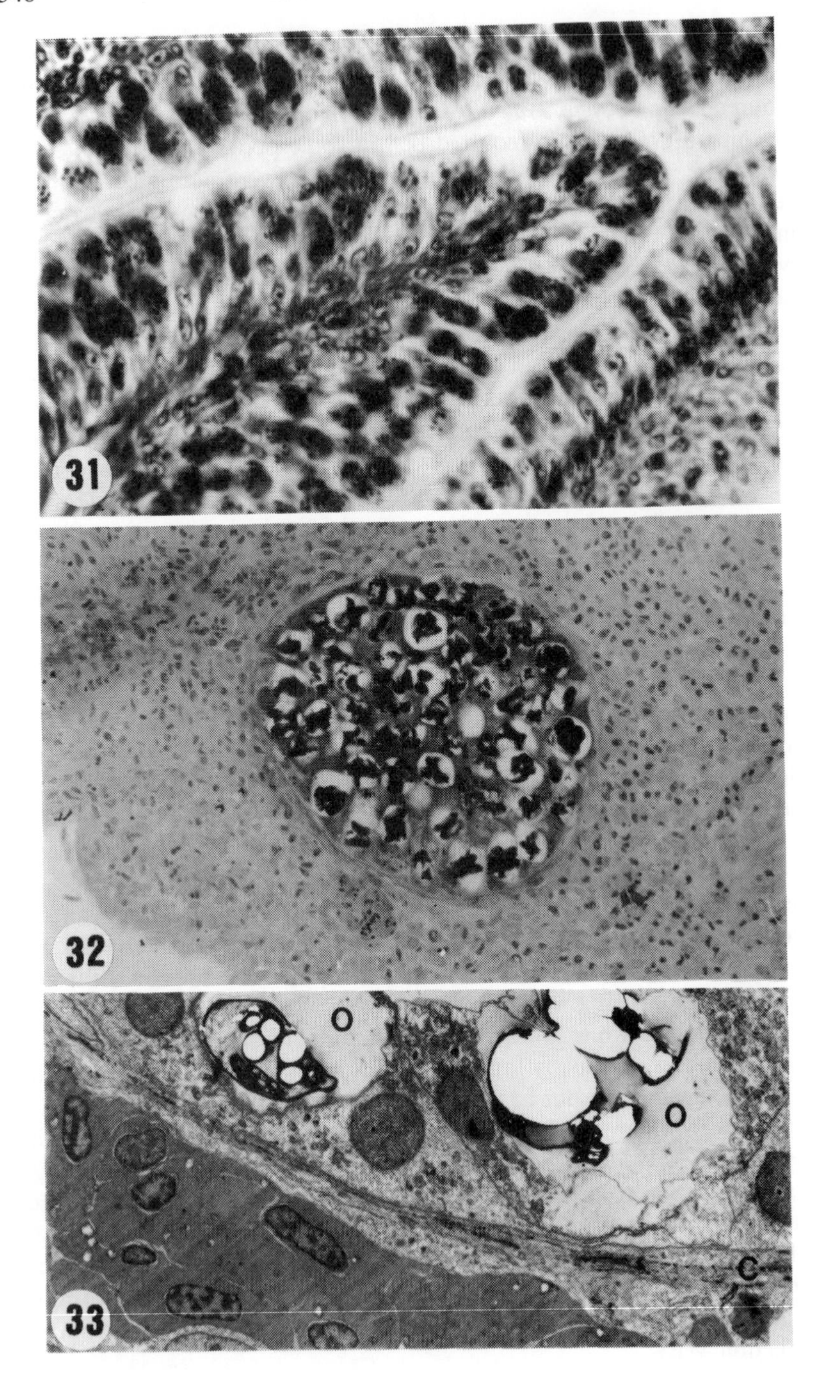
31
32
33
O
O
C

Pathological changes are also associated with *Epieimeria* species. *Epieimeria anguillae* from farmed eels, *Anguilla australis* and *Anguilla dieffenbachii*, is an example (Hine, 1975). In light infections, oocysts developed in the mucosal epithelium causing hypertrophy of adjacent cells and occasional ruture of the basement membrane. Oocysts in the submucosa caused vacuolation and destruction of the submucosal connective tissue. In heavy infections, oocysts compressed and destroyed columnar epithelial cells. Portions of mucosa became detached, exposing underlying connective tissue and, eventually, muscle layers. Extensive tissue destruction could result in emaciation and death of eels.

Paperna and Landsberg (1987) and Paperna (1990) observed a unique cell reaction in gut epithelial cells of juvenile *Oreochromis aureus* × *Oreochromis niloticus* hybrids infected with *Eimeria* (*s.l.*) *vanasi*. They found tubuli extending into the host cell cytoplasm from funnels located at the border of the parasitophorous vacuoles containing gamonts. The tubuli were fewer, vestigial or absent in cells with merogonic stages. Although these structures have not been described in other piscine coccidia, they are similar but not identical to the spine-like structures that protrude from the parasitophorous membrane of enterocytes of turkey and bobwhite quail infected with *Eimeria dispersa* (Long *et al.*, 1979; Millard and Lawn, 1982). The "spines" may consist of the skeletal filaments of the microvilli, assuming that the parasitophorous vacuole is formed from the invaginated cell membrane of the enterocyte (Millard and Lawn, 1982). It was also proposed that the function of the "spines" could be to anchor the parasitophorous vacuole membrane within the host cell (Long *et al.*, 1979).

2. *Extraintestinal stages*

Histopathological findings have also been reported for some coccidial infections in extraintestinal sites that include mainly liver, spleen, kidney, swim bladder and testes. In general, infection causes regressive changes in infected cells, their eventual destruction, and the development of an inflammatory reaction.

An inverse relationship was observed between intensity of infection of *Eimeria* species in the liver of blue whiting, *Micromesistius poutassou*, and

FIG. 31. *Goussia balatonica*. Light micrograph of section showing meronts infecting gut epithelium of white bream in a single nodule. Almost all epithelial cells are infected (× 300). (Kindly supplied by K. Molnár from Molnár, 1989, and reproduced by permission of the editor.) FIGS 32 and 33. *Goussia metchnikovi* in spleen of gudgeon. FIG. 32. Light micrograph of sporulated oocysts surrounded by a fibrotic capsule (× 240). FIG. 33. Transmission electron micrograph of capsule (c) and oocysts (o) (× 2300).

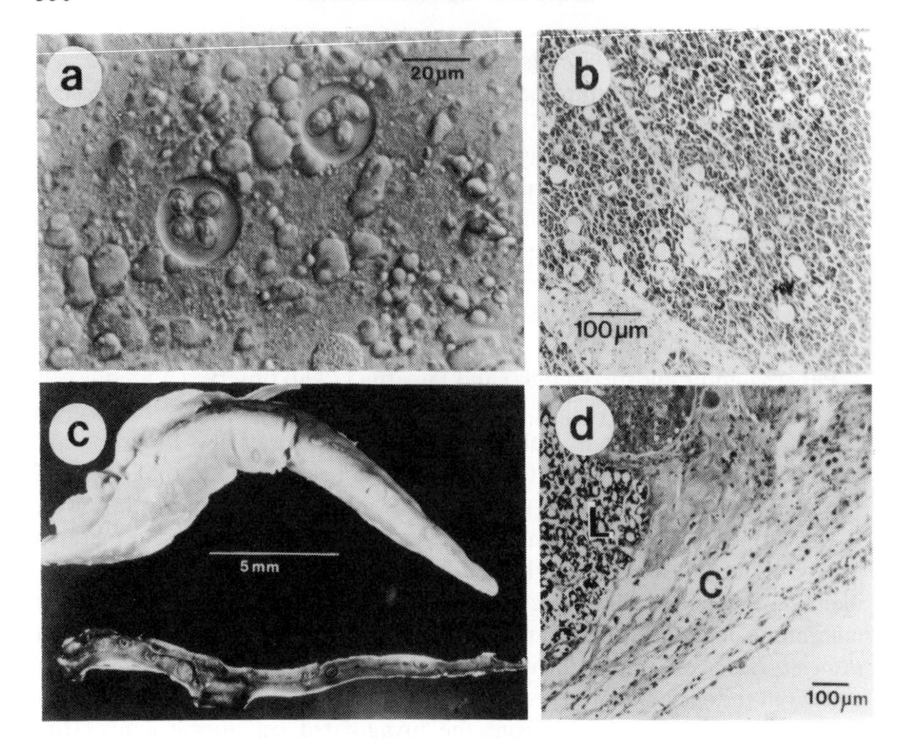

FIG. 34. Coccidial infection in the liver of blue whiting, *Micromesistius poutassou*. (a) *Eimeria* sp. oocysts in a fresh smear (photomicrograph, Nomarski optics). (b) Section of liver with medium intensity infection, showing oocysts scattered throughout the parenchyma and localized concentrations of oocysts in a matrix of collagen (photomicrograph, haematoxylin and eosin). (c) Liver with lesions (lower) compared with a liver from a fish of comparable length with a light *Eimeria* infection. (d) Section of the perimeter of a liver lesion; L, lesion tissue; C, fibrous capsule surrounding lesion (photomicrograph, haematoxylin and eosin). (Kindly supplied by K. MacKenzie from MacKenzie, 1981, and reproduced by permission of Blackwell Scientific Publications.)

host condition (MacKenzie, 1981). Infection was cumulative with increasing age of host. Coccidiosis resulted in extensive liver lesions of defined dark brown regions, and the author concluded that the *Eimeria* was the major factor responsible for poor condition in blue whiting (Fig. 34 a–d).

Infections of *Goussia clupearum* did not completely destroy the liver of herring, *Clupea harengus*, according to Morrison and Hawkins (1984), but these authors considered the damage caused and the replacement of tissue could stress the host. *Goussia thelohani* resembles *Goussia clupearum* and other coccidia in its site of development and pathological features (Blanc *et al.*, 1986; Daoudi *et al.*, 1988). Groups of oocysts found in livers of wrasse (*Symphodus tinca*) were surrounded by one or more fibrillar layers and necrotic amorphous material but gamonts were not encapsulated.

Oocysts are associated with melanomacrophage centres in the spleen of slender tuna (*Allothunnus fallai*), and in the liver of albacore (*Thunnus alalunga*). Jones (1990) examined infections of *Goussia auxidis* in these fish and in skipjack tuna (*Katsuwonus pelamis*), yellow tuna (*Thunnus albacares*) and *Scomber australasicus* from the western and central South Pacific Ocean. Host response to this coccidian appeared to be otherwise minimal, with slight proliferation of fibroblasts around the oocysts in some fish.

Heavy infections of *Calyptospora funduli* in killifishes caused hepatocyte degeneration and a gradual decrease or elimination of functional parenchyma (Solangi and Overstreet, 1980; Hawkins *et al.*, 1981). The host response involved infiltration of liver parenchyma by monocytes, which phagocytosed cellular debris, and infiltration of granular leucocytes into areas near developing macrogamonts.

Goussia degiustii infects the intestine, spleen, liver, and mesentery of common shiner and fathead minnow. In Ontario winter favours this development, and this may be linked to diminishing immune responses as temperature decreases (Paterson and Desser, 1982). Spleen seems to be the dominant site of infection with no apparent host reaction in wild populations of infected fish. Paterson and Desser (1982) reported that naturally infected fish caught at 10°C and then cooled to 4°C for 75 days had gamonts in the spleen. By contrast, warming fish to 24°C for 75 days elicited a host response producing encapsulated areas of spleen. In wild fish lymphocytes infiltrated around sporulated oocysts. Thin fibrous capsules developed around parasites and, within these, lymphocytes and cellular debris accumulated as oocysts were destroyed. Host reaction was pronounced at or near the surface of spleen, liver, or mesentery where clusters of oocysts were enveloped by capsules consisting of one to many layers. Additional fibrous layers enclosed smaller numbers of oocysts and this host reaction continued until the oocysts were obliterated. Similar oocyst nodules of *Goussia metchnikovi* were reported by Pellérdy and Molnár (1968), Dyková and Lom (1981) and Molnár (1981), from the spleen of gudgeon. Those seen in gudgeon in south-east England (Ball, 1983) are shown in Figs 32 and 33.

Coccidiosis of the kidneys also appears to be damaging. *Eimeria leucisci* infects the kidney of *Barbus barbus bocagei*. Parasitized epithelial cells of the renal tubuli were partially or completely destroyed, whereas unparasitized tubules were apparently normal (Alvarez-Pellitero and Gonzalez-Lanza, 1986). On some occasions fibrotic reactions were observed around infected tubules. In northern Greece, *Eimeria scardinii* in the kidney of roach, *Rutilus*

rutilus, showed a higher prevalence in June–August (22% approximately) than in September–November (1% approximately), with sporulated oocysts first seen in March and most abundant in summer (Athanassopoulou-Raptopoulou and Vlemmas, 1986). In heavy infections epithelial cells of tubules adjacent to oocysts were degenerate, glomeruli were shrunken and cells sloughed into the tubular lumen. Findings in infected rudd, *Scardinius erythrophthalmus*, were similar.

Odense and Logan (1976) carried out an extensive study on gadoid fish infected with *Goussia gadi* from Nova Scotia fishing banks. The average infection rate in the swim bladders of 2233 haddock, *Melanogrammus aeglefinus*, was 32%. No *Goussia gadi* infection was recorded in 92 cod (*Gadus morhua*), 56 cusk (*Brosme brosme*), 34 white hake (*Urophycis tenuis*), 34 silver hake (*Merluccius bilinearis*), or 33 pollock (*Pollachius virens*). The highest infection rates in the haddock were in late summer and autumn. There was no relationship between host sex and prevalence of infection, and the degree of infection had no apparent effect on the length/weight ratio of the fish. Although no tissue lesion could be attributed to *Goussia gadi*, the results indicated that the infection was fatal. The first sign of infection was a yellow waxy mass containing a few parasites and as the infection progressed the mass increased until the swim bladder was filled. This mass eventually occluded the swim bladder with, presumably, loss of buoyancy regulation capacity and impairment of sound production and resonation. Odense and Logan (1976) suggested that this would interfere with spawning, owing to difficulty with school formation and maintaining spawning position, and therefore ultimately affect survival.

Pinto (1957) and Pinto *et al.* (1961) examined the effects of *Eimeria sardinae* on the testes of the pilchard, *Sardina pilchardus*. In moderate infections, the host reaction was expressed by epithelial and connective cellular proliferation. Sperm production could proceed, although sex cells were parasitized, and numerous oocysts were found in the lumen of seminiferous canals. In heavy infections the testes showed softening and heterogeneous colour from marble-like to grey and rose. In heavy infections oocysts made up a large part of the gonad volume and there was little space left for sex cells. Pinto (1957) considered that this would adversely affect reproduction in the fish. Histological examination showed that the normal structure of testes was absent, sperm were lacking, and the testes were functionally sterile.

B. IMMUNITY

Little is known of the immune responses of fishes infected with coccidia, other than the cellular responses noted in the previous section. Non-specific responses are thought to occur in some fish, with humoral and cellular immune reactions taking place before clinical and anatomo-pathological symptoms of coccidial infection are manifest. For example, carp infected with *Goussia subepithelialis* show leucocytosis and eosinophilia, with increased phagocytic ability of neutrophils, increased myeloperoxidase activity, and serum lysozyme (Studnicka and Siwicki, 1990). Elevated natural antibody titre and high levels of coeruloplasmin probably result from cell and tissue injury, although when parasites reach the tissues, immunological activity apparently declines (Studnicka and Siwicki, 1990).

Premunition is thought to exist in infections with *Calyptospora funduli*, since fish with sporulated oocysts in the liver did not exhibit additional gamogony or sporogony when fed infective grass shrimp for as long as 30 days (Solangi and Overstreet, 1980). Cellular responses to *Calyptospora funduli* in the liver of killifish are marked, particularly in the early stages. At 18 d.a.i. the liver was infiltrated with many cells, including lymphocytes and macrophages, and eosinophils accumulated around gamonts, while collagen was laid down by fibroblasts (Solangi and Overstreet, 1980). By day 20 the inflammatory response had intensified, but was limited to foci near blood vessels and to deposition of yellow and brown pigment, which darkened with time, within fibrotic capsules. By day 30, when the oocyst walls were fully developed, inflammatory infiltrates had diminished or disappeared. Encapsulated material contained no iron but lipofuscins, argentaffin granules, polysaccharide and, later, melanins.

In contrast, Molnár (1981) suggested that several fish coccidia showed a high level of adaptation to their hosts, and as long as their intracellular development continued they were not immunogenic. This author believed that the defensive reaction began with the ageing of oocysts and with the disintegration of host cells. The host reaction, which was manifest as encapsulation or rejection of oocysts, depended on the host species and the location of the parasite.

VII. Conclusions

Dyková and Lom (1981) began the conclusion of their critical review of fish coccidia by asking a series of questions that they believed would be answered by experimental infections. We will conclude by examining (and, if possible, answering) Dyková and Lom's questions, and by posing a few of our own. In essence, we will attempt to establish areas where knowledge of fish coccidia has advanced significantly in the last 10–12 years, and to define areas where deficiency remains.

Some of Dyková and Lom's questions concerned paratenic hosts, "resting" oocysts, true intermediate hosts, and viability of sporozoites and oocysts. As a result of infection experiments, it is now known that some fish coccidia are transmitted directly, some require paratenic hosts for transmission, and at least one species is heteroxenous. In addition, some species may use both direct and indirect methods. Some oocysts require "overwintering" before they become infective, although there is little information on how long sporozoites and oocysts remain viable. As endogenous sporulation is so extensive among fish coccidia, are they also capable of direct transmission by autoinfection, as is *Cryptosporidium*? Fiebiger (1913) observed spontaneous opening of *Goussia gadi* sporocysts with sporozoite release *in situ*, and infection by *Goussia iroquoina* in the posterior portion of the intestine of common shiners in winter may result from autoinfection (Paterson and Desser, 1982).

Another question that Dyková and Lom posed concerned transmission of coccidia infecting extraintestinal organs. Knowledge of the mechanisms for transmission of coccidia infecting such sites is still fragmentary, and there is little understanding of how infections reach these sites. Are sporozoites transported there by leucocytes, as in some avian species of coccidia (Ball *et al.*, 1989)? Perhaps we should not be unduly influenced by work on homeotherms. As Desser (1981) noted, there is a tendency to consider fish coccidia as having life cycles very different from those of mammals and birds because species infecting the latter have been studied more extensively. Phylogenetically, fish coccidia are more primitive so that extraintestinal development may have been an early trend. Increased efficiency of the immune response in birds and mammals, compared to that of fish, may have acted against tissue stages, confining development mainly to the intestine of homeotherms. Would microscopical examination of gut contents and tissues of fossilized remains of ancient fishes and other animals be useful in determining phylogenetic relationships among coccidia?

Other questions that Dyková and Lom (1981) asked involved immunity to infection, host specificity and condition, and pathogenicity, especially of economically important species. Knowledge of immunity to fish coccidia has advanced very little, as is evident from the brevity of our section on this subject! Some fish coccidia are apparently host-specific, while others infect several different host species, but there is little to indicate why this is so. In contrast, the pathology of coccidiosis and its relationship to seasonality have received more attention, and experimental infections have helped much in this.

Some areas of particular confusion remain. This is true of the taxonomy of fish coccidia. *Crystallospora, Octosporella, Cryptosporidium*, and some recent additions to the genus *Calyptospora*, all require further study for reasons noted earlier. The separation of *Eimeria* from *Goussia* (see Dyková and Lom, 1981) seems entirely appropriate and most fish coccidia belong to one or other of these genera (Tables 1, 2 and 3). Dyková and Lom (1981) and Upton *et al.* (1984) thought that all species of fish coccidia lacking a Stieda body might eventually be shown to be species of *Goussia*. There are many fish coccidia currently classified as *Eimeria* but for which no Stieda body has been decribed (see Dyková and Lom, 1983), and at least one (*Eimeria* (*s.l.*) *vanasi* Landsberg and Paperna, 1987) that does not fit neatly into either *Eimeria* or *Goussia*. There may also be several instances of synonymity among *Eimeria* and *Goussia* species. Some species have been shown to be identical (e.g. *Goussia degiustii* and *Eimeria spleni*; *Goussia legeri* and *Eimeria stankovitchi*), but *Goussia carpelli* and *Eimeria cyprini* are listed both as synonymous (Dyková and Lom, 1983; Levine, 1988) and as separate species (Levine, 1988). Dyková and Lom (1981) provided a convincing argument for the establishment of the new genus *Epieimeria*, and this is widely accepted. If *Eimeria* that develop intracellularly but epiplasmally or extracytoplasmically are significantly different from those that develop intracytoplasmically, are *Eimeria* that develop within the nucleus also different? Should the genus *Nucleoeimeria* therefore gain wide acceptance? A similar argument applies to *Goussia*. Should the genera *Epigoussia* and *Nucleogoussia* be firmly established? Clearly the taxonomy of fish coccidia requires further refinement.

The development of the oocyst and sporocyst walls of fish coccidia also demands further study. Some oocyst walls are apparently thin enough to have been formed from macrogamete surface membranes, perhaps with endoplasmic reticulum additions. In other instances there is clear evidence that structures resembling wall-forming bodies are present and the involvement of some in oocyst wall formation seems probable. Sporocyst walls are usually substantial structures in fish coccidia. The origin of these is particularly intriguing and so far there is no entirely satisfactory explanation for their development.

As Dyková and Lom (1981) noted, infection experiments are probably the ideal way to solve many problems concerning fish coccidia. However, obtaining coccidia-free host fish is not always easy. This is particularly true of marine species that may be difficult to breed in captivity. In addition, coccidia may not lend themselves easily to transmission, and separating mixed infections is sometimes difficult. One useful development would be the establishment of pure lines of parasites from single oocysts, as has been done with avian coccidia. In addition, further knowledge might be gained by the use of cell culture, especially using coccidia from sterile sources such as internal organs.

The word "bewildering" was not used lightly at the beginning of this review. Fish coccidia are so diverse that they defy most attempts to bring them to order. Clearly, knowledge of the group has advanced significantly in the last few years and available information should provide a good basis for controlled experimentation to settle some outstanding questions.

References

Achmerov, A. C. (1959). Cited by Dogiel, V. A. and Achmerov, A. C. (1959).

Allamaratov, B. and Iskov, M. P. (1970). Some new species of parasitic protozoa of fishes in Surkhandarya. *Uzbekskii Biologicheskii Zhurnal* **4**, 43–46.

Alvarez-Pellitero, M. P. and Gonzalez-Lanza, M. C. (1985). *Goussia polylepidis* n. sp. (Apicomplexa: Sporozoea: Coccidia) from *Chondrostoma polylepis* (Pisces; Cyprinidae) of the Duero Basin (NW Spain). *Journal of Protozoology* **32**, 570–571.

Alvarez-Pellitero, M. P. and Gonzalez-Lanza, M. C. (1986). *Eimeria* spp. from cyprinid fish of the Duero Basin (north-west Spain). *Journal of Fish Diseases* **9**, 325–336.

Athanassopoulou-Raptopoulou, F. and Vlemmas, J. (1986). *Eimeria scardinii* Pellérdy and Molnár, 1968 in the kidneys of *Rutilus rutilus* (L.) and *Scardinius erythrophthalmus* (L.) from northern Greece. *Journal of Fish Diseases* **9**, 411–416.

Ball, S.J. (1983). *Eimeria metchnikovi* (Laveran, 1897) Reichenow, 1921 from the gudgeon, *Gobio gobio* L., in south-east England. *Journal of Fish Diseases* **6**, 201–203.

Ball, S. J. and Pittilo, R. M. (1990). Structure and ultrastructure. *In* "Coccidiosis of Man and Domestic Animals" (P. L. Long, ed.), pp. 17–41. CRC Press, Boca Raton, Florida.

Ball, S. J., Pittilo, R. M. and Long, P. L. (1989). Intestinal and extraintestinal life cycles of eimeriid coccidia. *Advances in Parasitology* **28**, 1–54.

Baska, F. and Molnár, K. (1989). Ultrastructural observations on different developmental stages of *Goussia sinensis* (Chen, 1955), a parasite of the silver carp (*Hypophthalmichthys molitrix* Valenciennes, 1844). *Acta Veterinaria Hungarica* **37**, 81–87.

Békési, L. and Molnár, K. (1991). *Calyptospora tucunarensis* n. sp. (Apicomplexa: Sporozoea) from the liver of tucunare *Cichla ocellaris* in Brazil. *Systematic Parasitology* **18**, 127–132.

Blanc, E., Daoudi, F., Marques, A. and Bouix, G. (1986). Evaluation statistique de l'action de coccidioses des poissons marins: étude realisée chez le maquereau. *European Mariculture Society Special Publication* **9**, 105–109.

Boulard, Y. (1977). Description d'*Eimeria quentini* n.sp., parasite intranucléaire du péritoine de la raie: *Aetobatis narinari* (Chondrichthyens, Myliobatidae) en Malaisie. *Protistologica* **13**, 529–533.

Chakravarty, M. and Kar, A. B. (1944). A new coccidian from the intestine of the fish *Notopterus notopterus* (Pallas). *Current Science* **13**, 51.

Chen, C.-L. (1956a). The protozoan parasites from four species of Chinese pond fishes. II. The protozoan parasites of *Mylopharyngodon piceus*. *Acta Hydrobiologica Sinica* **2**, 19–42. (Cited by Pellérdy, 1974.)

Chen, C.-L. (1956b). The protozoan parasites of four species of Chinese pond fishes: *Ctenopharyngodon idelius, Mylopharyngodon piceus, Aristichthys nobilis* and *Hypophthalmichthys molothrix*. III. The protozoan parasites of *Aristichthys nobilis* and *Hypophthalmichthys molothrix*. *Acta Hydrobiologica Sinica* **2**, 279–298. (Cited by Dyková and Lom, 1981.)

Chen, C.-L. (1960). See Chen, C.-L. (1973).

Chen, C.-L. (1962). See Chen, C.-L. (1973).

Chen, C.-L. (1964). Cited by Dyková and Lom (1983).

Chen, C.-L. (1973). "An Illustrated Guide to the Fish Diseases and Causative Pathogenic Fauna and Flora in the Hubei Province". Publishing House Science, China. (Cited by Dyková and Lom, 1983.)

Chen, C.-L. (1984). See Su, X.-Q. and Chen, C.-L. (1987).

Chen, C.-L. and Hsieh, S.-R. (1960). *Acta Hydrobiologica Sinica* **2**, 171–196. (Cited by Dyková, I. and Lom, J. 1983.)

Chen, C.-L. and Hsieh, S.-R. (1964). (Cited by Dyková, I. and Lom, J.,) (1983).

Cheung, P. J., Nigrelli, R. F. and Ruggieri, G. D. (1986). *Calyptospora serrasalmi* sp. nov. (Coccidia: Calyptosporidae) from liver of the black piranha, *Serrasalmus niger* Schomburgk. *Journal of Aquariculture and Aquatic Sciences* **4**, 54–57.

Chobotar, B. and Scholtyseck, E. (1982). Ultrastructure. *In* "The Biology of the Coccidia" (P. L. Long, ed.), pp. 101–165. University Park Press, Baltimore.

Conder, G. A., Oberndorfer, R. Y. and Heckmann, R. A. (1980). *Eimeria duszynskii* sp. n. (Protozoa: Eimeriidae), a parasite of the mottled sculpin, *Cottus bairdi* Girard. *Journal of Parasitology* **66**, 828–829.

Current, W. L. and Blagburn, B. L. (1990). *Cryptosporidium:* infections in man and domesticated animals. *In* "Coccidiosis of Man and Domestic Animals" (P. L. Long, ed.), pp. 155–185. CRC Press, Boca Raton, Florida.

Daoudi, F. (1987). "Coccidies et coccidioses de poissons méditerranéens: systématique, ultrastructure et biologie." Doctoral thesis, Montpellier University. (Cited by Molnár, 1989.)

Daoudi, F. and Marquès, A. (1986–1987). *Eimeria bouixi* n.sp. and *Eimeria dicentrarchi* n.sp. (Sporozoa—Apicomplexa), parasites from the fish *Dicentrarchus labrax* (Linné, 1758) in Languedoc. *Annales des Sciences Naturelles, Zoologie* (13e série) **8**, 237–242.

Daoudi, F., Radujković, B., Marquès, A. and Bouix, G. (1987). Nouvelles espèces de coccidies (Apicomplexa, Eimeriidae) des genres *Eimeria* Schneider, 1875, et *Epieimeria* Dyková et Lom, 1981, parasites de poissons marins de la baie de Kotor (Yougoslavie). *Bulletin du Muséum Nationale d'Histoire Naturelle, Paris* (4e série, section A) **9**, 321–332.

Daoudi, F., Radujković, B., Marquès, A. and Bouix, G. (1988). Pathogenicity of the coccidian *Goussia thelohani* (Labbé, 1896), in the liver and pancreatic tisues of *Symphodus tinca* (Linné, 1758). *Bulletin of the European Association of Fish Pathologists* **8**, 55–57.

Daoudi, F., Radujković, B., Marquès, A. and Bouix, G. (1989). Parasites des poissons marins du Monténégro: Coccidies. *Acta Adriatica* **30**, 13–30.

Davies, A. J. (1978). Coccidian parasites of intertidal fishes from Wales: systematics, development and cytochemistry. *Journal of Protozoology* **25**, 15–21.

Davies, A. J. (1990). Ultrastructural studies on the endogenous stages of *Eimeria variabilis* (Thélohan, 1893) Reichenow 1921, from *Cottus (Taurulus) bubalis* Euphrasen (Teleostei: Cottidae). *Journal of Fish Diseases* **13**, 447–461.

Davronov, O. (1987). [Coccidia from fishes of Uzbekistan.] *Parazitologiya* **21**, 115–120. (In Russian, English summary.)

Desser, S. S. (1981). The challenge of fish coccidia. *Journal of Protozoology* **28**, 260–261.

Desser, S. S. and Lester, R. (1975). An ultrastructural study of the enigmatic "rodlet cells" in the white sucker *Catostomus commersoni* (Lacépède) (Pisces: Catostomidae). *Canadian Journal of Zoology* **53**, 1483–1494.

Desser, S. S. and Li, L. (1984). Ultrastructural observations on the sexual stages and oocyst formation in *Eimeria laureleus* (Protozoa, Coccidia) of perch, *Perca flavescens*, from Lake Sasajewun, Ontario. *Zeitschrift für Parasitenkunde* **70**, 153–164.

Doflein, F. (1909). "Lehrbuch der Protozenkunde", 2nd edn. G. Fischer, Jena.

Dogiel, V. A. (1948). Parasitic protozoa of fishes from Peter the Great Bay. *Transactions of All-Union Science Research Institute of Lake and River Fisheries*, Leningrad **27**, 17–66. (Cited by Pellérdy, 1974.)

Dogiel, V. A. and Achmerov, A. C. (1959). [New species of parasitic protozoa in fishes from River Amur]. *Československá Parasitologie* **6**, 15–25. (In Russian.)

Dogiel, V. A. and Bykhovskii, B. E. (1939). "Parazity ryb Kaspiiskogo morya" ["The parasites in the Caspian Sea"], pp.1–149. Izdatel'stvo ANSSSR, Moskva & Leningrad. (Cited by Shul'man, S. S. and Shtein, G. A., 1962.)

Dujarric de la Rivière, R. (1914). Sur une coccidie de l'estomac de la perche (*Coccidium percae* nova species). *Comptes Rendus de la Société de Biologie* **76**, 493–494.

Duszynski, D. W., Solangi, D. W. and Overstreet, R. M. (1979). A new and unusual eimerian (Protozoa: Eimeriidae) from the liver of the gulf killifish *Fundulus grandis*. *Journal of Wildlife Diseases* **15**, 543–552.

Dyková, I. and Lom, J. (1981). Fish coccidia: critical notes on life cycles, classification and pathogenicity. *Journal of Fish Diseases* **4**, 487–505.

Dyková, I. and Lom, J. (1983). Fish coccidia: an annotated list of described species. *Folia Parasitologica* **30**, 193–208.

Dyková, I., Lom, J. and Grupcheva, G. (1983). *Eimeria branchiphilia* sp. nov. sporulating in the gill filaments of roach, *Rutilus rutilus* L. *Journal of Fish Diseases* **6**, 13–18.

Elmassian, M. (1909). Une nouvelle coccidie et un nouveau parasite de la Tanche, *Coccidium rouxi* nov. spec., *Zoomyxa legeri* nov. gen. nov. spec. *Archives de Zoologie Expérimentale et Générale*, 5e Série, **2**, 229–270.

Fantham, H. B. (1932). Some parasitic protozoa found in South Africa—XV. *South African Journal of Science* **29**, 627–640.

Fantham, H. B. and Porter, A. (1943). *Plasmodium struthionis*, sp.n., from Sudanese ostriches and *Sarcocystis salvelini*, sp.n., from Canadian speckled trout (*Salvelinus fontinalis*), together with a record of a *Sarcocystis* in the eel pout (*Zoarces angularis*). *Proceedings of the Zoological Society of London*, Series B, **113**, 25–30.

Fiebiger, J. (1913). Studien über die Schwimmblasencoccidien der Gadusarten (*Eimeria gadi* n.sp.) *Archiv für Protistenkunde* **31**, 95–137.

Fitzgerald, P. R. (1975). New coccidia from the spiny dogfish shark (*Squalus acanthias*) and great sculpin (*Myoxocephalus polyacanthocephalus*). *Journal of the Fish Research Board of Canada* **32**, 649–651.

Fournie, J. W. and Overstreet, R. M. (1983). True intermediate hosts for *Eimeria funduli* (Apicomplexa) from estuarine fishes. *Journal of Protozoology* **30**, 672–675.

Fournie, J. W., Hawkins, W. E. and Overstreet, R. M. (1985). *Calyptospora empristica* n.sp. (Eimeriorina: Calyptosporidae) from the liver of the starhead topminnow, *Fundulus notti*. *Journal of Protozoology* **32**, 542–547.

Fujita, T. (1934). Note on *Eimeria* of herring. *Proceedings of 5th Pacific Science Congress* **5**, 4135–4139.

Gadjadhar, A. A., Wobeser, G. and Stockdale, P. H. G. (1983a). Coccidia of domestic and wild waterfowl (Anseriformes). *Canadian Journal of Zoology* **61**, 1–24.
Gajadhar, A. A., Cawthorn, R. J., Wobeser, G. A. and Stockdale, P. H. G. (1983b). Prevalence of renal coccidia in wild waterfowl in Saskatchewan. *Canadian Journal of Zoology* **61**, 2631–2633.
Gauthier, M. (1921). Coccidies du chabot de rivière (*Cottus gobio* L.) *Compte Rendu Hebdomadaire des Séances de l'Académie des Sciences* **173**, 671–674.
Goodwin, M. A. (1989). Cryptosporidiosis in birds—a review. *Avian Pathology* **18**, 365–384.
Grassé, P. P. (1953). "Traité de Zoologie". Tome I: "Protozoaires: Rhizopodes, Actinopodes, Sporozoaires, Cnidosporidies". Fascicule 2. Masson, Paris.
Gregory, M. W. (1990). Pathology of coccidial infections. *In* "Coccidiosis of Man and Domestic Animals" (P. L. Long, ed.), pp. 235–261. CRC Press, Boca Raton, Florida.
Halawani, A. (1930). On a new species of *Eimeria* (*E. southwelli*) from *Aëtobatis narinari*. *Annals of Tropical Medicine and Parasitology* **24**, 1–3.
Hardcastle, A. B. (1943). A check list and host-index of the species of the protozoan genus *Eimeria*. *Proceedings of the Helminthological Society of Washington* **10**, 35–69.
Hardcastle, A. B. (1944). *Eimeria brevoortiana*, a new sporozoan parasite from the menhaden (*Brevoortia tyrannus*), with observations on its life history. *Journal of Parasitology* **30**, 60–68.
Hawkins, W. E., Solangi, M. A. and Overstreet, R. M. (1981). Ultrastructural effects of the coccidium, *Eimeria funduli* Duszynski, Solangi and Overstreet, 1979 on the liver of killifishes. *Journal of Fish Diseases* **4**, 281–295.
Hawkins, W. E., Solangi, M. A. and Overstreet, R. M. (1983a). Ultrastructure of the macrogamont of *Eimeria funduli*, a coccidium parasitizing killifishes. *Journal of Fish Diseases* **6**, 33–43.
Hawkins, W. E., Solangi, M. A. and Overstreet, R. M. (1983b). Ultrastructure of the microgamont and microgamete of *Eimeria funduli*, a coccidium parasitizing killifishes. *Journal of Fish Diseases* **6**, 45–57.
Hawkins, W. E., Fournie, J. W. and Overstreet, R. M. (1983c). Organization of sporulated oocysts of *Eimeria funduli* in the gulf killifish *Fundulus grandis*. *Journal of Parasitology* **69**, 496–503.
Hawkins, W. E., Fournie, J. W. and Overstreet, R. M. (1984). Ultrastructure of the interface between stages of *Eimeria funduli* (Apicomplexa) and hepatocytes of the longnose killifish, *Fundulus similis*. *Journal of Parasitology* **70**, 232–238.
Hine, P. M. (1975). *Eimeria anguillae* Léger and Hollande, 1922, parasitic in New Zealand eels. *New Zealand Journal of Marine and Freshwater Research* **9**, 239–243.
Hoffman, G. L. (1965). *Eimeria aurati* n.sp. (Protozoa: Eimeriidae) from goldfish (*Carassius auratus*) in North America. *Journal of Protozoology* **12**, 273–275.
Hoover, D. M., Hoerr, F. J., Carlton, W. W., Hinsman, E. J. and Ferguson, H. W. (1981). Enteric cryptosporidiosis in the naso tang, *Naso lituratus* Bloch and Schneider. *Journal of Fish Diseases* **4**, 425–428.
Jastrzebski, M. (1982). New species of intestinal coccidia of freshwater fish. *Bulletin de l'Académie Polonaise des Sciences* **30**, 1–12.
Jastrzebski, M. (1984). Coccidiofauna of cultured and feral fishes in fish farms. *Wiadomosci Parazytologiczne* **30**, 141–163.
Jastrzebski, M. (1989). Ultrastructural study on the development of *Goussia aculeati*,

a coccidium parasitizing the three-spined stickleback *Gasterosteus aculeatus*. *Diseases of Aquatic Organisms* **6**, 45–53.

Jastrzebski, M. and Komorowski, Z. (1990). Light and electron microscopic studies on *Goussia zarnowskii* (Jastrzebski, 1982): an intestinal coccidium parasitizing the three-spined stickleback, *Gasterosteus aculeatus* (L.). *Journal of Fish Diseases* **13**, 1–24.

Jastrzebski, M., Pastuszko, J., Kurska, E. and Badowska, M. (1988). Coccidia of the stickleback *Gasterosteus aculeatus* L. *Wiadomosci Parazytologiczne* **34**, 55–63.

Jones, J. B. (1990). *Goussia auxidis* (Dogiel, 1948) (Apicomplexa: Calyptosporidae) from tuna (Pisces: Scombridae) in the South Pacific. *Journal of Fish Diseases* **13**, 215–223.

Kent, M. L. and Hedrick, R. P. (1985). The biology and associated pathology of *Goussia carpelli* (Léger and Stankovitch) in goldfish *Carassius auratus* (Linnaeus). *Fish Pathology* **20**, 485–494.

Kent, M. L., Fournie, J. W., Snodgrass, R. E. and Elston, R. A. (1988). *Goussia girellae* n.sp. (Apicomplexa: Eimeriorina) in the opaleye, *Girella nigricans*. *Journal of Protozoology* **35**, 287–290.

Kent, M. L., Moser, M. and Fournie J. W. (1989). Coccidian parasites (Apicomplexa: Eucoccidorida) in hardy head fish, *Atherinomorus capricornensis* (Woodland). *Journal of Fish Diseases* **12**, 179–183.

Kulemina, I. V. (1969). New species of endoparasitic protozoa from the fry of the Lake Seliger. *Zoologicheskii Zhurnal* **48**, 1295–1298.

Labbé, A. (1893). Sur deux coccidies nouvelles, parasites des poissons. *Bulletin de la Société Zoologique de France* **18**, 202–204.

Labbé, A. (1894). Recherches zoologique et biologique sur les parasites endoglobulaires du sang des vertebrés. *Archives de Zoologie Expérimentale et Générale* (3e série) **2**, 55–258. (Cited by Hardcastle, 1943.)

Labbé, A. (1986). Recherches zoologiques, cytologiques et biologiques sur les coccidies. *Archives de Zoologie Expérimentale et Générale* (3e série) **4**, 517–654.

Lacey, S. M. and Williams, I. C. (1983). *Epieimeria anguillae* (Léger and Hollande, 1922) Dyková and Lom, 1981 (Apicomplexa: Eucoccidia) in the European eel, *Anguilla anguilla* (L.). *Journal of Fish Biology* **23**, 605–609.

Laguesse, E. (1906). Les "Stabchendrusenzellen" (M. Plehn) sont des sporozoaires parasites. *Anatomischer Anzeiger* **28**, 414–416.

Landau, I., Marteau, M., Golvan, Y., Chabaud, A. G. and Boulard, Y. (1975). Hétéroxénie chez les coccidies intestinales de poissons. *Compte Rendu Hebdomadaire des Séances de l'Académie des Sciences*, Série D **281**, 1721–1723.

Landsberg, J. H. and Paperna, I. (1985). *Goussia cichlidarum* n.sp. (Barrouxiidae, Apicomplexa), a coccidian parasite in the swimbladder of cichlid fish. *Zeitschrift für Parasitenkunde* **71**, 199–212.

Landsberg, J. H. and Paperna, I. (1986). Ultrastructural study of the coccidian *Cryptosporidium* sp. from stomachs of juvenile cichlid fish. *Diseases of Aquatic Organisms* **2**, 13–20.

Landsberg, J. H. and Paperna, I. (1987). Intestinal infections by *Eimeria* (*s.l.*) *vanasi* n.sp. (Eimeriidae, Apicomplexa, Protozoa) in cichlid fish. *Annales de Parasitologie Humaine et Comparée* **62**, 283–293.

Laveran, L. A. (1897). Sur une coccidie du goujon. *Comptes Rendus de la Société de Biologie* **49**, 925–927.

Lee and Chen, C.-L. (1964). See Chen (1973).

Léger, L. and Bory, T. (1932). *Eimeria pigra* n.sp. nouvelle coccidie juxtaépitheliale, parasite du gardon rouge. *Compte Rendu Hebdomadaire des Séances de l'Académie des Sciences* **194**, 1710–1712.

Léger, L. and Hesse, E. (1919). Sur une nouvelle coccidie parasite de la truite indigène. *Compte Rendu Hebdomadaire des Séances de l'Académie des Sciences* **168**, 904–906.

Léger, L. and Hollande, G. C. (1922). Coccidie de l'intestin de l'anguille. *Compte Rendu Hebdomadaire des Séances de l'Académie des Sciences* **175**, 999–1002.

Léger, L. and Stankovitch, S. (1921). L'enterite coccidienne des alevins de carpe. *Travaux du Laboratoire de Pisciculture de l'Université de Grenoble* **13**, 191–194.

Levine, N. D. (1973). Introduction, history and taxonomy. *In* "The Coccidia: *Eimeria, Isospora, Toxoplasma*, and related genera" (D. M. Hammond and P. L. Long, eds), pp. 1–22. University Park Press, Baltimore and Butterworth, London.

Levine, N. D. (1982). Taxonomy and life cycles of coccidia. *In* "The Biology of the Coccidia" (P. L. Long, ed.), pp.1–33. Edward Arnold, London.

Levine, N. D. (1983). The genera *Barrouxia, Defretinella* and *Goussia* in the coccidian family Barrouxiidae (Protozoa, Apicomplexa). *Journal of Protozoology* **30**, 542–547.

Levine, N. D. (1984a). The genera *Cryptosporidium* and *Epieimeria* in the coccidian family Cryptosporidiidae (Protozoa: Apicomplexa). *Transactions of the American Microscopical Society* **103**, 205–206.

Levine, N. D. (1984b). Taxonomy and review of the coccidian genus *Cryptosporidium* (Protozoa, Apicomplexa). *Journal of Protozoology* **31**, 94–98.

Levine, N. D. (1984c). The taxonomic position of the coccidian genus *Crystallospora* (Protozoa, Apicomplexa). *Journal of Fish Diseases* **7**, 317–318.

Levine, N. D. (1988). "The Protozoan Phylum Apicomplexa", Vol. 1. CRC Press, Boca Raton, Florida.

Levine, N. D. and Tadros, W. (1980). Named species and hosts of *Sarcocystis* (Protozoa: Apicomplexa: Sarcocystidae). *Systematic Parasitology* **2**, 41–59.

Li, L. and Desser, S. S. (1985a). The protozoan parasites of fish from two lakes in Algonquin Park, Ontario. *Canadian Journal of Zoology* **63**, 1846–1858.

Li, L. and Desser, S. S. (1985b). Three new species of *Octosporella* (Protozoa: Coccidia) from cyprinid fish in Algonquin Park, Ontario. *Canadian Journal of Zoology* **63**, 1859–1862.

Lom, J. (1971). Remarks on the spore envelopes in fish coccidia. *Folia Parasitologica* **18**, 289–293.

Lom, J. and Dyková, I. (1981). New species of the genus *Eimeria* (Apicomplexa: Coccidia) from marine fish. *Zeitschrift für Parasitenkunde* **66**, 207–220.

Lom, J. and Dyková, I. (1982). Some marine fish coccidia of the genera *Eimeria* Schneider, *Epieimeria* Dyková and Lom and *Goussia* Labbé. *Journal of Fish Diseases* **5**, 309–321.

Lom, J., Desser, S. S. and Dyková, I. (1989). Some little-known and new protozoan parasites of fish from Lake Sasajewun, Algonquin Park, Ontario. *Canadian Journal of Zoology* **67**, 1372–1379.

Long, P. L. and Joyner, L. P. (1984). Problems in the identification of species of *Eimeria*. *Journal of Protozoology* **31**, 535–541.

Long, P. L., Millard, B. J. and Lawn, A. M. (1979). An unusual local reaction to an intracellular protozoan parasite *Eimeria dispersa*. *Zietschrift für Parasitenkunde* **60**, 193–195.

Lukeš, J. and Dyková, I. (1990). *Goussia janae* n.sp. (Apicomplexa, Eimeriorina) in dace *Leuciscus leuciscus* and chub *L. cephalus*. *Diseases of Aquatic Organisms* **8**, 85–90.

Lukeš, J. Steinhagen, D. and Korting, W. (1991). *Goussia carpelli* (Apicomplexa, Eimeriorina) from cyprinid fish: field observations and infection experiments. *Angewandte Parasitologie* **32**, 149–153.

MacKenzie, K. (1981). The effect of *Eimeria* sp. infection on the condition of blue whiting, *Micromesistius poutassou* (Risso). *Journal of Fish Diseases* **4**, 473–486.

MacLean, S. A. and Davies, A. J. (1990). Prevalence and development of intraleucocytic haemogregarines from northwest and northeast Atlantic mackerel, *Scomber scombrus* L. *Journal of Fish Diseases* **13**, 59–68.

Mandal, A. K. and Chakravarty, M. M. (1965). A new coccidium, *Eimeria zygaenae* from hammer-headed shark, *Zygaena blockii*. *Science and Culture* **31**, 381–382.

Marinček, M. (1973a). Développement d'*Eimeria subepithelialis* (Sporozoa, Coccidia)—parasite de la carpe. *Acta Protozoologica* **12**, 195–216.

Marinček, M. (1973b). Les changements dans le tube digestif chez *Cyprinus carpio* à la suite de l'infection par *Eimeria subepithelialis* (Sporozoa, Coccidia). *Acta Protozoologica* **12**, 217–224.

Mayberry, L. F., Marchiondo, A. A., Ubelaker, J. E. and Kažić, D. (1979). *Rhabdospora thelohani* Laguessé, 1895 (Apicomplexa): new host and geographic records with taxonomic considerations. *Journal of Protozoology* **26**, 168–178.

Mayberry, L. F., Bristol, J. R., Sulimanović, D., Fijan, N. and Petrinec, Z. (1986). *Rhabdospora thelohani*: epidemiology of and migration into *Cyprinus carpio* bulbus arteriosus. *Fish Pathology* **21**, 145–150.

McConnell, E. E., Vos, A. J. de, Basson, P. A. and Vos, V. de (1971). *Isospora papionis* n.sp. (Eimeriidae) of the chacma baboon *Papio ursinus* (Ken 1792). *Journal of Protozoology* **18**, 28–32.

McGladdery, S. E. (1987). Potential of *Eimeria sardinae* Apicomplexa Eimeridae oocysts for distinguishing between spawning groups and between first and repeat-spawning Atlantic herring *Clupea harengus harengus*. *Canadian Journal of Fish and Aquatic Science* **44**, 1379–1385.

Mikhailov, T. K. (1975). [Parasites of fish of the Azerbaidzhan water bodies.] *Izdat. ELM Baku*, USSR (In Russian; cited by Dyková and Lom, 1983 and Levine, 1988.)

Millard, B. J. and Lawn, A. M. (1982). Parasite–host relationship during the development of *Eimeria dispersa* Tyzzer 1929, in the turkey (*Meleagris gallopavo gallopavo*) with a description of intestinal intra-epithelial leucocytes. *Parasitology* **84**, 13–20.

Molnár, K. (1976). Histological study of coccidiosis in silver carp and the bighead by *Eimeria sinensis* Chen, 1956. *Acta Veterinaria Academiae Scientiarum Hungaricae* **26**, 303–312.

Molnár, K. (1977). Comments on the nature and methods of collection of fish coccidia. *Parasitologia Hungarica* **10**, 41–45.

Molnár, K. (1978). Five new *Eimeria* species (Protozoa: Coccidia) from freshwater fishes indigenous in Hungary. *Parasitologia Hungarica* **11**, 5–11.

Molnár, K. (1979). Studies on coccidia of Hungarian pond fishes. *In* "Coccidia and Further Prospects of their Control", pp. 179–183. International Symposium, Prague.

Molnár, K. (1981). Some peculiarities of oocyst rejection of fish coccidia. Fish, pathogens and environment in European polyculture. In "Proceedings of an International Seminar, Szarvas, Hungary", pp. 170–183.

Molnár, K. (1982). Nodular coccidiosis in the gut of tench, *Tinca tinca* L. *Journal of Fish Diseases* **5**, 461–470.
Molnár, K. (1986). Occurrence of two new *Goussia* species in the intestine of the sterlet (*Acipenser ruthenus*). *Acta Veterinaria Hungarica* **34**, 169–174.
Molnár, K. (1989). Nodular and epicellular coccidiosis in the intestine of cyprinid fishes. *Diseases of Aquatic Organisms* **7**, 1–12.
Molnár, K. and Baska, F. (1986). Light and electron microscopic studies on *Epieimeria anguillae* (Léger and Hollande, 1922), a coccidium parasitizing the European eel, *Anguilla anguilla* L. *Journal of Fish Diseases* **9**, 99–110.
Molnár, K. and Fernando, C. H. (1974). Some new *Eimeria* (Protozoa, Coccidia) from freshwater fishes in Ontario, Canada. *Canadian Journal of Zoology* **52**, 413–419.
Molnár, K. and Hanek, G. (1974). Seven new *Eimeria* spp. (Protozoa: Coccidia) from freshwater fishes of Canada. *Journal of Protozoology* **21**, 489–493.
Molnár, K. and Rohde, K. (1988a). Seven new coccidian species from marine fishes in Australia. *Systematic Parasitology* **11**, 19–29.
Molnár, K. and Rohde, K. (1988b). New coccidians from freshwater fishes of Australia. *Journal of Fish Diseases* **11**, 161–169.
Moroff, T. and Fiebiger, J. (1905). Über *Eimeria subepithelialis* n.sp. *Archiv für Protistenkunde* **6**, 166–174.
Morrison, C. M. and Hawkins, W. E. (1984). Coccidians in the liver and testis of the herring *Clupea harengus* L. *Canadian Journal of Zoology* **62**, 480–493.
Morrison, C. M. and Odense, P. H. (1978). Distribution and morphology of rodlet cells in fish. *Journal of the Fisheries Research Board of Canada* **35**, 101–116.
Morrison, C. M. and Poynton, S. L. (1989a). A new species of *Goussia* (Apicomplexa, Coccidia) in the kidney tubules of the cod, *Gadus morhua* L. *Journal of Fish Diseases* **12**, 533–560.
Morrison, C. M. and Poynton, S. L. (1989b). A coccidian in the kidney of the haddock, *Melanogrammus aeglefinus* (L.). *Journal of Fish Diseases* **12**, 591–593.
Mukherjee, M. and Haldar, D. P. (1980). Observations on *Eimeria glossogobii* sp.n. (Sporozoa: Eimeriidae) from a freshwater teleost fish. *Acta Protozoologica* **19**, 181–185.
Obiekezie, A. I. (1986). *Goussia ethmalotis* n.sp. (Apicomplexa: Sporozoea), a coccidian parasite of the West African shad, *Ethmalosa fimbriata* Bowditch 1825 (Pisces: Clupeidae). *Zeitschrift für Parasitenkunde* **72**, 827–829.
Odense, P. H. and Logan, V. H. (1976). Prevalence and morphology of *Eimeria gadi* (Fiebiger, 1913) in the haddock. *Journal of Protozoology* **23**, 564–571.
Overstreet, R. M. (1981). Species of *Eimeria* in nonepithelial sites. *Journal of Protozoology* **28**, 258–260.
Overstreet, R. M., Hawkins, W. E. and Fournie, J. W. (1984). The coccidian genus *Calyptospora* n.g. and family Calyptosporidae n. fam. (Apicomplexa), with members infecting primarily fishes. *Journal of Protozoology* **31**, 332–339.
Paperna, I. (1987). Scanning electron microscopy of the coccidian parasite *Cryptosporidium* sp. from cichlid fishes. *Diseases of Aquatic Organisms* **3**, 231–232.
Paperna, I. (1990). Fine structure of the gamonts of *Eimeria* (*s.l.*) *vanasi*, a coccidian from the intestine of cichlid fishes. *Diseases of Aquatic Organisms* **9**, 163–170.
Paperna, I. and Cross, R. H. M. (1985). Scanning electron microscopy of gamogony and sporogony stages of *Goussia cichlidarum*, a coccidian parasite in the swimbladder of cichlid fishes. *Protistologica*, **21**, 473–479.
Paperna, I. and Landsberg, J. H. (1985). Ultrastructure of oogony and sporogony in *Goussia cichlidarum*, Landsberg and Paperna, 1985, a coccidian parasite in the

swimbladder of cichlid fish. *Protistologica*, **21**, 349–359.
Paperna, I. and Landsberg, J. H. (1987). Tubular formations extending from the parasitophorous vacuoles in the gut epithelial cells of cichlid fish infected by *Eimeria vanasi*. *Diseases of Aquatic Organisms* **2**, 239–242.
Paperna, I., Landsberg, J. H. and Feinstein, N. (1986). Ultrastructure of the macrogamont of *Goussia cichlidarum* Landsberg and Paperna, 1985, a coccidian parasite in the swimbladder of cichlid fish. *Annales de Parasitologie Humaine et Comparée* **61**, 511–520.
Paterson, W. B. and Desser, S. S. (1981a). Ultrastructure of macrogametogenesis, macrogametes and young oocysts of *Eimeria iroquoina* Molnar and Fernando, 1974 in experimentally infected fathead minnows (*Pimephales promelas*, Cyprinidae). *Journal of Parasitology* **67**, 496–504.
Paterson, W. B. and Desser, S. S. (1981b). An ultrastructural study of microgametogenesis and the microgamete in *Eimeria iroquoina* Molnar and Fernando, 1974, in experimentally infected fathead minnows (*Pimephales promelas*, Cyprinidae). *Journal of Parasitology* **67**, 314–324.
Paterson, W. B. and Desser, S. S. (1981c). An ultrastructural study of *Eimeria iroquoina* Molnar and Fernando, 1974, in experimentally infected fathead minnows (*Pimephales promelas*, Cyprinidae). 3. Merogony. *Journal of Protozoology* **28**, 302–308.
Paterson, W. B. and Desser, S. S. (1981d). *Rhabdospora thelohani* Laguesse, 1906 is not a member of the Apicomplexa. *Journal of Parasitology* **67**, 741–744.
Paterson, W. B. and Desser, S. S. (1982). The biology of two *Eimeria* species (Protista: Apicomplexa) in their mutual fish hosts in Ontario. *Canadian Journal of Zoology* **60**, 764–775.
Paterson, W. and Desser, S. S. (1984). Ultrastructural observations on fertilization and sporulation in *Goussia iroquoina* (Molnar and Fernando, 1974) in experimentally infected fathead minnows (*Pimephales promelas*, Cyprinidae). *Journal of Parasitology* **70**, 703–711.
Patnaik, M. M. and Acharya, B. N. (1972). *Eimeria ambassi* sp. nov. from a minnow (*Barbus ambassis*). *Indian Journal of Microbiology* **12**, 53–54.
Peirce, M. A. (1980). A simple staining technique to demonstrate the sporozoite refractile globule in coccidian parasites. *Parasitology* **80**, 551–554.
Pellérdy, L. P. (1974). "Coccidia and Coccidiosis", 2nd edn. Paul Parey, Berlin.
Pellérdy, L. P. and Molnár, K. (1968). Known and unknown eimerian parasites of fishes in Hungary. *Folia Parasitologica* **15**, 97–105.
Pellérdy, L. P. and Molnár, K. (1971). *Eimeria acerinae* sp.n. in the ruff (*Acerina cernua*). *Parasitologica Hungarica* **4**, 121–124.
Philbey, A. W. and Ingram, B. A. (1991). Coccidiosis due to *Goussia lomi* (Protista: Apicomplexa) in aquarium-reared Murray cod, *Maccullochella peeli* (Mitchell) (Percichthyidae). *Journal of Fish Diseases* **14**, 237–242.
Pinto, C. (1928). Synonymie de quelques espèces du genre *Eimeria* (Eimeridia, Sporozoa). *Comptes Rendus de la Société Biologique* **98**, 1564–1565.
Pinto, J. D. S. (1957). Parasitic castration in males of *Sardina pilchardus* (Walb.) due to testicular infestation by the coccidia *Eimeria sardinae* (Thélohan). *Revista da Faculdade de Ciencias de la Universidade de Lisboa*, Serie C **5**, 209–224.
Pinto, J. S., Barraca, I. F. and Assis, M. E. (1961). Nouvelles observations sur la coccidiose par *Eimeria sardinae* (Thélohan), chez les sardines des environs de Lisbonne, en 1961. *Notas e Estudos Instituto de Biologia Maritima, Portugal* **23**, 1–13.

Plehn, M. (1924). "Praktikum der Fischrankheiten". Stuttgart. (Cited by Levine, N. D., 1988.)

Reichenow, E. (1921). Die Coccidien. *In* "Handbuch der Pathogenen Protozoen" (S. von Prowazek and W. Noller, eds), pp. 1136–1277. J. A. Barth, Leipzig.

Reichenow, E. (1953). "Lehrbuch der Protozoenkunde". Teil II, 2. Hälfte. Fischer, Jena.

Romero-Rodriguez, J. (1978). Coccidiopatias de peces. Estudio del Protozoa Eimeriidae: *Eimeria carassiusaurati*, n.sp. *Revista Iberica de Parasitologia* **38**, 775–781.

Schneider, A. (1875). Note sur la psorospermies oviforme du poulpe. *Archives de Zoologie Expérimentale et Générale* **4** (notes et revue), XL–XLV.

Schneider, A. (1881). Sur les psorospermies oviformes ou coccidies. Espèces nouvelles ou peu connues. *Archives de Zoologie Expérimentale et Générale* **9**, 387–404.

Scholtyseck, E. (1973). Ultrastructure. *In* "The Coccida: *Eimeria, Isospora, Toxoplasma* and Related Genera" (D. M. Hammond and P. L. Long, eds), pp. 81–144. University Park Press, Baltimore and Butterworths, London.

Setna, S. B. and Bana, R. H. (1935). *Eimeria harpodoni* n.sp., a new coccidium from *Harpodon nehereus* (Ham. and Buch.). *Journal of the Royal Microscopical Society* **55**, 165–169.

Shul'man, S. S. (1962). Cited by Molnár, K., (1981).

Shul'man, S. S. and Shtein, G. A. (1962). Phylum Protozoa. *In* "Key to Parasites of Freshwater Fish of the USSR" (B. E. Bykhovskii, ed.), pp. 5–235. (English translation (1984) by Israel Programme for Scientific Translations, Jerusalem.)

Shul'man, S. S. and Zaika, V. E. (1962). Cited by Shul'man S. S. and Shtein, G. A. (1962).

Shul'man, S. S. and Zaika, V. E. (1964). Izvestiya sibirskogo otdeleniya. *Izvestiya Akademii Nauk SSSR, Ser. Biol.* **2**, 126–130.

Solangi, M. A. and Overstreet, R. M. (1980). Biology and pathogenesis of the coccidium *Eimeria funduli* infecting killifishes. *Journal of Parasitology* **66**, 513–526.

Stankovitch, S. (1920). Sur deux nouvelles coccidies parasites des poissons cyprinides. *Compte Rendu des Séances de la Société de Biologie* **83**, 833–835.

Stankovitch, S. (1921). Sur quelques coccidies nouvelles des poissons cyprinides. *Compte Rendu des Séances de la Société de Biologie* **85**, 1128–1130.

Stankovitch, S. (1924). *Eimeria misgurni* n.sp. et *Eimeria cobitis* n.sp., deux nouvelles coccidies des poissons d'eau douce. *Compte Rendus des Séances de la Société de Biologie* **9**, 255–258.

Steinhagen, D. (1991a). Ultrastructural observations on sporozoite stages of piscine coccidia: *Goussia carpelli* and *G. subepithelialis* from the intestine of tubificid oligochaetes. *Diseases of Aquatic Organisms* **10**, 121–125.

Steinhagen, D. (1991b). Ultrastructural observations on merogonic and gamogonic stages of *Goussia carpelli* (Apicomplexa, Coccidia) in experimentally infected common carp *Cyprinus carpio*. *European Journal of Protistology* **27**, 71–78.

Steinhagen, D. and Körting, W. (1988). Experimental transmission of *Goussia carpelli* (Léger and Stankovitch, 1921; Protista: Apicomplexa) to common carp, *Cyprinus carpio* L. *Bulletin of the European Association of Fish Pathologists* **8**, 112–113.

Steinhagen, D. and Körting, W. (1990). The role of tubificid oligochaetes in the transmission of *Goussia carpelli*. *Journal of Parasitology* **76**, 104–107.

Steinhagen, D., Körting, W. and van Muiswinkel, W. B. (1989). Morphology and biology of *Goussia carpelli* (Protozoa: Apicomplexa) from the intestine of experimentally infected common carp *Cyprinus carpio*. *Diseases of Aquatic Organisms* **6**, 93–98.

Steinhagen, D., Lukeš, J. and Körting, W. (1990). Ultrastructural observations on gamogonic stages of *Goussia subepithelialis* (Apicomplexa, Coccidia) from common carp *Cyprinus carpio*. *Diseases of Aquatic Organisms* **9**, 31–36.

Studnicka, M. and Siwicki, A. (1990). The nonspecific immunological response in carp (*Cyprinus carpio* L.) during natural infection with *Eimeria subepithelialis*. *Israeli Journal of Aquaculture—Bamidgeh* **42**, 18–21.

Stunkard, H.W. (1969). The sporozoa: with particular reference to infections in fishes. *Journal of the Fisheries Research Board of Canada* **26**, 725–739.

Su, X.-Q. (1987). A new species of family Barrouxiidae from freshwater fish of China (Eucoccidia: Barrouxiidae). *Acta Zootaxonomica Sinica* **12**, 119–120.

Su, X.-Q. and Chen, C.-L. (1987). Three new species of *Eimeria* parasite from freshwater fishes in Hubei province (Sporozoa: Eimeriidae). *Acta Zootaxonomica Sinica* **12**, 10–15.

Thélohan, P. (1890). Sur deux coccidies nouvelles, parasites de l'epinoche et de la sardine. *Compte Rendus des Séances de la Société de Biologie* **42**, 345–348.

Thélohan, P. (1892a). Sur quelques nouvelles coccidies parasites des poissons. *Compte Rendus des Séances de la Société de Biologie* **44**, 12–14.

Thélohan, P. (1892b). Sur quelques coccidies nouvelles parasites des poissons. *Journal de l'Anatomie et de la Physiologie Normales et Pathologiques de l'Homme et des Animaux* **28**, 151–162.

Thélohan, P. (1892c). Sur des sporozoaires indéterminés parasites des poissons. *Journal de l'Anatomie et de la Physiologie Normales et Pathologiques de l'Homme et des Animaux* **28**, 163–171.

Thélohan, P. (1893). Nouvelles recherches sur les coccidies. *Comptes Rendus Hebdomonadaires de l'Académie des Sciences* **117**, 247–249.

Upton S. J. and Duszynski, D. W. (1982). Development of *Eimeria funduli* in *Fundulus heteroclitus*. *Journal of Protozoology* **29**, 66–71.

Upton, S. J., Reduker, D. W., Current, W. L. and Duszynski, D. W. (1984). Taxonomy of North American fish Eimeriidae. *National Oceanic and Atmospheric Administration Technical Report, National Marine Fisheries Service* **11**, 1–18.

Upton, S. J., Bristol, J. R. Gardner, S. L. and Duszynski, D. W. (1986). *Eimeria halleri* sp.n. (Apicomplexa: Eimeriidae) from the round stingray, *Urolophus halleri* (Rajiformes: Dasyatidae). *Proceedings of the Helminthological Society of Washington* **53**, 110–112.

Upton, S. J., Gardner, S. L. and Duszynski, D. W. (1988). The round stingray, *Urolophus halleri* (Rajiformes: Dasyatidae), as a host for *Eimeria chollaensis* sp.nov. (Apicomplexa: Eimeriidae). *Canadian Journal of Zoology* **66**, 2049–2052.

Van den Berghe, L. (1937). On a new coccidium, *Eimeria raiarum* sp.n., with special reference to the biology of the oocysts. *Parasitology* **29**, 7–11.

Yakimoff, W. L. and Gousseff, W. F. (1935). Kokzidien bei Fishen (*Carassius carassius*). *Zeitschrift für Infektionskrankheiten* **48**, 149–150. [In German; English summary].

Yakimoff, W. L. and Gousseff, W. F. (1936). *Eimeria syngnathi* n.sp., a new coccidium from the great pipe fish (*Syngnathus nigrolineatus*). *Journal of the Royal Microscopical Society* (3rd Series) **56**, 376.

Zaika, V. E. (1966). K faune prosteishiki-parazitov ryb cernogo morya. *In* "Gel'mintofauna Zhivotnykh Yuzhnykh Morei", Vol. 13. Nauk, Dumka, Kiev, (Cited by Levine, N. D., 1988.)

The Sexuality of Parasitic Crustaceans

A. RAIBAUT

Laboratoire de Parasitologie Comparée, URA CNRS 698, Université Montpellier II, Sciences et Techniques du Languedoc, Place Eugène Bataillon, 34095 Montpellier Cédex 5, France

AND

J. P. TRILLES

Laboratoire d'Ecophysiologie des Invertébrés, Université Montpellier II, Sciences et Techniques du Languedoc, Place Eugène Bataillon, 34095 Montpellier Cédex 5, France

I. Introduction

As with the insects, the crustaceans constitute a world in their own right. Often extremely differentiated, they have populated all possible environ-

ADVANCES IN PARASITOLOGY VOL. 32
ISBN 0-12-031732-X

ments with an extraordinary diversity of biological forms. They are to be found in all of the seas, throughout the whole range of latitudes and at every possible depth from the sea shore to the abyssal zone. They occur in every river, lake and lagoon. Certain groups have even managed to adapt to a terrestrial life.

Numerous crustaceans also live in association with other living organisms, be it animals or plants (Table 1).

TABLE 1 *Groups of crustaceans including: * parasitic species only; ** parasitic and free-living species*

- Phylum: Arthropoda
  - Subphylum: Antennata (Mandibulata)
    - Superclass: Crustacea
      - Class: Maxillopoda
        - Subclass: Branchiura*
        - Subclass: Copepoda
          - Order: Harpacticoida**
          - Order: Cyclopoida**
          - Order: Poecilostomatoida*
          - Order: Siphonostomatoida*
          - Order: Monstrilloida*
        - Subclass: Cirripedia
          - Order: Ascothoracica*
          - Order: Rhizocephala*
        - Subclass: Tantulocarida*
      - Class: Malacostraca
        - Subclass: Eumalacostraca
          - Superorder: Peracarida
            - Order: Amphipoda**
            - Order: Isopoda**

The relationships established by certain crustaceans with other organisms are very loose, often appearing to be an oportunistic exploitation of an occasional encounter. For others, however, there is an obligate metabolic and trophic dependence on other organisms. These are the parasitic crustaceans that live on, or in, a host organism. The association of the parasitic crustacean with the host organism is essential to enable the parasite to live, reproduce and for its species to survive. The host species may itself, in some cases, be a crustacean.

The crustaceans are of particular interest to the marine pathologist because they constitute a family of the most extraordinary forms of parasitism. Although large numbers of parasites are to be found throughout the Crustacea, certain groups of crustaceans are particularly well endowed with parasitic species. These are the branchiurans, the copepods, the tantulocarids, the rhizocephalic cirripedes and the isopods.

All representatives of the branchiuran subclass are ectoparasites. Their hosts are generally bony fishes, mainly freshwater species. Some species, such as *Dolops ranarum* (Stuhlmann, 1891), are also capable of living on adult amphibians or their tadpoles. Although they are parasitic, the branchiurans have not lost their locomotory capacity, and are thus able to move from one host to another.

Amongst the parasitic copepods it is possible to find all stages from species living in simple associations, that are more or less definitive and that constitute the first fruit of budding unions capable of leading to parasitism, to endoparasitic species. This is the reason for the extreme diversity of their structural organisation (Huys and Boxshall, 1991) and the evolution of forms that are so vastly different from free-living types (the "morphological exuberance" of Kabata, 1979) that, without knowledge of the development of their nauplius larvae, it would be difficult to situate them within the copepods, or even to classify them as copepods at all. For example, one of the best modern copepodologists, J. H. Stock, described with Van der Spoel (1976), a parasite of a pteropod mollusc, with merely the suggestion that it might be a copepod.

The recently created subclass Tantulocarida (Boxshall and Lincoln, 1983), encompasses a number of curious crustaceans that are ectoparasites of other crustaceans, such as copepods (Harpacticoida, Misophrioida), ostracods (Myodocopida) and malacostracans (Tanaidacea, Isopoda, Amphipoda and Cumacea).

The rhizocephalans have undoubtedly evolved from cirripedes, such as *Tubincella*, which is literally screwed into the epidermis of whales, *Chelonobia*, which is firmly attached to the shell of marine turtles by ramifications situated at the base of the plates of its exoskeleton, *Anelasma*, which is completely covered over by the skin of the shark on which it lives, and *Rhizolepas*, which sinks a dense network of ramifications into the tissues of its annelid host. The rhizocephalans differ from these species only in their intense structural regression provoked by their parasitic way of life.

Finally, the isopods have greatly contributed to the history of marine parasitology, as well as to that of parasitology in general. This, probably very old, group has given rise to some remarkable examples of adaptive evolution. They are parasites of various invertebrate and vertebrate groups, but especially of crustaceans and fishes. Some parasitic isopods, such as the Gnathiidae and the Cymothoidae, show little morphological modification, whilst others, such as the Epicaridea, display a maximal complexity of parasitic adaptation.

Set against the diversity amongst the parasitic Crustacea is the comparable diversity of their host species that include not only members of all of the major invertebrate phyla (sponges, coelenterates, molluscs, annelids, echinoderms, tunicates and even crustaceans) but also of certain vertebrate groups, such as fish (branchiurans, copepods, isopods), amphibians (branchiurans, copepods) and cetacean mammals (cirripedes, copepods and also amphipods).

It is often difficult to establish how parasitism has come about. Nevertheless, it is probable that the evolution of crustacean parasitism dates back to early geological eras. The oldest known crustacean parasite, dating back 100 million years, is a copepod *Kabatarina pattersoni* (Cressey and Boxshall, 1989). Thus the origin of parasitic forms cannot be found with certainty amongst present-day, free-living forms, with which they are merely distant cousins.

It is, nevertheless, interesting to note that the crustaceans, owing to the multiplicity of their evolutionary levels, probably constitue one of the most favourable groups for the study of the evolutionary aspects of comparative parasitology. They can thus provide useful models of parasitic adaptations in fields such as sexuality.

II. Types of Sexuality

The copepods and isopods are the only crustacean groups to contain both free-living species, capable of moving about, and parasitic species. The free-living forms are essentially gonochoric. The non-parasitic species of cirripedes, on the other hand, are sessile and fixed to an inert surface. These species are generally hermaphroditic.

Parasitism has led to a multiplicity of sexual modes whose mechanisms are derived from ancestral models that have been conserved in the free-living or parasitic forms (gonochorism and hermaphrodism) or from original models specific to the parasitic way of life, such as pseudo-hermaphrodism or cryptogonochorism.

A. Gonochorism

Gonochorism (separate sexes) occurs amongst a number of parasitic species belonging to the copepods, isopods and branchiurans. The gonochorism of crustacean parasites can be considered as having two forms, depending on whether the parasitism is imaginal or protelian.

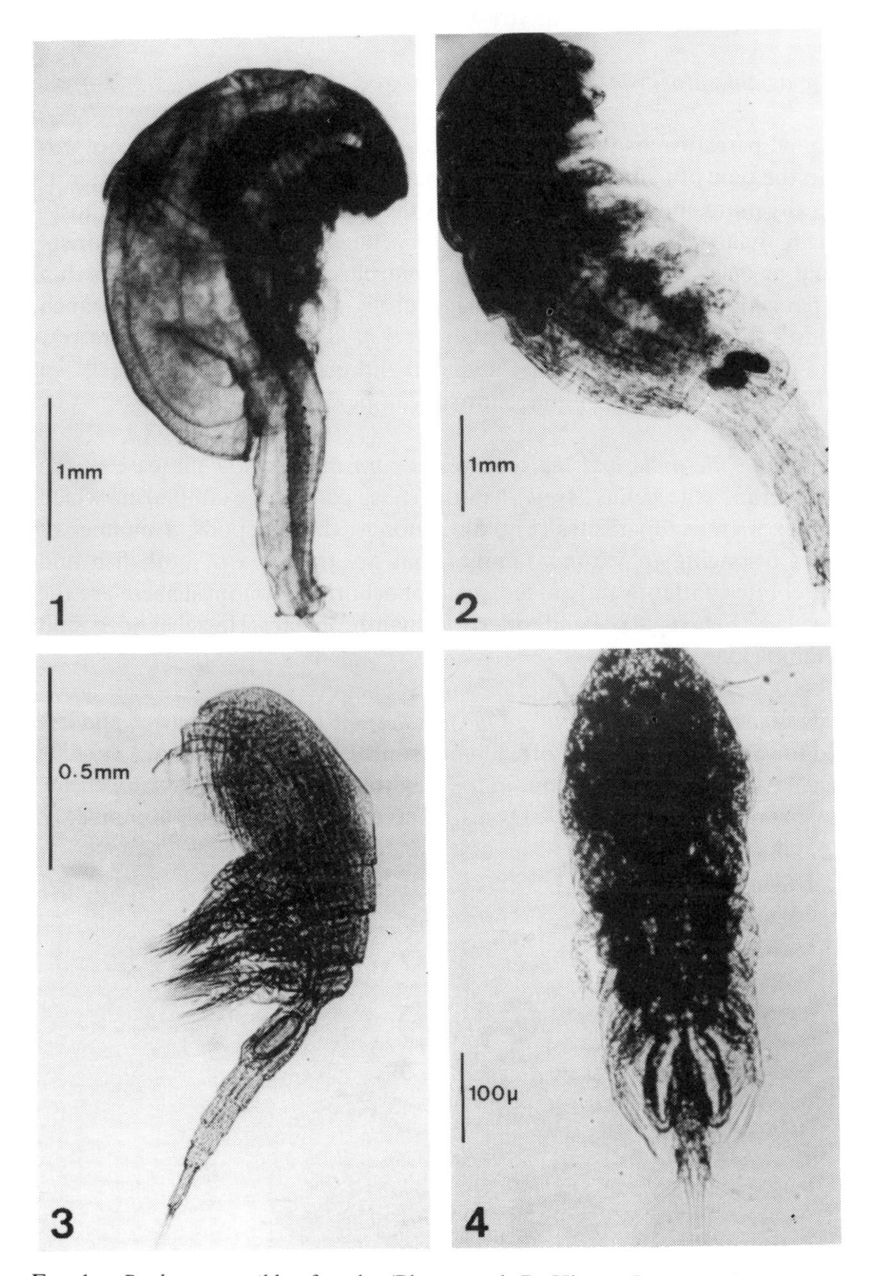

FIG. 1. *Pachypygus gibber* female. (Photograph R. Hipeau-Jacquotte.)

FIG. 2. *Pachypygus gibber* typical male. (Photograph R. Hipeau-Jacquotte.)

FIG. 3. *Pachypygus gibber* atypical male. (Photograph R. Hipeau-Jacquotte.)

FIG. 4. *Ergasilus lizae* male. (Photograph A. Raibaut and O. K. Ben Hassine.)

1. *Imaginal parasitism*

Imaginal parasitism is the most widespread and the sexuality varies according to the type of parasitism. Sexual dimorphism, which is discussed later, is generally more and more pronounced as the degree of intimacy of the host–parasite relationship increases. One very unusual type of gonochorism should be mentioned here. This is the trimorphism of certain copepods that are parasitic on ascidians (Hipeau-Jacquotte, 1978 a,b) or lamellibranch molluscs (Do and Kajihara, 1984; Do *et al.*, 1984), in which there are two males (one typical and the other atypical) and one female (Figs 1, 2, 3).

Three types of imaginal gonochorism can be distinguished:

(a) Both the male and the female are parasitic. This is the case in the branchiurans (the genus *Argulus* is the most common example) in which both sexes are ectoparasites (Fig. 5). Amongst the copepods, a number of species belonging to various families that are parasites of both fish and invertebrates, fall into this category of gonochorism. Amongst these are the Caligidae (Figs 6, 7), Pandaridae, Lernanthropidae, Hatschekiidae and Lichomolgidae.

(b) The male is a "parasite" of the female. The males are dwarfed and live fixed to various parts of the, often hypertrophic, body of the female. In some cases the male may move about in the vicinity of the female, the latter being sessile and permanently fixed. Dwarf males are encountered in a number of

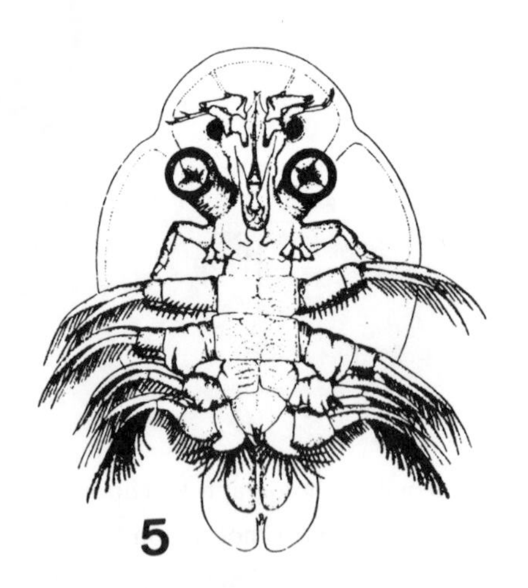

FIG. 5. *Argulus laticauda* male. (From Wilson, 1903.)

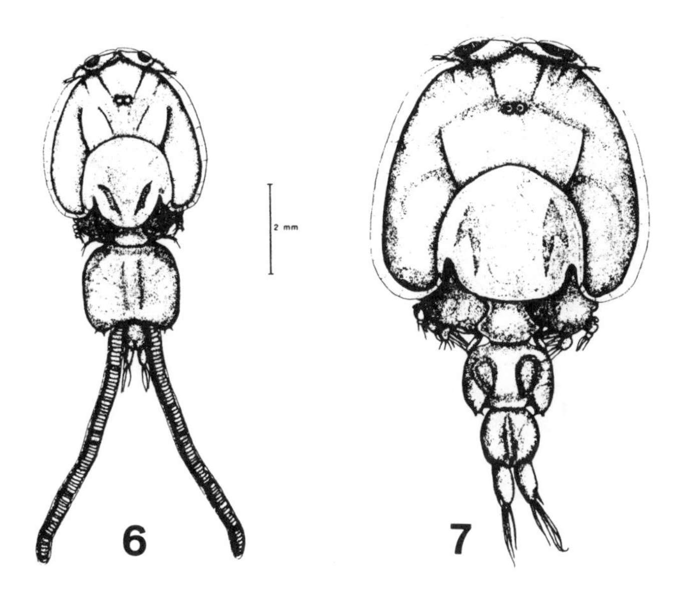

FIG. 6. *Caligus pageti* female. (From Ben Hassine, 1983.)

FIG. 7. *Caligus pageti* male. (From Ben Hassine, 1983.)

copepod parasites of invertebrates and fish. Amongst the latter, the Lernaeopodidae (Fig. 8), Naobranchiidae, Chondracanthidae (Fig. 9) and Sphyriidae contain the most striking examples of pronounced sexual dimorphism. Cases of male dwarfism are also to be found among the copepod parasites of invertebrates. This is especially the case among the Nicothoidae, Splanchnotrophidae, Herpyllobiidae, Echiurophilidae and Xenocoelomatidae. It is also the case in the epicarid Bopyrina (Figs 11 a,b,c) where the dwarf males are normally located on the female herself, for example between the pleopods in the bopyrians. The dwarf male is located in roughly the same place in the Dajidae. In certain cases, the male may lodge in a hemispherical ventral cavity located on the posterior portion of the female's body (e.g. *Notophryxus*). This cavity may contain a coiled cord onto which the male can fix itself (*Aspidophryxus*).

Often there is only one male, but sometimes there are several, especially amongst the Entoniscidae; in *Portunion maenadis*, the larger the female the greater the number of males. Up to 30 males have been found on one female, along with several cryptoniscian larvae (Veillet, 1945). These are found amongst the eggs in the incubatory cavity of the female. The males are also gregarious amongst the Dajidae and the Athelginae (dorsal, abdominal parasites of pagurid crabs) in which the males are also frequently

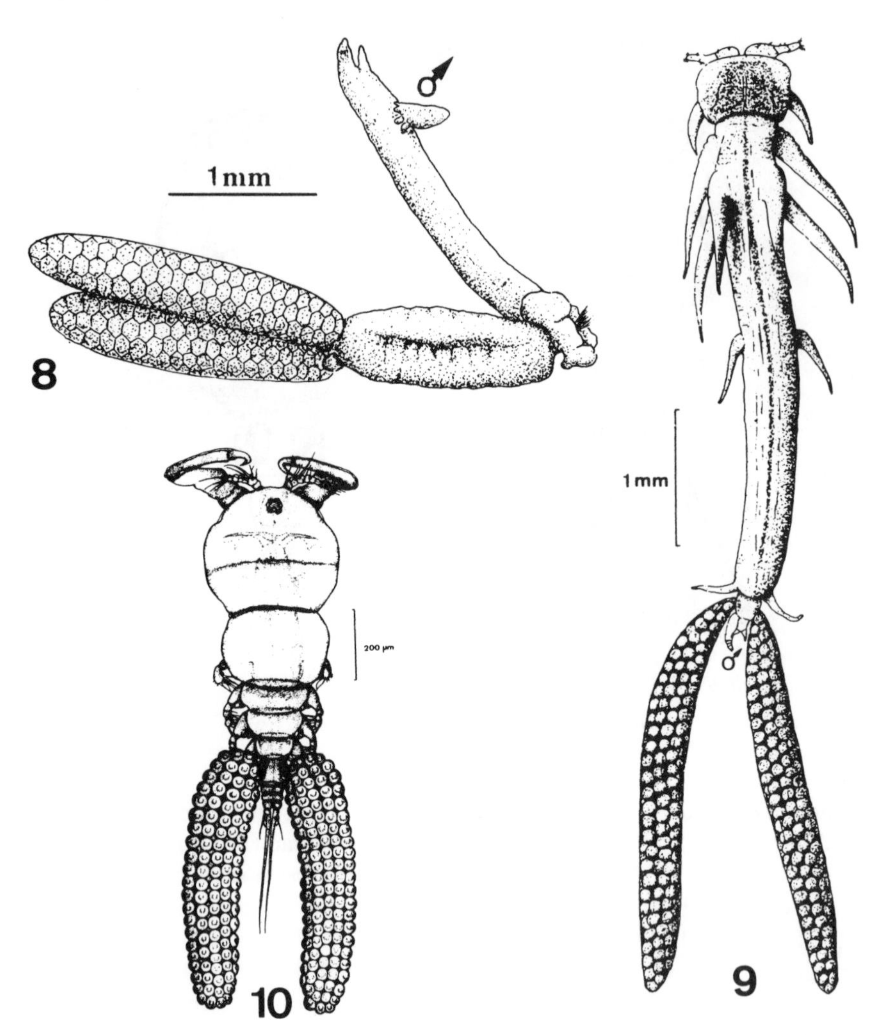

FIG. 8. *Alella macrotrachelus* female with dwarf male. (From Caillet and Raibaut, 1979.)

FIG. 9. *Chondracanthus angustatus* female with dwarf male. (From Raibaut *et al.*, 1971.)

FIG. 10. *Ergasilus lizae* female. (From Ben Hassine, 1983.)

FIG. 11. (A) *Cepon elegans* female in dorsal view with a male between the pleopods (ce, cephalogaster; o, incubation chamber; he, hepatic caeca; pl, pleopods). (After Giard and Bonnier, 1887, redrawn by J.-P. Trilles.) (B) *Cepon pilula* male, dorsal view (an_1, 1st antenna; an_2, 2nd antenna; r, rostre; chr, chromatophores; he, hepatic coeca; coe, heart). (After Giard and Bonnier, 1887, redrawn by J.-P. Trilles.) (C) *Ione thoracica* female in ventral view with a male between the pleopods. (After Reverberi and Pitotti, 1942 and Baer, 1951; redrawn by J.-P. Trilles.)

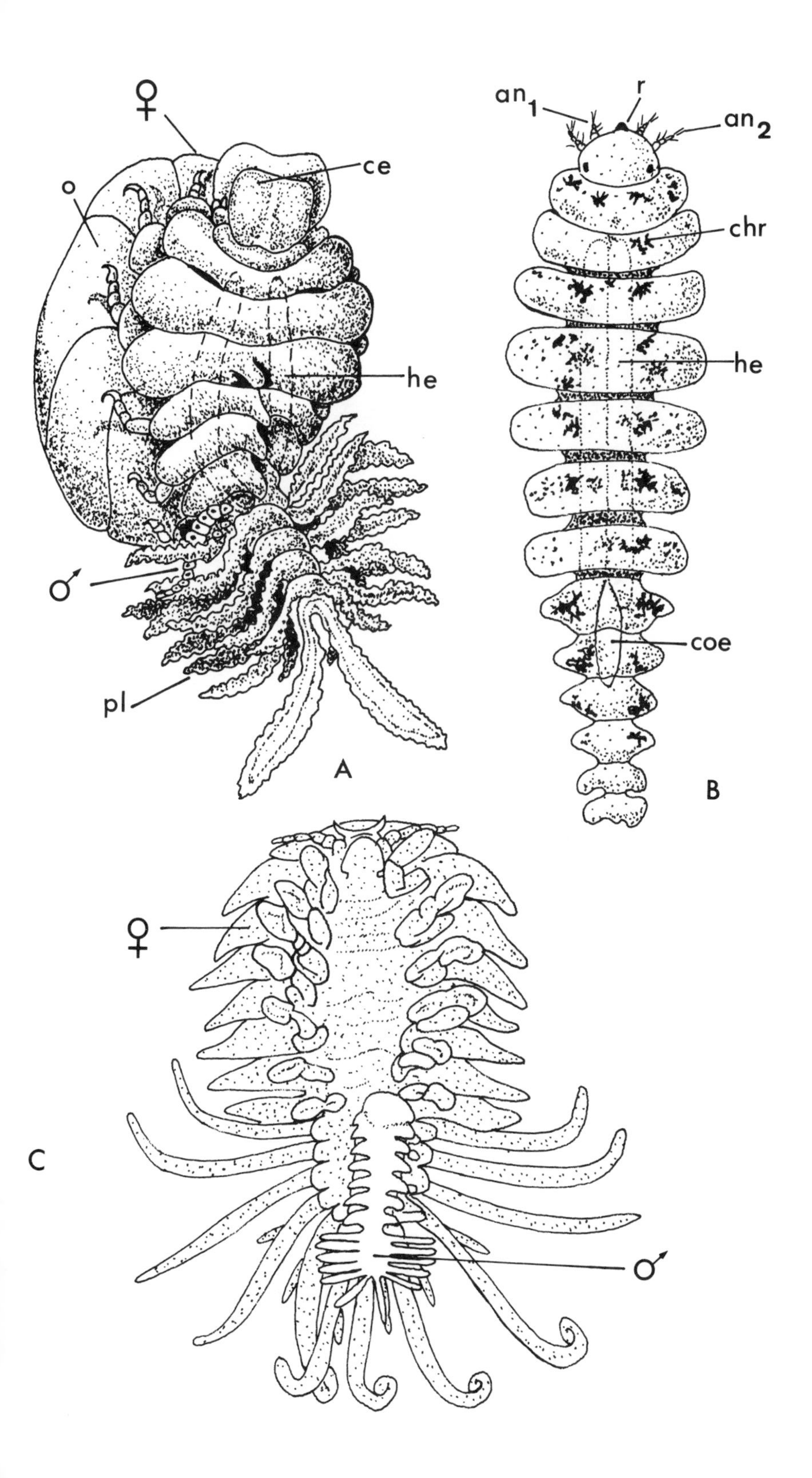

♀
o
ce
he
♂
pl
A
an1
r
an2
chr
he
coe
B
♀
C
♂

supernumerary. In the Athelginae, for example, the adult male occupies its usual position between the female's pleopods, whilst the other males, that are rarely well developed, are to be found in the incubatory cavity of the female that may or may not be full of eggs. Two or more males have also been described in a dozen species of branchial bopyrians.

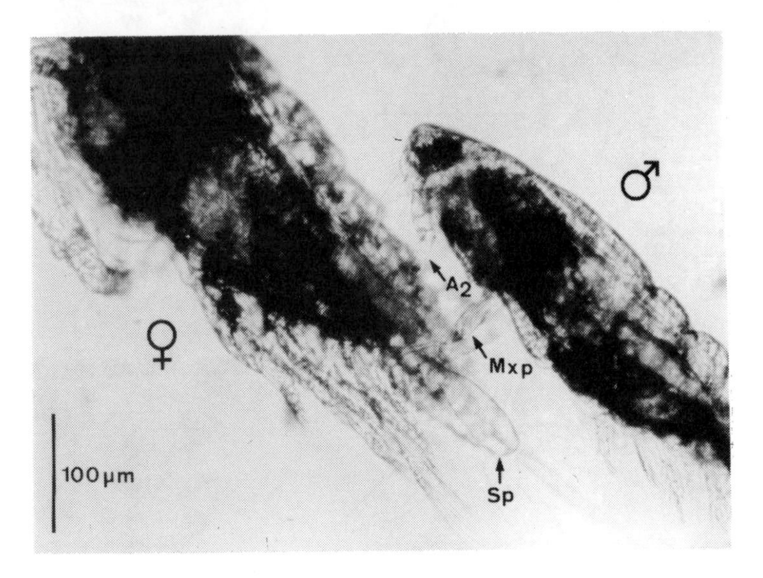

FIG. 12. *Ergasilus lizae* in precopulatory mate (A_2, second antenna; Mxp, maxilliped; Sp, spermatophore). (Photograph A. Raibaut and O. K. Ben Hassine.)

(c) The male is free-living, the female is parasitic. Amongst the copepods, the most representative example is the Ergasilidae, which are parasites of fish and in which only the female is parasitic (Figs 4, 10). Larval forms are free-living and fertilization occurs shortly before the female becomes attached to the host (Fig. 12), the male degenerating and dying (Zmerzlaya, 1972; Ben Hassine, 1983). This form of sexuality is universal amongst the tantulocarids (Boxshall and Lincoln, 1987; Boxshall, 1991; Huys, 1991). In both sexes, development starts with a tantulus larva (Figs 13A, 14A) which attaches itself to its crustacean host. The development of each male larva culminates in the formation of a single free-swimming adult (Fig. 13D♂♂). The ontogenesis of the latter occurs in an expanded trunk sac of the male tantulus (Fig. 13B♂, C♂, 14B). The trunk sac originates at, or near, the back of the thorax of the tantulus. The development of the female is very different. Just behind the head of the female tantulus (Fig. 13B♀) appears a swelling that enlarges into a trunk sac, whilst the larval trunk

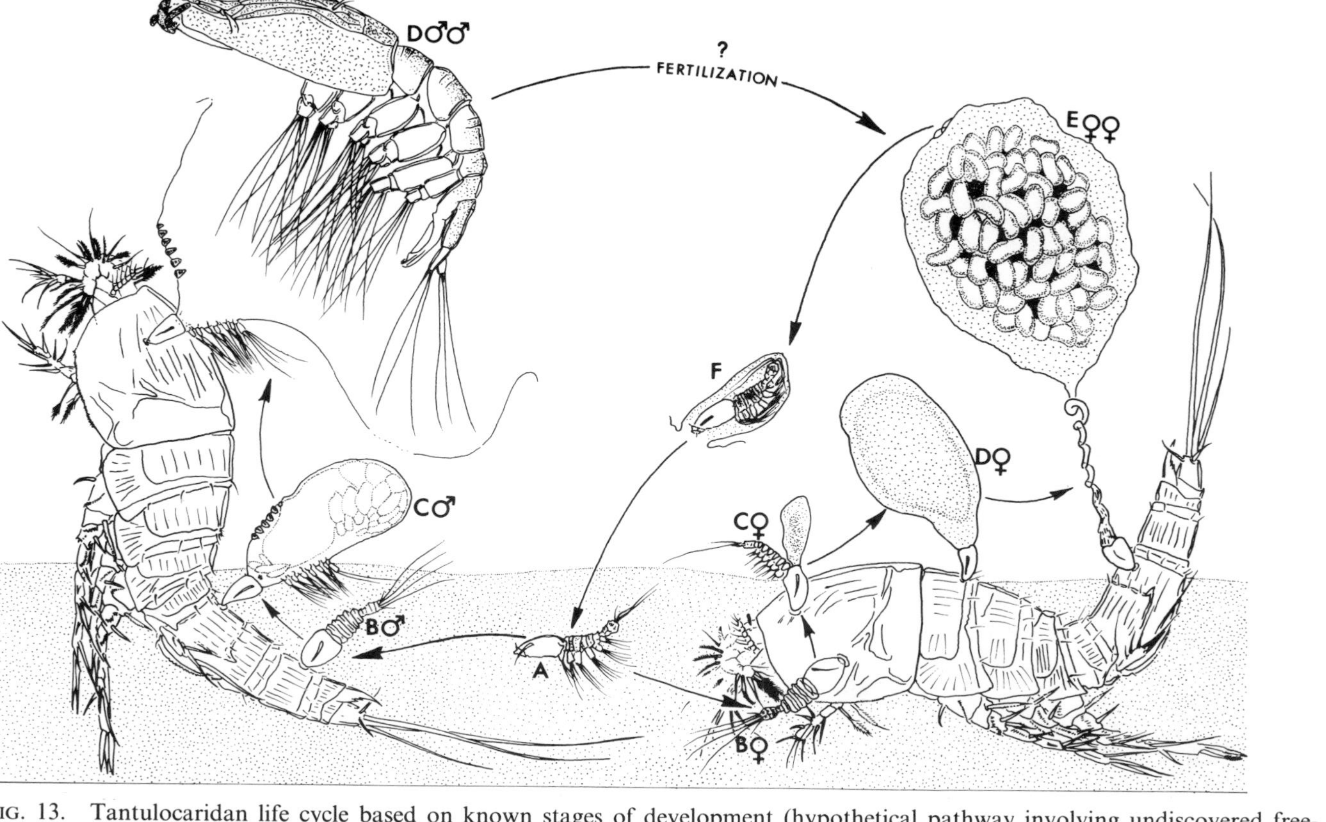

FIG. 13. Tantulocaridan life cycle based on known stages of development (hypothetical pathway involving undiscovered free-swimming female not illustrated). (A) tantulus larva; B♂, C♂, male development; D♂♂, free-swimming adult male; B♀, C♀, D♀, female development; E♀♀, adult female. (Redrawn from Huys, 1991 by N. Le Brun.)

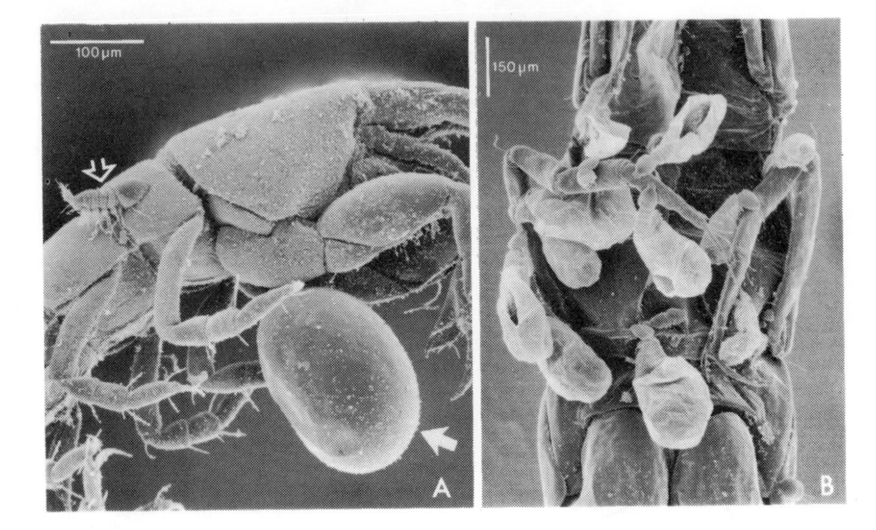

FIG. 14. Scanning electron micrographs of *Microdajus* species. (A) Lateral view of tanaid host, showing tantulus larva (hollow arrow) and adult female (solid arrow). (B) Ventral view of thorax of tanaid host showing numerous male tantuli with expanded trunk sacs containing metamorphosing males. (Photographs G. A. Boxshall.)

disintegrates (Fig. 13C♀, D♀). The adult female (Fig. 13E♀, 14A) contains eggs which develop directly into fully formed tantuli, without moulting (Fig. 13F).

2. *Protelian parasitism*

This type of parasitism occurs in species where the reproductive adult males and females cease to be parasitic. It is found amongst the gnathiid isopods and the monstrillid copepods.

(a) *Gnathiidae*. Only the icthyohaematophagous larvae (zuphea: thoracic segmentation entirely visible; praniza: some thoracic segments not apparent) are parasitic, whilst the adults are free living, inhabiting burrows dug by the males. The temporary nature of the parasitic association is very clear in *Gnathia maxillaris* where the larvae, alternately zuphea and praniza, feed at three separate occasions on the blood of fish, falling to the bottom after each meal to complete their digestion (Monod, 1926; Mouchet, 1928a,b; Stoll, 1962; Amanieu, 1963, 1969; Cals, 1978) (Figs 15, 16 A, B, C).

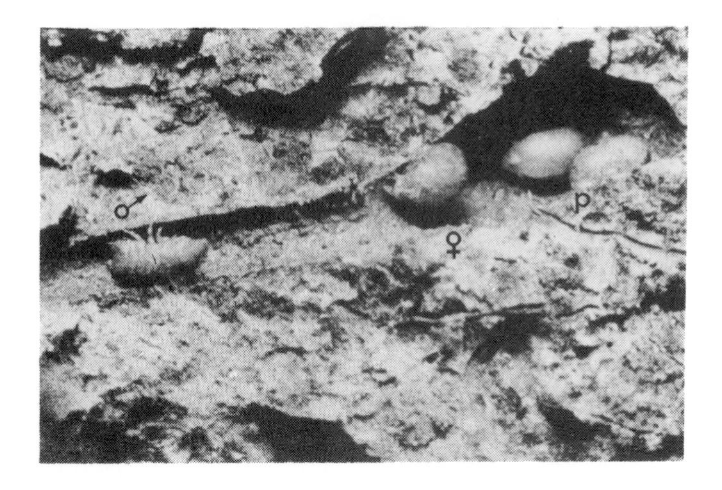

FIG. 15. A *Paragnathia formica* burrow dug in a marine sediment. On the right hand side is a female with two praniza (p) in the "nest" in the burrow. To the left is a male escaping via the access tunnel running from the "nest" to the sediment surface. (From Amanieu, 1969.)

(b) *Monstrilloida*. An extremely curious group of copepods, not only because parasitism concerns only the larvae, but especially because it is the nauplii that are infestive. This is a unique phenomenon amongst the parasitic copepods (Malaquin, 1901) (Fig. 17 A, B, C).

Protelian parasitism also occurs in the tantulocarids, but only in the male phase (Fig. 13).

It is remarkable that in both of these groups the adults are uniquely reproductive stages and no longer feed (Gnathiidae) or are indeed incapable of feeding (lack of mouthparts and gut in the Monstrilloida). This is also the case of the male tantulocarids that possess neither cephalic appendages nor a mouth. They live on reserves accumulated during the parasitic larval life.

Gonochorism is usually stable, though there are cases of labile, epigamic sexual differentiation, under the predominant influence of external factors. The larvae have the potentiality to become either sex. For example, in certain Epicarida, the first larva to become fixed to a host develops into a female, whilst other larvae that follow become males.

B. HERMAPHRODISM

Hermaphrodism is well represented in the order Isopoda.

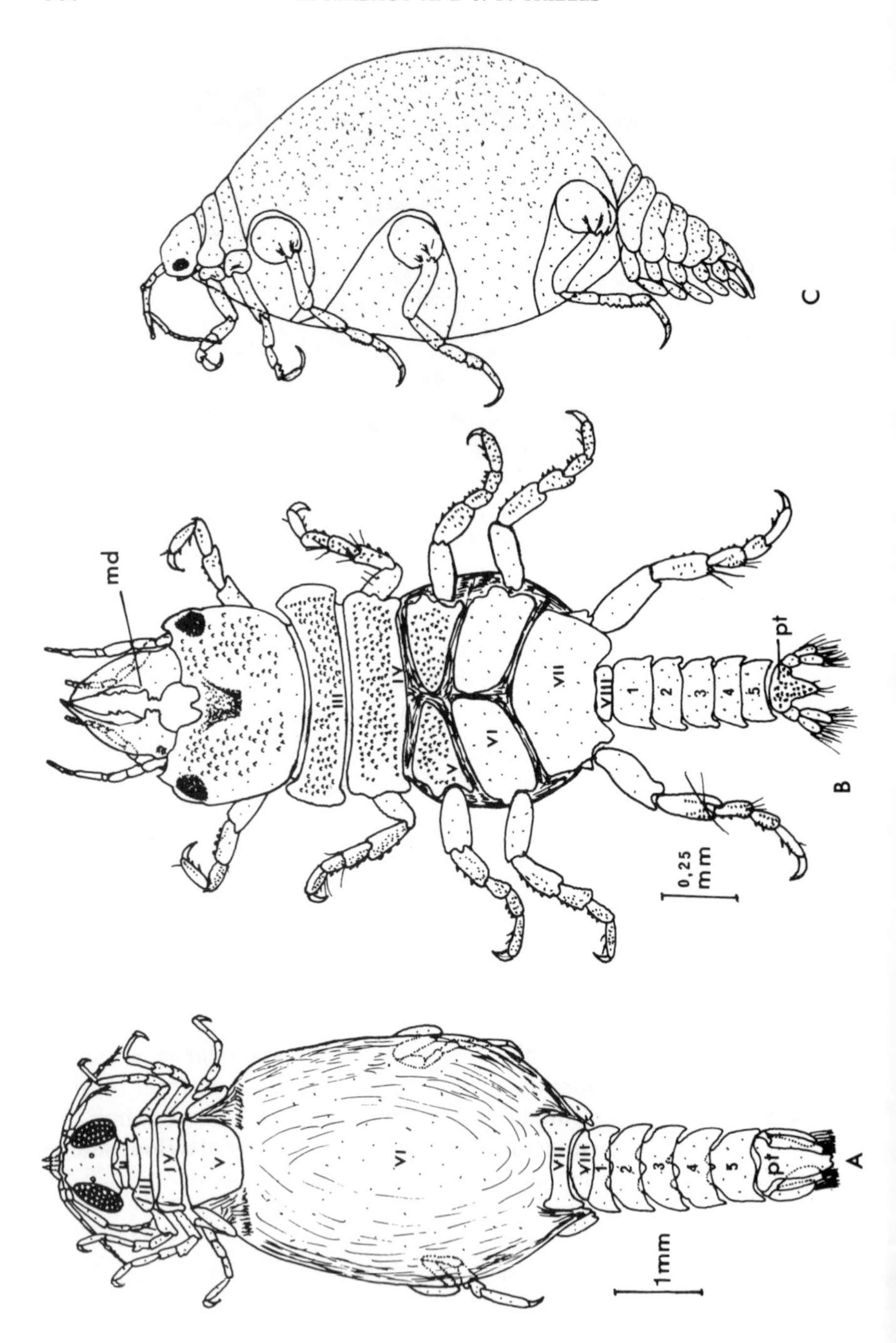
md
pt
III
IV
V
VI
VII
VIII
1
2
3
4
5
0,25 mm
1mm
A
B
C

There is no doubt about the existence of protandrous hermaphrodism in the Cymothoidae. These fish parasites (Fig. 18 A, B) pass through successive male and female sexual stages, this passage comprising a maximum of five stages: male puberty, post puberty or prolongation of male puberty, transition stage or sexual inversion, female puberty, and post puberty or prolongation of female puberty. Characteristic of sexual inversion are the slowing down of spermatogenesis (Figs 19, 20), that is often totally arrested, and the renewal and intensification of ovarian development (Trilles, 1968a,b, 1969).

Slight differences do occur, according to species, genus and especially ecophysiological characteristics. In the majority of cases, and particularly in the case of totally host-dependent species such as the buccal cymothoadians that rapidly and totally lose their swimming capacities, the five stages mentioned above can be distinguished. In certain species, however, especially those in which swimming capacities have been conserved for longer, thus permitting the parasites to exchange hosts, the prolongation of male puberty may be lacking. Finally, the sexual inversion may occur during a single moult. The possibility of changing sex occurs earlier or later according to the species; in a number of cases this is dependent upon an inhibitory influence of the female partner on the male of a couple of parasites attached to the same host fish. This phenomen is clearly observed in buccal cymothoadians that are unable to swim when adult and are therefore uncapable of leaving their host. This mechanism would obviously be less effective in species that conserve their swimming capacities for greater lengths of time and are, thus, capable of changing host.

The type of female phase reached is also conditioned by the degree of adaptation to parasitism. A distinct female phase is only clearly observed in the buccal parasites whilst other species, especially ectoparasites, show a condition closer to the hermaphrodism (a male phase followed by a simultaneous hermaphroditic phase) present, for example, in the free-living Oniscoidea of the family Rhyscotidae (Johnson, 1961): during a part, or the totality, of the female phase, male sexual characteristics, such as the *appendix masculina* or seminal vesicles full of spermatozoa, persist. Could this be linked to the possibility of self-fertilization? Results obtained so far are incapable of confirming this hypothesis.

FIG. 16. (A) *Praniza*, dorsal view (II–VIII, thoracic tergites; 1–5, free abdominal tergites; pt, pleotelson). (After Cals, 1978; redrawn by J.-P. Trilles.) (B) *Gnathia amboinensis* male, general view of the dorsal face (III–VIII, thoracic tergites—V and VI are divided into two tergites; 1–5, free abdominal tergites; pt, pleotelson; md, mandible). (After Cals, 1978; redrawn by J.-P. Trilles.) (C) *Paragnathia formica* female, lateral view. (After Monod, 1926, redrawn by J.-P. Trilles.)

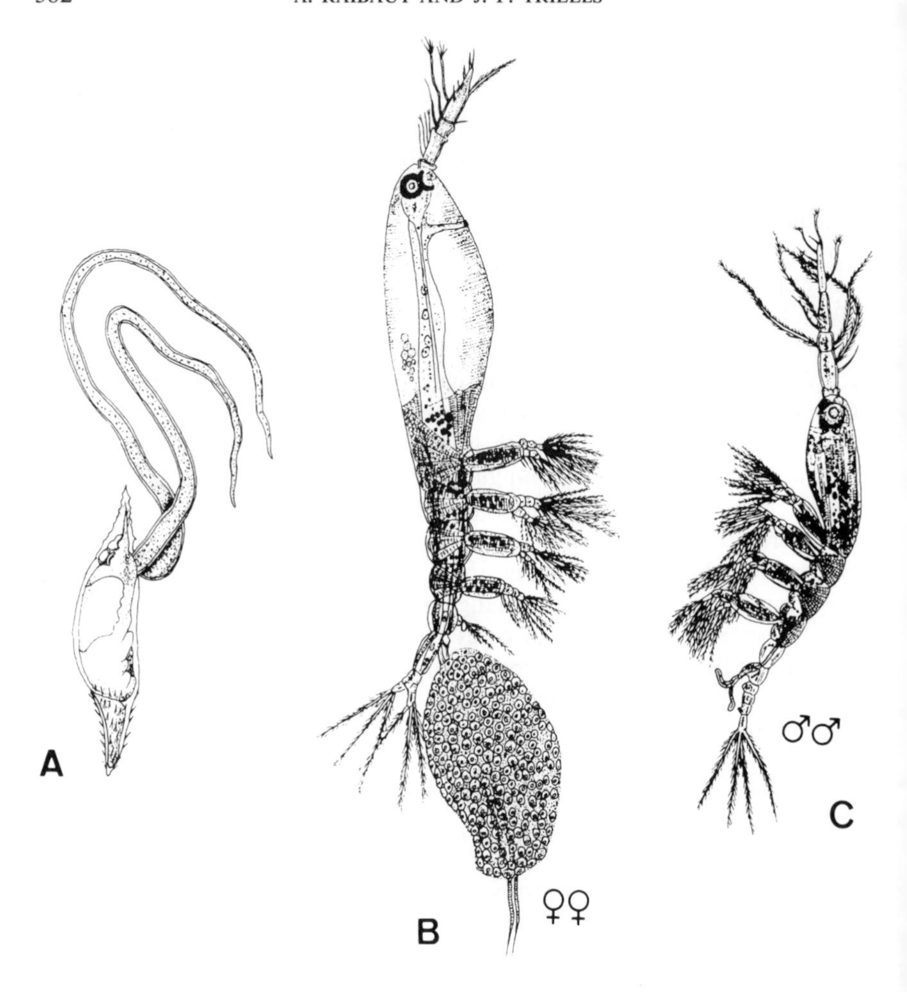

FIG. 17. *Haemocera danae*, monstrillid copepod parasite of the polychaete annelid *Salmacyna dysteri*. (A) embryonic coelomic endoparasite of the annelid. (B) Free-swimming gravid female. (C) Free-swimming adult male. (Redrawn from Malaquin, 1901 by A. Raibaut.)

This type of protandrous hermaphrodism is also encountered in the epicarid cryptoniscians. The larval cryptoniscians become male, acting for a while as a free-living male phase, before becoming fixed to their definitive host and metamorphosing into a parasitic female phase. It is of note that the undifferentiated cryptoniscian, that is male with functional testes, is capable of fertilizing either another larva at the same stage in which the ovaries have developed (*Crinoniscus*, in which the physiological inversion appears to

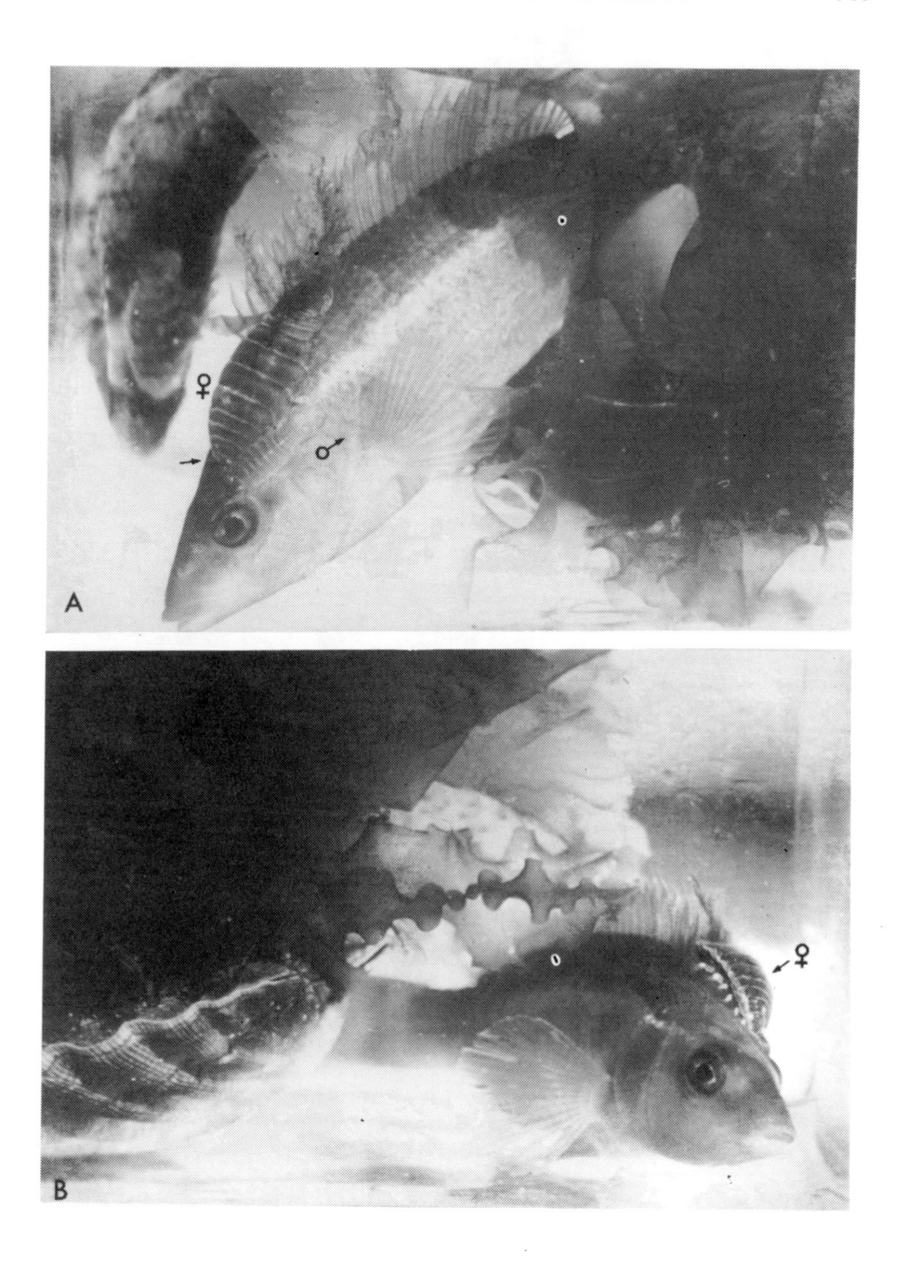

FIG. 18. Bony fishes of the family Labridae parasitized by *Anilocra frontalis* (Isopoda, Cymothoidae). (A) Dorsal view of a couple, male and female. (B) Side view of a female. (Original photographs by J.-P. Trilles.)

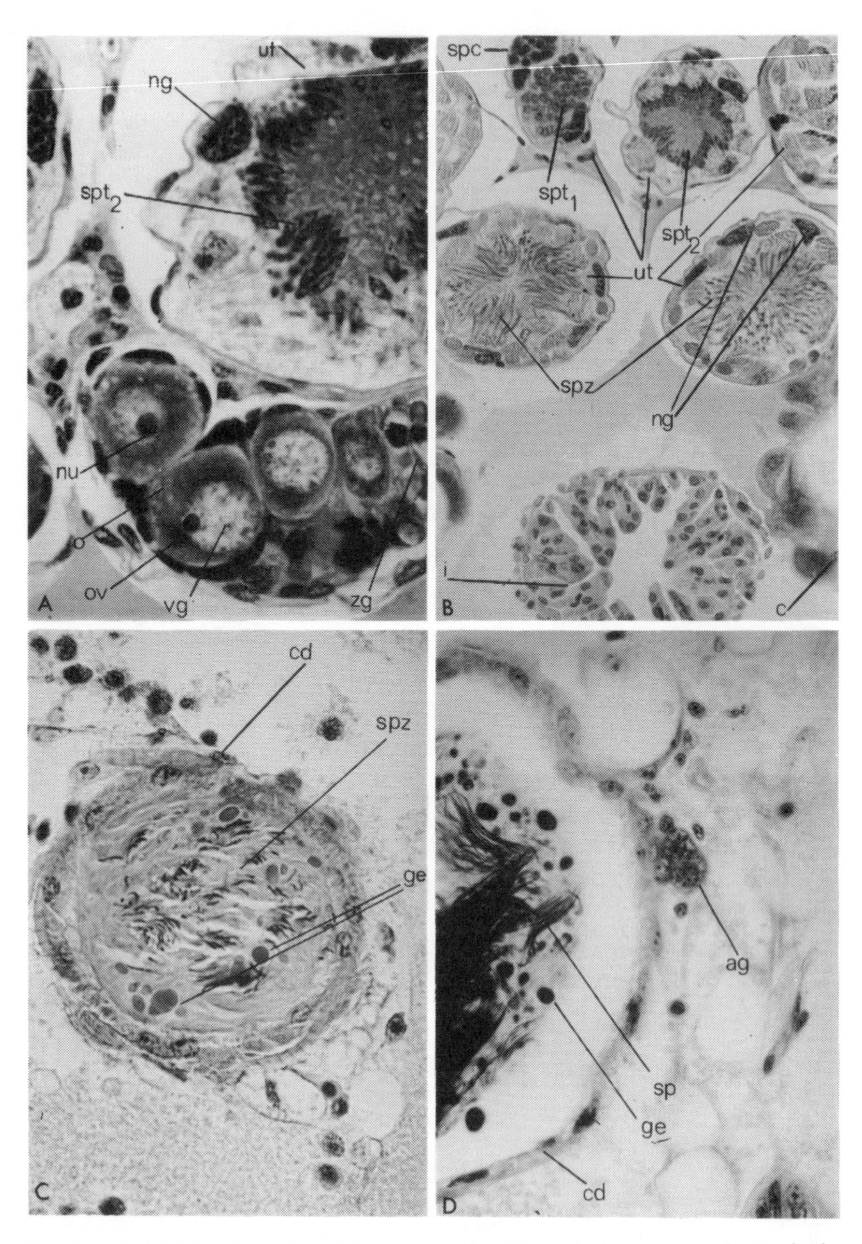

FIG. 19. (A) Male-phase *Nerocila*: cross-section through the ovary and a testicular utricle with spermatids. (B) Male-phase *Nerocila*: cross-section of six testicular utricles. (C) Male-phase *Nerocila*: cross-section of the vas deferens. (D) Male-phase *Nerocila*: portion of the androgenic gland lying against the vas deferens. ut, utricles; ng, giant nuclei; spt_{1-2}, stage 1 and 2 spermatids; nu, nucleolus; o, ovary; ov, oocytes; vg, germinal vesicle; zg, female germinative zone; spc, spermatocytes; i, intestine; c, digestive coeca; spz, spermatozoa; cd, vas deferens; ge, excretory granules; ag, androgenic gland masses. (From Trilles, 1968b.)

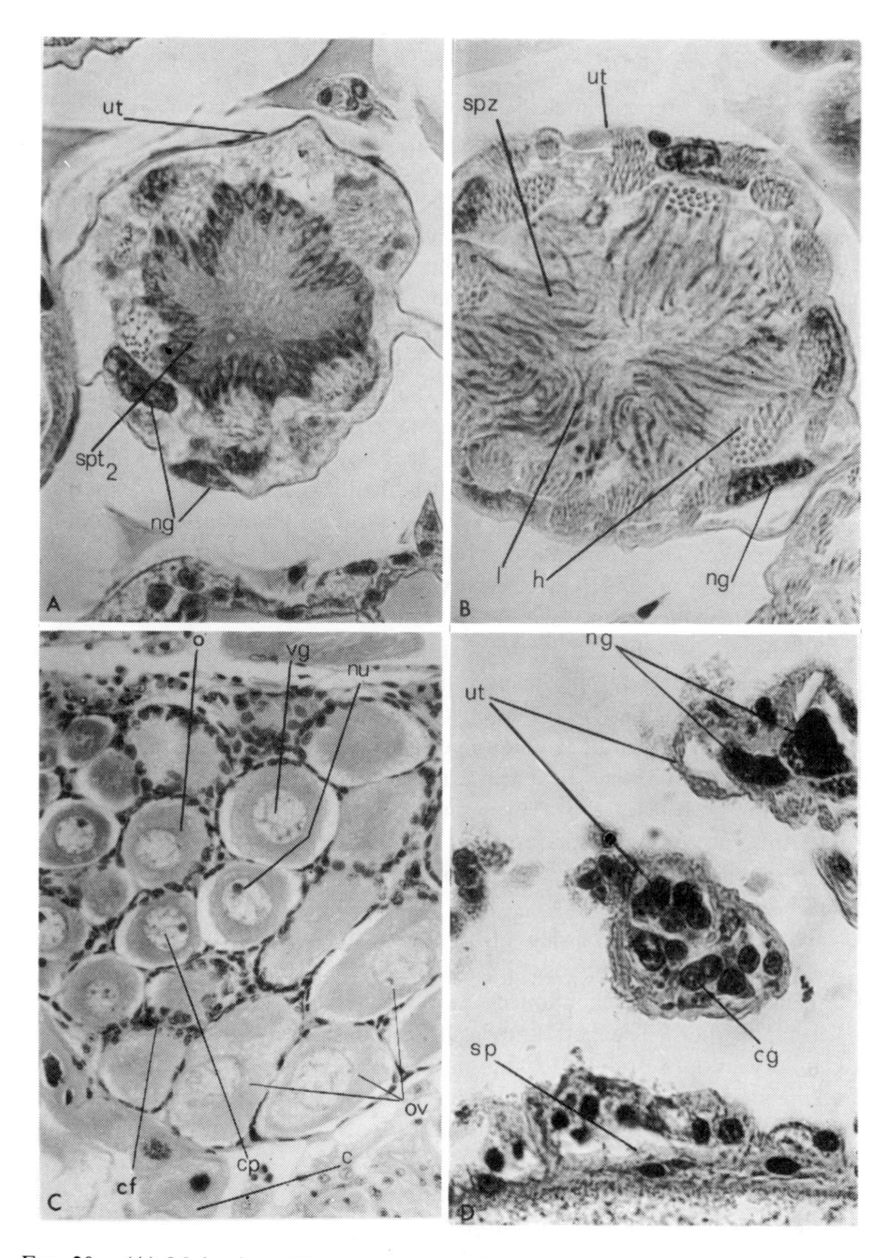

FIG. 20. (A) Male-phase *Nerocila*: cross-section through a testicular utricle with spermatids. (B) Male-phase *Nerocila*: cross-section of a testicular utricle showing a zone of spermatozoa. (C) Female-phase *Nerocila*: cross-section of the testicular utricles of an individual in the female sexual phase. ut, utricles; spt_2, stage 2 spermatids; ng, giant nuclei; spz, spermatozoa; l, flagellae of the spermatozoa; h, heads of the spermatozoa; o, ovary; ov, oocytes; nu, nucleolus; cf, follicular cells; vg, germinal vesicle; cp, plumate chromosomes; c, digestive coeca; cg, germinal cells; sp, pericardiac septum. (From Trilles, 1968b.)

precede the morphological inversion) (Bocquet-Védrine, 1974a,b), or a female (*Hemioniscus* and probably most of the other Cryptoniscina) (Goudeau, 1976).

The female morphology is always the result of the partial, or complete, loss of larval appendages and of the extension of certain parts of the body under the pressure of the ovaries and eggs. Indeed, with the exception of the ovaries, the internal organs become atrophic or disappear. The females become brood pouches that die when they liberate their larvae. Only one species (*Scalpelloniscus*) is perhaps gonochoric. The cryptoniscians of this species are morphologically and sexually differentiated into males and females; these possess the gonads characteristic of their respective sex and would thus appear to be gonochoric. It is not know, however, whether the sexes are genetically determined or whether this species is a protandrous hermaphrodite, as is the case with the other Cryptoniscina, with the morphological transformation from male to female occurring during a supplementary moult. In this case it would be halfway between the *Crinoniscus* and *Hemioniscus* models.

It should be added that the distinction between the modes of sexual differentiation in the Bopyrina and the Cryptoniscina is less radical than first appears. Thus, it has been demonstrated that, following the experimental elimination of the adult female of certain Bopyrina, the male remaining in direct contact with the host can transform into a new female. In consequence, in both cases, the cryptoniscians show an apparent sexual bipotency and, thus, according to species, all of the intermediate stages between a more or less labile gonochorism and an obligate protandrous hermaphrodism can be observed.

To terminate this overview of hermaphrodism in parasitic crustaceans, a few species of parasitic copepods, that infest for the most part polychaete annelids, should be mentioned. These have long been considered to be hermaphrodites. The best known case is doubtless *Xenocoeloma brumpti* Caullery and Mesnil, 1919 that is a parasite of the annelid *Polycirrus arenivorus* Caullery. It is a highly regressed copepod taking on the form of a cylindrical sack, suspended from its host, with no trace of segmentation or appendages (Fig. 21A). Caullery (1922) wrote of *X. brumpti*: "De l'organisme du Copépode, qui est assez volumineux (5–6 mm de longueur), il ne reste plus, en somme, que l'appareil génital, qui est comme greffé sur l'Annélide ... mais *Xenocoeloma* présente un nouveau paradoxe, c'est qu'à la différence de tous les Copépodes connus, il est hermaphrodite"* (Fig. 22).

*Authors' translation: "Of the original copepod organism, that is relatively voluminous (5–6 mm in length), all that remains is the genital apparatus that is grafted onto the annelid ... but *Xenocoeloma* creates a new paradox, in contrast to all other known copepods, it is a hermaphrodite"

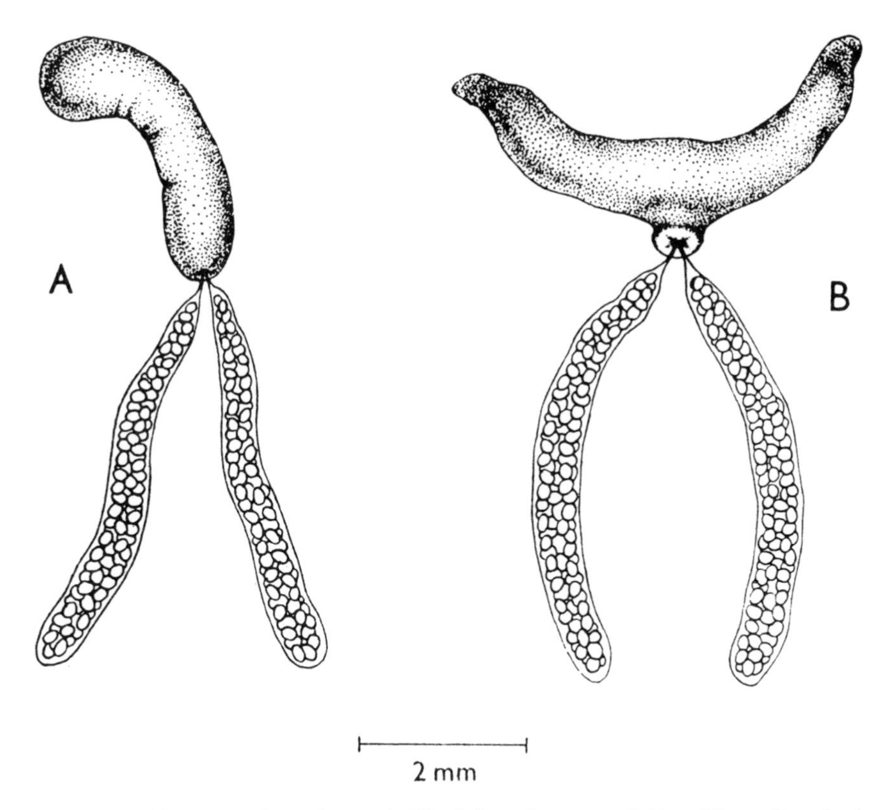

FIG. 21. (A) *Xenocoeloma brumpti*. (B) *Aphanodomus terebellae*. (From Bresciani and Lützen, 1960.)

Other species of copepods, such as *Aphanodomus terebellae* (Levinsen, 1878) (Fig. 21B) and *Flabellicola neapolitana* Gravier, 1918, both parasites of annelids, have been considered to be hermaphroditic, as have certain species of rhizocephalans, for example *Sacculina carcini* (Delage, 1884) and *Chthamalophilus delagei* (Bocquet-Védrine, 1957, 1961).

In reality, this is not the case, and as will be seen in the following section, the peculiar sexuality of these copepods, and most certainly that of all of the rhizocephalans, derives from a particular and specific form of parasitic life. This form of sexuality is cryptogonochorism.

C. CRYPTOGONOCHORISM

In 1960, Bresciani and Lützen described the morphology and anatomy of a curious species of copepod found on the west coast of Sweden and living as a

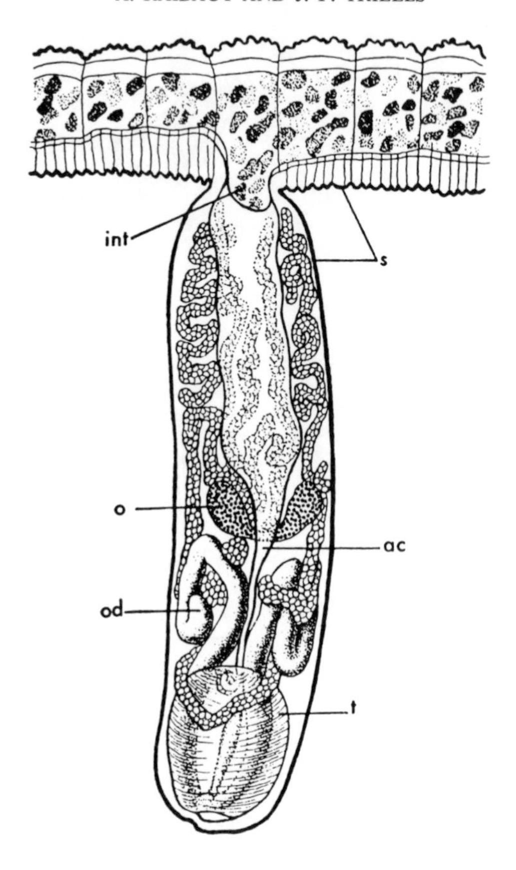

FIG. 22. *Xenocoeloma brumpti*—sagittal section of the parasite attached to its host. ac, axial cavity formed by the host's coelom; int, host's intestine; o, ovary; od, oviduct; s, skin of both the host and the parasite; t, testis. (From Caullery and Mesnil, 1919a.)

parasite of the ascidian *Ascidiella aspersa*. This species was *Gonophysema gullmarensis*, which is highly regressed and characterized notably by the joint presence of an ovary and a testis. The two authors concluded that the genus *Gonophysema* had affinities with two other genera, *Xenocoeloma* and *Aphanodomus*, comprised of species then considered to be hermaphrodites.

Whilst studying the larval development of *Gonophysema gullmarensis*, Bresciani and Lützen (1961) discovered in these copepods an unusual form of sexuality that they later (1972) named cryptogonochorism.

The developmental cycle of *Gonophysema gullmarensis* (Fig. 23) starts with the hatching of a nauplius larva which, after a short pelagic life, becomes a copepodite larva. This latter is the infectious stage, and having

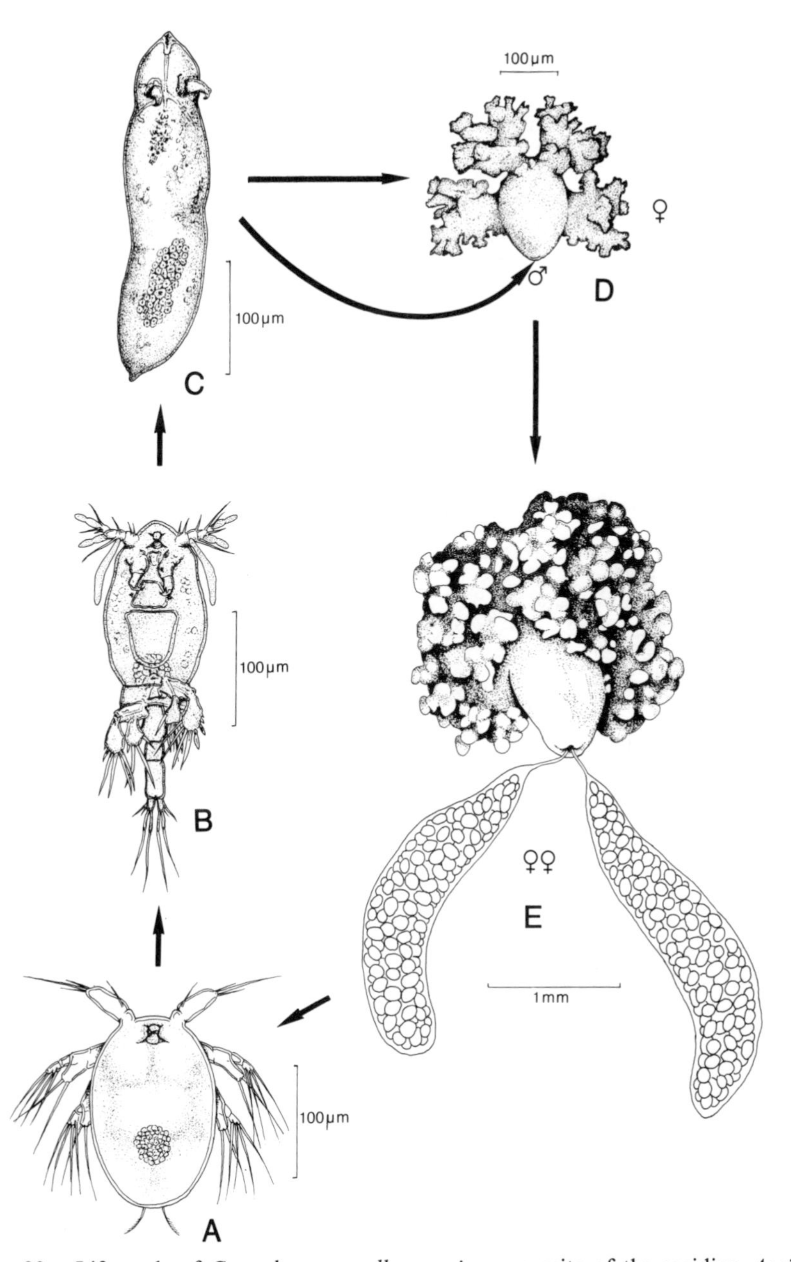

FIG. 23. Life cycle of *Gonophysema gullmarensis*, a parasite of the ascidian *Ascidiella aspersa*. (A) Nauplius. (B) Copepodite. (C) Onychopodite. (D) Young female. (E) Gravid adult female. (Redrawn and modified from Bresciani and Lützen, 1960, 1961 by A. Raibaut, 1985.)

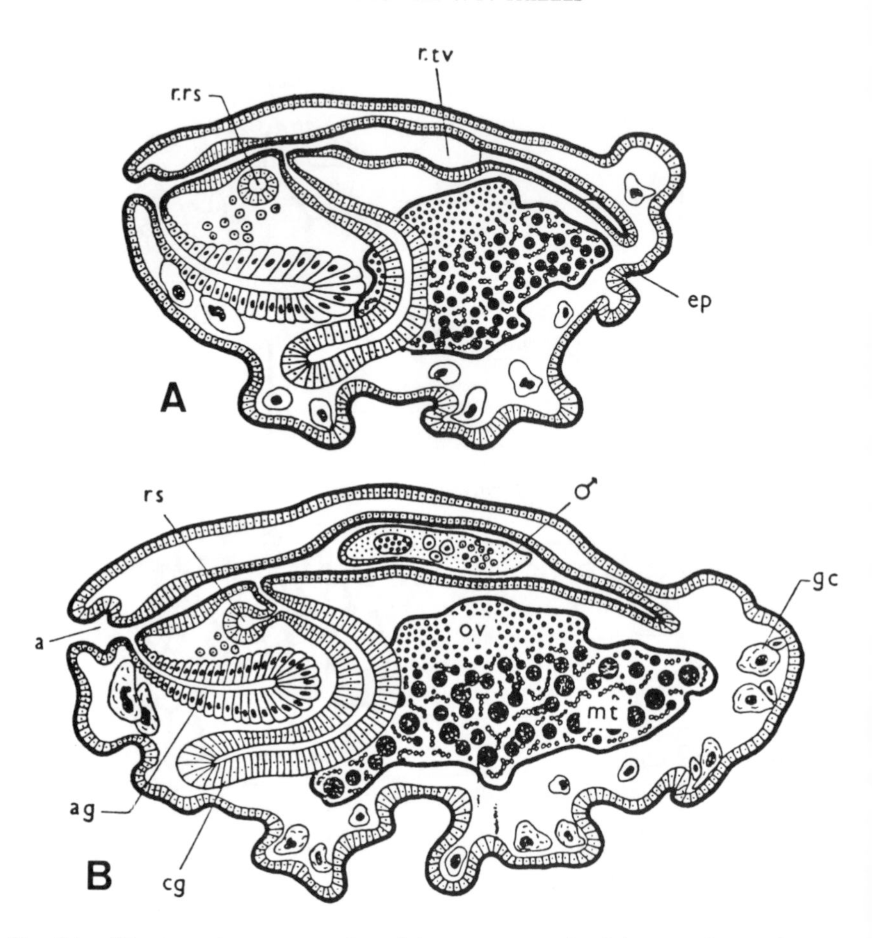

FIG. 24. Diagramatic representation of the organogenesis of the two phases of post-larval development of *Gonophysema gullmarensis* (median sections). (A) Young female without male. (B) Older female with a male in the testicular vesicle. a, atrium; ag, atrial gland; cg, cement gland; ep, epidermis; gc, giant cells; mt, part of ovary that has attained maturity; ov, ovary; rs, receptaculum seminis; r.rs, rudiments of receptaculum seminis; r.tv, rudiments of the testicular vesicle. (From Bresciani and Lützen, 1961.)

attained its host, moults to become an unusual type of larva named an onychopodite. This larva penetrates the tissues of the ascidian, and after a short time spent in the circulatory system of the host, becomes attached to the epithelium of the peribranchial cavity where it undergoes its final metamorphosis leading to the differentiation of a female. This female is then invaded by one or several other onychopodites (up to seven) that take up their place in a cavity originally called the testicular vesicle. Here they undergo a drastic regression. The male gonads of *Gonophysema gullmarensis* are in reality dwarf males (Fig. 24) reduced to their simplest possible form, namely testes. Thus, the phenomen of cryptogonochorism was discovered in *Gonophysema* and later extended to *Aphanodomus terebellae* (Bresciani and Lützen, 1972, 1974) and recently found by Nagasawa *et al.* (1988) in *Pectenophilus ornatus*, a remarkable copepod parasite of the Japanese scallop *Patinopecten yessoensis*.

In light of the results obtained concerning the sexuality of *Gonophysema gullmarensis* and *Aphanodomus terebellae* it might be considered that copepod hermaphrodism, especially that of *Xenocoeloma brumpti*, is unlikely. Only the study of the developmental cycle will provide a definitive answer to this problem. Nevertheless, the reproductive systems of *Gonophysema* and *Xenocoeloma* have common anatomical features, suggesting the same type of sexuality. These include the presence of egg sacs derived from a single median orifice, the presence of an atrium, and especially the fact that the male and female gonads are well separated. Moreover, Caullery and Mesnil (1919a) pointed out the unusual nature of *Xenocoeloma* hermaphrodism amongst the Crustacea in that male and female reproductive organs are derived from different germ lines. Finally, it should be pointed out that Bocquet *et al.* (1970) have demonstrated a testicular membrane distinct from the testicular vesicle in *Xenocoeloma alleni* (Brumpt, 1879).

At the present state of our knowledge, it can be concluded that the copepods are solely gonochoric.

It was at virtually the same time, particularly as a result of the studies of Reinhard (1942) and Veillet (1943) and especially the later studies of Ichikawa and Yanagimachi (1958), Yanagimachi (1961b) and Yanagimachi and Fujimaki (1967), that a comparable mechanism of cryptogonochorism to that described for certain copepods was demonstrated in the rhizocephalans. Up to this point, opinion had been divided.

In 1884, Delage concluded that *Sacculina carcini* was a hermaphrodite. Even though the studies of cypris behaviour and the observations of Lilljeborg (1861) for *Peltogaster* and Müller (1862, 1863) for *Peltogaster socialis* and *Lernaeodiscus porcellanae* worried him. In both cases the authors noted the presence of cypris larvae, or their empty carapaces, in a pericloacal position. Delage considered these cypris larvae to be "primordial males" (rather than complementary males). He excluded in the same hypothesis the opportunistic fixation of these larvae and suggested that they might fertilize the first eggs of the mature *Sacculina*. They would, thus, correspond to relics of primitive gonochorism; the author interpreting hermaphrodism in the rhizocephalans as a secondary adaptation, the principal rôle of which was to overcome the difficulties of fertilization inherent to the biological characteristics of these parasites.

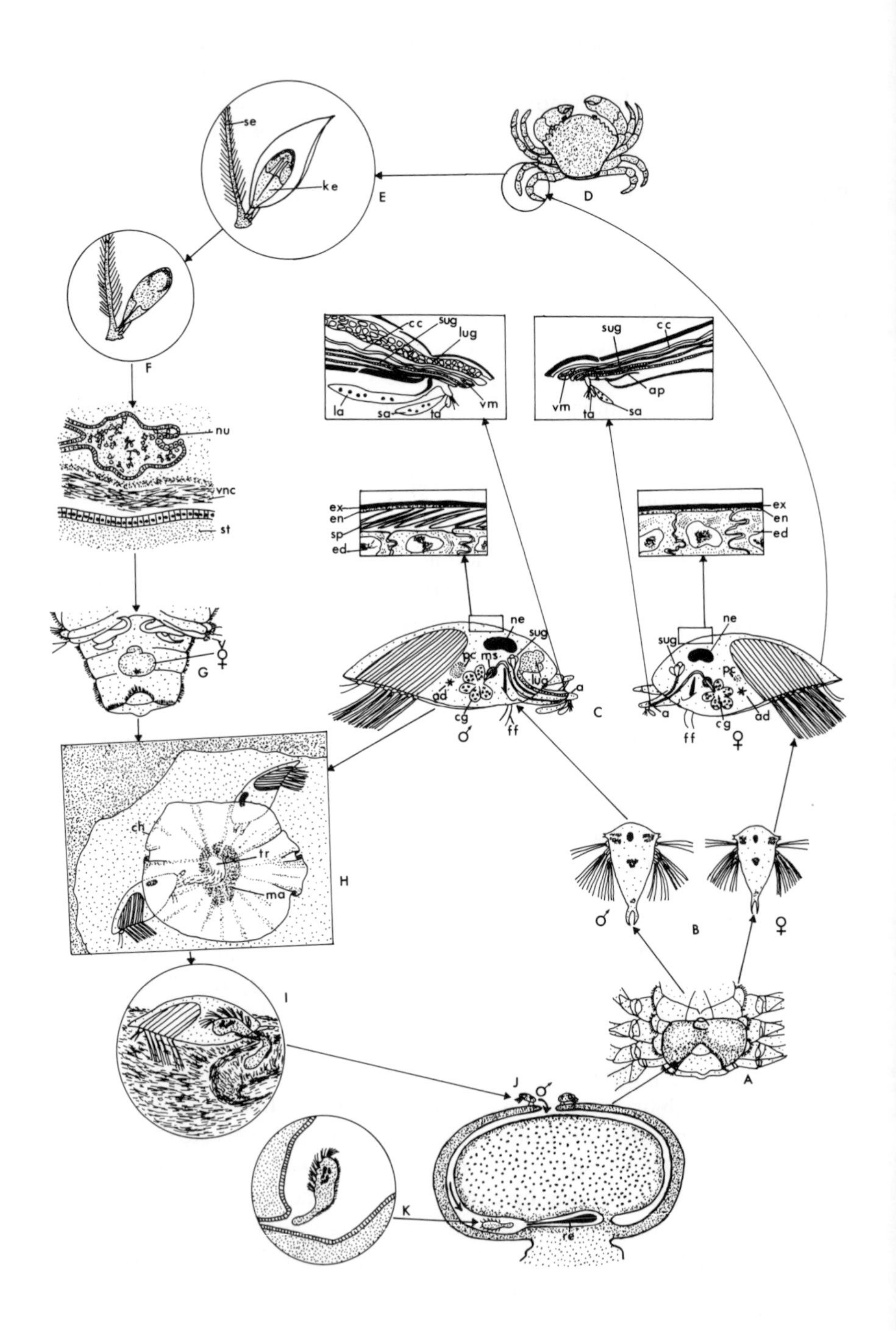
se
ke
E
D
F
cc
sug
lug
la
sa
ta
vm
sug
cc
ap
vm
ta
sa
nu
vnc
st
ex
en
sp
ed
ex
en
ed
G
ne
sug
pc
ms
lug
ad
cg
a
ff
C
ne
sug
pc
a
cg
ad
ff
ch
tr
ma
H
B
I
J
A
K
re

On the contrary, in 1906, Smith suggested these cypris larvae to be non-functional, and self-fertilization to be the general rule in *Sacculina* and *Peltogaster*. They merely represented the remnants of primitive protandrous hermaphrodism.

It was Reinhard (1942) who invalidated Smith's hypothesis by pointing out that the male cypris larvae were functional and that they penetrated the cavity of the visceral mass of a young external parasite and injected a small clump of undifferentiated cells, certain of which would develop into male gametes. But Reinhard did not question the occurrence of hermaphrodism in the rhizocephalans. He only conceded, as did Delage earlier, that only the first eggs were fertilized by these "foreign" spermatozoa. In parallel to hermaphrodism, there existed, therefore, complementary, functional males.

As with his predecessors, however, Reinhard (1942) did not notice any morphological differences enabling him to distinguish between eventual male and hermaphroditic cypris larvae. He hypothesized that sexual determinism was linked to the site of fixation, as in *Bonellia* for example, and accepted that there was only one type of undifferentiated larva that would develop into either a hermaphrodite following fixation to the crustacean host, or a complementary male following fixation to the visceral mass of a rhizocephalan of the same species.

The studies of Veillet (1943) on *Lernaeodiscus galatheae* marked a turning point in the search for a solution to this enigma. Indeed, this author discovered the existence of two types of larvae, that are morphologically similar but of different size. Whilst conserving the idea of a larval male, he nevertheless conceded the existence of early sexual determination, comparable to that observed in the scalpellid cirripeds with their dwarf male. He, thus, totally excluded epigenetic sexual determinism induced by fixation site.

FIG. 25. Life cycle of the kentrogonid rhizocephalan *Sacculina carcini*. (A) Adult externa. (B) Nauplii. (C) Cyprids. (D) Settlement on crab. (E) Cypris-kentrogon metamorphosis. (F) Stylet penetration and injection of internal parasite. (G) Emergence of virgin female externa. (H) Male cypris, implantation of trichogon into female externa. (I) Male cypris-trichogon metamorphosis. (J) Settlement of male cyprids on virgin female externa. (K) Implantation of trichogon into a receptacle. a, antennules; ad, carapace adductor muscle; ap, apodeme; cc, cement canal; cg, cement gland; ch, cuticular hood covering the mantle aperture; ed, epidermis; en, endocuticle; ex, exocuticle; ff, frontal filaments; Ke, kentrogon; la, large aesthetasc; lug, large unicellular gland; ma mantle aperture; ms, muscular sac of cement gland; ne, nauplius eye; nu, nucleus; pc, postganglion cells; re, receptacle; sa, small aesthetasc; Se, seta; Sp, spines of epicuticle; St, sternum; sug, small unicellular glands; ta, terminal aesthetasc on fourth antennular segment; tr, trichogon; vm, valve muscle of cement canal; vnc, ventral nerve cord. (Compiled and redrawn by J. P. Trilles from Hoeg, 1987a,b, 1990.)

More recently, the Japanese authors Ichikawa and Yanagimachi (1958, 1960), Yanagimachi (1961b) and Yanagimachi and Fujimaki (1967) have attempted to analyse their reproduction and development, principally in *Peltogasterella gracilis* (Boschma) (= *Peltogasterella socialis* Krüger), but also in *Peltogaster paguri* and *Sacculina senta* (Yanagimachi and Fujimaki, 1967). These studies have led to a radically gonochoric conception of rhizocephalan sexuality, at least for the Peltogasteridae, that is, nevertheless, similar to the cryptogonochorism described in parasitic copepods. The visceral mass, which is not hermaphroditic, emanates from a female cypris; the testicular part results from the colonization of a special receptacle by a clump of cells derived from male cypris larvae. The term "cypris cell receptacle" has been applied to this sterile pseudo-testis that would appear to play exclusively the role of seminal receptacle. Besides this, biometric and morphological studies have permitted the authors to distinguish a dimorphism in the eggs and the larvae. They have observed that an adult female produces either large eggs (160–170 μm in diameter) or small eggs (140–150 μm in diameter), that develop respectively into large or small larvae. At the nauplius stage, the larvae differ only in size. At the cypris stage, on the other hand, their structure is also different; the most noteworthy distinction being in the morphology of the antennules that are the fixation organs.

The large cypris larvae are those that fix themselves in the vicinity of the mantle orifice of the young female parasitic *Peltogasterella* and function as male larvae. The smaller cypris larvae attach themselves to the base of a seta of a pagurid crab and become adult females.

The cryptogonochorism of *Peltogasterella* was no longer open to doubt, although it remained uncertain whether this model could be applied to the totality of the rhizocephalans, especially when taking into account the doubts forwarded by Bocquet-Védrine (1972), in particular concerning *Sacculina senta* studied by the Japanese authors but also from personal and unpublished observations of *Sacculina carcini*.

More recent and very detailed studies led us to hypothesize that cryptogonochorism is a generalization amongst the rhizocephalans.

Besides the Peltogasteridae, differences in the sizes of male and female cypris larvae have also been observed in the Lernaeodiscidae (Ritchie and Hoeg, 1981) and perhaps certain Sacculinidae (Rubiliani, 1984). As far as cypris morphology is concerned, sexual dimorphism has been observed in the Lernaeodiscidae, Peltogasteridae and the Sacculinidae (suborder Kentrogonida). The big, broad aesthetasc of the third antennular segment is present uniquely in the male cypris larvae of these rhizocephalans; as for the aesthetasc of the fourth antennular segment, though it is present in both species, it is longer in the male cypris larvae than in the females (Fig. 25). The spiny process, of the third antennular segment, that is sometimes visible on the attachment disc is not always a male characteristic. Though it is such a characteristic in the cypris larvae of the Sacculinidae, Peltogasteridae and perhaps also *Mycetomorpha vancouverensis*, it is totally absent in the Lernaeodiscidae (Glenner *et al.*, 1989). No sexual dimorphism has been observed in the cypris larvae of the Chthamalophilidae, Clistosaccidae or Sylonidae. With the exception of *Mycetomorpha vancouverensis* (Mycetomorphidae), the male and female cypris larvae of the suborder Akentrogonida would appear characteristically to be morphologically identical. This is probably related to their specific means of penetration, without passing through an infesting kentrogon stage. The three species *Clistosaccus paguri*, *Sylon hippolytes* and *Chthamalophilus delagei*, and probably all other species of Akentrogonida, however, are none the less gonochoric (cryptogonochoric) (Lützen, 1981a; Hoeg, 1982, 1985a).

Following the example of the Japanese authors, particularly their study of *Peltogasterella gracilis*, the developmental cycles *per se* of a number of other species have been studied in greater or lesser detail. Amongst these species are *Boschmaella japonica*, a parasite of *Chthamalus challengeri* and *Balanus amphitrite amphitrite*, *Chthamalophilus delagei*, a parasite of *Chthamalus stellatus*, *Clistosaccus paguri*, a parasite of *Pagurus bernhardus*, *Lernaeodiscus porcellanae*, a parasite of *Petrolisthes cabrilloi*, *Ptychascus barnwelli* and *P. glaber*, parasites of semiterrestrial crabs (*Uca*, *Arcatus* and *Sesarma*), *Sacculina carcini*, a parasite of *Carcinus maenas*, *Sylon hippolytes*, a parasite of *Spirontocaris lilljeborgi*, *Thompsonia littoralis*, a parasite of *Leptodius exaratus*, *T. reinhardi*, a parasite of *Discorsopagurus schmitti* and *T. dofleini*, a parasite of *Portunus pelagicus*.

The results of these studies can be regrouped into two distinct fundamental models (Hoeg, 1993):

(a) In the Kentrogonida, for example *Sacculina carcini* (Fig. 25), the externa is attached to the ventral surface of the host's abdomen. It liberates nauplii that develop into morphologically distinct male and female cypris larvae. Indeed, in the Kentrogonida, with the exception of a small number of species (*Ptychascus*, *Sesarmaxenos*, *Typhosaccus*) that hatch as cypris larvae, the larvae are liberated at the nauplius stage, that are directly implicated in dispersion. The female cypris larvae attach themselves to the base of a plumate seta on the host crab, preferably on one of the legs. The female larva metamorphoses into a kentrogon that introduces, or sinks, a hollow stylet into the host through which it injects parasite cells into the crab's haemocoel. This very young internalized primordium spreads away from the point of penetration and the female parasite emerges externally on the ventral face of the host's abdomen. At this stage it is a virgin externa, around the mantle opening of which will fix the male cypris larvae. The latter metamorphose into nonsegmented larvae that possess no appendages and that are covered in spines. These are the trichogon larvae that emerge *via* one of the antennules of the cypris larvae and then move around in the mantle cavity of the externa until they reach the cypris cell receptacles where they implant themselves. The trichogon larva is maintained here throughout the life of the female parasite, producing spermatozoa that fertilize the ova produced by the adult externa (Delage, 1884; Rubiliani *et al.*, 1982; Lützen, 1984; Walker, 1985; Hoeg, 1987a,b, 1990; Andersen *et al.*, 1990).

It is of course a similar cycle that is observed in *Peltogasterella gracilis* (Yanagimachi, 1961b), but also, for example, in *Lernaeodiscus porcellanae* (Ritchie and Hoeg, 1981; Hoeg and Ritchie, 1985, 1987). It should be added, however, that in the latter case it is the male that triggers the sexual maturation of the female. This phenomenon has also been described in certain parasitic copepods, such as *Lernaea cyprinacea* in which the premetamorphic females that have not been fertilized have their development blocked at this stage (Bird, 1968). The female cypris larva of *Lernaeodiscus porcellanae* penetrates into the branchial cavity of the host crab and attaches itself to a gill filament. The transformation into a kentrogon and the rest of the developmental cycle are classic. To start off with, the virgin externa does not possess an orifice giving access to the internal cavity of the mantle. This appears only after a moult. The externa then becomes very attractive to the male cypris larvae that attach themselves in the proximity of the orifice, sometimes in large numbers (from 3 to 10). The rest of the cycle is classic, though it is essential that the externa be "fertilized" by a male cypris larva before it can attain sexual maturity.

(b) In the Akentrogonida, for example *Clistosaccus paguri* (Fig. 26), it is the cypris larva itself that penetrates the host without metamorphosing into a kentrogon. The externa of *Clistosaccus paguri* liberates cypris larvae

---

FIG. 26. Life cycle of the akentrogonid rhizocephalan *Clistosaccus paguri*. (A) Adult externa: (B) Cyprids. (C) Settlement on a hermit crab; (D) Penetration and injection of the internal parasite. (E) Young interna. (F) Settlement above a late interna. (G) Penetration into the interna and injection of the spermatogonia. (H) Emergence of externa and migration of spermatogonia into the receptacle. a, antennules; ad, carapace adductor muscle; c, cement; cc, cement canal; cg, cerebral ganglion; cgl, cement gland; cu, cuticle of host crab; ec, embryonic cells; ep, epidermis; he, haemocoel; ic, inclusion-filled cells; ma, mantle tissue; ms, muscular sac; nu, nucleus; re, receptacle; sug, small unicellular glands; th, thorax. (Compiled and redrawn by J. P. Trilles from Hoeg, 1990.)

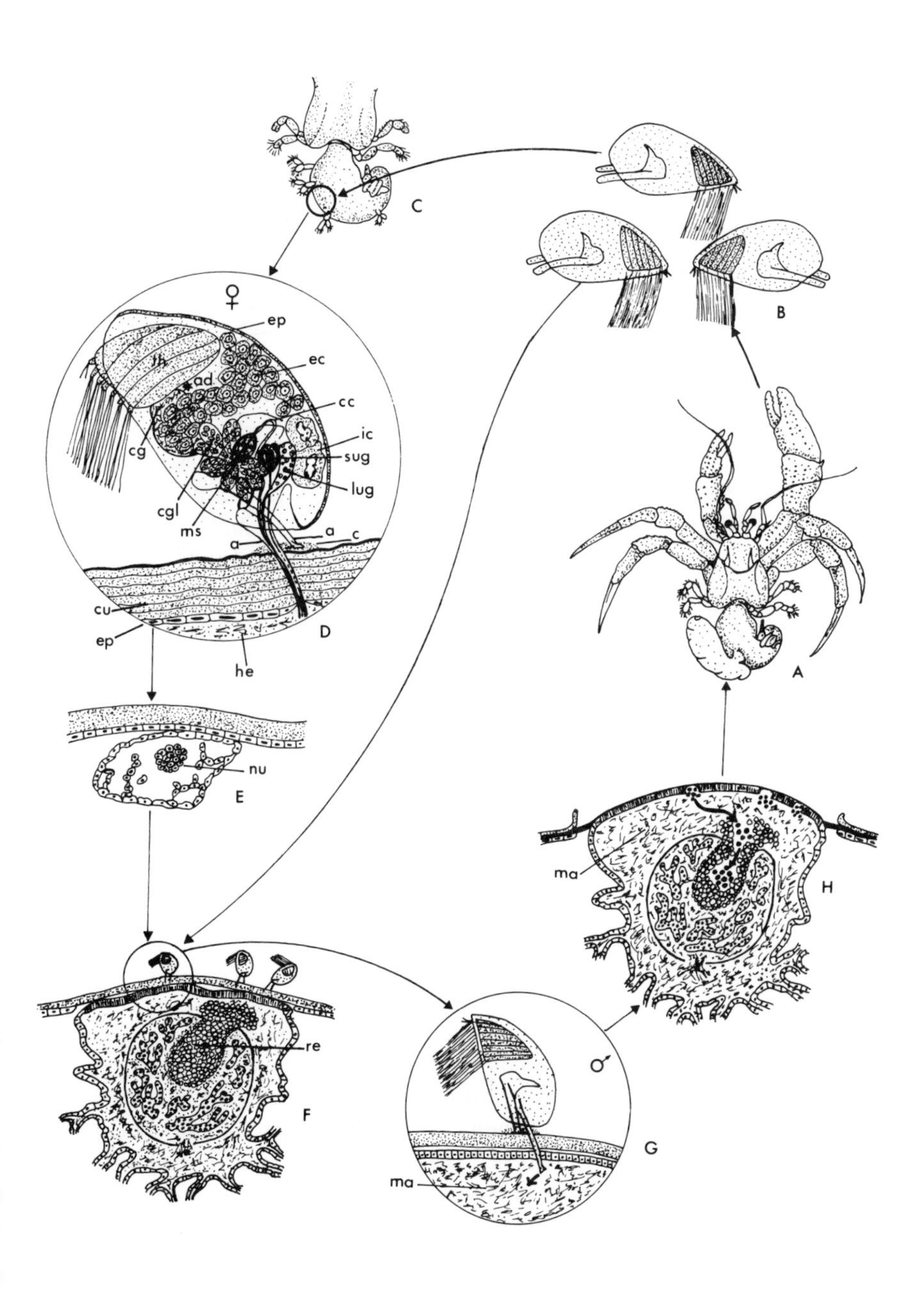
C
B
A
♀
ep
ec
th
ad
cc
ic
cg
sug
lug
cgl
ms
a
a
c
cu
ep
he
D
nu
E
ma
H
re
F
♂
G
ma

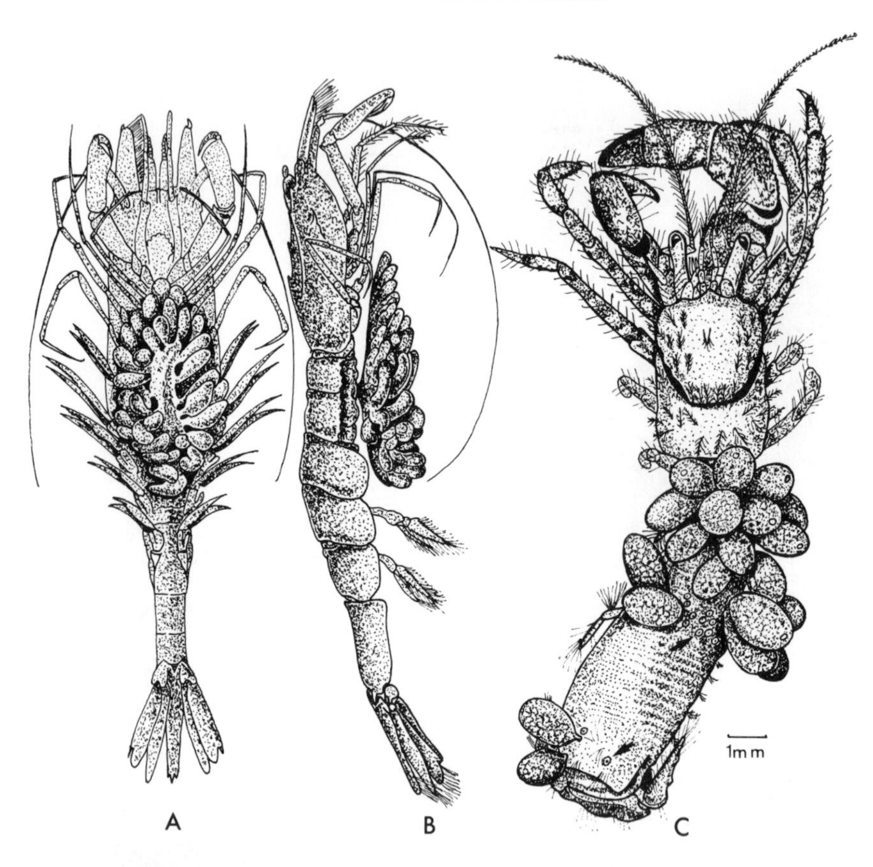

FIG. 27. (A) and (B) *Mycetomorpha vancouverensis* Potts on *Crangon communis* Rathb. (From Potts, 1912 and Baer, 1951; redrawn by J.P. Trilles.) (C) *Thompsonia reinhardi* parasitic on *Discorsopagurus schmitti*; numerous externae and scars (two scars indicated by arrowheads.) (From Lützen, 1992; redrawn by J.P. Trilles.)

which, as we have seen, display no sexual dimorphism. In all species of Akentrogonida, the larvae hatch at the cypris stage. Such abbreviated larval development must obviously entail a decrease in the ability to disperse, while on the other hand more larvae may actually succeed in settling (Hoeg, 1992). A female cypris larva fixes onto the abdomen of a hermit crab, and sinks an antennule into the crab's tegument, through which it injects the internal parasitic cells into the crab's haemocoel. The internal parasite develops directly under the tegument around the point of penetration. The male cypris larvae attach themselves to the tegument of the crab immediately above a female interna that is at an advanced stage of maturity. These too sink in an antennule, piercing both the host's

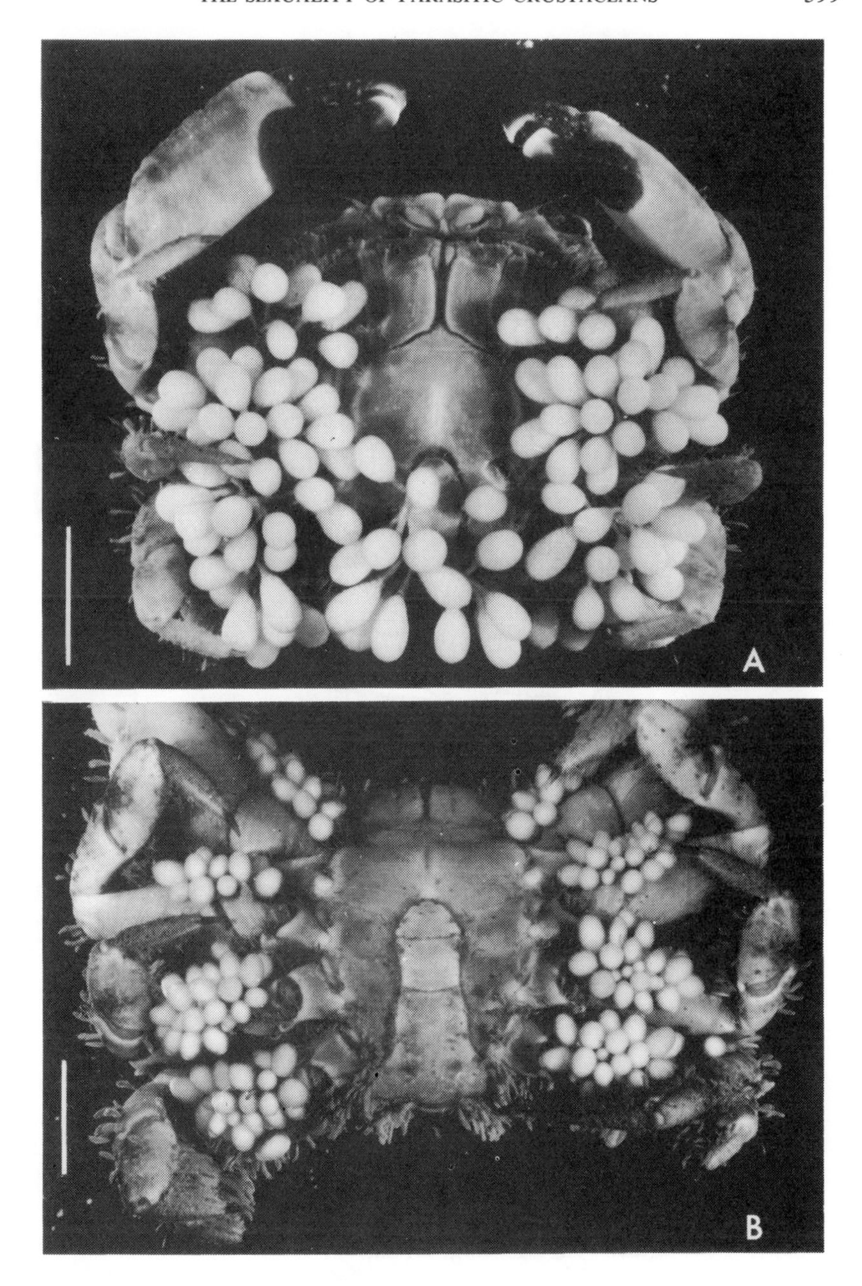

Fig. 28. *Thompsonia littoralis* on its host, *Leptodius exaratus*. (A) Gravid externae. (B) Immature externae. Scale bar = 3 mm. (From Lützen and Jespersen, 1990.)

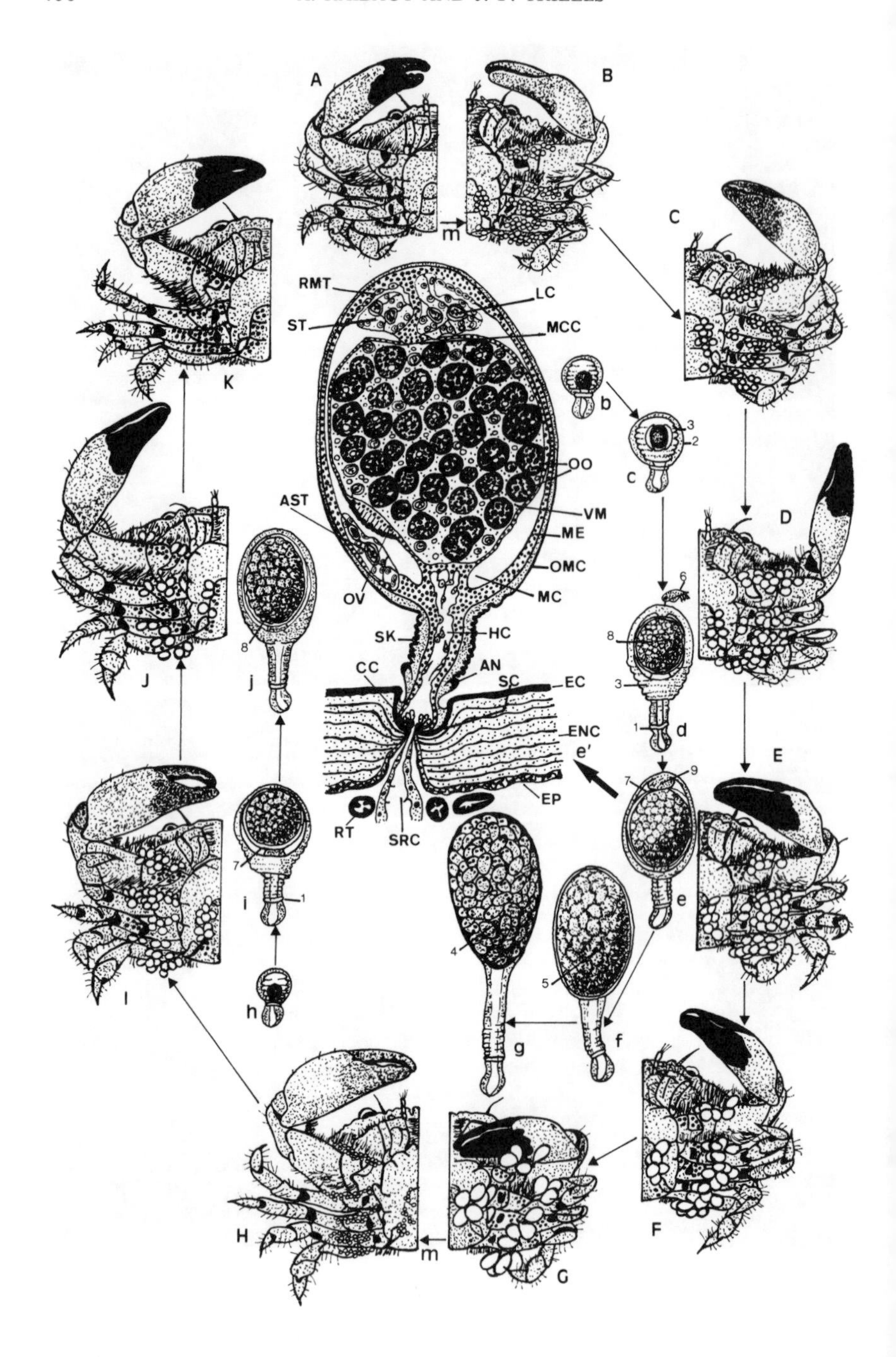
A
B
m
C
D
E
F
G
H
m
I
J
K
RMT
LC
ST
MCC
b
3
2
c
OO
VM
AST
ME
OMC
OV
MC
SK
HC
CC
AN
SC
EC
ENC
e'
EP
RT
SRC
6
8
3
1
d
7
9
e
8
j
7
i
1
h
4
5
g
f

tegument and the interna beneath. They inject their spermatogonia that migrate through the connective tissue of the interna's mantle to take up position in a single, central receptacle. From these spermatogonia will develop the spermatozoa that will fertilize the ova of the adult female externa. It is also possible that, in some cases, the male cypris larvae may attach themselves to virgin externa that have recently erupted. It would be interesting to know how the male cypris larvae recognize a virgin externa or how they manage to locate an internalized female parasite. Whatever the answer, the possibility of the implantation of male cells in the primordia (first stages of growth of the female in the host) increases the length of time during which the virgin female parasite may receive male cypris larvae. This constitutes an obvious advantage in cases where the cypris larvae are rare (Hoeg, 1982, 1985b, 1990, 1992). This type of fertilization by sperm injection, which is common in certain heteropteran insects and which can be characterized as traumatic, probably occurs in the tantulocarids. The male possesses an abdominal penis consisting of an intromissable stylet. Given the size of this penis (100 μm) compared to that of the tantulus larva (130 μm), the possibility of mating occurring at this stage may be excluded. Since no free-swimming, non-feeding female stage has yet been discovered the only possibility left for fertilization is the injection of sperm through the body wall of the attached female that lacks gonopores (Boxshall, 1991; Huys, 1991).

An identical cycle occurs in *Sylon hippolytes* (Lützen, 1981a; Bower and Boutillier, 1990; Hoeg, 1990, 1991), but in this case there is no seminal

---

FIG. 29. Part of the life cycle of *Thompsonia littoralis* on its host, *Leptodius exaratus*. (A–K) Stages in a fertile (b–g) and an abortive generation (h–j) and longitudinal section through externa before ovulation (e′). Following moulting in an internally infested crab (A), a primary generation of externae emerges (B, 25 days) and grows to sexual maturity (C, 35 days; D, 45 days; E, 70 days; F, 100 days; G, 130 days) after transfer of male cells by cypris larvae (d, D). Following the disappearance of the externa, a second generation appears when the host moults again (H, 18 days); in the absence of male cyprids, these externae fail to develop spermatogenic tissue and to ovulate and drop off prematurely (I, 45 days; J, 70 days) resulting in a scarred crab (K, 100 days). Moulting in the externae between c and d, and h and i; oviposition between e and f. (1, annulus; 2, cuticle 1; 3, cuticle 2; 4, cyprids; 5, embryos; 6, male cypris; 7, mantle cavity; 8, ovary; 9, spermatogenic tissue; AN, annulus; AST, accessory spermatogenic tissue; CC, cuticular collar; EC, exocuticle of host; ENC, endocuticle of host; EP, epidermis of host; HC, haemolymphatic canal; LC, large cells in spermatogenic tissue; m, moulting; MC, mantle cavity; MCC, fused mantle cavity cuticles; ME, mantle; OMC, outer mantle cuticle; OO, oocyte, OV, oviduct; RMT, rounded mantle cells; RT, root; SC, socket in host cuticle; SK, stalk; SRC, sinusoid root canal; ST, spermatogenic tissue; VM, visceral mass). (Compiled and redrawn by J. P. Trilles from Lützen and Jespersen, 1992.)

receptacle and the implanted spermatogonia develop directly within the ovarian lobules, and probably at the expense of certain oocytes. *Sylon* would appear to spawn only once.

An apparatus homologous with a seminal receptacle does not exist amongst the other Akentrogonida. Spermatogenesis takes place in one or several spermatogenic islets (= ilôts spermatogènes of Bocquet-Védrine, 1961), which are hollow, epithelia-delimited capsules situated either in the mantle tissue or floating freely in the mantle cavity. The origin of the male cells has not been directly witnessed in any of the species concerned, but strong circumstantial evidence indicates that they are implanted by functional male cypris in a manner comparable to the Clistosaccidae (Hoeg, 1992).

In *Mycetomorpha vancouverensis* and *Thompsonia* (Fig. 27), the spermatogenic islets are situated in the mantle. This is the case, for example, in *Thompsonia littoralis* (Figs 28, 29) which, by associating asexual and sexual reproduction, develops several successive generations of multiple externae, originating from a single, common root system. The roots spread out, from the central nervous system, along the principal nerves (Fig. 39C) and, when the crab moults, a number of externae emerge that are symmetrically disposed on the abdomen and at the base of the thoracic appendages. The number of externae progressively decreases as certain break off, leaving a scar in their place. The remaining externae reach sexual maturity. They contain spermatogonia arranged in one, or several, spermatogenic islets in the mantle. These were probably injected by male cypris larvae. Such externae develop within the space of a few months and liberate young cypris larvae. With each generation, the number of externae increases, but, in the absence of male cypris larvae, certain generations may completely abort (Lützen and Jespersen, 1992).

FIG. 30. (A) The colonial rhizocephalan *Thompsonia dofleini* infesting the ventral side of a mature male *Portunus pelagicus* caught near Tuas, Singapore. (B) External life cycle of *Thompsonia dofleini*: A–H, stages in the development of E-externae. (A) Before moulting. (B) Formation of mantle cavity. (C–E) Growth of ovary and vitellogenesis. (F) Newly divided eggs in ovary. (G) Spawned externa. (H) Basal and mid portion of externa with cypris larvae. (I) Male cypris attached to externa, inoculating spermatogonia. (J–K) Early and late stages in the development of S-externae. a, annulus; bm, basal membrane around ovary; cr, connecting root; cy, cypris larva; C1, cuticle 1; C2, cuticle 2; do, degenerating oocytes; e, eggs in early division; em, embryos; en, endocuticle of host; epidermis of host; m, mantle; mc, mantle cavity; mcy, male cypris; o, ovary; rs, root system; sc, scar; so, spent ovary; sp, spermatogonia; ss, sperm and spermatids; tc, thickened cuticle of mantle cavity; y, yolk granules. (Drawn by B. Beyerholm, after Jespersen and Lützen, 1992.)

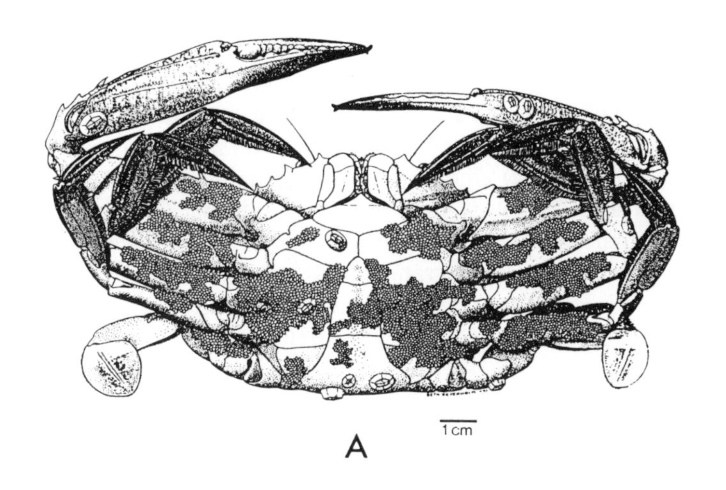
1 cm
A

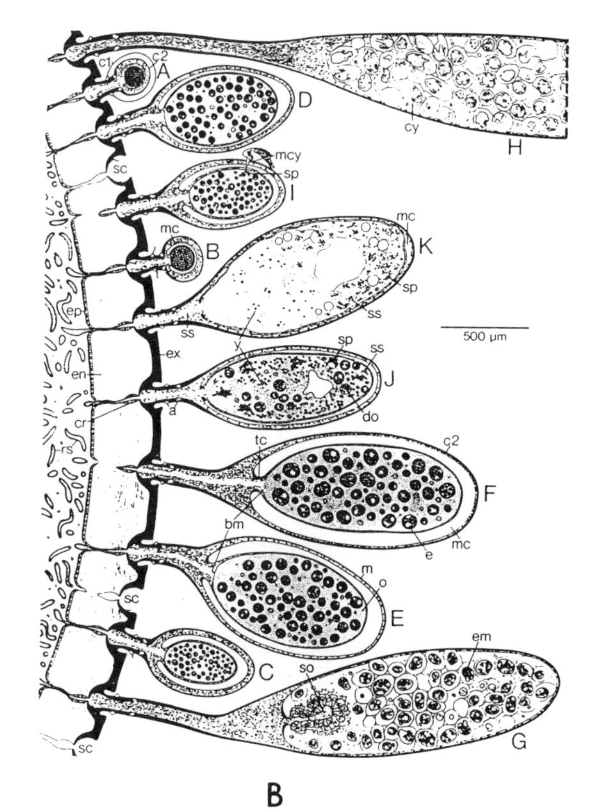
c1
c2
A
D
cy
H
sc
mcy
sp
I
mc
mc
B
K
sp
ep
ss
ss
500 µm
y
sp
ss
ex
en
J
a
do
cr
rs
tc
c2
F
bm
e
mc
m
o
sc
E
em
C
so
G
sc
B

It is of note, however, that in *Thompsonia dofleini* (Fig. 30), there is a developmental cycle that bears some resemblance to that of *Sylon* and which corresponds to a supplementary level in the evolution of sexuality in the rhizocephalans. This species is indeed extremely interesting, as spermatozoa production is limited to the visceral mass of a small number of externae (S-externae) amongst a vast population (several thousands). The spermatogonia, which were probably introduced into the young externae by male cypris larvae, multiply and develop at the expense of the oocytes which rapidly disintegrate, and finally disappear. From here, the spermatozoa are transferred to the other externae (E-externae) via the common root system (Jespersen and Lützen, 1992).

In the Duplorbidae, the Chthamalophilidae and *Pirusaccus*, the mature spermatogenic islets float freely in the mantle cavity. They are, however, derived from an evagination of the mantle, probably as a result of the implantation of male cells, through the tegument, by cypris larvae (Bocquet-Védrine, 1961; Bocquet-Védrine and Bourdon, 1984; Lützen, 1985; Hoeg, 1992). A similar development is also observed in *Boschmaella japonica*, in which a single, primary spermatogenic islet evaginates from the mantle and subsequently fragments into several, distinct, secondary islets in which spermatogenesis occurs (Hoeg *et al.*, 1990).

In the Akentrogonida, the position of the spermatogenic islets is directly related to the morphology of the antennules. Thus, in the male cypris larvae of the Clistosaccidae, which possess relatively long and thin antennules, the spermatogonia can be injected deep enough to reach the central seminal vesicle (*Clistosaccus*) or the ovary (*Sylon*). Only the mantle may be attained by the shorter antennules of the male cypris larvae of the Chthamalophilidae.

III. Sexual Dimorphism

Sexual dimorphism in crustacean parasites is often related to the different constraints imposed upon the males and the females by the parasitic way of life. The females are often sedentary and fixed and by consequence the most highly transformed in order to maintain the fixation. It is also the females that have the highest energetic demands.

A. Size

Amongst the crustacean parasites, the exaggeration of sexual dimorphism is almost always constituted by the gigantism of the female with respect to the male. As described above, there are a number of examples of this amongst the bopyrian epicarid isopods, as well as in the copepods. This is also what is observed in the protandrous hermaphrodites, such as the cryptoniscians; during the male phase, the individual is dwarfed and lives on a giant female, into which it may be transformed once having functioned as a male.

There are, however, examples where the differences in size of the males and females are slight. This is the case in a number of copepods and cymothoadians, where there exists, nevertheless, a slight size advantage in favour of the females. There are a certain number of exceptions to this rule. Thus, in certain lichomolgid copepods, belonging to the species *Indomolgus brevisetosus, Rhynchomolgus corallophilus* and *Temnomolgus eurynotus*, it is the male that is bigger than the female (Humes and Stock, 1973). Furthermore in certain species, "giant" males, derived from males that have undergone a supplementary moult, have been observed. This is the case in *Caligus pageti*, a copepod parasite of mullet, in which the males are approximately twice the size of the females (Ben Hassine and Raibaut, 1981).

B. MORPHOLOGY

An important form of sexual dimorphism is, theoretically, that which concerns the general form of the body and appendages. In many cases, this dimorphism is slight. Examples of significant dimorphism are to be found, however, amongst the epicarids, the rhizocephalans and certain copepods.

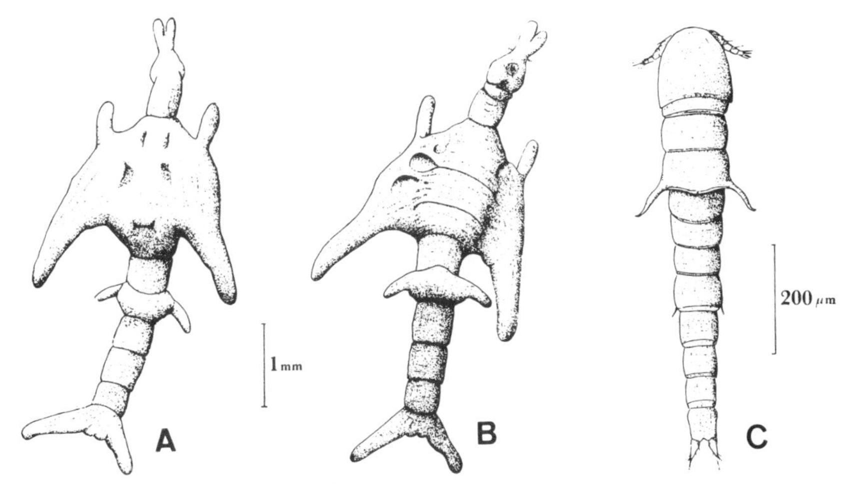

FIG. 31. *Colobomatus mugilis*. (A) Female, dorsal view. (B) Ventral view. (C) Male, dorsal view. (From Raibaut *et al.*, 1978.)

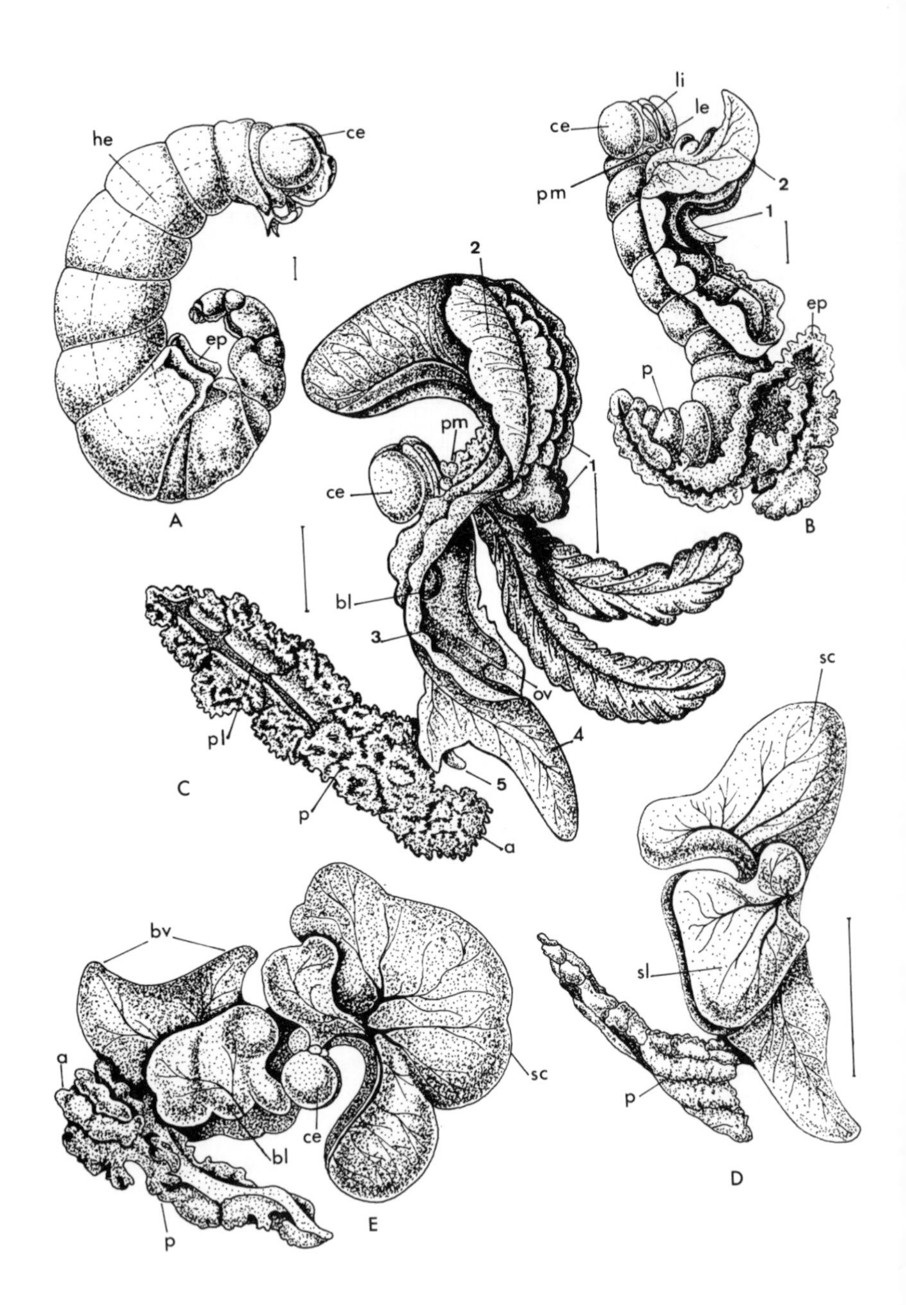
he
ce
ep
A
li
le
ce
pm
2
1
ep
p
B
2
pm
ce
1
bl
3
ov
4
5
pl
C
p
a
sc
sl
p
D
bv
a
bl
ce
sc
p
E

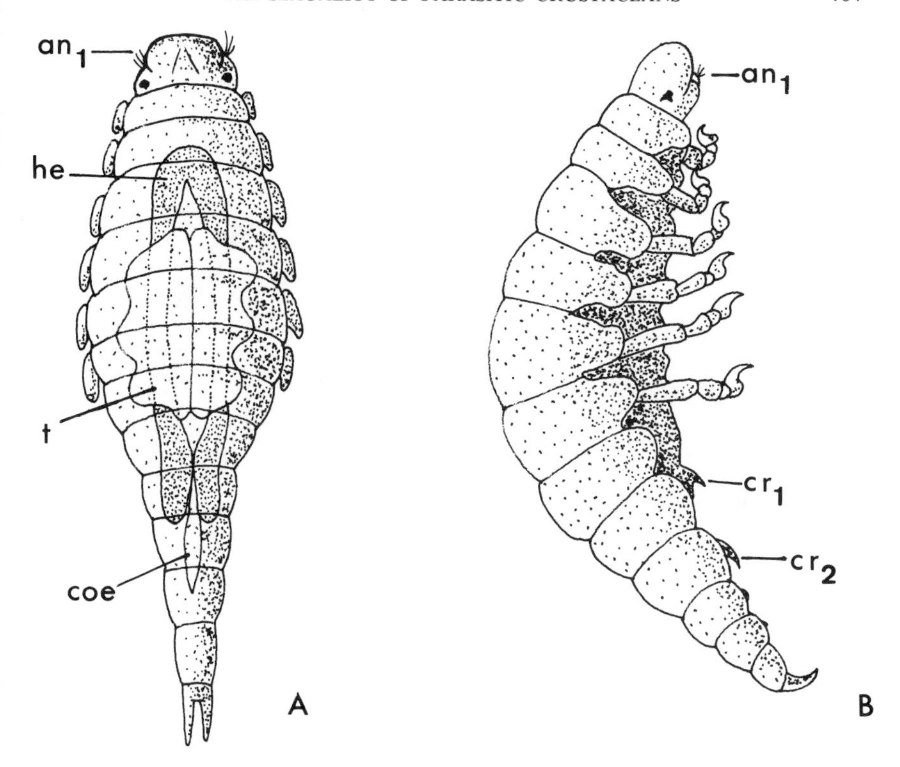

FIG. 33. (A) *Portunion maenadis*, male, dorsal view (an_1, 1st antenna; coe, heart; he, hepatopancreatic coeca; t, testis). (B) *Portunion kossmanni*, male, dorsal view (an_1, 1st antenna; cr_1, cr_2, ventral hooks). (Redrawn by J. P. Trilles from Giard and Bonnier, 1887.)

FIG. 32. (A) (B) (C) and (D) *Portunion maenadis* Giard. Development of a young female into a gravid female, showing the extraordinary development of the oostegites and transformation of the pleopods into respiratory organs. (A) very young female, arched ventrally. (B) Older female whose incubatory cavity has not yet closed and that has begun to arch dorsally. (C) Adult female before the first spawning with an incubatory cavity that is partially open along the medio-ventral line and the abdomen of which has been turned to expose the underside. (D) Adult female post spawning. Scales, actual sizes. (E) *Portunion kossmanni*. Adult female with incubatory cavity full of eggs. a, pleural plate of the first abdominal segment; bl, lateral hump; bv, ventral humps; ce, cephalogaster; ep, pleural plates of the first abdominal segments; he, hepatic coecum; le, 2nd antenna; li, 1st antenna; p, pleon; pl, pleopod; pm, maxillipeds; sc, cephalic sac or anterior hood of the incubatory cavity; sl, lateral sac of this same cavity; 1–5, first to fifth pairs of incubatory plates. (Compiled and redrawn by J. P. Trilles from Giard and Bonnier, 1887.)

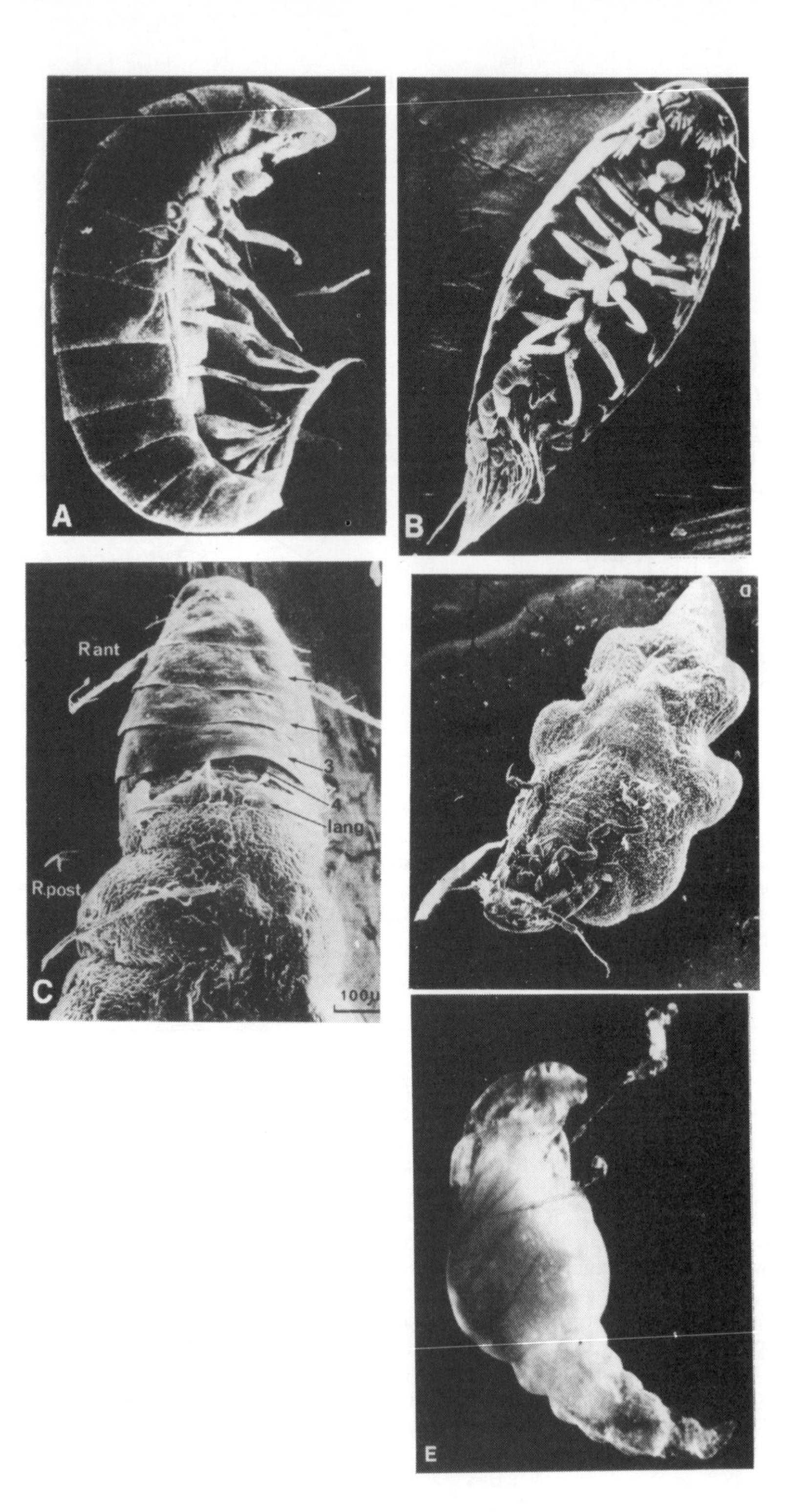

A
B
R ant
3
4
lang
R.post
C
100μ
E

In the Cymothoidae, sexual dimorphism can be observed, although it is little pronounced; the male and the female have both conserved the appearance of a typical isopod, and differ only in their degree of femininity (individuals in the female phase are relatively larger) that encompasses the morphology of certain appendages, or simply their presence (appendix masculinae in males, oostegites in females, form of the maxillipeds, chaetotaxy etc.). In certain cases, the female phase may be recognized by the asymmetry of corresponding individuals (for example, in the branchial species); there is sometimes a dimorphism in the pigmentation, certain males being highly pigmented when the corresponding females are whitish and very pale (Trilles, 1968b, 1969).

The majority of female copepod parasites are more highly modified than the respective males, the latter conserving a primitive aspect owing to the fact that they are frequently less "specialized" than their partners and are usually more mobile. Thus, for example, female Philichthyidae are characterized by a highly modified body, whereas the males show a clear segmentation and have swimming appendages (Fig. 31). Sexual dimorphism may be limited to certain parts of the body, or to certain appendages. In the first case this often concerns the genital segment, which is relatively more highly developed in the female owing to the presence of numerous eggs being evacuated into the ovigerous sacs. The appendages that are most often affected by sexual dimorphism are the antennae and the maxillipeds. In male Caligidae, these appendages are more bulky and complex than in the females. The males of certain species, such as *Scambicornus poculiferus* and *Aspidomolgus stoichactinus*, which are parasites of a sea cucumber of the genus *Synapta* and a sea anemone of the genus *Actinia* respectively, have antennae armed with suckers that are lacking in the females (Humes, 1969). More rarely, the sexual dimorphism may concern certain thoracic appendages, i.e. P1, P2 and P5. Finally, in certain cases a particular appendage may be present in only one of the sexes. Thus, the males of *Ergasilus lizae* possess

---

FIG. 34. (A) *Hemioniscus balani*, male stage, lateral view. (B) *Hemioniscus balani*, male stage, ventral view. (C) *Hemioniscus balani*, antero-dorsal view of an individual that has just undergone the sexual inversion half moult (lang, cuticular strip corresponding to the posterior part of the 4th free thoracic tergite, in the process of moulting; 1, 2 and 3, 1st, 2nd and 3rd free thoracic tergites; 4, reduced 4th free thoracic tergite; R ant, anterior region of the young female; R post, posterior region of the young female). (D) *Hemioniscus balani*, female, ventral view. (E) *Crinoniscus equitans*, female after a half meal (12–16 h after the sexual inversion moult), artificially removed from the ovigerous plate of *Balanus*, to which were attached the "appendages" of its second thoracic segment. (A, B, C and D from Goudeau, 1976, 1977; E from Bocquet-Védrine and Bocquet, 1972b.)

a pair of maxillipeds that are lacking in the females. It is, moreover, with the help of these appendages that the males seize the females to form couples in precopulatory attitude (Ben Hassine, 1983) (Fig. 12).

Sexual dimorphism can reach exceptional limits in the epicarids and rhizocephalans. Sexual dimorphism is already very marked in the Bopyrina Entoniscidae (Figs 32, 33). The females are deeply modified, manifestly as a result of the characteristics of these endoparasites that are sited in the visceral cavity of brachyurans or anomurans and are enveloped in a membrane that is of host origin.

But this is nothing in comparison to the situation observed in the majority of Cryptoniscina, where individuals in the female sexual phase are enormously regressed and often transformed into simple embryo-filled sacs that die when the embryos are released. They have most disconcerting forms that often bear no resemblance to isopods, or even Crustacea. It is often difficult, and even hazardous, to describe them accurately.

These females, that are often highly deformed, nevertheless conserve certain more or less important attributes of the cryptoniscian males, which serve in particular in the anchorage to the host.

In *Hemioniscus balani*, the female is divided into two remarkably disparate regions, the dissimilarity of which becomes more and more accentuated throughout its lifetime: the anterior part is similar to the corresponding region of the male; the posterior region, however, loses all traces of segmentation and appendages and undergoes a progressive hypertrophy (Fig. 34). The unchanged anterior region comes to resemble a minute tubercle placed on a multilobed mass; this region has not moulted and includes the cephalon and nearly all of the pereionites I to IV. On the contrary, two ring systems allow the posterior region to continue moulting, thus leading to the distal hypertrophy. This system permits the parasite to remain permanently anchored to the host tegument, facilitating continuous nutritional intake. The incubatory pocket only invades the posterior region (Goudeau, 1977).

In *Crinoniscus* a highly characteristic anchorage ring results from a very unusual sexual inversion moult, since the casting off of the anterior cuticle (cephalon and the first four pereionites) precedes that of the posterior region (in general, it is the inverse in isopods). The resulting young female, which has conserved a normal metamerization, has lost her pereiopods and pleopods. A composite formation remains, however, that is composed of pereiopod II of the male and a flexible cord inserted on pereionite II. This apparatus is employed by *Crinoniscus* to hold on to the barnacle host while it lacerates the latter's eggs on which it feeds. Whilst feeding, the female swells, almost visibly, and the thoracic metamerization quickly fades away. The replete animal takes on a globular form, due mainly to the dilation of

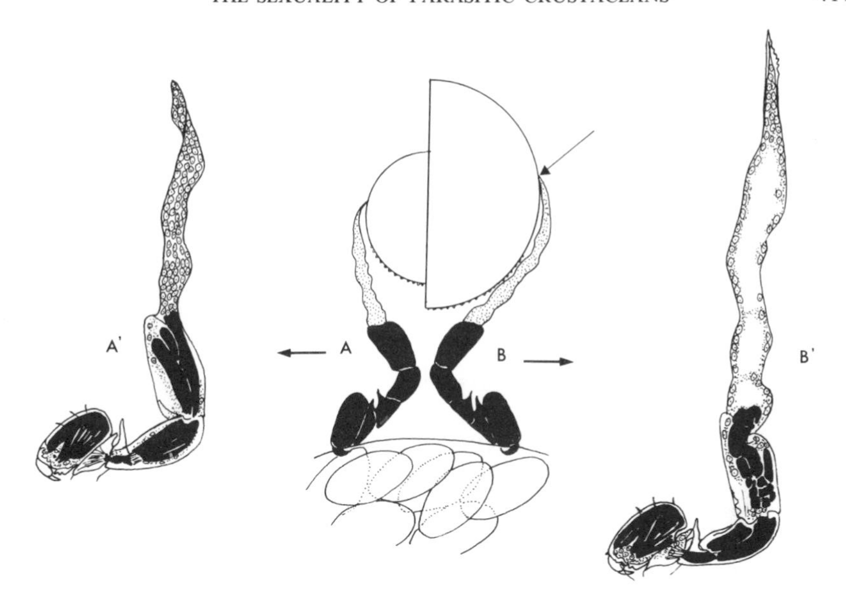

FIG. 35. Diagrammatic half sections of the second free thoracic segment of female *Crinoniscus equitans* attached to the ovigerous plate of *Balanus*. (A) Very young female. (B) Female after a half meal. The arrow in (B) indicates the position at which the cord will break at the end of the repletion meal. "Appendages" of the second thoracic segment of *Crinoniscus equitans*—A′, very young female, just after the sexual inversion moult; B′, female having taken approximately half a meal. (Compiled and redrawn by J. P. Trilles from Bocquet-Védrine and Bocquet, 1972b.)

pereionites III, IV and V. Following this, and during another highly unusual moult, the female undergoes another spectacular transformation to take on the definitive and typical fleur-de-lis form (Bocquet-Védrine and Bocquet, 1972a,b; Bocquet-Védrine, 1976) (Figs 34, 35).

In other cases, there exist neoformed anchorage devices that, in the females, constitute the only differentiated elements in otherwise excessively degraded organisms. They consist of trunk-like organs or peduncles that are derived from the anterior region of the parasites. *Lernaeocera* (copepod) (Fig. 36) and *Danalia* (isopod) (Fig. 37) are amongst the most representative examples (Altes, 1962; Kabata, 1979).

In certain other Cryptoniscina, such as *Ancyroniscus* (Fig. 38), it is the posterior region of the parasite that is intraviscerally internalized in the host, and as such constitutes a simple anchorage device for an organism that is itself extraordinarily simplified. The female that has laid eggs occupies the totality of the incubatory chamber of the host *Dynamene* and consists of a transparent sac full of eggs and embryos. This organism is lobed: four large, finger-shaped lobes are intravisceral and four others are intramarsupial. A

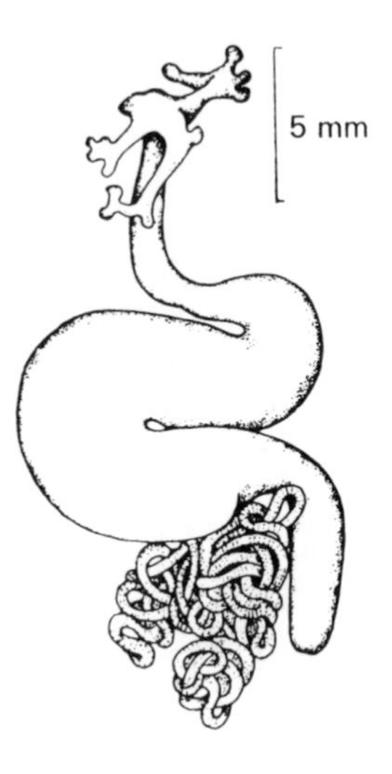

FIG. 36. Adult female *Lernaeocera branchialis*, a blood-feeding, pennellid copepod parasite of the branchial cavity of gadid fishes. (From Kabata, 1979.)

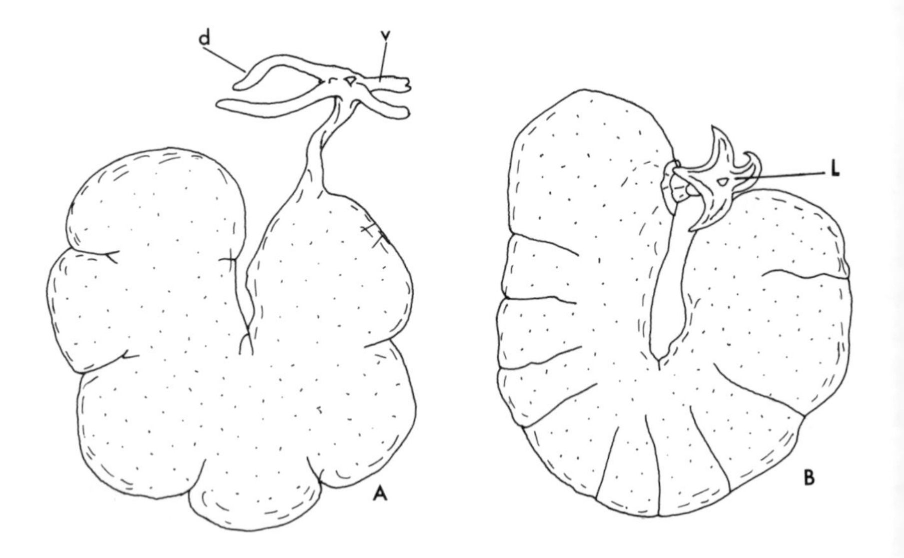

FIG. 37. (A) Adult female *Danalia curvata* (liriopsid isopod). d, v, dorsal and ventral lobes of the neck. (From Caullery, 1907 and Altes, 1981.) (B) Adult female *Danalia gypsilon*. L, lobes of the neck. (From Wimpenny, 1927 and Altes, 1981.)

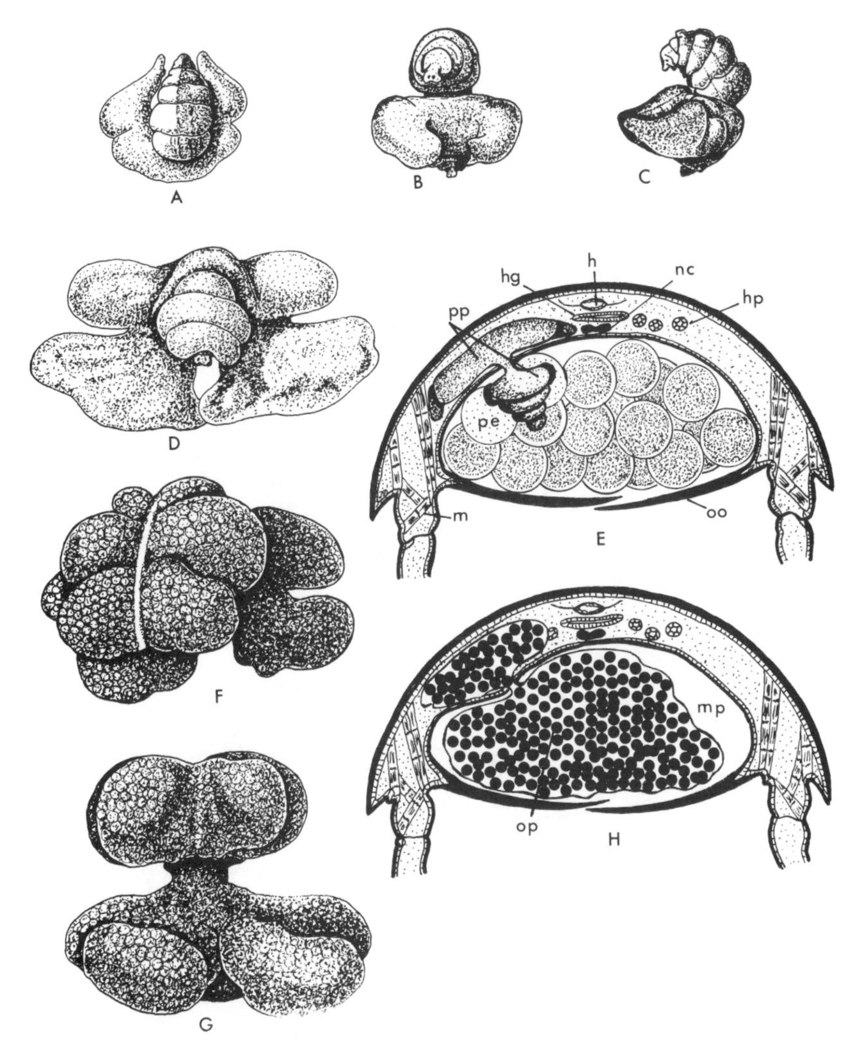

Fig. 38. (A) (B) and (C) Young female *Ancyroniscus bonnieri*. (A) dorsal view. (B) Ventral view. (C) lateral view. (From Caullery and Mesnil, 1920.) (D) Female *Ancyroniscus bonnieri* at an advanced stage, post spawning, viewed from above. (from Caullery and Mesnil, 1920.) (E) Diagrammatic representation of an ovigerous female *Dynamene bidentata* (Adams) in transverse section showing a pre-ovigerous stage of *Ancyroniscus bonnieri* with female parasite partially situated in the host's marsupium and absorbing the brood. h, heart; hg, hindgut; hp, hepatopancreatic caeca; m, muscle; nc, nerve cord; oo, oostegite; pe, partially absorbed egg; pp, preovigerous parasite. (From Holdich, 1975.) (Compiled and redrawn by J. P. Trilles.) (F) and (G) Female *Ancyroniscus bonnieri* post-spawning. (F) Viewed from above, the upper surface of the figure is that which is apparent when the parasite is viewed *in situ* on the host. (G) Figure demonstrating the two parts of the parasite, internal and external, and the isthmus that joins them. (From Caullery and Mesnil, 1920.) (H) Diagrammatic representation of an ovigerous-stage female *Ancyroniscus bonnieri* parasitizing *Dynamene bidentata*, in transverse section. mp, marsupium of host; op, ovigerous parasite. (From Holdich, 1975.)

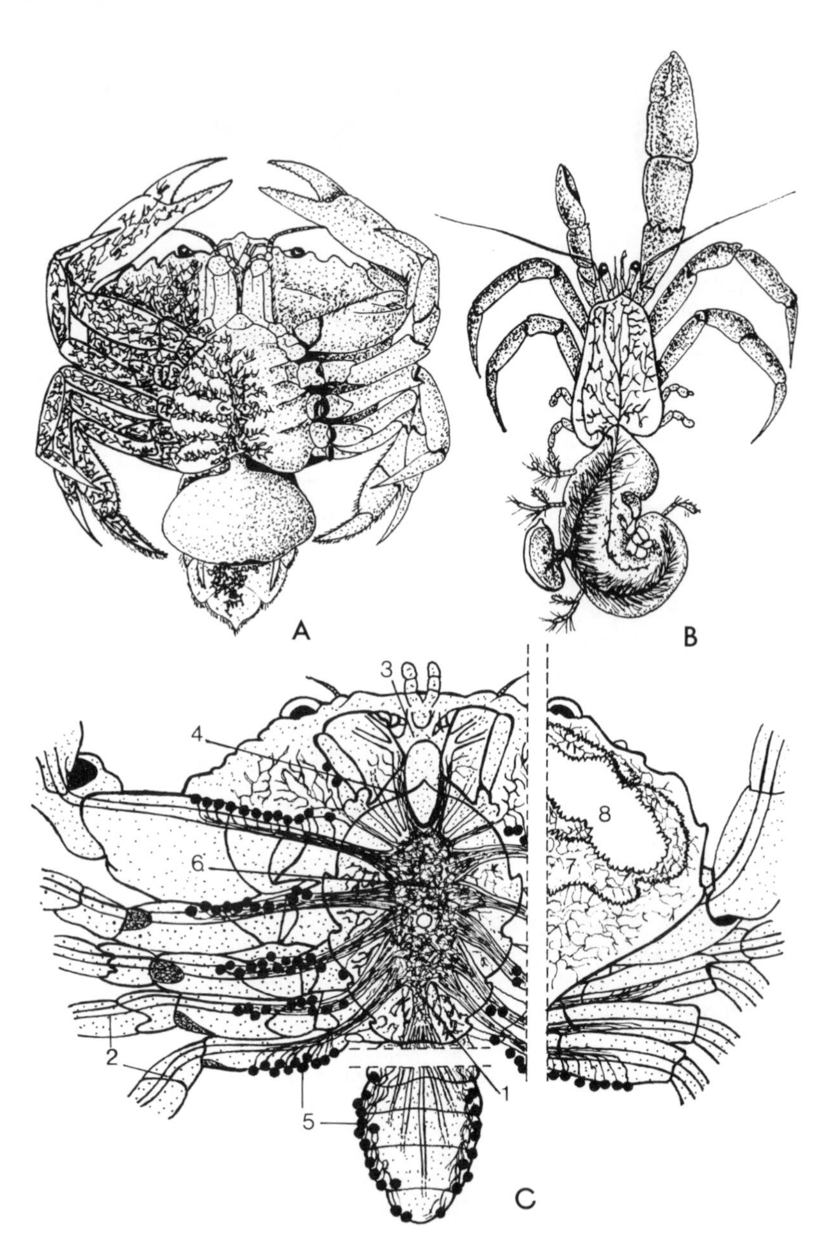

FIG. 39. (A) Crab parasitized by *Sacculina*; the root system is shown on the left-hand side of the figure. (From Boas in Caullery, 1950.) (B) *Peltogaster paguri* Rathke; the internal root system is shown in the pagurid host. (From Perez, 1932 in Baer, 1946.) (C) Distribution of the root system of *Thompsonia littoralis* in a female of the common littoral crab, *Leptodius exaratus*; ventral and (abdomen and right part of illustration) dorsal aspect. 1, abdominal nerve; 2, ambulatory nerve; 3, cerebral ganglion; 4, oesophageal connective; 5, externae; 6, fused thoracic ganglia; 7, hepatopancreas; 8, ovary. (From Lützen and Jespersen, 1992.) (Compiled and redrawn by J. P. Trilles.)

more or less narrow isthmus joins the two parts and none of the organs specific to the cephalic region subsist (Caullery and Mesnil, 1919b; Holdich, 1975).

It is well known that in the rhizocephalans also, for example in *Sacculina* and *Peltogaster*, the regression of the female body is extraordinary and of such magnitude that it is only via their larval forms that it is possible to classify them as crustaceans, let alone as cirripedes. During its development, the female parasite acquires a very simple, highly regressed organization consisting of an internalized, absorbent root system (interna) serving its nutritional requirements, and an external reproductive system (externa), that is sometimes multiple, e.g. *Thompsonia* (Figs 39, 40). *Chthamalophilus*, however, is unique among the Rhizocephala in that the root system is replaced by a slightly lobed bladder-like structure, which is alleged to be enveloped by the infolded and, at least partly, continuous epidermis of the barnacle host (Bocquet-Védrine, 1961; Hoeg, 1992) (Fig. 41).

It is also worth recalling the Ascothoracica, which are less well known, but possess none the less a considerable sexual dimorphism and a fundamental, prodigiously deformed type.

C. BEHAVIOUR

A sexual dimorphism is often observed in behaviour. In certain gnathiids, such as *Paragnathia formica*, the male praniza appear more resistant than the female praniza to the high temperatures in summer. This would explain the fact that male metamorphoses may occur throughout the year, whereas those of the female are seasonal (Stoll, 1962; Charmantier, 1980, 1982). In the Cymothoidae, it is sometimes observed that the female has a greater specificity than the male, that is often accompanied by the two sexes having different positions on the host fish; in a number of species, and especially the branchial and buccal species, the position on the host of individuals in female phase is highly characteristic, whereas that of the males is much more variable (Trilles, 1968b, 1969) In a certain fashion this is what is also observed in the Epicarida. Although the larval male of the *Cryptoniscina* is often encountered on, or in the vicinity of, the female (*Ancyroniscus*, *Clypeoniscus*, *Cabirops*, etc), he does not take up residence. He remains vagrant and may pass from one female to another, as is easily observed in *Hemioniscus*, for example. As with the Cymothoidae, it would appear the male *Clypeoniscus* may sometimes remain on its host and become female when its partner dies (Sheader, 1977).

Within the framework of the sexual dimorphism of behaviour should be included the phenomenon of chemotaxy, involving the secretion of sexual

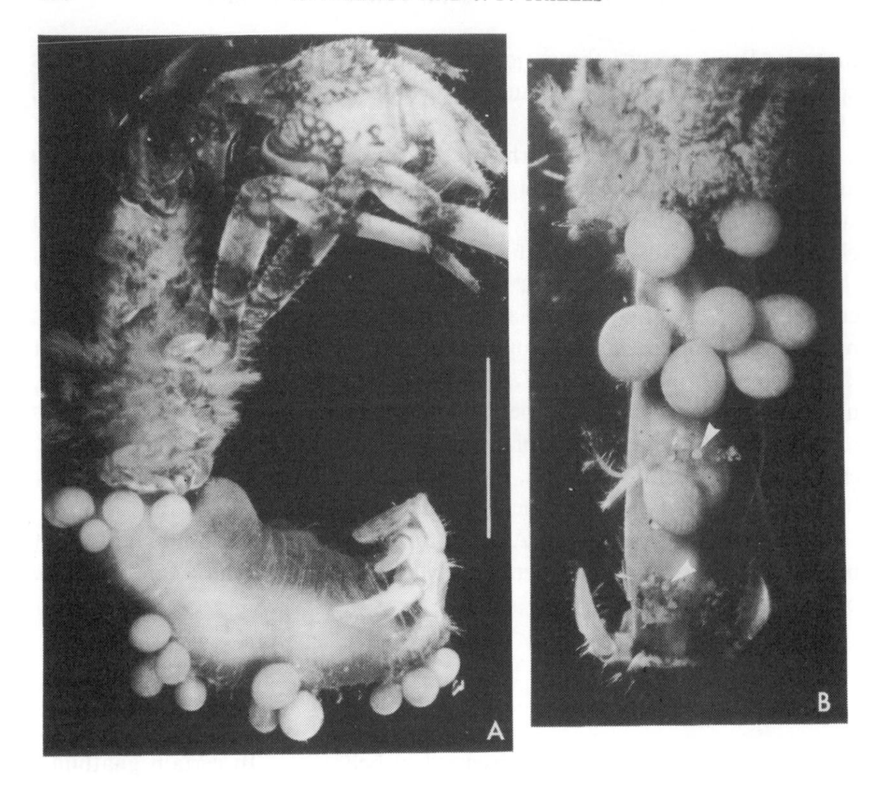

FIG. 40. *Thompsonia reinhardi* on its host, *Discorsopagurus schmitti*. (A) Lateral view of host with several externae. (B) Dorsal view of host with 8 externae and several scars (2 of which are indicated by arrowheads. Scale represents 3 mm. (From Lützen, 1992.)

pheromones by the females in order to attract males. This phenomenon is known in several aquatic crustaceans (Meusy and Payen, 1988). Such attraction of males by females has been described for a number of parasitic species belonging to a variety of groups (Cymothoidae, Epicarida, Rhizocephala, Copepoda). In the Epicarida, this attraction would also appear to be effective as it is rare to find an adult female that is totally devoid of males. In certain species, the absence of males has been noted in autumn. Along with an attractive substance emitted by the host, there must exist a special factor produced by the females. In the entoniscians, this substance must be produced throughout the life of the female; indeed the latter is accompanied by males of all sizes and ages (Veillet, 1945). This is also what certainly occurs in the Dajidae and the Athelginae, but is more exceptional amongst the Bopyridae. It would seem on the other hand that the male-attracting

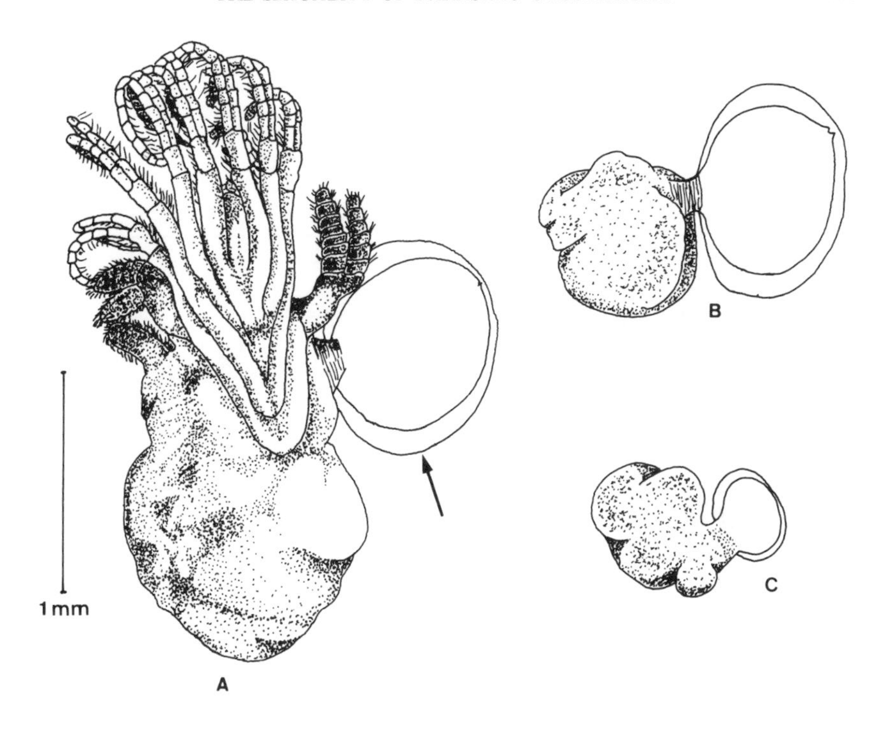

FIG. 41. (A) Adult *Chthamalophilus delagei* on its host, *Chthamalus stellatus* (Poli). The mantle of the barnacle host has been removed. (B) and (C) *Chthamalophilus delagei*, at two stages of the development of the external sac, completely removed from their hosts. In both drawings, the anchor region is shaded. Each individual possesses a double cuticular layer around the external sac. (Compiled and redrawn by J. P. Trilles from Bocquet-Védrine, 1961.)

substance is secreted, or becomes operative, only at a given stage of the development of the female, this stage varying from one species to another in the Bopyridae. The attraction phenomenon associated with the females is also known to occur in parasitic copepods. Thus, Caillet and Raibaut (1979) have demonstrated experimentally that larval males are strongly attracted to a young female *Alella macrotrachelus* (Lernaeopodidae) (Fig. 42). In a number of species of free-living (mainly harpacticoid) and parasitic copepods, the male, once having localized a female, grasps the latter for an extended period of precopulatory mate guarding. This phase has been described for a number of copepod groups (Boxshall, 1990) and terminates by the deposition of two spermatophores, following a postural change of the two partners that take up an *in copula* position. The modalities of this sexual behaviour have been remarkably well studied by Anstensrud (1989a: 1990a,b,c; 1992) in two species of fish parasite, namely *Lernaeocera bran-*

chialis and *Lepeophtheirus pectoralis*, both in experimental conditions, with the interesting technique of individual marking with vital stains (Anstensrud, 1989b), and in natural conditions.

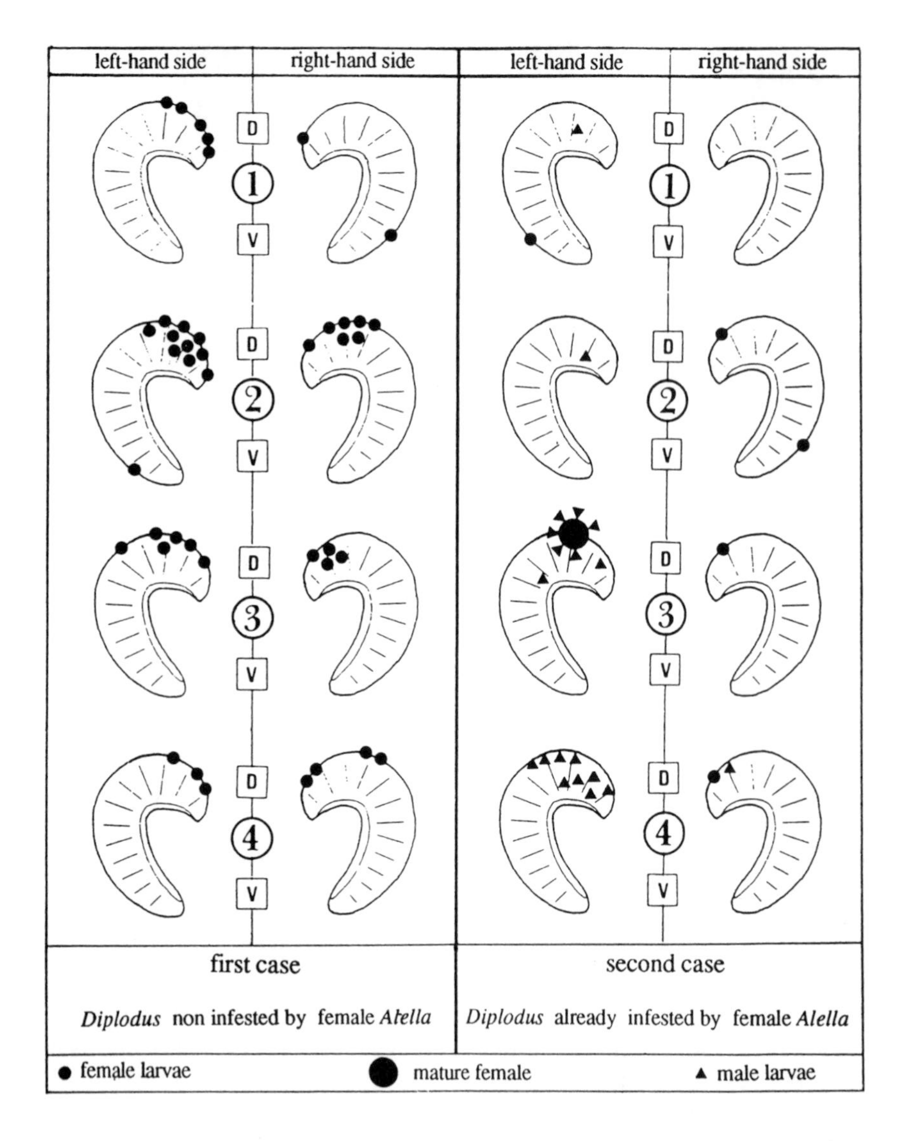

FIG. 42. Experimental infestation of white bream *Diplodus sargus* (Bony fish, Sparidae) by infective copepodites of the lernaeopodid *Alella macrotrachelus*. (From Caillet and Raibaut, 1979.)

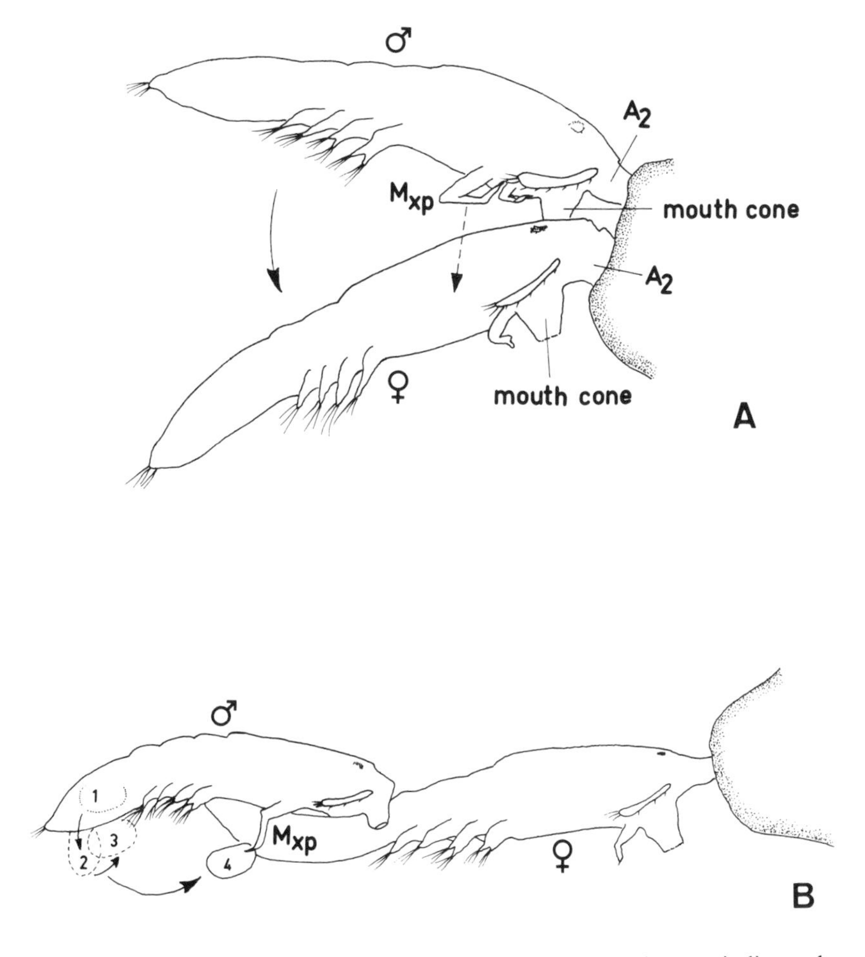

FIG. 43. (A) *Lernaeocera branchialis* in precopula position. Arrows indicate the male's reaction towards other males. (B) *Lernaeocera branchialis* in copula position. 1 spermatophore still inside the male's genital complex; 2, 3, spermatophore at different stages of expulsion; 4, spermatophore attached to the female's vulva; A_2, second antennae; M_{xp}, maxilliped. (Redrawn by N. Le Brun from Anstensrud, 1989a.)

The life cycle of *Lernaeocera branchialis* is heteroxenic. Gravid females are attached to the gill cavity of the final host fish, a gadid species such as cod (*Gadus morhua* (L.)) and whiting (*Merlangius merlangus* (L.)) (Kabata, 1979). The larva hatches as a nauplius which, after a short period of pelagic life, and following a moult, becomes an infective copepodite (Kabata, 1958). A second nauplius stage sometimes occurs (Capart, 1948; Sproston, 1942;

Anstensrud, 1989a). The copepodite attaches itself to a first host, a flounder (*Platichthys flessus* (L.)), and moults into a chalimus larva which is attached by a frontal filament to the gill of its host. After four successive chalimus stages, the larvae develop into either adult males or young premetamorphic females. Copulation occurs on the flatfish host that cannot, therefore, be referred to as an intermediate host, but rather as a fertilization host (Tirard and Raibaut, 1989). The premetamorphic females, whether they are fertilized or not, leave their flatfish host and search for a gadid final host.

Once a male localizes a female, still attached to the host by the frontal filament (chalimus IV), he adopts a precopulatory position. This comprises his attachment, by his second antennae, to, or near, the female's frontal filament or second antennae (Fig. 43A). When a male in precopula position is disturbed by other males, it responds by holding on to the female chalimus larva with the maxillipeds (Fig. 43A). This reaction is accompanied by rapid strokes with the swimming legs which make the male vibrate. This reaction, however, does not prevent other males from forming clusters around precopulating pairs. Immediately after female ecdysis, the male crawls backwards on the dorsal side of the female using its second antennae, and the maxillipeds are moved back and forth over the ventral surface of the female. This backwards movement is continuous, without any obvious interruptions, until the male reaches its final position and the second antennae grasp the female's trunk just behind the last leg-bearing thoracic segment (Fig. 43B, 44). Two spermatophores are simultaneously transferred between the mature male and the premetamorphic female in copulatory position. These spermatophores are expelled after a series of contractions of the male's genital complex (Fig. 43B). The males are highly polygynous. Anstensrud (1992) has demonstrated that the adult male *Lernaeocera branchialis* can take up a precopulatory position with all of the larval stages attached to the flatfish host, but there is a marked preference for the female chalimus IV or the premetamorphic virgin females. In addition, the males will not generally attach themselves to females that have already been fertilized, and even prefer a copepodite to a female chalimus IV larva that is occupied by another male. This would appear to indicate a change in the pheromone signals according to the sexual state of the chalimus IV or premetamorphic females (virgin, coupled, fertilized) culminating in a repulsive signal associated with the mass of sperm in the *receptaculum seminis* of the fertilized female.

Lepeophtheirus pectoralis, a parasite of the flounder, has a different mating strategy (Anstensrud, 1990a). During precopulation, the male clasps the female, just in front of the genital segment, with his second antennae, sometimes assisted by his maxillipeds (Fig. 45A). This is comparable to the precopulatory mate of *Ergasilus lizae* (Fig. 12) described by Ben Hassine

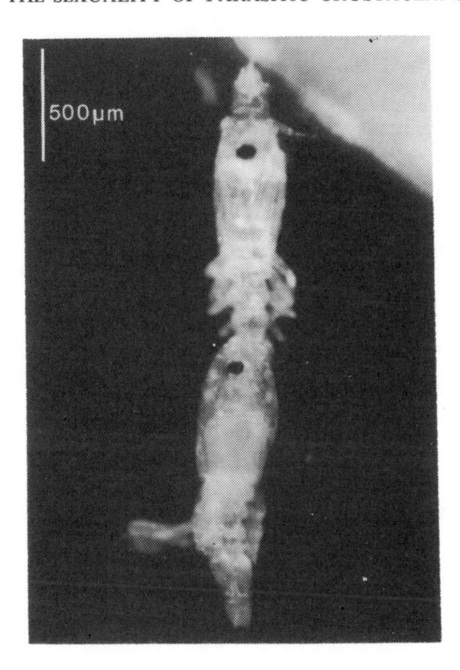

FIG. 44. *Lernaeocera lusci* in copula position. (Photograph A. Raibaut and C. Tirard.)

(1983). When the female *Lepeophtheirus pectoralis* moults to become an adult, the male rotates around the female to position himself ventrally (Fig. 45B). Copulation then occurs. The male moves his maxillipeds from the orifice of the *receptaculum seminis* towards the anterior part of the female's genital complex, thus bending the complex up and away from himself (Fig. 45C). In this position, the male bends his own genital complex towards the long setae of the second swimming legs and two spermatophores are simultaneously expelled (Fig. 45D). The males copulate only once with each female. Following the deposition of the spermatophores, a plug obstructs the vulvae of the female for about 1 month. Here again, the male distinguishes the sexual state of the female and it is probable that chemical cues are initially responsible for the male's discrimination between preadult and adult females. It is interesting to note, however, that males discriminate virgin females with an expanded genital complex, even though spinsters have been shown to be receptive and fertile (Anstensrud, 1990c). The discrimination against spinsters suggests that the size (or shape) of the genital complex, and thus mechanoreception, is of crucial importance for mate recognition by male *Lepeophtheirus pectoralis* (Anstensrud, 1992).

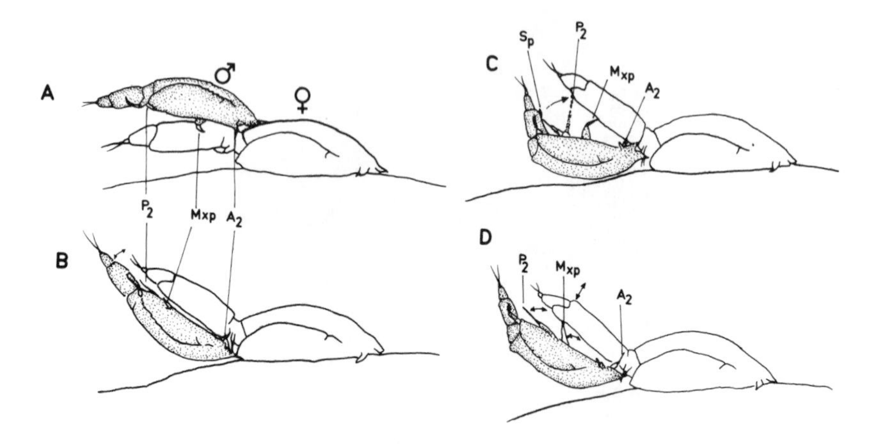

FIG. 45. Mating positions of *Lepeophtheirus pectoralis*. Maxillae, first and third swimming legs are omitted from the drawings. (A) Precopula position. (B)–(D) Copula position. M_{xp}, maxilliped; A_2, second antennae; P_2, second swimming leg; Sp, spermatophores. (Redrawn by N. Le Brun from Anstensrud, 1990a.)

D. EFFECTS ON THE HOST

A parasitic crustacean feeds despoiling its host, lysing its tissues, drawing its blood, blocking its reproduction, etc ... and therefore by inducing a number of disorders of varying gravity. The overall burden will obviously depend upon the number of parasites on each host, but also on the sex of the parasite(s). It is evident that in some cases, such as infestation by cymothoadians, the two sexes affect the host in the same way, although this effect is slightly greater for the females (Romestand and Trilles, 1975, 1976; Romestand *et al.*, 1977; Romestand, 1979). In other parasitic associations the effect of the female is very great while that of the males is negligible. This difference, when it exists, is obviously related to a difference in feeding behaviour. In the Epicarida, the feeding female induces metabolic and haematological effects in the host, and has an impact on its growth, moulting, gonads, secondary sexual characters, egg laying and fecundity, etc. In *Sacculina*, it is of no surprise that it is the female that has an action on the endocrine glands and nervous system of the host (Rubiliani, 1984), and that also induces a variable decrease in the total number of hyaline and granular haemocytes and often perturbs the host's protein metabolism (Sanviti *et al.*, 1981). Equally, it is only the female *Peltogaster* that induces a decrease in the lipid content of pagurid crabs of either sex. It is well known that the female *Sacculina*, once the visceral sac is externalized, inhibits the moulting of its

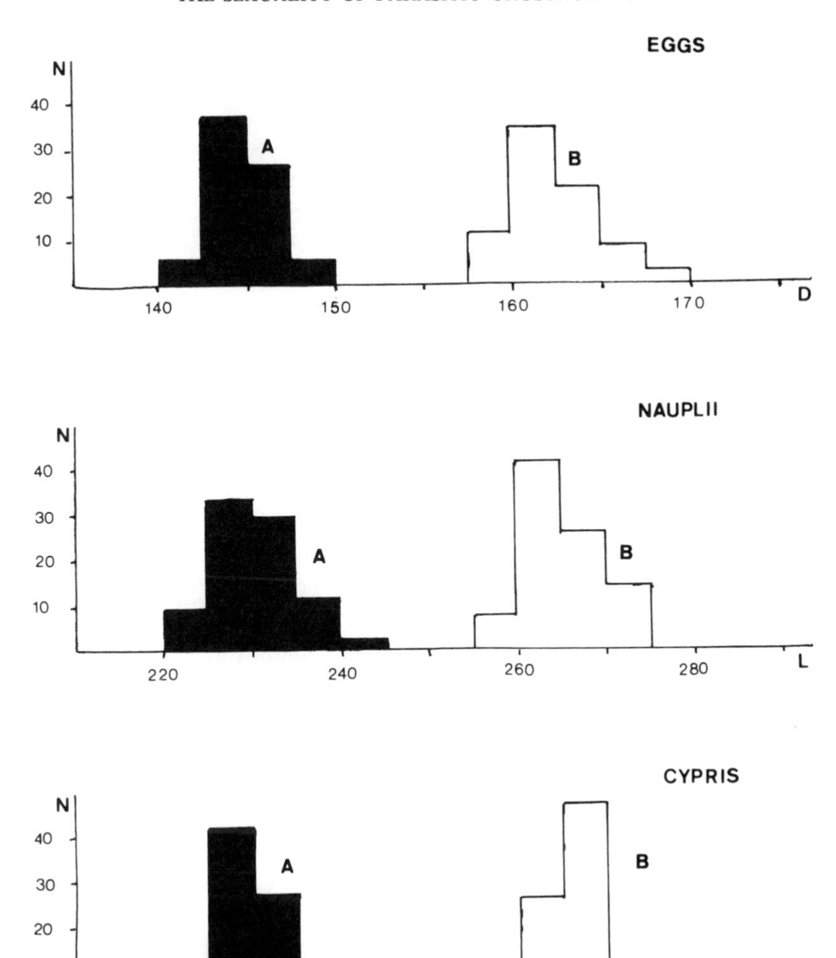

FIG. 46. Range in size of eggs, nauplii and cypris derived from two different females, A and B, of *Peltogasterella gracilis*. N, number; D, diameter in μm; L, length in μm. (From Yanagimachi, 1961b, modified.)

host. This is also the case with other parasites, such as *Aparobopyrus*, which diminishes the moulting frequency and the growth of its host *Pachycheles* (Van Wyck, 1982). As for the effects of female parasites on the gonads and secondary sexual characters of the host, there are a great number of examples, especially amongst the Epicarida and Rhizocephala.

The local or general pathological effects of parasitic copepods on their hosts are, in most cases, caused by the female parasites on account of the considerable energy requirements of reproduction. In a number of cases most adaptive modifications in the male converge to leave it the unique and indispensible rôle of insemination and a minimal trophic requirement. There is of course the phenomenon of cryptogonochorism. Can one really speak of nutrition for a male that is reduced to a testis? There is also the possibility of mating before the parasitic life that concerns only the female. Finally, there is the extraordinary appearance of heteroxeny in the pennellid copepods that include certain blood-sucking mesoparasites that inflict severe wounds on their hosts. A unique occurrence in the world of parasites is that this heteroxeny only affects one sex, the females. The males remain on a single host (invertebrate or fish), their food intake being reduced (Tirard, 1991).

IV. Sexual Determinism

In the parasitic crustaceans, as in other members of the class and indeed all multicellular organisms, the genetic determination of sex is syngamic, resulting from a mechanism comprising an equilibrium between male and female genes. The great majority of crustaceans possess an XX/XY or ZZ/ZW system. In a certain number of cases, however, it may be concluded that there exists a polyfactorial (polygenic) system. Though studies of the genetic determinism of sex are rare for parasitic crustaceans, it might be supposed that it is similar to the free-living forms. It should also be added that in addition to mechanisms of genetic determinism of sex, the crustaceans demonstrate very clear phenomena of epigenetic control of sexual differentiation. Among these, two major classes may be distinguished, depending on whether the epigenetic intervention is only at the phenotype level or directly affecting the sexual differention *per se*. As with the other invertebrates, in the crustaceans, such variation in the expression of genes can be correlated with certain specific life styles, as is the case of parasitism.

Although they are now relatively old, a certain number of studies have attempted to produce caryotypes of representatives of several orders of crustaceans. The generally large number of chromosomes, as well as their small size, have often hampered these studies and the results are often disappointing. A certain number of results, however, have confirmed the presence of heterochromosomes in some species and their almost certain absence in others.

Amongst the parasitic crustaceans, sexual chromosomes have only been observed in two species. In *Mytilicola intestinalis*, a copepod parasite of *Mytilus*, Orlando (1973) has demonstrated a heterogametic ZW determinism of the female sex (the males are in consequence homogametic ZZ). In the rhizocephalan *Peltogasterella gracilis* (= P. *socialis*), a parasite of *Pagurus lanuginosus*, Yanagimachi (1961b) has also demonstrated a phenomenon appearing to involve the existence of sexual chromosomes. As discussed above, in this species the author has distinguished, by their respective sizes, two types of eggs, nauplii and cypris larvae (Fig. 46), the latter also displaying certain structural differences, involving in particular the antennules. Each of these two types is produced separately by a distinct category of female that are themselves morphologically undistinguishable. The smaller larvae develop into females, and the larger ones function as males. On the metaphase plate of the first division in the maturation of the large, male-producing eggs, Yanagimachi (1961b) has demonstrated the presence of 15 bivalent chromosomes, whilst in the smaller, female-producing eggs, there are 15 bivalent chromosomes plus a small univalent chromosome designated X (Fig. 47). Thus, in this species there is a syngamic genetic determinism of sex with the following chromosome complement: ♀ 30A + X and ♂ 30A + 0. This does not mean that certain females cannot exceptionally produce both males and females. In the case where a female produces numerous females and a few males, the phenomenon might be explained by the occasional loss of the X chromosome during the meiotic anaphase. The case of females that produce numerous males and few females, which have been demonstrated to still have the 30A + 0-type caryotype, is far more difficult to explain. Bresciani and Lützen (1962) have studied a species of Cyclopoid, *Thespesiopsyllus paradoxus*, a gut parasite of the brittle star, *Ophiopholis aculeata*. They noted the presence of two types of nauplius in the stomach of the host. One type is large, green in colour and develops into a female copepodite. The Scandinavian authors consider that the others, that are small and pinkish in colour, become male copepodites. In a study of the infestation of white bream (a teleostean fish of the family Sparidae) by the lernaeopodid *Alella macrotrachelus*, Caillet and Raibaut (1979) consider that the infesting free-swimming copepodites have separate sexes, despite the fact that they are morphologically identical. As soon as the copepodite anchors itself to the fish by a frontal fixation filament, the future chalimus stage can be seen by transparence, the latter having a distinct sexual morphological differentiation (pupal stage). In addition, on a single gill, it is possible to observe a male and a female chalimus side by side (Fig. 42).

In certain parasitic crustaceans, therefore, there is no doubt about the genetic determination of sex, as the existence of well defined sexual chromosomes has been demonstrated. There is, today, for example, unequivocal evidence that the male and female cypris larvae of *Peltogasterella* are genetically determined.

Nevertheless, in certain species of parasitic crustaceans there appears to

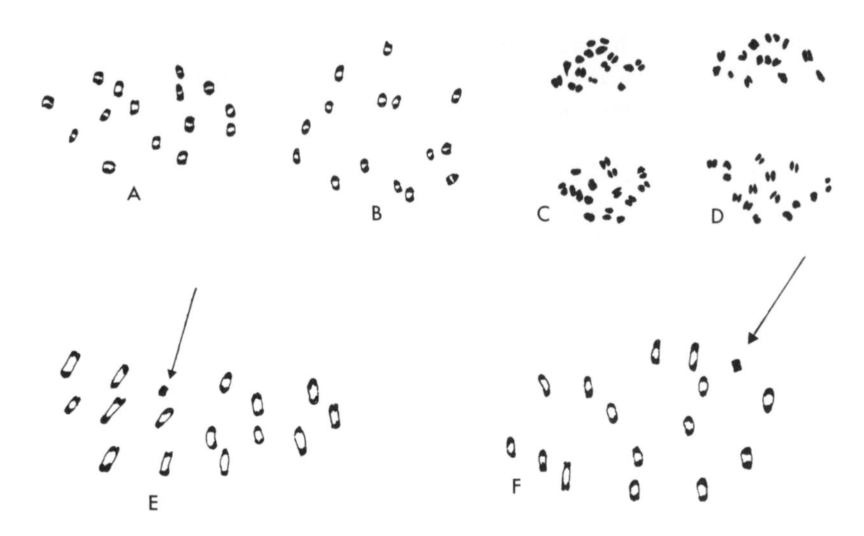

FIG. 47. Chromosomes of the first maturation division of the large and small eggs of *Peltogasterella gracilis*. (Magnification: × 1500. Camera lucida drawings.) (A)–(B) Chromosomes of the first maturation division metaphase of the large male-producing egg. (C)–(D) Chromosomes of the first maturation division anaphase of the large male-producing egg. (E)–(F) Chromosomes of the first maturation division metaphase of the small female-producing egg. A univalent chromosome (indicated by an arrow) is found beside 15 bivalent chromosomes. (Redrawn by Trilles J.-P. from Yanagimachi, 1961b, modified.)

exist a polyfactorial determinism of sex. It is, doubtless, in this latter case that the expression of genes might be modulated by epigenetic factors present in the parasite's environment (host or host's environment). Here may lie the explanation of the establishment of phenotypic sexual determination in certain copepod parasites of ascidians, for example, or the sexual determinism in certain cases of labile gonochorism (such as in certain epicarids) or of certain forms of hermaphrodism, such as that of the cymothoadians.

The copepod *Pachypygus gibber* is an excellent example of such an epigenetic influence on the differentiation of the sexual phenotype, as in this trimorphic species there exist typical females, typical males and atypical males (Figs 1, 2, 3). The morphology, the life cycle and the sexuality of individuals of the three categories appear to be closely linked to the physiology of the host *Ciona* (Hipeau-Jacquotte, 1978a,b,c, 1980, 1984, 1988). The atypical males are only found in very young specimens of the host, where females are absent, whereas the typical males develop in older *Ciona* in the presence of females (Figs 48, 49). But this action does not concern the reproductive system, as it has recently been shown that, in both

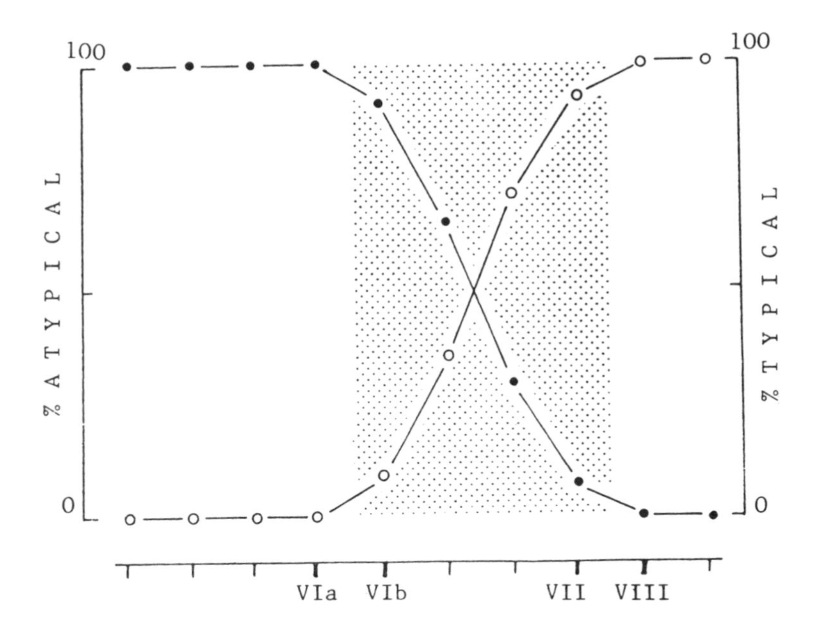

FIG. 48. Relationship between the relative percentage of the type of 3rd copepodite of *Pachypygus gibber* and the development of the *Ciona* host. Counts of 3rd copepodites (150 specimens) have been used to determine the first stage at which the differentiation between typical and atypical forms is apparent. The relative percentage of each form was determined for each stage of *Ciona*. The transitional phase during which both forms were found is indicated by the shaded frame. (From Hipeau-Jacquotte, 1988.)

the typical and the atypical males, the various stages of spermatogenesis are the same and lead to a single type of spermatozoa and a single type of fertilized egg (Hipeau-Jacquotte and Coste, 1989). This phenomenon is therefore epigenetic, as far as the phenotype is concerned, and results from the interaction of genetic determinants and environmental factors (metagamic determination according to Hipeau-Jacquotte, 1988). For the atypical males, Hipeau-Jacquotte (1984) first proposed the intervention of a "young host factor" implicated in the atypical morphogenesis and that is in fact related to a limited food supply, which is in agreement with the reasoning of Charnov and Bull (1977). It should be added that there are interactions between complementary sexes.

The case of *Pachypygus* is comparable to that of *Pseudomyicola spinosus*. In this species, Do *et al.* (1984) have demonstrated the existence of two types of male according to the method of culture (isolated atypical male; associated typical male; isolated or associated female). But there is no mention of age, host size or available food resources as factors influencing sexual

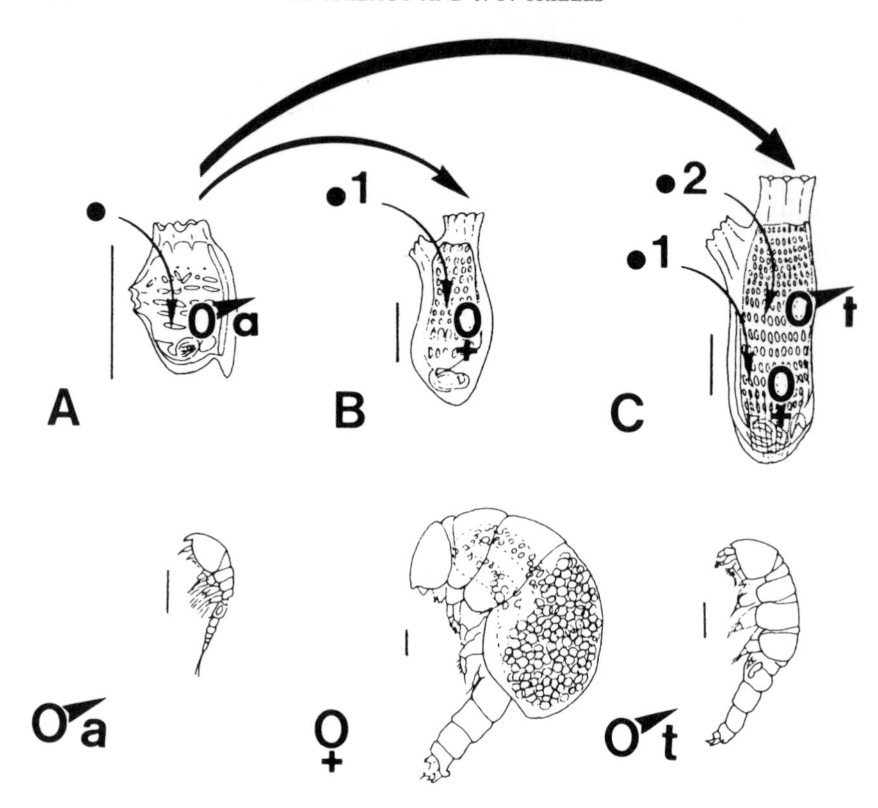

FIG. 49. Morphogenesis of the infective larva (●) of *Pachypygus* in relation to the developmental stage of *Ciona* (A)–(C) and to previous infestation of *Ciona*. (A) Pre-critical phase (uni-or multi-infestation): atypical male (s). (B) Just after the critical phase (uni-infestation): female. (C) Post-critical phase (uni-or multi-infestation): female (s) and typical male (s). t, typical male; a, atypical male; 1,2, chronology of infestation. Note the additional free phase of atypical male (thick arrows). Scales: (A) 0.1 cm; (B)(C) 1 cm and for three copepod phenotypes: 400 μm. (From Hipeau-Jacquotte, 1988.)

differentiation. Also comparable is *Pachypygus macer* in which a female dimorphism is related to the size of the ascidian host, and therefore to the quantity of food available (Monniot, 1986). Such influences are equally known in certain free-living copepods, the factors concerned being essentially temperature and salinity. The trophic factor is, on the other hand, important in the larval morphogenesis of *Artemia* free-living Anostraca.

In other cases, the epigenetic influence is more important and controls the totality of sexual differentiation, including the reproductive system.

There are, thus, on record, cases of labile gonochorism with the preponderant effector of sexual differentiation being the host environment. Thus, at the turn of the century, Malaquin (1901) noted, during his remarkable study of the Monstrillida, that individual sex was influenced by the abundance of larvae in the host. When an annelid is parasitized by several nauplii, the development of the latter is towards maleness, whereas a single larva in a host becomes female or male. Bacci *et al.* (1958), in a study of the sex ratio of *Mytilicola intestinalis*, noted that the number of males increased the heavier was the infestation: Nevertheless, the Italian biologists, having observed, on one hand, that when the parasites were isolated in their hosts the sex ratio was not far from unity, and on the other hand that the presence of adult females did not inhibit the differentiation of larvae into females, concluded "per una forte componente genetica nel determinisimo dei sessi cui si sovrappone une influenza ambientale di natura per ora imprecisabilile"†. As pointed out above, the presence of heterochromosomes in *Mytilicola intestinalis* was unknown at this time. Heegaard (1947) hypothesized that "substances" secreted by the females induced copepodite larvae to develop into males, the larvae not submitted to this influence developing into females. Bresciani and Lützen (1961) suggested that such phenomena were likely to occur in *Gonophysema gullmarensis*: "Our observations can easily be explained if we suppose that of the onychopodids which collect round a female a number will invade her by chance and develop into males, while the remaining part is prevented from doing so for some reason or other, and will develop into females in the tissue of the ascidian. Unfortunately, we cannot prove experimentally that the sex determination in *Gonophysema* is not genetically determined".

But, the best known and most thoroughly studied examples are members of the Epicarida, and in particular *Ione thoracica*, a parasite of the branchial cavity of *Callianassa subterranea* and *C. tyrrhena* (Kossmann, 1881), that has been studied by Reverberi and Pitotti (1942) (Fig. 50) and Reverberi (1947), and more recently *Stegophryxus hyptius*, an abdominal parasite of *Pagurus longicorpus* (Reinhard, 1949). In the two cases, the observations are practically identical, and the only recognizable differences are those that one would expect of two genera, as far as behaviour and habitat are concerned (Owens and Glazebook, 1985).

The larval *Ione* anchors itself to tufts of setae on the abdomen of *Callianassa*; here it metamorphoses into a bopyridium (= the first post larval stage) and, whilst developing undergoes a slow migration to the host's

---

† Authors' translation: There is an important genetic component in the sexual determination, onto which is superimposed an environmental influence, the nature of which is at present unknown.

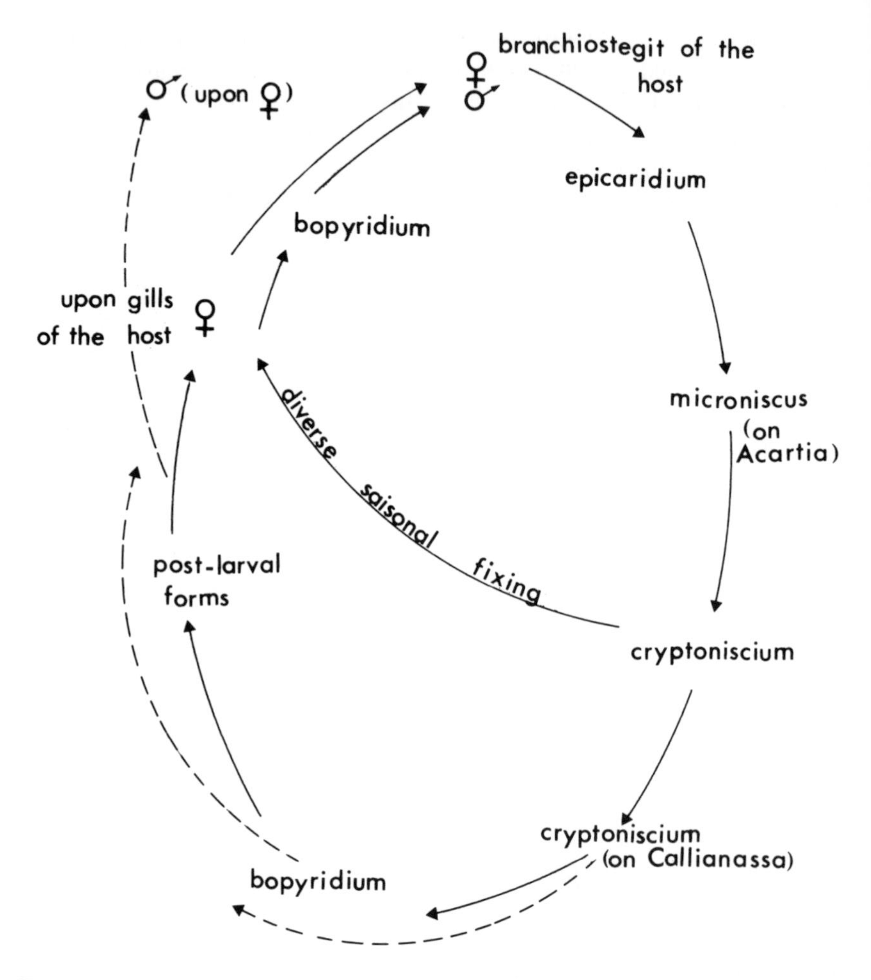

FIG. 50. *Ione thoracica*: Life cycle. (After Reverberi and Pitotti, 1942; Bourdon, 1969.)

branchial cavity. Once here the larva develops into a female and, having reached a certain age (5–6 months), it secretes a substance that attracts another planktonic cryptoniscian directly into the branchial cavity. This latter fixes to the female and rapidly becomes a male. If several larvae attach themselves simultaneously to the female, they all start their development into males, but only one will remain. On the majority of adult *Callianassa* that are already infested with a large *Ione*, a number of cryptoniscians are to be found hanging especially from the tufts of setae and between the pleopods. These develop into bopyridium and even more advanced stages, but only migrate to the branchial cavity if the male already in place disappears. Most of the time, these post larval forms eventually fall from the host's abdomen and die (Bourdon, 1969). In other experiments, Reverberi and Pitotti (1942) and Reverberi (1947) removed young males, at the beginning of their metamorphosis, from the female *Ione*, to which they were attached and transplanted them into the branchial cavity of unparasitized *Callianassa*. They rapidly attached themselves to the new host and took up residence. In the majority of cases they became female, demonstrating the existence of an undeniable inversion of sex. In one case, Reverberi also obtained a sexual inversion in the opposite sense, a young individual, that was already developing into a female, becoming male. He removed an adult female and a very young female, and placed the young individual on the older female in a small dish. The association persisted, the young female feeding on the older one. When the older female died, the author replaced her with another, and so forth, during several months until Reverberi finally obtained, in one case, a typical male with testes and spermatozoa. These experiments served to demonstrate, therefore, not only the trophic relationship linking the associated individuals, but also the influence of such an association on the sexual determinism of the younger "partner".

The results obtained by Reinhard (1949) with *Stegophryxus* are, in all aspects, virtually superimposable with those of Reverberi. It should be pointed out, however, that in this case the female is not attached to the gills, as with *Ione*, but to the host's abdomen. Under these conditions, the larvae have the choice of only two substrates; the abdomen of the host, or that of the female parasite herself. Presumptive females, that were still very young although already attached to the abdomen of the pagurid, were removed and transplanted onto older females. They became transformed into males. Reinhard, however, was unable to obtain the reverse transformation (from presumptive male into female), although this was perhaps feasible as the author was unable to obtain the attachment of young males to the host. It is of course conceivable that, though the males, once they are differentiated, can undergo sexual inversion, the differentiated females lose the capacity to transform into males. This contradicts the observation of Reverberi on *Ione*, though this observation concerned a single case that logically requires confirmation.

The definitive sex of the Bopyridae and the Phryxidae is, therefore, apparently labile and results from the specific conditions presiding over the development of the individual after the cryptoniscian stage. Reverberi and Pitotti (1942) originally accepted that the nutritional conditions were predominant in this phenomenon; the larvae directly absorbing abundant quantities of host blood (gills) becoming females; if the quantity of available food was poorer, on the other hand (fixation to the female as opposed to fixation to the host), the appearance of complementary males occurs. This is a conclusion that Reinhard did not exclude for *Stegophryxus*: the larvae that feed directly on the host become females. But Reverberi (1947) later accepted that it was more likely that the larvae that fixed themselves to the female parasites were masculinized by a masculinizing substance produced by the female, rather than by a less abundant food supply. This elegant conclusion was shared by Reinhard. The case of complementary males fixed onto the cuticle of *Callianassa* effectively suggests that the two factors, greater or lesser abundance of food and a masculinizing substance (that has no effect at a distance) are both involved.

Amongst the Crustacea, such epigenetic conditions may also take effect during the development of certain forms of hermaphrodism, such as the protandrous hermaphrodism of the cymothoadians discussed above. Certain events occurring in this case are comparable to events, described above concerning the Epicarida. Both cases are almost certainly cases of polyfactorial sexual determinism, with, in the case of the cymothoadians, the particularity of a preponderant action of genes encoding for masculinity at the beginning of differentiation.

Either a male parasite, e.g. *Anilocra*, that has developed on one host fish and is seeking to mate, or a larva that, to start with, will develop towards maleness, might attach itself to a host fish already infested with a parasite in the female sexual phase. In most species of Cymothoidae, as in *Anilocra*, an inhibitory epigenetic influence, dependent upon the presence of a female partner, will act upon the sex inversion of the male partner that will remain in the male phase as long as the female is present on the same fish. At the death of the female, this inhibitory influence is eliminated and the male immediately undergoes sexual inversion. Having in turn become a female, this parasite will now await the arrival of a male, or a larva that will potentially develop into a male, in order to mate. The ecophysiological conditions controlling this phenomenon have how been more or less completely elucidated. It is known that the change of sex is dependent upon the involution of the androgenic glands of the male; these latter degenerate at the moment of sexual inversion. This regression is under the control of a neurohormonal action, that is itself dependent upon the presence of a female. The individual in female sexual phase most certainly secretes a sexual pheromone, transported by the blood of the host (both partners are blood feeders), which acts on the neurosecretory system of the male, inducing it to secrete a masculinizing hormone that, in its turn, acts upon the androgenic glands, maintaining their activity. It is evident that when the female dies, this whole system of control disappears, permitting the sexual inversion to occur. Here, once again, it is a typical case of epigenetic control of sexuality (Trilles, 1963, 1968b).

V. Conclusion

Parasitism having appeared amongst free-living ancestors, it is of interest to note possible modifications of sexuality brought about by the acquirement of a parasitic life, wherever living representatives, that are at least close to the ancestral free-living forms, are available. The parasitic way of life creates the problem of mate encounter whenever one of the partners is definitively fixed, and even more so when both partners are fixed.

As a rule, the gonochorism inherited from free-living forms has persisted in the parasitic copepods. The permanent fixation of one of the sexes, that is always the female, whose energetic requirements are considerable on account of her extraordinary fecundity, has led to the adoption of two adaptive strategies to overcome the problem of mate encounter. The first, in a way, retraces evolution, as mating and fertilization occur during the phase when both partners are still free living. The second involves the two partners living permanently close to each other. In all cases, this involves the female attaching and developing first. In a second phase, the male, which is more often than not miniscule (pygmy males), attaches himself close to the female, on the female's body, in a groove on the female's body or even in the female's body. The last case, cryptogonochorism, is an exacerbation of sexual dimorphism since the male has lost all of his, now useless, somatic attributes and has become a simple testicle.

The Tantulocarida, the members of which are all parasitic, at least at the larval stage, have a reproductive strategy similar to that observed in certain copepods, such as the Ergasilidae, with a free-swimming male and a parasitic female. The method of insemination, however, is more like that of the akentrogonid rhizocephalans in which the male cypris injects spermatogonia into the female via an antennule. In the case of the Tantulocarida, the male has an abdominal penis that pierces the female's tegument.

Amongst the cirripeds, in which the fixation to an inert substrate has already entailed a constraint on sexual encounter, hermaphrodism with cross-fertilization is generalized in highly gregarious species such as the barnacles and goose-barnacles. In species, such as *Scalpellum*, where the population densities tend to be drastically reduced, cross-fertilization becomes difficult and the appearance of dwarf, complementary males is observed (Svane, 1986). These latter might be considered as taking over the role of the hermaphroditic partner in cross-fertilization. This leads to a reduction, or even a disappearance, of the male territory of the hermaphrodite in favour of gonochorism with an exaggeration of sexual dimorphism, as in the copepods. With the acquisition of a parasitic way of life, the ultimate stage of this process is reached, in which the male becomes internalized within the female, thus offering all the advantages of hermaphrodism with self-fertilization (it should be recalled that examples of this are extremely rare in the living world) without the inconvenience of the absence of genetic mixing. This stage is reached in the Rhizocephala (*Sacculina, Peltogasterella, Clistosaccus and Lernaeodiscus*) but with modalities that differ between species. Amongst these, it is interesting to recall the case of *Thompsonia dofleini*, the sexual reproduction of which associates an extraordinary increase in the efficiency of the male with no modification of the sex ratio.

The free-living isopods are fundamentally gonochoric with, however, a few isolated cases of protandrous (land-dwelling Rhyscotidae) or proterogynous (certain Anthuridae) hermaphrodism. Hermaphrodism, on the other hand, has been largely exploited by the parasitic isopods and is the predominant form of reproduction, even though the epicarid Bopyridae resort to gonochorism with dwarf males. It has been demonstrated experimentally that these latter conserve the potentiality for protandrous hermaphrodism. Finally there is the case of the Gnathiiidae that, not being parasitic at the adult stage (protelian parasitism), have conserved the gonochorism typical of the free-living forms. This phenomenon is perfectly reflected by the monstrillid copepods.

The sexuality of the Crustacea has become remarkably well adapted to the parasitic way of life. A number of forms have taken advantage of their general genetic plasticity (polyfactorial determinism) leading to two different strategies. The first concerns the epigenetic determinism typified by *Ione* (Bopyrina) or *Pachypygus* (Copepoda) facilitating the formation of couples and complementary "spare" males. It is of note that this type of determinism already exists in certain free-living species. The second corresponds to the generalization of the phenomenon of protandrous hermaphrodism (Cymothoidae and Cryptoniscina) that rarely occurs in free-living forms. In the Rhizocephalans, on the other hand, the maintenance of hermaphrodism is certainly not shown although it is constant among species fixed to an inert substrate.

For all other parasitic crustaceans with strict genetic determinism, two possibilities occur:

(a) Certain ones have evolved very little with respect to free-living forms. This is the case of the Gnathiidae in which the reproductive adults do not feed and the larvae show adult behaviour like that of the Cirolanidae or the Aegidae.

(b) In others, the solution has been the dwarfism of the males that live as parasites on the females. This solution is translated by the exaggeration of sexual dimorphism leading to its paroxysm, cryptogonochorism.

Finally, nothing is known of the sexual determinism of the Tantulocarida, on account of their recent discovery and of the difficulty encountered in observing them *in vivo*, and more particularly under experimental conditions.

Acknowledgements

The authors thank the following for their precious help in the preparation of this paper: Mme J. Bocquet-Védrine (Gif-sur Yvette, France), M. Goudeau (Université de Nice, France), R. Hipeau-Jacquotte (Université de Marseille, France), M. Amanieu (Université de la Réunion), R. Bourdon (Station Biologique de Roscoff, France), G. A. Boxshall (British Museum, London), J. T. Hoeg, A. Jespersen and J. Lützen (Institute of Cell Biology and Anatomy, University of Copenhagen, Denmark), R. Huys (University of Gent, Belgium), N. Le Brun (Laboratoire de Parasitologie Comparée, Université Montpellier II, France), I. Robbins (Station de Biologie Marine, Sète, France).

References

Altes, J. (1962). Sur quelques parasites et hyperparasites de *Clibanarius erythropus* (Latreille) en Corse. *Bulletin de la Société Zoologique de France* **87** (1), 88–97.

Altes, J. (1981). Les Liriopsides. *Bulletin de la Société d'Histoire Naturelle d'Afrique du Nord, Alger* **69** (3 & 4), 3–35.

Amanieu, M. (1963). Evolution des populations de *Paragnathia formica* (Hesse) au cours d'un cycle annuel. *Bulletin de l'Institut Océanographique de Monaco* **60**, 1261, 1–12.

Amanieu, M. (1969). Recherches écologiques sur les faunes des plages abritées de la région d'Arcachon. *Helgoländer wissenschaftliche Meeresuntersuchungen* **19**, 455–557.

Andersen, M. L., Bohn, M., Hoeg, J. T. and Jensen, P. G. (1990). Cypris ultrastructure and adult morphology in *Ptychascus glaber* and *P. barnwelli* n. sp. (Cirripedia: Rhizocephala) parasites on semiterrestrial crabs. *Journal of Crustacean Biology* **10** (1), 20–28.

Anstensrud, M. (1989a). Experimental studies of the reproductive behaviour of the parasitic copepod *Lernaeocera branchialis* (Pennellidae). *Journal of the Marine Biological Association of the United Kingdom* **69**, 465–476.

Anstensrud, M. (1989b). A vital stain for studies behaviour and ecology of the parasitic copepod *Lernaeocera branchialis* (Pennellidae). *Marine Ecology-Progress Series* **53**, 47–50.

Anstensrud, M. (1990a). Moulting and mating in *Lepeophtheirus pectoralis* (Copepoda: Caligidae). *Journal of the Marine Biological Association of the United Kingdom* **70**, 269–281.

Anstensrud, M. (1990b). Mating strategies of two parasitic copepods (*Lernaeocera branchialis* (L.)) (Pennellidae) and *Lepeophtheirus pectoralis* (Müller) (Caligidae) on flounder: polygamy, sex-specific age at maturity and sex-ratio. *Journal of Experimental Marine Biology and Ecology* **136**, 141–158.

Anstensrud, M. (1990c). Effects of mating on female behaviour and allometric growth in the two parasitic copepods *Lernaeocera branchialis* (L. 1767) (Pennellidae) and *Lepeophtheirus pectoralis* (Müller, 1776) (Caligidae). *Crustaceana* **59**, 245–258.

Anstensrud, M. (1992). Mate guarding and mate choice in two copepods *Lernaeocera branchialis* (L.) (Pennellidae) and *Lepeophtheirus pectoralis* (Müller) (Caligidae) parasitic on flounder. *Journal of Crustacean Biology* **12** (1), 31–40.

Bacci, G., Balata, M. and Romani, M. L. (1958). Rapporti numerici dei sessi in tre popolazioni di *Mytilicola intestinalis* Steuer. *Rendiconti del' Accademia Nazionale di Lincei* **25**, 557–563.

Baer, J. G. (1946). Le parasitisme. *Masson et Cie Editeurs, Paris*, 1–235.

Baer, J. G. (1951). "Ecology of Animal Parasites". pp. 1–224. The University of Illinois Press, Urbana.

Ben Hassine, O. K. (1983). "Les Copécodes Parasites de Poissons Mugilidae en Méditerranée Occidentale (Côtes Françaises et Tunisiennes). Morphologie, Bio-écologie, Cycles Évolutifs". Thèse d'Etat, Université Montpellier II. Sciences et Techniques du Languedoc, 1–452.

Ben Hassine, O. K. and Raibaut, A. (1981). Le développement larvaire expérimental de *Caligus pageti* Russel, 1925, Copépode parasite de poissons Mugilidae en Méditerranée. *Bulletin de l'office national des Pêches de Tunisie* **5** (2), 175–201.

Bird, N. T. (1968). Effects of mating on subsequent development of a parasitic copepod. *Journal of Parasitology, U.S.A.* **54** (6), 1194–1196.

Bocquet, C., Bocquet-Védrine, J. and L'Hardy, J. P. (1970). Contribution à l'étude du développement des organes génitaux de *Xenocoeloma alleni* (Brumpt), Copépode parasite de *Polycirrus caliendrum* Claparède. *Cahiers de Biologie Marine* **11**, 195–208.

Bocquet-Védrine, J. (1957). *Chthamalophilus delagei* nov. gen., nov. sp, Rhizocéphale nouveau, parasite de *Chthamalus stellatus*. *Comptes Rendus Hebdomadaires des Séances de l'Académie des Sciences* (sér. D) **244**, 1545–1548.

Bocquet-Védrine, J. (1961). Morphologie de *Chthamalophilus delagei* J. Bocquet-Védrine, Rhizocéphale parasite de *Chthamalus stellatus* (Poli). *Cahiers de Biologie Marine*, **2** (5), 455–593.

Bocquet-Védrine, J. (1967). Un nouveau Rhizocéphale parasite de Cirripède: *Microgaster balani* n. gen., n. sp. *Comptes Rendus Hebdomadaires des Séances de l'Académie des Sciences* (sér. D) **265**, 1630–1632.

Bocquet-Védrine, J. (1972). Les Rhizocéphales. *Cahiers de Biologie Marine* **13** (5), 615–626.

Bocquet-Védrine, J. (1974a). Chronologie du développement chez *Crinoniscus equitans* Perez (Isopode Cryptoniscien). *Archives de Zoologie Expérimentale et Générale* **115** (2), 197–204.

Bocquet-Védrine, J. (1974b). Parenté phylogénétique des Isopodes Cryptonisciens rangés jusqu'ici dans les familles des Liriopsidae et des Crinoniscidae. *Recherches Biologiques Contemporaines*, Imprimerie Vagner, Nancy, 73–78.

Bocquet-Védrine, J. (1976). Les voies d'absorption de l'eau au cours de l'acquisition de la forme adulte chez le Crustacé Isopode épicaride *Crinoniscus equitans* Perez. *Archives de Zoologie Expérimentale et Générale* **117**, 423–433.

Bocquet-Védrine, J. and Bocquet, C. (1972a). Réalisation de la forme définitive chez *Crinoniscus equitans* Perez, au cours de l'étape femelle du cycle de cet Isopode Cryptoniscien. *Comptes Rendus Hebdomadaires des Séances de l'Académie des Sciences, Paris* **275**, 2009–2011.

Bocquet-Védrine, J. and Bocquet, C. (1972b). La ceinture d'attache de la femelle juvénile de *Crinoniscus equitans* Perez (Isopode Cryptoniscien) et son importance adaptative. *Comptes Rendus Hebdomadaires des Séances de l'Académie des Sciences, Paris* **275**, 2235–2238.

Bocquet-Védrine, J. and Bourdon, R. (1984). *Cryptogaster cumacei* n. gen. n. sp.; premier Rhizocéphale parasite d'un Cumacé. *Crustaceana* **41**, 261–270.

Bonnier, J. (1900). Contributions à l'étude des Epicarides: les Bopyridae. *Travaux de la Station Zoologique de Wimereux* **8**, 1–475.

Bourdon, R. (1969). "Systématique des Céponiens et Biologie des *Cancricepon*". Thèse d'Université, Nancy, 1–17.

Bower, M. B. and Boutillier, J. A. (1990). *Sylon* (Crustacea, Rhizocephala) infections on the shrimps in British Columbia. *In* "Pathologie in Marine Science" (F. O. Perkins and T. C. Cheng, eds), pp. 267–275. Academic Press, New York.

Boxshall, G. A. (1990). Precopulatory mate guarding in copepods. *Bijdragen tot de Dierkunde* **60** (3/4), 209–213.

Boxshall, G. A. (1991). A review of the biology and phylogenetic relationships of the Tantulocarida, a subclass of Crustacea recognized in 1983. *Verhandlungen der Deutschen Zoolgischen Gesellschaft* **84**, 271–279.

Boxshall, G. E. and Lincoln, R. J. (1983). Tantulocarida, a new class of Crustacea ectoparasitic on other crustaceans. *Journal of Crustacean Biology* **3**, 1–16.

Boxshall, G. E. and Lincoln, R. J. (1987). The life cycle of the Tantulocarida (Crustacea). *Philosophical Transactions of the Royal Society, London* B **315**, 267–303.

Bresciani, J. and Lützen, J. (1960). *Gonophysema gullmarensis* (Copepoda parasitica): An anatomical and biological study of an endoparasite living in the Ascidian *Ascidiella aspersa*. I. Anatomy. *Cahiers de Biologie Marine* **1** (2), 157–184.

Bresciani, J. and Lützen, J. (1961). *Gonophysema gullmarensis* (Copepoda parasitica): An anatomical and biological study of an endoparasite living in the Ascidian *Ascidiella aspersa*. II. Biology and development. *Cahiers de Biologie Marine* **2** (4), 347–372.

Bresciani, J. and Lützen, J. (1962). Parasitic Copepods from the West coast of Sweden including some new or little known species. *Videnskab Elige Meddelelser fra Dansk Naturhistorisk Forening* **124**, 368–405.

Bresciani, J. and Lützen, J. (1972). The sexuality of *Aphanodomus* (Parasitic Copepoda) and the phenomenon of cryptogonochorism. *Videnskab Elige Meddelser fra Dansk Naturhistorisk Forening* **135**, 7–20.

Bresciani, J. and Lützen, J. (1974). On the biology and development of *Aphanodomus* Wilson (Xenocoelomidae), a parasitic Copepod of the Polychaete *Thelepus cincinnatus*. *Videnskab Elige Meddelser fra Dansk Naturhhistorisk Forening* **137**, 25–63.

Caillet, C. and Raibaut, A. (1979). Observations expérimentales sur la sexualité du Copépode Caligide *Clavellodes macrotrachelus* (Brian, 1906), parasite branchial du Sar *Diplodus sargus* (Linné, 1758). *Comptes Rendus Hebdomadaires des*

Séances de l'Académie des Sciences, Paris **288**, 223–226.

Cals, P. (1978). Expédition Rumphius II (1975). Crustacés parasites, commensaux, etc. (Th. Monod & R. Serène, éd.) IV. Crustacés Isopodes, Gnathiides. Particularités systématiques et morphologiques. Appareil piqueur de la larve hématophage. *Bulletin du Muséum National d'Histoire Naturelle, Paris* (3ème sér.) **520** (Zool. 356), 479–516.

Capart, A. (1948). Le *Lernaeocera branchialis* (Linné, 1758). *La Cellule* **52** (2), 159–212.

Caullery, M. (1907). Sur les Liriopsides, Crustacés Isopodes (Epicarides) parasites de Rhizocéphales. *Comptes Rendus Hebdomadaires des Séances de l'Académie des Sciences, Paris* **144**, 100–102.

Caullery, M. (1950). Le parasitisme et la symbiose. 2ème edn. G. Doin et Cie Editeurs. pp. 1–358.

Caullery, M. and Mesnil, F. (1919a). *Xenocoeloma brumpti* C. et M., Copépode parasite de *Polycirrus arenivorus* C. *Bulletin Biologique de France et de Belgique* **53**, 161–233.

Caullery, M. and Mesnil, F. (1919b). Sur un nouvel Epicaride (*Ancyroniscus bonnieri* n. g., n. sp.) parasite d'un Sphéromide (*Dynamene bidentata* Mont.). *Comptes Rendus Hebdomadaires des Séances de l'Académie des Sciences, Paris* **169**, 1430–1432.

Caullery, M. and Mesnil, F. (1920) *Ancyroniscus bonnieri* C. & M. Epicaride parasite d'un Sphéromide (*Dynamene bidentata* Mont.). *Bulletin Biologique de France et de Belgique* **53**, 1–36.

Charmantier, G. (1980). Etude écophysiologique des Crustacés Isopodes Gnathiidae: osmorégulation et résistance à la dessication des mâles de *Paragnathia formica* (Hesse, 1864). *Journal of Experimental Marine Biology and Ecology* **43**, 161–171.

Charmantier, G. (1982). Les glandes céphaliques de *Paragnathia formica* (Hesse, 1864) (Isopoda, Gnathiidae): localisation et ultrastructure. *Crustaceana* **42** (2), 179–193.

Charnov, E. L. and Bull, J. (1977). When is sex environmentally determined? *Nature* **266**, 828–830.

Cressey, R. and Boxshall, G. A. (1989). *Kabatarina pattersoni*, a fossil parasitic copepod (Dichelesthiidae) from a lower Cretaceous fish. *Micropaleontology* **35** (2), 150–167.

Delage, Y. (1884). Evolution de la Sacculine (*Sacculina carcini* Thomps.) Crustacé endoparasite de l'ordre nouveau des Kentrogonides. *Archives de Zoologie Expérimentale et Générale* **2**, 417–736.

Do, T. T. and Kajihara, T. (1984). Sex determination and atypical male development in a Poecilostomatoid Copepod, *Pseudomyicola spinosus* (Raffaele and Monticelli, 1885). *Syllogeus* **58**, 283–287.

Do, T. T., Kajihara, T. and Ho, J. S. (1984). The life history of *Pseudomyicola spinosus* (Raffaele and Monticelli, 1885) from the blue mussel, *Mytilus edulis galloprovincialis* in Tokyo Bay, Japan, with notes on the production of atypical male. *Bull. Ocean Res. Instit., Univ. Tokyo* **17**, 1–65.

Giard, A. and Bonnier, J. (1887). Contribution à l'étude des Bopyriens. *Travaux de la Station Zoologique de Wimereux* **5**, 1–272.

Glenner, H., Hoeg, J. T., Klysner, A. and Brodin Larsen, B. (1989). Cypris ultrastructure, metamporphosis and sex in seven families of Rhizocephalan barnacles (Crustacea: Cirripedia: Rhizocephala). *Acta Zoologica, Stockholm* **70** (4), 229–242.

Goudeau, M. (1976). Contribution à la biologie d'un Crustacé parasite: *Hemioniscus balani* Buchholz, Isopode Epicaride. Nutrition, mues et croissance de la femelle et des embryons. *Thèse Université Pierre et Marie Curie, Paris* VI, 1–56 (+8 publications et un texte non publié).

Goudeau, M. (1977). Contribution à la biologie d'un Crustacé parasite: *Hemioniscus balani* Buchholz, Isopode Epicaride: nutrition, mues et croissance de la femelle et des embryons. *Cahiers de Biologie Marine* **28**, 201–242.

Heegaard, P. (1947). Contribution to the phylogeny of the Arthropods. Copepoda. *Spolia Zoologica Musei Haunienssis* **8**, 1–236.

Hipeau-Jacquotte, R. (1978a). Développement post-embryonnaire du copépode ascidicole Notodelphyidae *Pachypygus gibber* (Thorell, 1859). *Crustaceana* **34**, 155–194.

Hipeau-Jacquotte, R. (1978b). Existence de deux formes sexuelles mâles chez le copépode ascidicole Notodelphyidae *Pachypygus gibber* (Thorell, 1859). *Comptes Rendus Hebdomadaires des Séances de l'Académie des Sciences, Paris* D **287**, 253–256.

Hipeau-Jacquotte, R. (1978c). Relation entre âge de l'hôte et type de développement chez un copépode ascidicole Notodelphyidae. *Comptes Rendus Hebdomadaires des Séances de l'Académie des Sciences, Paris* D **287**, 1207–1210.

Hipeau-Jacquotte, R. (1980). La forme mâle atypique du copépode ascidicole Notodelphyidae *Pachypygus gibber* (Thorell, 1859): description et synonymie avec *Agnathaner minutus* Canu, 1891. *Bulletin du Muséum d'Histoire Naturelle de Paris* A **4**, 455–470.

Hipeau-Jacquotte, R. (1984). A new concept in the evolution of the Copepoda: *Pachypygus gibber* (Notodelphyidae), a species with two breeding males. *Crustaceana* suppl. **7**, 60–67.

Hipeau-Jacquotte, R. (1988). Environmental sex determination in a crustacean parasite. *International Journal of Invertebrate Reproduction and Development* **14**, 11–24.

Hipeau-Jacquotte, R. and Coste, F. (1989). Reproductive system of the parasitic copepod *Pachypygus gibber*: spermatogenesis and spermatophore formation in dimorphic males, and discharge in female tracts. *Journal of Crustacean Biology* **9** (2), 228–241.

Hoeg, J. T. (1982). The anatomy and development of the rhizocephalan barnacle *Clistosaccus paguri* Lilljeborg and relation to its host *Pagurus bernhardus* (L.). *Journal of Experimental Marine Biology and Ecology* **58**, 87–125.

Hoeg, J. T. (1985a). Cypris settlement, Kentrogen formation and host invasion in the parasite barnacle *Lernaeodiscus porcellanae* (Müller) (Crustacea: Cirripedia: Rhizocephala). *Acta zoologica, Stockholm* **66**, 1–45.

Hoeg, J. T. (1985b). Mâle cypris settlement in *Clistosaccus paguri* Lilljeborg (Crustacea: Cirripedia: Rhizocephala). *Journal of Experimental Marine Biology and Ecology* **89**, 221–235.

Hoeg, J. T. (1987a). Mâle cypris metamorphosis, and a new male larval form the trichogon, in the parasitic barnacle *Sacculina carcini* (Crustacea: Cirripedia: Rhizocephala). *Philosophical Transactions of the Royal Society of London* B **317**, 47–63.

Hoeg, J. T. (1987b). The relation between cypris ultrastructure and metamorphosis in male and female *Sacculina carcini* (Crustacea Cirripedia). *Zoomorphology* **107**, 299–311.

Hoeg, J. T. (1990). "Akantrogonid" host invasion and an entirely new type of life cycle in the Rhizocephalan parasite *Clistosaccus paguri* (Thecostraca: Cirripedia) *Journal of Crustacean Biology* **10** (1), 37–52.

Hoeg, J. T. (1991). Functional and evolutionary aspects of the sexual system in the Rhizocephala (Crustacea: Thecostraca: Cirripedia). *In* "Crustacean Sexual Biology" (R. Bauer and J. Martin, eds), pp. 208–227. Columbia University Press, New York.

Hoeg, J. T. (1993). Rhizocephala. *In* "Microscopic Anatomy of Invertebrates" Vol. 9 "Crustacea" (F. Harrison, ed.), pp. 315–345. Wiley-Liss.

Hoeg, J. T. and Ritchie, L. E. (1985). Male cypris settlement and its effects on juvenile development in *Lernaeodiscus porcellanae* Müller (Crustacea: Cirripedia: Rhizocephala). *Journal of Experimental Marine Biology and Ecology* **87**, 1–11.

Hoeg, J. T. and Ritchie, L. E. (1987). Correlation between cypris age, settlement rate and anatomical development in *Lernaeodiscus porcellanae* (Cirripedia Rhizocephala). *Journal of Marine Biological Association of United Kingdom* **67**, 65–75.

Hoeg, J. T., Kapel, C. M., Thor, P. and Webster, P. (1990). The anatomy and sexual biology of *Boschmaella japonica*, an akentrogonid rhizocephalan parasite on barnacles from Japan (Crustacea Cirrepedia Rhizocephala). *Acta Zoologica (Stockh.)* **71**, 177–188.

Holdich, D. M. (1975). *Ancyroniscus bonnieri* (Isopoda, Epicaridea) infecting British populations of *Dynamene bidentata* (Isopoda, Sphaeromatidae). *Crustaceana* **28** (2), 145–151.

Humes, A. G. (1969). *Aspidomolgus stoichactinus* n. gen., n. sp. (Copepoda, Cyclopoida) associated with Actinarian in the West Indies. *Crustaceana* **16** (3), 225–242.

Humes, A. G. and Stock, J. H. (1973). A revision of the family Lichomolgidae Kossmann, 1877, Cyclopoid Copepods mainly associated with marine Invertebrates. *Smithsonian Contribution to Zoology* **127**, 1–368.

Huys, R. (1991). Tantulocarida (Crustacea: Maxillopoda): A New Taxon from the Temporary Meiobenthos. *Marine Ecology* **12** (1), 1–34.

Huys, R. and Boxshall, G. A. (1991). "Copepod Evolution". *The Ray Society, London* **159**, 1–468.

Ichikawa, A. and Yanagimachi, R. (1958). Studies on the sexual organization of the Rhizocephala. I. The nature of the "testis" of *Peltogasterella socialis* Krüger. *Annotationes Zoologicae Japonenses* **31**, 82–96.

Ichikawa, A. and Yanagimachi, R. (1960). Studies on the sexual organization of the Rhizocephala. II. The reproductive function of the larva (cypris) male of *Peltogaster* and *Sacculina*. *Annotationes Zoologicae Japonense* **33**, 42–56.

Jerpersen, A. and Lützen, J. (1992). *Thompsonia dofleini* Häfele, a colonial akentrogonid rhizocephalan with dimorphic, ova-or sperm producing, externae (Crustacea: Cirripedia). *Zoomorphology*, **112**, 105–116.

Johnson, G. (1961). Contribution à l'étude de la détermination du sexe chez les Oniscoïdes: phénomène d'hermaphrodisme et monogénie. *Bulletin Biologique de France et de Belgique* **95**, 177–267.

Kabata, Z. (1958). *Lernaeocera obtusa* n. sp. Its biology and its effects on the haddock. *Marine Research* **3**, 1–26.

Kabata, Z. (1979). "Parasitic Copepoda of British Fishes". *The Ray Society* **152**, 1–468.

Kossmann, R. (1881). Studien über Bopyriden III. *Ione thoracica und Cepon portuni. Mitteilungen aus der Zoologischen Station zu Neapel* **3**, 149–169.

Lilljeborg, W. (1861). Supplément au mémoire sur les genres *Liriope* et *Peltogaster*, H. Rathke. *Nova Acta Reg. Soc. Sci. Upsala* **3** (3), 73–102.

Lützen, J. (1981a). Observations on the Rhizocephalan barnacle *Sylon hippolytes* M. Sars parasitic on the prawn *Spirontocaris lilljeborgi* (Danielssen). *Journal of Experimental Marine Biology and Ecology* **50**, 231–254.

Lützen, J. (1981b). Field studies on regeneration in *Sacculina carcini* Thompson (Crustacea: Rhizocephala) in the isefjord, Denmark. *Journal of Experimental Marine Biology and Ecology* **53**, 241–249.

Lützen, J. (1984). Growth, reproduction and life span in *Sacculina carcini* Thompson (Cirripedia: Rhizocephala) in the Isefjord, Denmark. *Sarsia* **69**, 91–106.

Lützen, J. (1985). Rhizocephala (Crustacea, Cirripedia) from the deep sea. *Galathea Report* **16**, 99–112.

Lützen, J. (1992). Morphology of *Thompsonia reinhardi*, new species (Cirripedia: Rhizocephala), parasitic on the northeast pacific hermit crab *Discorsopagurus schmitti* (Stevens). *Journal of Crustacean Biology* **12** (1), 83–93.

Lützen, J. and Jespersen, A. (1990). Records of *Thompsonia* (crustacea: Cirripedia: Rhizocephala) from Singapore, including description of two new species, *T. littoralis* and *T. pilodiae*. *Raffles Bulletin of Zoology* **38** (2), 241–249.

Lützen, J. and Jespersen, A. (1992). A study of the morphology and biology of *Thompsonia littoralis* (Crustacea: Cirripedia: Rhizocephala). *Acta Zoologica* **73**, 1–23.

Malaquin, A. (1901). Le parasitisme évolutif des Monstrillides. *Archives de Zoologie Expérimentale et Générale* **9**, 81–232.

Meusy, J. J. and Payen, G. G. (1988). Female reproduction in malacostracean Crustacea. *Zoological Science* **5**, 217–265.

Monniot, C. (1986). Présence en Guadeloupe de deux phénotypes femelles du copépode ascidicole *Pachypygus macer* Illg, 1958. *Systematic Parasitology* **8**, 151–162.

Monod T. (1926). "Les Gnathiidae. Essai Monographique (Morphologie, Biologie, Systématique)." *Mémoires de la Société des Sciences Naturelles et Physiques du Maroc* **13**, 1–667.

Mouchet, S. (1928a). Note sur le cycle évolutif des Gnathiidae. *Bulletin de la Société Zoologique de France* **53**, 392–400.

Mouchet, S. (1928b). Contribution à l'étude de la digestion chez les Gnathiidae. *Bulletin de la Société Zoologique de France* **53**, 442–452.

Müller, F. (1862). Die Rhizocephalen, eine neue Gruppe schmarotzender Kruster. *Archiv fur Naturgeschichte, Berlin* **28**, 1–9.

Müller, F. (1863). Die zweite entwickelungsstufe der Wurzelkrebse (Rhizocephalen). *Archiv fur Naturgeschichte, Berlin* **29**, 24–33.

Nagasawa, K., Bresciani, J. and Lützen, J. (1988). Morphology of *Pectenophilus ornatus*, new genus, new species, a Copepod Parasite of the Japanese Scallop *Patinopecten yessoensis*. *Journal of Crustacean Biology* **8** (1), 31–43.

Orlando, E. (1973). Heterochromosomes in *Mytilicola intestinalis* Steuer (Copepoda). *Genetica* **44**, 244–248.

Owens, L. and Glazebook, J. S. (1985). Sex determination in the Bopyridae. *Journal of Parasitology* **71** (1), 134–135.

Perez, C. (1932). Sur les racines des Rhizocéphales parasites des pagures. *Archivi di Zoologia Italiana* **16**, 1315–1318.

Potts, F. A. (1912). Mycetomorpha, a new Rhizocephalan. *Zoologische Jahrbucher, Abeteilung fur Systematik, Okologie und Geographie der Tiere* **33**, 575–594.

Raibaut, A. (1985). Cycles évolutifs des Copépodes parasites et les modalités de l'infestation. *L'Année Biologique* (4ème sér.) **24** (3), 233–274.

Raibaut, A., Ben Hassine, O. K. and Maamouri, K. (1971). Copépodes parasites des Poissons de Tunisie (première série). *Bulletin de l'Institut d'Océanographie et de Pêche Salammbô* **2** (2), 169–197.

Raibaut, A., Caillet, C. and Ben Hassine, O. K. (1978). *Colobomatus mugilis* n. sp. (Copepoda, Philichthyidae) parasite de Poissons Mugilidés en Méditerranée occidentale. *Bulletin de la Société Zoologique de France* **103** (4), 449–457.

Reinhard, E. G. (1942). The reproductive role of the complemental males of *Peltogaster*. *Journal of Morphology* **70** 389–402.

Reinhard, E. G. (1949). Experiments on the determination and differentiation of sex in the Bopyrid *Stegophryxus hyptius* Thompson. *Biological Bulletin* **96**, 17–31.

Reverberi, G. (1947). Ancora sulla trasformazione sperimentale del sesso nei Bopiridi. La trasformazione delle femine giovanili in maschi. *Pubblicazioni della Stazione Zoologica di Napoli* **21** (1), 81–91.

Reverberi, G. and Pitotti, M. (1942). Il ciclo biologico e la determinazione fenotipica del sesso di *Ione thoracica* Montagu, Bopiride parassita di *Callianassa laticauda* Otto. *Pubblicazioni della Stazione Zoologica di Napoli* **19**, 111–184.

Ritchie, L. E. and Hoeg, J. T. (1981). The life history of *Lernaeodiscus porcellanae* (Cirripedia: Rhizocephala) and co-evolution with its porcellanid host. *Journal of Crustacean Biology* **1**, 334–347.

Romestand, B. (1979). Etude écophysiologique des parasitoses à Cymothoadiens. *Annales de Parasitologie Humaine et Comparée* **54** (4), 423–448.

Romestand, B., and Trilles, J. P. (1975). Les relations immunologiques "hôte–parasite" chez les Cymothoidae (Isopoda, Flabellifera). *Comptes Rendus Hebdomadaires des Séances de l'Académie des Sciences, Paris* (Sér D) **280**, 2171–2173.

Romestand, B. and Trilles, J. P. (1976). Au sujet d'une substance antithrombinique mise en évidence dans les glandes latéro-oesophagiennes de *Meinertia oestroides* (Risso, 1826) (Isopoda, Flabellifera, Cymothoidae: parasite de Poisson) *Zeitschrift fur Parasitenkunde* **50**, 87–92.

Romestand, B., Janicot, M. and Trilles, J. P. (1977). Modifications tissulaires et réactions de défense chez quelques Téléostéens parasités par les Cymothoidae (Crustacés Isopodes hématophages). *Annales de Parasitologie Humaine et Comparée* **52** (2), 171–180.

Rubiliani, C. (1984). "Les Relations Hôte–Parasite Chez les Crustacés: Développement et Modalités d'Action des Rhizocephales Sacculinidae sur la Reproduction des Crabes". Thèse de Doctorat d'Etat, Paris VI (Université Pierre et Marie Curie), 1–282, I–XLV.

Rubiliani, C., Turquier, Y. and Payen, G. C. (1982). Recherche sur l'ontogénèse des Rhizocéphales: I. Les stades précoces de la phase endoparasitaire chez *Sacculina carcini* Thompson. *Cahiers de Biologie Marine* **23**, 287–297.

Sanviti, G., Romestand, B. and Trilles J. P. (1981). Les Sacculines (*Sacculina carcini* Thompson, 1836) de *Carcinus mediterraneus* et *Pachygrapsus marmoratus*: comparaison immunochimique; étude comparée de leur influence sur la composition protéique de l'hémolymphe des deux hôtes. *Zeitschrift fur Parasitenkunde* **64**, 243–251.

Sheader, M. (1977). The breeding biology of *Idotea pelagica* (Isopoda: Valvifera) with notes on the occurrence and biology of its parasite *Clypeoniscus hanseni* (Isopoda: Epicaridea). *Journal of the Marine Biological Association of the United Kingdom* **57**, 659–674.

Smith, G. W. (1906). The Rhizocephala. *Fauna und Flora des Golfes von Neapel und der angrezenden Meeresabschnitte* **29**, 1–123.

Sproston, N. G. (1942). The developmental stages of *Lernaeocera branchialis* (Linn.). *Journal of the Marine Biological Association of the United Kingdom* **25**, 441–466.

Stock, J. H. and Van Der Spoel, S. (1976). *Pteroxena papillifera* n. gen., n. sp., an endoparasitic organism (Copepoda?) from the Gymnosomatous Pteropod, *Notobranchaea. Bull. Zool. Mus. Amsterdam* **5** (21), 177–180.

Stoll, C. (1962). Cycle évolutif de *Paragnathia formica* (Hesse) (Isopode-Gnathiidae). *Cahiers de Biologie Marine* **3**, 401–402.

Svane, J. (1986). Sex determination in *Scalpellum scalpellum* (Cirripedia: Thoracica: Lepadomorpha) a hermaphrodite goose barnacle with dwarf males. *Marine Biology* **90**, 249–253.

Tirard, C. (1991). "Biodiversité et Biogéographie évolutive dans les Systèmes Hôtes–Parasites. Le Modèle Gadiformes (Téléotéens)—Copépodes et Monogènes". Thèse Université Montpellier II, 1–132.

Tirard, C. and Raibaut A. (1989). Quelques aspects de l'écologie de *Lernaeocera lusci* (Bassett-Smith, 1896), Copépode parasite de Poissons Merlucciidae et Gadidae. *Bulletin d'Ecologie* **20**, 289–294.

Trilles, J. P. (1963). Mise en évidence d'une action du complexe céphalique neuroséc-rétoire sur la glande androgène et les gonades de *Nerocila orbignyi* Schioedte et Meinert (Isopoda, Cymothoïdae). *Comptes Rendus Hebdomadaires des Séances de l'Académie des Sciences, Paris* **257**, 1811–1812.

Trilles, J. P. (1968a). "Recherches sur les Isopodes Cymothoidae des Côtes Françaises. I. Systématiques et Faunistiques". Thèse Doctorat ès-Sciences Montpellier, 1–181.

Trilles, J. P. (1968b). "Recherches sur les Isopodes Cymothoidae des Côtes Fran çaises. Vol I: Bionomie et Parasitisme. Vol II: Biologie Générale et Sexualité". Thèse Doctorat ès-Sciences Montpellier, 1–793.

Trilles, J. P. (1969). Recherches sur les Isopodes Cymothoidae des côtes Françaises. Aperçu général et comparatif sur la bionomie et la sexualité de ces Crustacés. *Bulletin de la Société Zoologique de France* **94** (3), 433–445.

Van Wyk, P. M. (1982). Inhibition of the growth and reproduction of the Porcellanid Crab *Pachycheles rudis* by the Bopyrid Isopod *Aparobopyrus muguensis*. *Parasitology* **85**, 459–473.

Veillet, A. (1943). Note sur le dimorphisme des larves de *Lernoeodiscus galatheoe* Norman et Scott et sur la nature des "mâles larvaires" des Rhizocéphales. *Bulletin de l'Institut Oceanographique de Monaco* **841**, 1–4.

Veillet, A. (1945). Recherches sur le parasitisme des Crabes et des Galathées par les Rhizocéphales et les Epicarides. *Annales de l'Institut Oceanographique de Monaco* **22**, 193–341.

Walker, A. (1985). The cypris larvae of *Sacculina carcini* Thompson (Crustacea, Cirripedia, Rhizocephala). *Journal of Experimental Marine Biology and Ecology* **93**, 131–145.

Wilson, C. B. (1903). North American parasitic copepods of the family Argulidae, with a bibliography of the group and a systematic review of all known species. *Proceeding of the United States National Museum* **25**, 635–742, pls. 8–27.

Wimpenny, R. S. (1927). Observations sur *Danalia ypsilon* Smith. *Bulletin de l'Institut Océanographique de Monaco* **496**, 1–8.

Yanagimachi, R. (1961a). The use of cetyl alcohol in the rearing of Rhizocephalan larvae. *Crustaceana* **2**, 37–39.

Yanagimachi, R. (1961b). Studies on the sexual organization of the Rhizocephala III. The mode of sex-determination in *Peltogasterella. Biological Bulletin* **120**,

272–283.
Yanagimachi, R. and Fujimaki, N. (1967). Studies on the sexual organization of Rhizocephala. IV, on the nature of the "testis" of *Thompsonia. Annotationes Zoologicae Japonenses* **40**, 98–104.
Zmerzlaya, E. I. (1972). *Ergasilus sieboldi* Nordmann, 1832, its development, biology and epizooytic significance. *Izvjesca Gosud Nauchno-Issledovate Instituta Ozern. Rechn. Rybnogo Khozjajstva* **80**, 132–177.

Note added in Proof

In a communication to the First European Crustacean Conference, Paris, August 31–September 5, 1992, Huys R., Boxshall G. and Lincoln R., following their discovery of a free-swimming female ectoparasite on a Pacific deep water harp, state that tantulocarids exhibit a dual life cycle combining a sexual phase with a multiplicative parthenogenetic phase.

Index